T0215891

Lecture Notes in Computer Science   9573

Commenced Publication in 1973
Founding and Former Series Editors:
Gerhard Goos, Juris Hartmanis, and Jan van Leeuwen

Roman Wyrzykowski · Ewa Deelman
Jack Dongarra · Konrad Karczewski
Jacek Kitowski · Kazimierz Wiatr (Eds.)

Parallel Processing and Applied Mathematics

11th International Conference, PPAM 2015
Krakow, Poland, September 6–9, 2015
Revised Selected Papers, Part I

 Springer

Editors
Roman Wyrzykowski
Czestochowa University of Technology
Czestochowa
Poland

Ewa Deelman
Department of Computer Science
University of Southern California
Marina Del Rey, CA
USA

Jack Dongarra
Electrical Engineering and Computer
Science
University of Tennessee
Knoxville, TN
USA

Konrad Karczewski
Institute of Computer and Information
Science
Czestochowa University of Technology
Czestochowa
Poland

Jacek Kitowski
Department of Computer Science
AGH University of Science and Technology
Krakow
Poland

Kazimierz Wiatr
AGH University of Science and Technology
Krakow
Poland

ISSN 0302-9743   ISSN 1611-3349   (electronic)
Lecture Notes in Computer Science
ISBN 978-3-319-32148-6   ISBN 978-3-319-32149-3   (eBook)
DOI 10.1007/978-3-319-32149-3

Library of Congress Control Number: 2016934929

LNCS Sublibrary: SL1 – Theoretical Computer Science and General Issues

Printed on acid-free paper

This Springer imprint is published by Springer Nature
The registered company is Springer International Publishing AG Switzerland

Preface

This volume comprises the proceedings of the 11th International Conference on Parallel Processing and Applied Mathematics – PPAM 2015, which was held in Krakow, Poland, September 6–9, 2015. It was organized by the Department of Computer and Information Science of the Częstochowa University of Technology together with with the AGH University of Science and Technology, under the patronage of the Committee of Informatics of the Polish Academy of Sciences, in cooperation with the ICT COST Action IC1305 "Network for Sustainable Ultrascale Computing (NESUS)." The main organizer was Roman Wyrzykowski.

PPAM is a biennial conference. Ten previous events have been held in different places in Poland since 1994. The proceedings of the last six conferences have been published by Springer in the *Lecture Notes in Computer Science* series (Nałęczów, 2001, vol. 2328; Częstochowa, 2003, vol. 3019; Poznań, 2005, vol. 3911; Gdańsk, 2007, vol. 4967; Wrocław, 2009, vols. 6067 and 6068; Toruń, 2011, vols. 7203 and 7204; Warsaw, 2013, vols. 8384 and 8385).

The PPAM conferences have become an international forum for exchanging ideas between researchers involved in parallel and distributed computing, including theory and applications, as well as applied and computational mathematics. The focus of PPAM 2015 was on models, algorithms, and software tools that facilitate efficient and convenient utilization of modern parallel and distributed computing architectures, as well as on large-scale applications, including big data problems.

This meeting gathered more than 190 participants from 33 countries. A strict refereeing process resulted in the acceptance of 111 contributed presentations, while approximately 43 % of the submissions were rejected. Regular tracks of the conference covered important fields of parallel/distributed/cloud computing and applied mathematics such as:

- Numerical algorithms and parallel scientific computing
- Parallel non-numerical algorithms
- Tools and environments for parallel/distributed/cloud computing
- Applications of parallel computing
- Applied mathematics, neural networks, evolutionary computing, and metaheuristics

The plenary and invited talks were presented by:

- David A. Bader from the Georgia Institute of Technology (USA)
- Costas Bekas from IBM Research — Zurich (Switzerland)
- Pete Beckman from the Argonne National Laboratory (USA)
- Christopher Carothers from the Rensselaer Polytechnic Institute (USA)
- Barbara Chapman from the University of Houston (USA)
- Willem Deconinck from the European Centre for Medium-Range Weather Forecast (UK)
- Geoffrey C. Fox from Indiana University (USA)

- Dieter Kranzlmueller from the Ludwig-Maximilians-Universität München (Germany)
- Vladik Kreinovich from the University of Texas at El Paso (USA)
- Alexey Lastovetsky from the University College Dublin (Ireland)
- Carlos Osuna from ETH Zurich (Switzerland)
- Srinivasan Parthasarathy from the Ohio State University (USA)
- Enrique S. Quintana-Orti from the Universidad Jaime I (Spain)
- Thomas Rauber from the University of Bayreuth (Germany)
- Daniel Reed from the University of Iowa (USA)
- Rizos Sakellariou from the University of Manchester (UK)
- Boleslaw K. Szymanski from the Rensselaer Polytechnic Institute (USA)
- Manuel Ujaldon from Nvidia
- Jeffrey Vetter from the Oak Ridge National Laboratory and Georgia Institute of Technology (USA)
- Richard W. Vuduc from the Georgia Institute of Technology (USA)
- Torsten Wilde from the Leibnitz Supercomputing Centre (LRZ) (Germany)

Important and integral parts of the PPAM 2015 conference were the workshops:

- Minisympsium on GPU Computing organized by José R. Herrero from the Universitat Politecnica de Catalunya (Spain), Enrique S. Quintana-Orti from the Universidad Jaime I (Spain), and Robert Strzodka from Heidelberg University (Germany).
- The Third Workshop on Models, Algorithms and Methodologies for Hierarchical Parallelism in New HPC Systems organized by Giulliano Laccetti and Marco Lapegna from the University of Naples Federico II (Italy), and Raffaele Montella from the University of Naples Parthenope (Italy).
- Workshop on Power and Energy Aspects of Computation organized by Jee Choi from the IBM T.J. Watson Research Center (USA), Piotr Luszczek from the University of Tennessee (USA), Leonel Sousa from the Technical University of Lisbon (Portugal), and Richard W. Vuduc from the Georgia Institute of Technology (USA).
- Workshop on Scheduling for Parallel Computing — SPC 2015 organized by Maciej Drozdowski from the Poznań University of Technology (Poland).
- The 6th Workshop on Language-Based Parallel Programming Models — WLPP 2015 organized by Ami Marowka from the Bar-Ilan University (Israel).
- The 5th Workshop on Performance Evaluation of Parallel Applications on Large-Scale Systems organized by Jan Kwiatkowski from the Wrocław University of Technology (Poland).
- Workshop on Parallel Computational Biology — PBC 2015 organized by Bertil Schmidt from the University of Mainz (Germany) and Jarosław Żola from the University at Buffalo (USA).
- Workshop on Applications of Parallel Computations in Industry and Engineering organized by Raimondas Čiegis from the Vilnius Gediminas Technical University (Lithuania) and Julius Žilinskas from the Vilnius University (Lithuania).

- Minisymposium on HPC Applications in Physical Sciences organized by Grzegorz Kamieniarz and Wojciech Florek from the A. Mickiewicz University in Poznań (Poland).
- The Second Workshop on Applied High-Performance Numerical Algorithms in PDEs organized by Piotr Krzyżanowski and Leszek Marcinkowski from Warsaw University (Poland) and Talal Rahman from Bergen University College (Norway).
- Minisymposium on High-Performance Computing Interval Methods organized by Bartłomiej J. Kubica from the Warsaw University of Technology (Poland).
- Workshop on Complex Collective Systems organized by Paweł Topa and Jarosław Wąs from the AGH University of Science and Technology (Poland).
- Special Session on Efficient Algorithms for Problems with Matrix and Tensor Decompositions organized by Marian Vajtersic from the University of Salzburg (Austria) and Gabriel Oksa from the Slovak Academy of Sciences.
- Special Session on Algorithms, Methodologies, and Frameworks for HPC in Geosciences and Weather Prediction organized by Zbigniew Piotrowski from the Institute of Meteorology and Water Management (Poland) and Krzysztof Rojek from the Częstochowa University of Technology (Poland).

The PPAM 2015 meeting began with four tutorials:

- Scientific Computing with GPUs, by Dominik Göddeke from the University of Stuttgart (Germany), Robert Strzodka from Heidelberg University (Germany), and Manuel Ujaldon from the University of Malaga (Spain) and Nvidia.
- Advanced Scientific Visualization with VisNow, by Krzysztof Nowiński, Bartosz Borucki, Kerstin Kantiem, and Szymon Jaranowski from the University of Warsaw (Poland).
- Parallel Computing in Java, by Piotr Bała from the Warsaw University of Technology (Poland) and Marek Nowicki, Łukasz Górski, Magdalena Ryczkowska from the Nicolaus Copernicus University (Poland).
- Introduction to Programming with Intel Xeon Phi, by Krzysztof Rojek and Łukasz Szustak from the Częstochowa University of Technology (Poland).

An integral part of the GPU Tutorial was the CUDA quiz with participants challenged to maximize the performance on a common GPU model. The winner was Miłosz Ciżnicki from the Poznan Supercomputing and Networking Center. The winner received the prize of a Tesla K40 GPU generously donated by Nvidia for the conference given its role of PPAM sponsor. The second and third prizes were granted, respectively, to Michał Antkowiak and Łukasz Kucharski, both from the A. Mickiewicz University in Poznań.

Nvidia also donated another prize, GeForce GTX480 GPU, for the authors of the best paper presented at the Minisymposium on GPU Computing. This prize was awarded to Jan Gmys, Mohand Mezmaz, Nouredine Melab, and Daniel Tuyttens from the University of Mons, who presented the paper "IVM-Based Work Stealing for Parallel Branch-and-Bound on GPU."

Special Session on Algorithms, Methodologies, and Frameworks for HPC in Geosciences and Weather Prediction: Contemporary and future applications of numerical weather prediction, climate research, and studies in geosciences demand multidisciplinary advancements in computing methodologies, including the use of multi-/manycore processors and accelerators, scalable and energy-efficient frameworks, and big data strategies, as well as new or improved numerical algorithms. This includes, for example, development of scalable, high-resolution methods for integration of fluid PDEs and efficient iterative solvers, highly optimized ports to modern hardware (CPU, GPU, Xeon Phi), code development and portability strategies, and libraries for handling geophysical datasets.

The special session served as a multidisciplinary forum for the discussion of state-of-the-art research and development toward the next-generation geophysical fluid solvers and weather/climate prediction applications.

The special session featured a number of invited and contributed talks, covering recent advances in numerical algorithms, accelerator methodologies, energy-efficent computing, and large dataset managements, including:

- Algorithms and tools for the extreme-scale numerical weather prediction (invited plenary talk by Willem Deconinck et al.)
- Adaptation of COSMO Consortium weather and climate numerical models to hybrid architecures (invited plenary talk by Carlos Osuna et al.)
- Highly efficient port of the GCR solver using high-level stencil framework on multi- and many-core architectures (by M. Ciżnicki et al.)
- Autotuned scheduler for time/energy optimization for a fully three-dimensional MPDATA advection scheme on the hybrid CPU-GPU clusters (by K. Rojek et al.)
- Parallel alternating direction implicit preconditioners for all-scale atmospheric models (by Z. Piotrowski et al.)

The organizers are indebted to the PPAM 2015 sponsors, whose support was vital to the success of the conference. The main sponsor was Intel Corporation and the other sponsors were: Nvidia, Action S.A., and Gambit. We thank all the members of the international Program Committee and additional reviewers for their diligent work in reviewing the submitted papers. Finally, we thank all of the local organizers from the Częstochowa University of Technology and the AGH University of Science and Technology, who helped us run the event very smoothly. We are especially indebted to Grażyna Kołakowska, Urszula Kroczewska, Łukasz Kuczyński, Adam Tomaś, and Marcin Woźniak from the Częstochowa University of Technology; and to Krzysztof Zieliński, Kazimierz Wiatr, and Jacek Kitowski from the AGH University of Science and Technology.

We hope that this volume will be useful to you. We would like everyone who reads it to feel invited to the next conference, PPAM 2017, which will be held during September 10–13, 2017, in Lublin, the largest Polish city east of the Vistula River.

January 2016

Roman Wyrzykowski
Jack Dongarra
Ewa Deelman
Konrad Karczewski
Jacek Kitowski
Kazimierz Wiatr

Organization

Program Committee

Jan Węglarz	Poznań University of Technology, Poland, (Honorary Chair)
Roman Wyrzykowski	Częstochowa University of Technology, Poland, (Program Chair)
Ewa Deelman	University of Southern California, USA, (Program Co-chair)
Francisco Almeida	Universidad de La Laguna, Spain
Pedro Alonso	Universidad Politecnica de Valencia, Spain
Peter Arbenz	ETH, Zurich, Switzerland
Cevdet Aykanat	Bilkent University, Ankara, Turkey
Piotr Bała	Warsaw University, Poland
David A. Bader	Georgia Institute of Technology, USA
Michael Bader	TU München, Germany
Olivier Beaumont	Inria Bordeaux, France
Włodzimierz Bielecki	West Pomeranian University of Technology, Poland
Paolo Bientinesi	RWTH Aachen, Germany
Radim Blaheta	Institute of Geonics, Czech Academy of Sciences, Czech Republic
Jacek Błażewicz	Poznań University of Technology, Poland
Pascal Bouvry	University of Luxembourg
Jerzy Brzeziński	Poznań University of Technology, Poland
Marian Bubak	AGH Kraków, Poland, and University of Amsterdam, The Netherlands
Tadeusz Burczyński	Polish Academy of Sciences, Warsaw
Christopher Carothers	Rensselaer Polytechnic Institute, USA
Jesus Carretero	Universidad Carlos III de Madrid, Spain
Raimondas Čiegis	Vilnius Gediminas Technical University, Lithuania
Andrea Clematis	IMATI-CNR, Italy
Zbigniew Czech	Silesia University of Technology, Poland
Jack Dongarra	University of Tennessee and ORNL, USA, and University of Manchester, UK
Maciej Drozdowski	Poznań University of Technology, Poland
Mariusz Flasiński	Jagiellonian University, Poland
Tomas Fryza	Brno University of Technology, Czech Republic
Jose Daniel Garcia	Universidad Carlos III de Madrid, Spain
Pawel Gepner	Intel Corporation, USA
Domingo Gimenez	University of Murcia, Spain

Mathieu Giraud   LIFL and Inria, France
Jacek Gondzio   University of Edinburgh, Scotland, UK
Andrzej Gościński   Deakin University, Australia
Laura Grigori   Inria, France
Adam Grzech   Wroclaw University of Technology, Poland
Inge Gutheil   Forschungszentrum Juelich, Germany
Georg Hager   University of Erlangen-Nuremberg, Germany
José R. Herrero   Universitat Politecnica de Catalunya, Barcelona, Spain
Ladislav Hluchy   Slovak Academy of Sciences, Bratislava, Slovakia
Sasha Hunold   Vienna University of Technology, Austria
Florin Isaila   Universidad Carlos III de Madrid, Spain
Ondrej Jakl   Institute of Geonics, Czech Academy of Sciences,
                                 Czech Republic
Emmanuel Jeannot   Inria, France
Bo Kagstrom   Umea University, Sweden
Christos Kartsaklis   Oak Ridge National Laboratory, USA
Eleni Karatza   Aristotle University of Thessaloniki, Greece
Ayse Kiper   Middle East Technical University, Turkey
Jacek Kitowski   Institute of Computer Science, AGH, Poland
Joanna Kołodziej   Cracow University of Technology, Poland
Jozef Korbicz   University of Zielona Góra, Poland
Stanislaw Kozielski   Silesia University of Technology, Poland
Dieter Kranzlmueller   Ludwig Maximillian University, Munich, and Leibniz
                                 Supercomputing Centre, Germany
Henryk Krawczyk   Gdańsk University of Technology, Poland
Piotr Krzyżanowski   University of Warsaw, Poland
Krzysztof Kurowski   PSNC, Poznań, Poland
Jan Kwiatkowski   Wrocław University of Technology, Poland
Jakub Kurzak   University of Tennessee, USA
Giulliano Laccetti   University of Naples Federico II, Italy
Marco Lapegna   University of Naples Federico II, Italy
Alexey Lastovetsky   University College Dublin, Ireland
Joao Lourenco   University Nova of Lisbon, Portugal
Tze Meng Low   Carnegie Mellon University, USA
Hatem Ltaief   KAUST, Saudi Arabia
Emilio Luque   Universitat Autonoma de Barcelona, Spain
Vyacheslav I. Maksimov   Ural Branch, Russian Academy of Sciences, Russia
Victor E. Malyshkin   Siberian Branch, Russian Academy of Sciences, Russia
Pierre Manneback   University of Mons, Belgium
Tomas Margalef   Universitat Autonoma de Barcelona, Spain
Svetozar Margenov   Bulgarian Academy of Sciences, Sofia
Ami Marowka   Bar-Ilan University, Israel
Ricardo Morla   INESC Porto, Portugal
Norbert Meyer   PSNC, Poznań, Poland
Jarek Nabrzyski   University of Notre Dame, USA
Raymond Namyst   University of Bordeaux and Inria, France

Contents – Part I

Parallel Non-numerical Algorithms

Tools and Environments for Parallel/Distributed/Cloud Computing

Minisymposium on GPU Computing

**Special Session on Efficient Algorithms for Problems with Matrix
and Tensor Decompositions**

Contents – Part II

The 6th Workshop on Language-Based Parallel Programming Models (WLPP 2015)

The 5th Workshop on Performance Evaluation of Parallel Applications on Large-Scale Systems

Workshop on Parallel Computational Biology (PBC 2015)

Workshop on Applications of Parallel Computation in Industry and Engineering

Minisymposium on HPC Applications in Physical Sciences

Workshop on Complex Collective Systems

**Special Session on Algorithms, Methodologies and Frameworks
for HPC in Geosciences and Weather Prediction**

Parallel Architectures and Resilience

Exploring Memory Error Vulnerability
for Parallel Programming Models

Isil Oz[1][(✉)], Marisa Gil[2], Gladys Utrera[2], and Xavier Martorell[2]

[1] Computer Engineering Department, Marmara University,
Goztepe Campus, Kadikoy, 34722 Istanbul, Turkey
isil.oz@marmara.edu.tr
[2] Computer Architecture Department, Universitat Politecnica de Catalunya (UPC),
Jordi Girona, 3, 08034 Barcelona, Spain
{marisa,gutrera,xavim}@ac.upc.edu

Abstract. Transistor size reduction and more aggressive power modes in HPC platforms make chip components more error prone. In this context, HPC applications can have a diverse level of tolerance to memory errors that may change the execution in different ways. As the tolerance to memory errors depends on write frequency and access patterns, different programming models may exhibit a different behavior in the rate of failures and alleviate the performance loss caused by the overhead of fault-tolerance mechanisms. In this paper, we explore how tolerant to memory errors are two main parallel programming models, message-passing and shared memory: we perform a memory vulnerability analysis and also conduct error propagation experiments to observe the effect of memory errors through program flow. Our results show the need for soft error resiliency methods based on memory behavior of programs, and the evaluation of the tradeoffs between performance and reliability.

Keywords: Memory errors · Reliability · SDC · Programming models

1 Introduction

High performance computing platforms are increasing the number of processing elements, rising to exascale and petascale systems. This higher number of cores has the effect of increasing the likelihood of memory errors. In addition, transistor size reduction and more aggressive power modes in HPC platforms make chip components more error prone, and thus transient error rate increases [13,16].

Memory errors, which result from a fault in a single bit, are expected to increase. While some memory errors are masked by an overwrite and do not affect the program execution, the rest may change the execution in different ways, including program crash or incorrect output generation [16]. Current research considers incorporating fault-tolerance mechanisms to detect these errors and recover from them in order to make the system more robust [6,8,11,19]. In terms of performance, these error-detection techniques incur at slowdown on

© Springer International Publishing Switzerland 2016
R. Wyrzykowski et al. (Eds.): PPAM 2015, Part I, LNCS 9573, pp. 3–11, 2016.
DOI: 10.1007/978-3-319-32149-3_1

each memory access due to the additional circuitry they need, and once the error is detected, recovering from it.

As the tolerance to memory errors depends on write frequency and access patterns [11], different programming models may exhibit a different behavior in the rate of failures due to memory, and thus the time spent in recovering can vary by improving performance. Besides, due to the propagation effect, in a parallel program one faulty value may be distributed along data flow to multiple threads or processes through shared memory or message-passing system.

In this paper, we perform a memory vulnerability analysis for a set of programming model implementations by conducting fault injection experiments. We also conduct error propagation analysis to observe the effect of memory errors through program flow. We focus on data memory that represents around 90 % of the total memory in HPC solvers; in addition a memory error on code produces a crash in the execution, so there is no way to recover or detect bad results. We can summarize the main contributions of this work as follows:

- We design an instrumentation-based fault injection framework, which emulates bit-flip memory errors, to quantify fault behavior of parallel programs, and perform a high-level fault analysis to track error propagation through data flow.
- We conduct an experimental study for miniFE benchmark [1] and analyze the memory error vulnerability of different programming models.
- We compare memory vulnerability rates, and perform a performance-reliability analysis by discussing the impact of programming models on both execution time and SDC (Silent Data Corruption) rates.

Our results show that programming models have an impact on memory error vulnerability. However, we also observe that a better understanding of the algorithmic flow and data structures is needed to improve performance by gaining reliability. Our error propagation analysis demonstrates that the propagation effect depends on where the error is introduced.

2   Error Vulnerability Analysis

Data in a memory location is vulnerable to soft errors if an error on that memory location can affect the execution [11]. The vulnerability is defined between Write-Read and Read-Read operations [14,18]. A memory location is vulnerable during the lifetime of a value that starts with the first write operation and ends with the last read operation. The vulnerability of one location is the sum of vulnerable intervals (unsafe durations). If a value is not read by the program until it is overwritten (between consecutive Write operations), this time interval is not vulnerable.

We track memory access patterns to get vulnerable intervals in a program. We follow memory store (Write) and load (Read) operations, and calculate vulnerability rate (VR) as follows:

$$VR(M_i) = \frac{\sum Vinterval}{Ttime},$$

where $V interval$ represents the vulnerable time intervals that M_i memory location is vulnerable, i.e., time between Write-Read and Read-Read operations, and $Ttime$ represents total execution time of the program. The VR of a program can be calculated by taking the average vulnerability rates of all memory locations mapped by the program.

3 Fault Injection Methodology

In this work, we consider bit-flip transient faults, which are transient transitions of bit values due to external factors such as particle strikes, electrical noise, and cosmic rays [13,16]. Although our analysis includes results for single-bit errors, it can also be extended to multi-bit upsets. We assume that the execution is correct if the application terminates successfully and produces correct results. We classify the faulty execution as SDC (Silent Data Corruption) if the computation results are different from the results without soft errors, and ABORT if the execution terminates with an error code due to some program error such as segmentation fault, floating-point exception, division by zero.

We design and implement a fault injection framework based on the Pin system, which performs dynamic instrumentation for programs [9,10]. We implement a thread-level Pintool to trace all instructions and memory operations of a parallel program. We ensure the synchronization of threads by using Pin's lock mechanism and inject our code into analysis routines protected by locks for each thread.

First, our tool determines a fault injection point by generating one random instruction, one random execution unit (thread number for OpenMP, process number for MPI), and one random faulty bit. Then, instrumentation for program execution starts by tracking instructions and memory operations. When the program reaches the fault instruction point (the instruction), it randomly selects one memory location previously accessed by the program and flips the fault bit. Then it continues instrumentation by tracking the program execution, and reports the execution outcome if it is a crash, incorrect output, or correct execution. By repeating 1,000 injections to achieve statistical significance for each programming model, we evaluate SDC rates to determine vulnerability behavior.

4 Experimental Platform

We conduct our experiments on our cluster, which is composed of 128 nodes, each one with: 2x Intel Xeon E5649 6-Core at 2.53 GHz. For this work, we use a homogeneous single-node in order to execute the programs in the same conditions and analyze the memory behavior.

In this work, we use gcc compiler version 4.6.1, with OpenMP 3.0 and OmpSs 1.99.4. For MPI, we use Intel MPI 4.0.2.003. The programming models we use are: OpenMP [4], a shared-memory fork/join model, OmpSs [3] a shared-memory task-based model, and MPI [2], a message-passing programming model.

In all three cases, the algorithm is the same, except for the minimum changes required by the programming model in use.

For our evaluation purposes we choose the miniFE 2.0 rc3 benchmark from Mantevo suite [1]. MiniFE solves a linear-system using unpreconditioned conjugate gradients (CG), which carries out the SpMV and represents the most significant fraction of the miniFE total run-time. The Mantevo package includes MPI and OpenMP implementations, and for the task-based model we use a previous proposal using the OmpSs programming model [15].

5   Experimental Results

We track memory operations and calculate vulnerable intervals, as explained in Sect. 2. For MPI, we track processes separately to monitor accessed memory locations, since there is separate address space for each process. We observe that memory access pattern for each process is similar, so we report the results for one process. Figure 1 shows memory vulnerability values calculated by tracking vulnerable times during the lifetime of the values. For all programming models, the values are high, but OpenMP and MPI values are larger than OmpSs. Since the OmpSs runtime generates several copies of data structures, many more distinct memory locations are accessed, and data store/load operations are preformed more frequently. Thus, the vulnerability of each location is relatively smaller (not much time between Write-Read) when compared to OpenMP and MPI, which explains the lower vulnerability values.

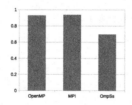

Fig. 1. Memory vulnerability rate (VR) values for programming models.

5.1   Fault Injection Results

We perform fault injection experiments for 4-core, 8-core, and 12-core executions in a single node by using our fault injection framework. One can observe in Fig. 2 that memory errors have more effect on MPI than on OpenMP and OmpSs versions. While OpenMP and OmpSs exhibit similar behavior for 4-core and 8-core executions, OmpSs has fewer erroneous cases for 12-core execution. The trend for both ABORT and SDC rates is similar, and the aggregate results demonstrate the common behavior.

If we look at the SDC rates, which is a more common vulnerability metric, we can see that OpenMP and MPI exhibit similar behavior, and OmpSs has

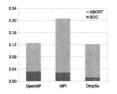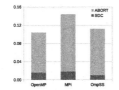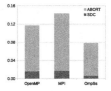

Fig. 2. ABORT and SDC results for OpenMP, MPI and OmpSs on 4, 8 and 12 cores

smaller values, as in the case of vulnerability values (See Fig. 1). In order to see the tradeoff between reliability and performance, we collect SDC rates as the vulnerability metric, and the number of floating point operations performed per time unit (gigaflops per second) as the performance metric. Figure 3 presents both results including normalized values for 4-core, 8-core, and 12-core execution scenarios. We can see that there is a clear tradeoff between reliability and performance for different programming models. While MPI has the largest SDC rates (except the 4-core case), its gigaflop values are also the largest among programming models, which shows its better performance. While the order of versions is $MPI < OpenMP < OmpSs$ in terms of reliability (the lower the SDC, the more reliable the program), the performance is in the opposite order $OmpSs < OpenMP < MPI$ (the larger the Gflops, the better performance).

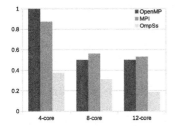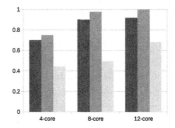

Fig. 3. Reliability and performance behavior of OpenMP, MPI, OmpSs (left: normalized SDC rates; right: normalized Gflops) on 4, 8 and 12 cores

5.2   Error Propagation

We also investigate the propagation of memory errors through program execution. Since MiniFE calculations (functions) depend on the previous calculations (functions), we analyze how the errors spread through data flow. A major computation in miniFE is the CG solver, which performs a sparse matrix vector multiplication. There are six consecutive operations performed by the CG solver, dot_r2, $daxpby$, $matvec$, dot, $daxpby$, and $daxpby$ (For the sake of better understanding, we call these operations dot_r2, 1_daxpby, $matvec$, dot, 2_daxpby, and 3_daxpby when we present results). Figure 4 shows flow graphs for each programming model. The most time-consuming part is the sparse matrix vector product ($matvec$ function).

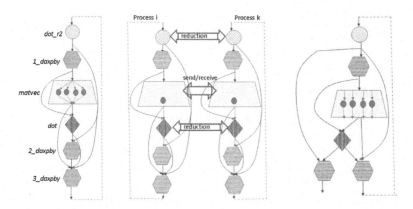

Fig. 4. Flow graphs for different programming models (OpenMP, MPI, OmpSs).

The other operations are dot products. At *dot_r2* and *dot* functions, a reduction operation is also performed to sum the dot product values resulting from all the threads. They open parallelism inside each function. The figure shows the case of the *matvec* function as an example. In this example, we see the OpenMp and OmpSs programming models when working with 4 PEs (threads), and the MPI programming model with 2 PEs (processes). In OpenMP and OmpSs all threads work on the same copy of the structures. In OmpSs, the flow of the execution is controlled by data dependencies, so after the execution of the *dot* function, both *2_daxpby* and *3_daxpby* can start execution as they do not depend on each others' output. In MPI, each process has its own copy of the data and the exchange of data is performed at specific points through send/receive operations (*matvec* function) or through global reduction operations (*dot* and *dot_r2*).

In order to see the error propagation effect through data flow, we corrupt the input vectors of each operation and check the output values. In Fig. 4, the input-output dependence flow between functions is shown by the flow direction of the arrows. We perform high-level fault injections by modifying a random index of the input vectors. We store the correct execution results generated by each operation, then compare them with the results obtained by faulty executions, and report the incorrect value rate for each case. We only report the effect of errors on *1_daxpby*, *matvec*, *2_daxpby*, and *3_daxpby* operations by specifying incorrect value rates. We perform experiments for 12-core execution. We collect the number of incorrect values calculated by each function through iterations. For a matrix size of $100 \times 100 \times 100$, one can observe that miniFE converges to a valid result before 300 iterations, so we limit the erroneous execution to 300 iterations. We modify the values (random input vector element) at different functions as well as at different iterations (iteration 50, 100, 150, 200, and 250) to determine the effect of the errors at different execution points. Then, we compare the output of each function at the end of the iterations 100, 200, and 300 to observe the propagation effect through iterations. We modify the input vectors by adding a small value (0.1) to one random element at iteration 50.

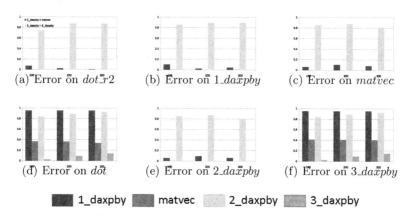

Fig. 5. Propagation error effect by modifying input vectors in the specified function at iteration 50. The results are shown at iteration 100, 200 and 300

One may observe that errors introduced at iteration 50 produce almost the same effects at all three programming models, so we include only one set of results (specifically collected by MPI execution, but infinitely small differences for OpenMP and OmpSs). Figure 5 presents the incorrect value rates (on the y axis) at each observed function, and the iterations (on the x axis) at which the result is observed. For instance, in Fig. 5(a), errors are introduced in dot_r2 and 1_daxpby and 2_daxpby outputs show changes of 6 % and 74 %, respectively at the end of iteration 100. The corresponding values at the end of iteration 200 and 300 show that the error propagating to 1_daxpby decreases along iterations, but increases for 2_daxpby. For the rest of the functions there is no significant propagation effect.

One can observe that the final result (i.e. the output vector of 2_daxpby, the lighter shaded column) is always significantly affected by corrupted data: above 75 %. Closing consideration of the propagation effect for errors introduced in dot and 3_daxpby, reveals that all functions are clearly affected in their output values, and these incorrect values are accumulated to corrupt the final result by a larger amount. In particular, at the iterations 200 and 300, the corruption is always more than 90 %. So, these two functions are eligible for applying fault tolerance techniques to improve soft-error reliability in an efficient way.

6   Related Work

Luo et al. [11] present a memory error vulnerability analysis for data-intensive applications running on large-scale datacenters. They investigate program segments and memory regions in a program, and compare the soft error tolerance across program and memory regions. In this work, we implement a similar approach for soft error emulation framework which simulates single-bit memory error in a random location in a program. Our framework uses binary instrumentation instead of software debuggers.

Shantharam et al. [12] present both theoretical and empirical analysis for the impact of a soft error on sparse matrix vector multiplication kernel operations. The work conducts a high-level soft error insertion methodology and demonstrates the propagation through a sequence of operations. Our analysis investigates the propagation effect through both functions and iterations.

Bronevetsky et al. [5] investigates the soft error vulnerability of iterative linear algebra methods by conducting high-level fault injection experiments. The authors analyze all data structures, and evaluate some types of error detection and error tolerance methods. Our fault injection methodology to analyze the soft error vulnerability introduces single-bit flips directly to the memory locations by using binary instrumentation.

Fang et al. [7] proposes a methodology to quantify the error resilience of OpenMP applications. The work extends a recent fault injection framework [17] to inject faults into OpenMP programs, and provides error resilience characteristics for a set of benchmarks. While the paper considers the faults in computational elements, our fault model focuses on memory elements and data values.

To the best of our knowledge, this is the first work that evaluates fault behavior and compares the memory error vulnerability of parallel programming models by presenting both soft error characteristics and error propagation through data flows.

7   Conclusions

In this work, we present a memory error vulnerability analysis for different programming models including MPI, OpenMP, and OmpSs. Our results show that programming models have an impact that varies the memory error vulnerability, so a better understanding about data and error patterns can improve the program performance by choosing the best programming model to make the most of silent errors. We also observe that there is a need to modify either the algorithmic flow or data structures to acquire a significant amount of reliability gain. Our error propagation analysis demonstrates that the propagation effect depends on the functions and data structures into which the error is introduced.

In future work, we plan to evaluate other applications, using different data types as well as different access patterns, including several data types. Furthermore, as we are evaluating message-passing models and HPC applications, we plan to evaluate how errors are propagated among nodes, even when using heterogeneous architectures.

Acknowledgments. The authors acknowledge the support of the BSC (Barcelona Supercomputing Centre). We would like to thank the anonymous reviewers for their comments, which helped us to improve the manuscript. This work has been supported by the European Commission through the HiPEAC-3 Network of Excellence (FP7/ICT-217068), the Spanish Ministry of Education (TIN2012-34557), and the Generalitat de Catalunya (2014-SGR-1051).

References

1. Mantevo benchmark suite. http://mantevo.org/
2. Mpi. http://www.mcs.anl.gov/research/projects/mpi
3. Ompss. http://pm.bsc.es/ompss
4. Openmp forum. http://openmp.org/forum/
5. Bronevetsky, G., de Supinski, B.: Soft error vulnerability of iterative linear algebra methods. In: International Conference on Supercomputing (ICS) (2008)
6. Cappello, F., Geist, A., Gropp, B., Kale, L., Kramer, B., Snir, M.: Toward exascale resilience. Int. J. High Perform. Comput. Appl. **23**(4), 374–388 (2009)
7. Fang, B., Pattabiraman, K., Ripeanu, M., Gurumurthi, S.: Evaluating the error resilience of parallel programs. In: Fault Tolerance for HPC at eXtreme Scale (FTXS) Workshop (2014)
8. Goloubeva, O., Rebaudengo, M., Reorda, M.S., Violante, M.: Software-Implemented Hardware Fault Tolerance. Springer, New York (2006)
9. Li, D., Vetter, J.S., Yu, W.: Classifying soft error vulnerabilities in extreme-scale scientific applications using a binary instrumentation tool. In: Proceedings of the International Conference on High Performance Computing, Networking, Storage and Analysis (2012)
10. Luk, C.-K., Cohn, R., Muth, R., Patil, H., Klauser, A., Lowney, G., Wallace, S., Reddi, V.J., Hazelwood, K.: Pin: building customized program analysis tools with dynamic instrumentation. In: Proceedings of the Conference on Programming language Design and Implementation (2005)
11. Luo, Y., Govindan, S., Sharma, B., Santaniello, M., Meza, J., Kansal, A., Liu, J., Khessib, B., Vaid, K., Mutlu, O.: Characterizing application memory error vulnerability to optimize datacenter cost via heterogeneous-reliability memory. In: Proceedings of the International Conference on Dependable Systems and Networks (2014)
12. Shantharam, M., Srinivasmurthy, S., Raghavan, P.: Characterizing the impact of soft errors on iterative methods in scientific computing. In: International Conference on Supercomputing (ICS) (2011)
13. Shivakumar, P., Kistler, M., Keckler, S., Burger, D., Alvisi, L.: Modeling the effect of technology trends on the soft error rate of combinational logic. In: Proceedings of the International Conference on Dependable Systems and Networks (2002)
14. Sridharan, V., Kaeli, D.R.: Eliminating microarchitectural dependency from architectural vulnerability. In: Proceedings of IEEE 15th International Symposium on High Performance Computer Architecture (HPCA) (2009)
15. Utrera, G., Gil, M., Martorell, X.: Analyzing the impact of programming models for efficient communication overlap in high-speed networks. In: International Conference on High Performance Computing and Simulation (HPCS) (2014)
16. Weaver, C., Emer, J., Mukherjee, S.S., Reinhardt, S.K.: Techniques to reduce the soft error rate of a high-performance microprocessor. In: 31st Annual International Symposium on Computer Architecture (ISCA) (2004)
17. Wei, J., Thomas, A., Li, G., Pattabiraman, K.: Quantifying the accuracy of high-level fault injection techniques for hardware faults. In: International Conference on Dependable Systems and Networks (DSN) (2014)
18. Yan, J., Zhang, W.: Evaluating instruction cache vulnerability to transient errors. In: Proceedings of the 2006 Workshop on MEmory Performance: DEaling with Applications, Systems and Architectures (MEDEA) (2006)
19. Zhang, W.: Computing and minimizing cache vulnerability to transient errors. IEEE Des. Test **26**(2), 44–51 (2009)

An Approach for Ensuring Reliable Functioning of a Supercomputer Based on a Formal Model

Alexander Antonov, Dmitry Nikitenko, Pavel Shvets$^{(\boxtimes)}$, Sergey Sobolev,
Konstantin Stefanov, Vadim Voevodin$^{(\boxtimes)}$, Vladimir Voevodin,
and Sergey Zhumatiy

Research Computing Center, Lomonosov Moscow State University, Moscow, Russia
{asa,dan,shpavel,sergeys,cstef,vadim,voevodin,serg}@parallel.ru

Abstract. In this article we describe the Octotron project intended to ensure reliability and sustainability of a supercomputer. Octotron is based on a formal model of computing system that describes system components and their interconnections in graph form. The model determines relations between data describing current supercomputer state (monitoring data) under which all components are functioning properly. Relations are given in form of rules, with the input of real monitoring data. If these relations are violated, Octotron registers the presence of abnormal situation and performs one of the predefined actions: notification of system administrators, logging, disabling or restarting faulty hardware or software components, etc. This paper describes the general structure of the model, augmented with details of its realization and evaluation at supercomputing center in Moscow State University.

Keywords: Reliability · Autonomous operating · Model of supercomputer · Monitoring · Supercomputing

1 Introduction

A supercomputing center maintenance practice determines a set of strict requirements for the technologies and facilities supporting supercomputer operation: maintaining high productivity of the supercomputer, constant monitoring of all potential sources of emergency situations and the performance of all critical components, automatic decision-making by maintenance and support systems, and guaranteed operator notification about the supercomputers current status, among other requirements. Until all of these requirements are met, neither the efficient operation of the supercomputer nor the safety of its hardware can be guaranteed.

The reasons for such requirements are clear. Supercomputers are expensive and therefore downtime is unallowable. The demand for supercomputers is high, and that means the maximum operational equipment should be available to the users. Supercomputers require high power consumption, which means their status needs to be monitored closely to avoid an equipment loss. And the complexity of the maintenance problem grows extremely fast with the vast number

R. Wyrzykowski et al. (Eds.): PPAM 2015, Part I, LNCS 9573, pp. 12–22, 2016.
DOI: 10.1007/978-3-319-32149-3_2

of components in any modern supercomputer. Fulfillment of these requirements sets the foundation for the Octotron system. Lets make a formal supercomputer operation model and make it available to the Octotron as an input data. The Octotron will constantly observe the current state of supercomputer using the monitoring systems and compare it to the model. If actual supercomputer state does not correspond with the model, Octotron can perform one of the preprogrammed actions, such as notifying the administrator via SMS, disabling the malfunctioning device, and so on. With this approach we, first, guarantee complete control over the situation; second, guarantee compliance between the expected and actual behavior of the supercomputer; and third, will be confident that we will find out about any event that deserves our attention.

This approach leads to a number of useful outcomes. In particular, it allows us to not only guarantee the reliable operation of the existing fleet of systems within a supercomputing center, but also to ensure continuity in maintenance when moving to a new generation of machines. Indeed, once an emergency situation arises, it is reflected in the model along with the causes and traceable features, and an adequate reaction is programmed into the model.

The paper is organized as follows. In Sect. 2 we describe interesting existing solutions for the discussed and related problems. In Sect. 3 we briefly state what our goals for the Octotron system being created are. Section 4 contains detailed description of the structure of our system, with the focus on one of its most important parts supercomputer model. This structure description is continued in Sect. 5 with the explanation how our system operates. In Sect. 6 we show how the Octotron system is being used on real supercomputer systems in our university. Finally, Sect. 7 contains conclusions and acknowledgments.

2   Background and Related Works

The work to ensure reliable supercomputer operations has been going on for a while, and a broad range of materials and methods has been accumulated. The following approaches to handling emergency situations are provided [1]: forecasting potential failures and their consequences; preventing failures; reducing the number of errors and their impact on system health; ensuring resilience against failures.

In global practices, the resilience for a supercomputer is primarily viewed in the context of ensuring reliable application execution. The methods available to support the execution of a large application in a potentially unreliable environment are based on creating checkpoints and logging communications, which allows the application to recover from failures in computing system components [2]. However, these mechanisms do not address equipment safety issues. Maintaining reliable operations and monitoring the status of a supercomputing system can be done using proprietary vendor hardware and software solutions (HP BTO Software [3], xCAT [4]). An alternative option is to install and configure one of the freely distributed monitoring systems (Nagios [5], Zabbix [6], Ganglia [7]) and writing a set of scripts to respond to specific subsets of potentially dangerous situations.

Currently, the aforementioned systems do not use a coherent model or description of a computer as input data. In our opinion, lack of model usage makes very difficult, if not impossible, to analyze and react properly on complex global fault situations, which concern not only one component such as one node or server, but a variety of cluster components. The powerful Zenoss [8] monitoring system offers the target system modeling concept, but that is only understood as automatic identification of the system configuration Failure detection based on a system of rules has been implemented in ClustrX Safe/AESS (Automated Notification/Equipment Shutdown System) [9] by the T-Platforms company. However, this system is focused solely on infrastructure monitoring and does not affect the supercomputers computational and software parts. It is also the vendors proprietary solution.

Another interesting recent development is the Iaso [10] system designed by NUDT University in China. This system supports the autonomous operation of the Tianhe-2 supercomputer, the current leader of the Top500 rating. It is an integrated piece of software that addresses all of the issues of automatic fault detection and elimination within system components, locating the root causes of failures, performing self-diagnostics and self-testing of the computing system, and recovering applications after failures. The ideas implemented in Iaso partially correlate to the ideas of this project. Even though Iaso is declared to be a universal software complex, it requires a modified Linux kernel on the clusters computing nodes with its own client modules installed. Iaso receives a description of the target system as input, but this is mainly used for monitoring and controlling the network infrastructure. Iaso is not publicly available at the moment.

Therefore, currently there have not been found any open system that uses formal model for maintaining reliability of a supercomputer.

It should be mentioned that Octotron system is intended to work with existing monitoring systems like collectd, Nagios, or Zabbix as data source thus avoiding unnecessary duplication of data collecting agents.

3   Requirements for Octotron System

The primary functional requirement for the Octotron system has in fact been formulated: the system must allow a supercomputer to independently control its own operation by comparing its current state to a predefined model. Since this task is not a trivial one, and Octotron must operate in a complex supercomputer environment, an additional set of requirements has been formulated which the system must meet:

1. be able to control all key failure causes in a supercomputers hardware and software components;
2. allow the monitored area to be expanded to include any components that were not originally controlled by the system but caused a failure;
3. be able to react to emergencies independently from the operator by performing a set of predefined actions to eliminate a failure or to notify support team;

4. support the current generation of teraflops and petaflops supercomputers and be ready to work with the next generation of supercomputing systems, where the number of components will increase by an order of magnitude;
5. verify the integrity of its own operation and evaluate the adequacy of information it has received on the state of the supercomputer;
6. accumulate experience from previous supercomputing system support, minimizing the number of repeated failures on both existing and prospective supercomputers.

Considering these requirements, an architecture was proposed for the Octotron system based on a formal model of supercomputer operation, which is described in the next section of this paper.

4   The Supercomputer Model

Octotron represents the supercomputer model in the form of a graph [11] (Fig. 1), which is used to describe all typical modern supercomputer components and relations between them. The model is accompanied by a set of rules and reactions. The rules help the supercomputer to register a failure or emergency, and the reactions describe what actions need to be taken once a rule is triggered.

Vertices in the graph correspond to physical or logical components of the supercomputer that need to be monitored: computing nodes, UPS modules, job queues, software components, licenses, etc. The criteria are simple: everything that the efficient operation of a supercomputer depends on must be reflected in the model. The graph edges correspond to the relationships between components, e.g. consists of, provides power to, connected with Infiniband. Each vertex in the graph is associated with a set of attributes which describe that components status: processor temperature, amount of memory, number of jobs in a queue, etc. The Octotron system updates attribute values through the supercomputers own monitoring systems (like collectd or Nagios) or directly via external interfaces on the components. The latter method is used, in particular, to work with engineering infrastructure over the SNMP protocol, or to interact with a GSM modem, or to get the current status of software licenses.

The core of Octotron is written in Java programming language. The system generates operative graph in the memory, while using Neo4j database as a long-term storage. All write requests are executed on both graphs so that we can keep current supercomputer state description up-to-date, while read requests use only memory-stored graph for improved performance. Neo4j can be used independently from the Octotron system, serving as a standardized interface for side tools, such as visualization, analysis, debugging and so on. Database support is optional and can be disabled, but in case of termination or failures all data will be lost.

Python language was chosen as the primary language for model description. We use the Jython interpreter, which executes the code on a Java virtual machine and allows classes from a Java code to be used in a Python program. Since Python is a rather simple and clear language, even an untrained person can create a model, following the examples and documentation for the Octotron system. Heres a sample of code describing the model shown on Fig. 1.

```
# creating basic components
room = CreateObject()
chiller = CreateObject()

# creating components with attributes
ups = CreateObject({"sensor" : {"load" : Long()}})
air_cond = CreateObject({"sensor" : {"fluid_temp" : Long()}})
hot_aisle = CreateObject({"sensor" : {"air_temp" : Long()}})
rack = CreateObjects(3, {"sensor" : {"temp" : Long()}})

# creating contains edges
OneToOne(room, ups, "contains")
OneToOne(room, air_cond, "contains")
OneToOne(room, hot_aisle, "contains")
OneToEvery(room, rack, "contains")

# creating power edges
OneToOne(ups, air_cond, "power")
OneToEvery(ups, rack, "power")

# creating chill edges
OneToOne(chiller, air_cond, "chill")
OneToOne(air_cond, hot_aisle, "chill")
OneToEvery(hot_aisle, rack, "chill")
```

Once created, such model can be rather easily updated in case supercomputer structure is needed to be changed (for example, computational core upgrade or some topology modifications is going to be made) by simply modifying this model description and running model creation process again. In this case all attributes from the older version of the model will persist if its object still exists in the new model.

5   Octotron System Operation

After a supercomputing system model has been created, we need, first, to add actual data on the structure of supercomputer components; second, to organize verification of abnormal statuses for the components (i.e. rules); and third, to define the respective system reactions to these situations. All of these activities are implemented in the Octotron system.

The supercomputer model is accompanied with a set of rules, which define deviations in the supercomputers behavior from what is set in the model. Each attribute in the model is linked to a set of rules that is triggered with every change in the value of the given attribute. In particular, the rules can be used

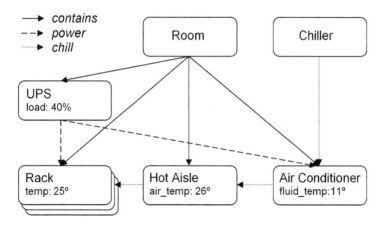

Fig. 1. General idea of the Octotron supercomputer model: vertices (computing nodes, UPS modules, job queues, software components, licenses, etc.), edges (relationships), attributes

to determine the rate at which attribute values change, which helps control situations similar to this: a rapid growth in the error rate on network interfaces can signal about a network failure. The rules are assigned to vertices and edges but can access attributes not only from the same vertex but also from adjacent vertices with the given relationship type.

Most commonly used templates of rules and their compositions available in our system are the following:

1. compare an attribute value with a constant or another attribute. Example: identify when a component temperature exceeds predefined limits;
2. aggregate values of several attributes and check the result. Example: identify when temperature exceeds the predefined limits for several sensors in a hot aisle;
3. check attribute values in two adjacent (according to the graph) vertices. Example: different operational modes on a pair of ports directly connected to an Ethernet network;
4. convert an attribute value for a further usage with other rule. Example: convert an absolute value to percentage value;

In addition, reaction modifiers allow extending checks in two ways:

1. check if reaction condition is maintained over a certain period of time. Example: LoadAVG value (the number of OS processes ready for execution) at a node can awhile go beyond ordinary limits, but if the value stays high for a long time, this can indicate problems on the node;
2. check if a reaction condition repeats several times in a row. Example: a node fails an SSH access check more than three times in a row.

If a rule registers an abnormal operation of a component, the corresponding reaction is triggered. A reaction represents a set of actions, such as recording data in the log file, notifying system administrators by e-mail or SMS, or invoking a custom command. Rules and reactions, as well as the model itself, are written in Python using a special library.

Set of rules to check and report number of available nodes is shown below. It writes a message to a log file if more than 20 % of nodes are unavailable (ping failed).

```python
def NodeGroupModule(nodes_count):
  return {
    "var" : {
      # calculate the number of neighborhood objects with
      # attribute "ping" equals to "false"
      "failed_ping" : ASoftMatchCount(False,
        EDependencyType.OUT, "ping"),

      # converts the number to a percentage value,
      # basing on provided total nodes count
      "failed_pct_ping" : ToPct("failed_ping",
        nodes_count),

      # returns true if the percentage is below 20 or
      # false otherwise
      "failed_pct_ping_ok" :
        UpperThreshold("failed_pct_ping", 20),
    },

    "react" : {
      # reaction for an object is triggered when the
      # attribute "failed_pct_ping_ok" becomes "false"
      Equals("failed_pct_ping_ok", False):

        # reaction writes the message with category
        # "Danger" to a log file
        ( Danger("tag", "SYSTEM").Msg("descr",
          "too many unavailable nodes")

        # when the "failed_pct_ping_ok" attribute turns
        # back to "true" -- another entry will be added to
        # a log file
        , Recover("tag", "SYSTEM").Msg("descr",
          "nodes are available again"))
    }
  }
}
```

Data on a supercomputers status is imported into Octotron from external sources: a monitoring system or directly via component interfaces. Octotron is not tied to any specific monitoring system, but can work with any one through a custom import module. Data from various supercomputer subsystems can be imported into Octotron at various intervals, depending on the necessary reaction speed to failures in each respective component.

Users and administrators can interact with the system with a web browser through HTTP requests. All requests are divided into few categories: view requests, modification requests and control requests, with an option of a separate authentication. View requests allow users to query all information about objects, attributes, rules and reactions. Modification requests allow to import data and change the model. Control requests are used by administrators to perform such actions as terminating model, suppressing all reactions and accessing system performance metrics.

Data import into Octotron system is made by import modules a set of tools that convert output of different monitoring system or methods to required format. Import modules do not have a strict structure; the only requirement for module is to provide a valid import request. We are planning to provide a default modules for typical monitoring tools and methods, such as SNMP, collectd, ping and ssh checks, written on Bash language and Python.

The Octotron system also features several levels of built-in diagnostics and self-diagnostics. To verify the appropriate operation of the system, an independent service graph with basic needed functionality is added to the system database. Octotron processes the specific rule for this service graph that involves this functionality, which in turn is verified by the external process. If the check passes, the system is considered to be operating correctly.

Another check is related to the frequency by which attribute values associated with the actual supercomputer components are updated. Each sensor attribute can be assigned a timeout, during which its value should be received from the source. If the value is not updated within that time, it indicates that either the respective component is faulty, or there is an error in its monitoring status. Similarly a timeout can be specified during which the value must change or, on the contrary, remain stable.

The last verification level is the monitoring of one Octotron system with another Octotron system: two independent copies of Octotron that are monitoring two supercomputers are being monitored by the third one. This third instance provides two types of diagnostics: (1) checks if Ocrotron processes are alive and (2) send requests to service graph described earlier to verify that it is handled correctly.

6 Evaluation of the Octotron System at Moscow State University

Octotron system is deployed on two Moscow State University supercomputers: Chebyshev (60 TFlops performance peak, 625 nodes, 5,000 cores, 42 racks and 1 hot aisle) and Lomonosov (1.7 PFlops peak, 5,000 nodes, 82,000 cores, 115 racks and 5 hot aisles). Chebyshev supercomputers model contains 10,228 vertices, 24,698 edges and 205,044 attributes. Lomonosov supercomputers model contains about 116,000 vertices, 332,000 edges and 2,400,000 attributes. Models reflect the following supercomputer components: a power supply system (UPS, battery modules); a cooling system (chillers, in-row air conditioners, environment monitoring); a management component (access nodes, job queues); a computing component (chassis, nodes, disks, memory); a shared file system; Ethernet network (switches, ports); and an Infiniband network (switches, network manager). The following relationships exist between the said components: contains, chill, connected via Ethernet, connected via service network, connected via Infiniband network, includes and provides power to.

As an example, we describe how Octotron operates on the Chebyshev supercomputer mentioned above. The main supplier of operational data for this supercomputers computing components is the monitoring system based on collectd. The hardware infrastructure supplies data via SNMP.

Every 10 min, the Chebyshev supercomputer hosts report the following data: the number of active SSH user sessions; the number of active software licenses; the number of jobs in each of the six partitions of the supercomputer (total number, jobs queued, jobs being executed, completed jobs); the number of processors (total, available for job execution, blocked); and the GSM modem account balance for sending emergency SMS notifications.

The following data is collected from all computing nodes at a higher frequency (every minute): the temperature inside the node; each processors temperature; ID of a job being executed; the status of the file system; memory status (total/free/occupied memory); Infiniband card status (sent/received packet counter, errors); Ethernet card status (errors); average node load, number of zombie processes, and other system data. SMART information on HDD status is additionally collected from nodes equipped with local hard disks.

The following data is collected from the shared file system every 10 min: free/occupied space, the performance, as well as the status and load of each blade module. Data from Ethernet switches are collected at the same intervals.

Information on the supercomputers climate control system is collected every minute. It includes data from several indoor temperature and humidity sensors and the status of each of 8 in-row air conditioners (air/coolant temperature before/after the air conditioner, and a number of various alerts). More than 60 parameters are recorded at the same interval from each of the five UPS units: the status of grid power and the UPS operating module, battery status, etc.

About 160 rules are used to control the operation of the Chebyshev supercomputer. Some of them are: GSM modem account balance is close to the deactivation limit; failures in the operation of two or three chillers; substantial increase in the error rate on network interfaces; number of user sessions at the host is below threshold.

In both cases of Lomonosov and Chebyshev supercomputers, overhead of monitoring systems is very low: $<1\%$ of computation load of nodes and $<1\%$ of communication network bandwidth is used. Octotron instance for Lomonosov supercomputer works on one dedicated server (2x Xeon E5450, 32 GB RAM), uses $<10\%$ of CPU and 6 GB of physical memory (20 GB of virtual). Chebyshev instance requires much less resources.

Within the evaluation period Octotron system is being used together with ClustrX Safe/AESS (mentioned in Sect. 2) on Lomonosov supercomputer. In this case Octotron only notifies about found failures but do not perform automatic actions like shutting down hardware. We still had few opportunities to fully check functionality of our system, since in most cases failures are local and not very hard to find. The most common failures are: node went down; temperature of HW component is rising; IB card failed; etc. A few less common cases that were discovered: not expected RAM memory volume on several nodes; chiller failure; not enough disk space on utility server; password attack on utility server. One interesting case we'd like to mention here. During summer outside temperature was getting very high, and at the same time small part of cooling hardware was down. Supercomputer became to overheat, and Octotron system notified administrators that this temperature problem is getting critical within the whole machine.

7   Conclusion

Currently, Octotron is used to support the reliable autonomous operation of the Chebyshev and Lomonosov supercomputer at MSU. In the near future, it will be deployed on the new Lomonosov-2 supercomputer.

The architecture and implementation of the Octotron system meets all of the criteria developed earlier. Potential sources of supercomputer failure are identified, thanks to a complete model of the target system and automatic monitoring of each component. Octotron is easily scalable for any supercomputer with petaflops performance, as confirmed by early experiments with the Lomonosov supercomputer. There are interesting areas for development, too. For example, further scalability can be achieved by breaking down the supercomputer model into a set of smaller models and launching several independent copies of Octotron, each monitoring its own part of the system. The overheads of the current version of Octotron are small and have no serious impact on the supercomputers performance.

Now, the Octotron system is following several directions of development. One is related to developing a shared bank of potential failures. Furthermore, interactive model visualization tools are being developed which allow the model not only to be viewed, but also promptly modified. Automated model creating tools will be expanded as well. Another interesting and promising area is an in-depth analysis of the flow of events taking place inside a supercomputer. This is aimed both at locating the root cause of failures and emergencies, and at forecasting such situations in the future. It is important that all of this functionality can be implemented by developing new modules and linking them to the already fully functioning Octotron kernel.

It is also noteworthy that the supercomputer model gets a value of its own as the key repository of knowledge on its structure. Moreover, the supercomputer model can be effective for educational purposes, since it can easily be used to demonstrate key components of a supercomputer and the relationships between them.

Octotron is available under an open MIT license [12,13].

Acknowledgments. This material is based upon work supported by the Ministry of Education and Science of the Russian Federation (Agreement N14.607.21.0006, unique identifier RFMEFI60714X0006).

References

1. Snir, M., Wisniewski, R.W., Abraham, J.A., Adve, S.V., Bagchi, S., Balaji, P., Belak, J., Bose, P., Cappello, F., Carlson, B., Chien, A.A., Coteus, P., Debardeleben, N.A., Diniz, P., Engelmann, C., Erez, M., Fazzari, S., Geist, A., Gupta, R., Johnson, F., Krishnamoorthy, S., Leyffer, S., Liberty, D., Mitra, S., Munson, T.S., Schreiber, R., Stearley, J., Hensbergen, E.V.: Addressing failures in exascale computing. Int. J. High Perform. Comput. Appl. **28**(2), 129–173 (2014)
2. Cappello, F., Geist, F., Gropp, W., Kale, S., Kramer, B., Snir, M.: Toward exascale resilience: 2014 update. Supercomput. Front. Innovations **1**(1), 5–28 (2014)
3. HP, BTO Software (part of HP Software Division). http://www8.hp.com/us/en/software/enterprise-software.html
4. Ford, E., Elkin, B., Denham, S., Khoo, B., Bohnsack, M., Turcksin, C., Ferreira, L.: Building a Linux HPC Cluster with xCAT. An IBM Redbooks Publication. http://www.redbooks.ibm.com/abstracts/sg246623.html?Open
5. Nagios monitoring system description. http://www.nagios.org/
6. Zabbix system. http://www.zabbix.com/
7. Massie, M.L., Chun, B.N., Culler, D.E.: The ganglia distributed monitoring system: design, implementation, and experience. Parallel Comput. **30**(7), 817–840 (2004)
8. Zenoss - Open Source Network Monitoring and Systems Management. http://www.zenoss.org/
9. Clustrx SAFE (AESS) software. http://www.t-platforms.com/products/hpc/software/clustrxproductfamily/clustrx-safe.html
10. Lu, K., Wang, X., Li, G., et al.: Iaso: an autonomous fault-tolerant management system for supercomputers. Front. Comput. Sci. **8**(3), 378–390 (2014)
11. Antonov, A.S., Vad, V., Voevodin Vl, V., Voevodin, S.A. Zhumatiy, D.A. Nikitenko, S.I. Sobolev, K.S. Stefanov, P.A. Shvets: Securing of reliable, efficient autonomous functioning of supercomputers: basic principles and system prototype. Vestnik UGATU (Scientific Journal of Ufa State Aviation Technical University) 18(2(63)), 227–236 (2014)
12. Octotron core repository. https://github.com/srcc-msu/octotron_core
13. Octotron framework: modeling and monitoring of complex computer systems. https://github.com/srcc-msu/octotron

Sparse Matrix Multiplication on Dataflow Engines

Vladimir Simic$^{(\boxtimes)}$, Vladimir Ciric, Nikola Savic, and Ivan Milentijevic

Faculty of Electronic Engineering, University of Nis, Aleksandra Medvedeva 14,
P.O. Box 14, 18000 Nis, Serbia
vsimic@elfak.ni.ac.rs

Abstract. In this paper, a novel architecture for sparse matrix multiplication is proposed. The architecture is suitable for implementation in specific environments such as dataflow engines. In order to avoid multiple streaming of elements from the host, we propose the architecture which buffers the elements from the input stream in on-chip memory in the form of pages. In the case of sparse matrices, the architecture processes only pages with non-zero elements. The proposed architecture allows replication of its blocks in order to parallelize the computation. The architecture is implemented on Maxeler dataflow engine based on Virtex 5 FPGA. The implementation results are given.

Keywords: Sparse matrices · Matrix multiplication · Maxeler dataflow engine

1 Introduction

Matrices are powerful mathematical tool used in many engineering and science fields and research activity is constantly directed toward implementation of high performance algorithms for matrix computations [1]. Matrices are crucial in modeling of dynamic systems, stock market prediction, solving of large systems of linear equations, etc. In many of such problems extremely large data sets emerge, which implies using of very large matrices. Increasing the size of matrices brings new challenges for efficient storage and high performance processing [2–6].

Sparse matrix multiplication is well known computation and it is in the core of many different engineering and science problems [5]. Irregularity of sparse matrix data representation makes them extremely unsuitable for multiplication on dataflow engine, which demands highly regular data streams.

FPGA platforms are widely used for acceleration of computations [7]. The Maxeler company introduced lately new concept, as well as programming language for designing of FPGA architectures. With high irregularities of sparse matrices and relatively small computational demands, it is a challenge to design streaming dataflow engine with acceptable performances.

In this paper, a novel architecture for sparse matrix multiplication, based on paging and efficient exploitation of streamed elements, is proposed. The architecture is suitable for implementation in specific environments such as dataflow

© Springer International Publishing Switzerland 2016
R. Wyrzykowski et al. (Eds.): PPAM 2015, Part I, LNCS 9573, pp. 23–30, 2016.
DOI: 10.1007/978-3-319-32149-3_3

engines. In order to avoid multiple streaming of elements from the host, we propose the architecture which buffers the elements from the input stream in on-chip memory in the form of pages. The architecture processes only pages with non-zero elements, which makes it suitable for multiplication of sparse matrices. The proposed architecture allows replication of its blocks in order to parallelize the computation. The architecture will be implemented on Maxeler dataflow engine based on Virtex 5 FPGA. The implementation results will be given. Although sparse matrices irregularities make them less suitable for highly regular deeply pipelined architectures, it will be shown that the proposed architecture can achieve the speedup of about 10 times compared to equivalent CPU based implementation.

2   Maxeler Dataflow Computation

To support dataflow processing based on FPGA technology, Maxeler brings a software architecture with Java based programming environment for end users. Main programming model is Data-Flow Engine (DFE), with Manager and Kernel as central building blocks Fig. 1 [8]. Kernel implements actual computations of the DFE and manager orchestrates data streams [8].

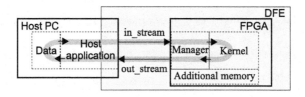

Fig. 1. Maxeler programming model

The Host application shown in Fig. 1 is executed on the host PC and manages data flow configuration, data preparation, streaming data to manager and one or more computational kernels. Maxeler provides flexibility in defining the source of dataflow, and for the single FPGA designs it can be either the host application or additional on-chip memory (Fig. 1). Furthermore, Maxeler brings great flexibility in defining dataflow movement in more complex designs.

3   Design of Architecture for Sparse Matrix Multiplication

Let $A = [a_{i,j}]$ and $B = [b_{i,j}]$ be two matrices with dimensions $M \times L$ and $L \times N$, respectively. The result of multiplication of matrix A and B is matrix $C = [c_{i,j}]$, with dimension $M \times N$ and elements

$$c_{i,j} = \sum_{k=1}^{L} a_{i,k} \cdot b_{k,j}, \tag{1}$$

where $i = 1, 2, \ldots, M$ and $j = 1, 2, \ldots, N$. The matrix A is referred as left operand, or left matrix, and matrix B as right matrix.

In order to design a streaming processor for matrix multiplication, the straightforward solution is to feed the architecture with left and right matrix multiplication operands using separate input streams, and to have one stream for the resulting matrix [1]. Having in mind the target platform, there are two concerns that should be addressed in that case. The first is balancing of two input streams over the same data path from the host. The second, and the most obvious is the fact that each matrix element is used several times during the computation. From (1) it can be seen that the element $a_{1,1}$ is multiplied with elements $b_{1,1}, b_{1,2}, \ldots, b_{1,n}$ in order to obtain $c_{1,1}, c_{1,2}, \ldots, c_{1,n}$, respectively, where $1 \leq n \leq N$. In order to avoid multiple streaming of elements from host CPU, we propose the architecture which will buffer one input stream in on-chip memory. Once when the operand is buffered, it is easier to optimize the execution in case of sparse matrices. The second stream will be used to fetch the corresponding prebuffered element in order to perform the multiplication.

The proposed architecture is shown in Fig. 2a. The architecture has one input data path which is used to multiplex left and right operand in the form of data streams ("in_stream_A" and "in_stream_B" in Fig. 2a). In order to achieve multiplexing, one input stream has to be held while the other stream flows, and vice versa. This functionality is provided by blocks $a\mathbf{H}$ and $b\mathbf{H}$ (Fig. 2a), where $\mathbf{H}$ stands for 'hold', and a and b are the number of clock cycles. On-chip memory is a temporary cache for elements $a_{i,j}$ of the left operand A.

After buffering the elements $a_{i,j}$ of the left operand A, the elements of the right operand $B = [b_{i,j}]$ are streamed into the architecture. Indices from the elements of the input stream $B = [b_{i,j}]$ are used to address on-chip memory to obtain corresponding element for multiplication (Fig. 2a). The architecture accumulates all intermediate results and generates resulting element $c_{i,j}$ to output stream when the computation is done. This process is driven by the control logic shown in Fig. 2a.

Let us consider the size of on-chip memory. The straightforward solution is to provide enough on-chip memory to store all elements $A = [a_{i,j}]$. However, this is not applicable for large matrices. The other solution is to store one row of matrix A at the time. In that case, the size of on-chip memory depends on matrix dimension which is not implementation-friendly. Furthermore, the dimension of the matrix which can be processed is still limited by the architecture. Thus, we propose the architecture with fixed size on-chip memory block. If the number of elements in one row is less or equal than available on-chip memory block size, all elements from one row will be in the memory at the same time. If the number of elements in one row is greater than available memory size, only a subset of elements $a_{i,n_p \cdot p + 1}, a_{i,n_p \cdot p + 2}, \ldots, a_{i,(n_p+1) \cdot p}$, will be stored in on-chip memory, where i stands for the row, while p is referred as "page size" $(1 \leq p \leq N)$, and n_p is the page number, such that $0 \leq n_p \leq N_p - 1$, where N_p is the number of pages per row. Thus, we introduced the input stream paging. Let us note that $N_p = \frac{N}{p}$.

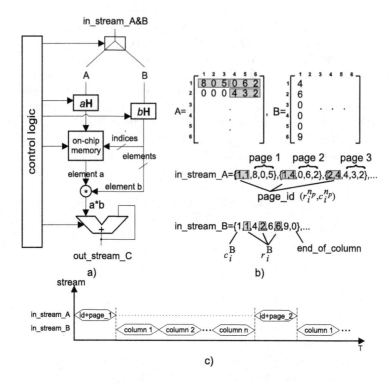

Fig. 2. The proposed architecture: (a) the kernel of streaming processor, (b) example of input data streams, (c) multiplexing of input streams

The example of input stream formats is given in Fig. 2b. The multiplexing of stream of pages is shown in Fig. 2c. Let us assume that the architecture multiplies two 6×6 matrices A and B (dimensions $L = M = N = 6$ in Fig. 2b). In order to illustrate the formation of data streams "in_stream_A" and "in_stream_B" only the first two rows from matrix A and the first column from matrix B are shown in Fig. 2b. The chosen page size is $p = 3$. Thus, there are two pages per row ($N_p = \frac{N}{p} = 2$). Pages of matrix A are shaded in Fig. 2b.

The "in_stream_A" is formed by adding the page identificator (page_id in Fig. 2b) before the first element of every page in the stream. Page identificator is a pair of elements: the page row index ($r_i^{n_p}$), and the column index of the first element from the page ($c_i^{n_p}$), where n_p is the page index ($1 \le n_p \le N_p$). Thus, for the example shown in Fig. 2b, the stream "in_stream_A" for the page 2 is $\{1,4,0,6,2\}$, where $r_i^2 = 1$, $c_i^2 = 4$, while set $\{0,6,2\}$ represents the elements. The "in_stream_B" is formed by adding the column identificator at the beginning (c_i^B), followed by pairs (r_i^B, b), where r_i^B is row index, and b is the value of the element (Fig. 2b). The pair (c_i^B, r_i^B) from the stream "in_stream_B" is used to address the on-chip memory in order to fetch element from the matrix A that corresponds to the current element b. For that purpose the control logic checks if the condition $c_i^{n_p} \le r_i^B \le c_i^{n_p} + p - 1$ holds. The next page will be fetched when the elements from all columns are multiplied with all elements from the current page (Fig. 2c).

The architecture is optimized in such manner that the "in_stream_A" doesn't contain the pages with all-zero elements as it is shown in Fig. 2b. The total number of pages in matrix $A_{N \times N}$ is $N_p \cdot N$. The number of non-zero pages can be significantly lower in the case of sparse matrices. We will denote the average number of non-zero pages per row as N_p^{avg}, where $N_p^{\text{avg}} \leq N_p$.

The parallelization of the computation is possible in the case when the throughput between host and the FPGA platform is enough to support several parallel streams. The proposed architecture can be replicated as a whole in order to enable parallel computation. By replicating kernel functional blocks and increasing the size of on-chip memory buffer using available resources of FPGA, it is possible to process several pages at the same time. The parallel architecture with K parallel kernels is shown in Fig. 3, where each kernel represents one architecture from Fig. 2a.

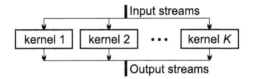

Fig. 3. Multi-kernel architecture for parallel processing of multiple pages

4   Implementation Results

The proposed architecture is implemented and evaluated on a Maxeler MAX2C card equipped with Virtex 5 FPGA. The host is Intel based PC with i5 2500 CPU with 4 GB of RAM running Centos 6.5 x64 operating system. All results are obtained for integer based arithmetic kernels.

The proposed architecture is evaluated for different matrix dimensions and different average number of pages per row. Multiplication is performed on basic architecture ($K = 1$) and two multi-kernel architectures for $K = 2$ and $K = 4$ (Figs. 2a and 3). By replicating the basic computational kernel two and four times, as it is shown in Fig. 3, two additional architectures are implemented. All three implemented architectures have on-chip memory blocks of the same size $p = 1000$ elements (Fig. 2a). As the implemented size of one integer element is 4 B, the total size of one on-chip memory block is 4 kB.

The results of implementation are given in Table 1. For each architecture ($K = \{1, 2, 4\}$) three columns are given, which show occupied FPGA resources for implementation of the architecture. The target Maxeler FPGA device has the total of 149760 Look-Up Tables (LUTs), 149760 Flip-Flops (FFs), and 516 Block RAM (BRAMs), 36kbits each, available for the user application. The 36 kbits BRAM block is programmable from $32\,\text{k} \times 1\,\text{b}$ to $512 \times 72\text{b}$, in various depth and width configurations. In our implementation, due to the size of the integer, BRAM block for on-chip memory is configured as $1\,k \times 32\,b$. Thus, roughly one out of 516 available BRAMs is required for storing of one page.

From Table 1 it can be seen that the percentage of occupied resources varies from about 6.5 %, for $K = 1$, up to about 11 % for $K = 4$. Four times replicated kernel doesn't occupy four times more resources than the basic architecture. This is due to the fact that we used relatively small size of on-chip memory which occupies only 3 BRAMs more for $K = 4$ than it is required for $K = 1$. The majority of difference between 35 used BRAMs for $K = 1$ and 51 for $K = 4$ goes to the Finite State Machine of the control logic, accumulation of intermediate results, and other architecture logic.

Table 1. FPGA resources occupation for $K = 1$, $K = 2$ and $K = 4$

Implementation	$K = 1$			$K = 2$			$K = 4$		
FPGA resources	LUTs	FFs	BRAMs	LUTs	FFs	BRAMs	LUTs	FFs	BRAMs
Occupied	10000	13337	35	10696	14471	39	11944	16685	51
%	6.68	8.91	6.78	7.14	9.66	7.56	7.98	11.14	9.88

To evaluate the performances of implemented architectures, six different sets of square sparse matrices $A_{N \times N}$ and $B_{N \times N}$ are prepared. The results are shown in Table 2. Three different dimensions are chosen ($N = \{2000, 5000, 10000\}$) for two different matrix densities, i.e. six different tests in total (Tests # 1 to 6 in Table 2). As the size of on-chip memory block is fixed ($p = 1000$), the number of pages per row is $N_p = 2$ for $N = 2000$ and $N_p = 5$ and $N_p = 10$ for $N = 5000$ and $N = 10000$, respectively.

However, more important is to vary the number of non-zero pages in test matrix A. Thus, we divided the tests in two groups regarding the number and distribution of non-zero pages in the matrix A. The first group (Tests # 1, 2 and 3) stands for the tests where the matrix A has majority of pages with all-zero elements. Precisely, one out of twenty pages has some non-zero elements in the case of Tests # 1 and 3 ($N_p^{\text{avg}} = \frac{2}{20} = 0.1$ and $N_p^{\text{avg}} = \frac{10}{20} = 0.5$), and one out of ten pages has some non-zero elements in the case of Test # 2 ($N_p^{\text{avg}} = \frac{5}{10} = 0.5$). On the other hand, in Tests # 4 to 6 matrix A is still sparse, but distribution of elements is such that there are no pages with all-zero elements (Table 2), i.e. $N_p = N_p^{\text{avg}}$. The total number of pages ($N_p \cdot N_p^{\text{avg}}$) in the stream "in_stream_A" for all tests is given in bold (Table 2). For all tests, sparse matrix B is chosen to contain $|b_{i,j}|$ arbitrary distributed non-zero elements. Test matrices A and B are prepared by the host application prior to streaming to the architectures.

The tests are executed for $K = \{1, 2, 4\}$. Bottom half of Table 2 lists total execution times obtained for the three architectures. In order to compare execution times of the proposed architectures on Maxeler platform with the execution of the same algorithm on CPU, additional row is given. The CPU in Table 2 refers to equivalent algorithm executed on the CPU. Execution times are given in seconds. Let us note that the time needed for the preparation of input data is not included in the results given in Table 2.

From Table 2 it can be seen that increasing the number of non-zero pages of matrix A and number of elements in matrix B the total execution time is increased for each of the implementations. Faster execution of Test # 4 compared with Test # 3 is due to the fact that the number of non-zero elements in the stream "in_stream_B" ($|b_{i,j}|$) in Test # 4 is much lower than the number of non-zero elements in Test # 3. The execution is faster due to the fact that the architectures have to multiply much less number of elements for the matrix B in that case.

The maximum obtained speed-up for given results is about 10 times in the case of $K = 4$ for Test # 6. Graphical illustration of the results from Table 2 is shown in Fig. 4. Due to the large range of evaluated values for both the number of pages and achieved execution times, the both axes of the graphic from Fig. 4 are given in logarithmic scale.

Table 2. Time required for matrix multiplication on architectures with $K = \{1, 2, 4\}$

Test #		1	2	3	4	5	6
	N	2000	5000	10000	2000	5000	10000
A	N_p	2	5	10	2	5	10
	N_p^{avg}	0.1	0.5	0.5	2	5	10
	$N \cdot N_p^{avg}$	200	2496	4992	4000	25000	100000
B	$|b_{i,j}|$	524288	4196304	8388608	524288	4494304	16777216
Time [s]	$K = 1$	0.61	14.51	44.64	10.77	301.62	1376.05
	$K = 2$	0.58	13.43	41.60	9.36	264.23	1339.33
	$K = 4$	1.03	8.94	20.91	12.60	193.77	549.94
	CPU	0.46	46.30	162.86	9.21	888.56	6298.77

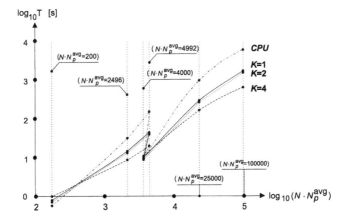

Fig. 4. Graphical representation of the time required for matrix multiplication

5   Conclusion

In this paper, a novel architecture for sparse matrix multiplication is proposed. The architecture is suitable for implementation in specific environments such as dataflow engines. In order to avoid multiple streaming of elements from the host, we propose the architecture which buffers the elements from the input stream in on-chip memory in the form of pages. In the case of sparse matrices, the architecture processes only pages with non-zero elements. The proposed architecture allows replication of its blocks in order to parallelize computation. The architecture is implemented on Maxeler dataflow engine based on Virtex 5 FPGA. Three architectures with different number of parallel kernels are implemented and evaluated. The achieved speedup for $K = 4$ parallel kernels, and matrix dimension $N = 10^5$, is about 10 times, compared to equivalent CPU implementation.

Acknowledgment. The research was supported in part by the Serbian Ministry of Education, Science and Technological Development (Project TR32012).

References

1. Milovanovic, I., Bekakos, M.P., Tselepis, I.N., Milovanovic, E.I.: Forty-three ways of systolic matrix multiplication. Int. J. Comput. Math. **87**(6), 1264–1276 (2010)
2. Smith, T.M., Van De Geijn, R., Smelyanskiy, M., Hammond, J.R., Van Zee, F.G.: Anatomy of high-performance many-threaded matrix multiplication. In: 28th IEEE International Parallel and Distributed Processing Symposium, pp. 1049–1059. IEEE (2014)
3. Matam, K., Indarapu, S., Kothapalli, K.: Sparse matrix-matrix multiplication on modern architectures. In: 19th International Conference on High Performance Computing (HiPC), pp. 1–10. IEEE (2012)
4. Saule, E., Kaya, K., Çatalyürek, Ü.V.: Performance evaluation of sparse matrix multiplication kernels on Intel Xeon Phi. In: Wyrzykowski, R., Dongarra, J., Karczewski, K., Waśniewski, J. (eds.) PPAM 2013, Part I. LNCS, vol. 8384, pp. 559–570. Springer, Heidelberg (2014)
5. Weifeng, L., Vinter, B.: An efficient GPU general sparse matrix-matrix multiplication for irregular data. In: IEEE 28th International Parallel and Distributed Processing Symposium, pp. 370–381. IEEE (2014)
6. Yavits, L., Morad, A., Ginosar, R.: Sparse matrix multiplication on an associative processor. IEEE Trans. Parallel Distrib. Syst. (2014)
7. Ciric, V., Cvetkovic, A., Simic, V., Milentijevic, I.: Tropical algebra based framework for error propagation analysis in systolic arrays. Elsevier Appl. Math. Comput. **225**, 512–525 (2013)
8. Maxeler Technologies, MaxCompiler documentation, Version 2014.1, Maxeler Technologies

Energy Efficient Calculations of Text Similarity Measure on FPGA-Accelerated Computing Platforms

Michał Karwatowski[1,2(✉)], Paweł Russek[1,2], Maciej Wielgosz[1,2],
Sebastian Koryciak[1,2], and Kazimierz Wiatr[1,2]

[1] AGH University of Science and Technology, Mickiewicza Av. 30,
30-059 Krakow, Poland
{mkarwat,russek,wielgosz,koryciak,wiatr}@agh.edu.pl
[2] ACC Cyfronet AGH, Nawojki 11, 30-950 Krakow, Poland

Abstract. This paper presents an impact of the customized hardware accelerator on the overall performance of the text similarity computing system. The hardware processing module that is presented in the paper is a building block of the processing engine in the search system of related documents. The engine is used in the phase of preliminary retrieval of similar documents. The TF-IDF weighting scheme and cosine similarity metric are used by the module. Evaluation boards equipped with Xilinx's Field Programmable Gate Array (FPGA) were utilized as a hardware platforms for implementation of the selected time-consuming operations. The series of tests was conducted, and the results of the hardware-accelerated solutions were compared against the standard software implementation. The two different FPGA-enabled platforms were employed in the experiments. The low-power and the high-performance platform were used to compare the metrics of different hardware solutions. We provide the adequate results and conclusions that present that the energy and speed metrics of the text similarity calculations can be improved thanks to the hardware accelerator. Consequently, the cluster of FPGA-enabled nodes is proposed for the large scale processing.

Keywords: FPGA · Reconfigurable systems · Custom computing · Document similarity

1 Introduction

Text processing has become the significant operation that is performed in today's computing centers. The necessary algorithms can be very complex and require substantial computing power [1]. The primary impulse to perform the experiment that is presented in this article was an idea to use newly emerging semiconductor devices to one of the most common tasks of today's ICT systems *i.e.* data browsing and searching. The hybrid FPGA SoC devices that integrate the multi-processor subsystem and FPGA structure on a single silicon chip become

© Springer International Publishing Switzerland 2016
R. Wyrzykowski et al. (Eds.): PPAM 2015, Part I, LNCS 9573, pp. 31–40, 2016.
DOI: 10.1007/978-3-319-32149-3_4

available now. These devices target a market of mobile devices and for that reason they feature low-power consumption. The same time, low-power calculations is a goal of big computing centers where reduction of power dissipation is a major issue to operate and thrive. The presence of the FPGA structure in a computing node provides green computing capability that is possible thanks to the advancements in semiconductor technology but also thanks to the method of computing with the aid of the custom processing architectures.

This paper presents the performance gain that can be achieved thanks to the use of CPU-FPGA heterogeneous computing node as a platform, which allow a system designer to implement the most computationally demanding parts of the algorithm in the customized hardware. Most computing platforms consists of processors, which compared to FPGA, are flexible to perform a wider range of various tasks. However, FPGAs offers the ability for the adoption of the processing architecture for the required algorithm [2]. Importantly, the hardware solution involves fine grain parallelism and pipelining of operations. This customization boosts the performance while lowering power consumption. Additionally, the computing nodes can be combined to form the computing cluster. The idea to aggregate nodes of reduced power (both computational and electric power) to build High Throughput Systems (HTC) is not new, as such systems are already commercially available. This approach is justified for search operations because they are IO-bound rather than compute-bound problems [3].

Our custom accelerator relies on the efficient and parallel operation of sparse vector multiplication. Acceleration of sparse matrix-based calculations has been discussed in many papers. Wang *et al.* [4] designed the complete system for mobile devices that accelerated matrix operations. They composed their system of multiple operation specific accelerators. Overall performance matched some state-of-the-art processors maintaining low energy consumption. Dorrance *et al.* [5] designed scalable kernel for sparse matrix-vector multiplication. In their work they proved that FPGAs can reach higher performance comparing to their theoretical maximum than GPUs or CPUs. Jamro *et al.* [6] designed scalable kernel for sparse matrix multiplication for High-Performance Computing on FPGAs. The authors proposed the custom architecture and presented the study how the matrices sparsity factor influences the ratio of expensive floating-point arithmetic operations to cheap integer comparison operations.

However, our work is concentrated on a task that is specific to the document similarity search engine. The papers [4,6] focused on optimization of a single sparse matrix operation, while the system described in this paper is able to perform many simultaneous calculations for a large set of sparse vectors.

2 Text Processing

Our text similarity search method requires that all text words (n-grams) are converted to their numerical representation. In the case of implementation that is presented here, a single word is used as the document term. Consequently, the text document can be treated as a vector. The Vector Space Model (VSM) has already been successfully used as a conventional method for text representation in previous works [7]. This model represents a document as a vector of features. In this scheme, a set of documents may be presented as the two-dimensional term/document matrix. Separate documents are represented as vectors in an N-dimensional vector space that is built upon the N different terms that occur in the considered text corpus *i.e.* a reference set of documents. The comparison and matching of the texts can be performed by using the *cosine similarity* measure in VSM. The cosine measure can be calculated according to the equation

$$cosine\ similarity(\mathbf{u}, \mathbf{v}) = \frac{\sum_{i=0}^{N}(u_i \cdot v_i)}{\|\mathbf{u}\| \cdot \|\mathbf{v}\|}, \tag{1}$$

where $v = (v_0, v_1, \ldots, v_N)$ and $u = (u_0, u_1, \ldots, u_N)$ are the text document vectors that are sparse in practice as they contain many zeros.

The most common algorithm for weighing coefficients in VSM is the method called *Term Frequency-Inverted Document Frequency* (TF-IDF). TF-IDF is a numerical statistic that reflects how important is a word/term to represent the document in the context of the whole documents collection. TF-IDF is often used as a weighting process by the information retrieval and text mining algorithms. There are also many others weighing schemes and similarity measures [8], and they can be easily adopted by our FPGA accelerator.

3   Computing with the FPGA-Enabled Computer Cluster

This paper is the continuation and further development of the work that was presented in [9–11]. In our system, we wanted to combine two elements: the FPGA accelerator [9] and a distributed processing model [10]. All development had to be done with the MapReduce model and accelerator oriented processing in mind. The architecture of FPGA custom processor was proposed and developed, and the efficient text data representation for 'map'/'reduce' methods that are suitable for documents vectors processing were projected. Finally, Java classes for Hadoop were developed that could delegate data execution to the hardware accelerator.

In a sparse vector format, the documents are stored as a list of couples of the integer index and non-zero numerical value. The list is ordered by the index value. The index is a 22-bit unsigned integer, and TF-IDF value is a 10-bit fixed-point number. Those bit-sizes constitute 32 bits in total that fits the bit-width of the accelerator interface. Additionally, each vector is normalized so that during the calculations every vector can be considered of the same length. The vectors are sent to the accelerator as the streams of couples to calculate their similarity. To calculate *cosine similarity* the accelerator reads vector data from the separate streams; it compares the indexes and when they match coefficients are multiplied, and the similarity value is updated.

Nowadays, MapReduce data processing model is a standard for data processing of large data sets in distributed databases. The first and most popular framework for MapReduce calculations is Hadoop. Hadoop's sequence file format

(SequenceFileAsBinaryInputFormat class) was used to store the vectors. They were kept in a binary key-value pairs. The 'key' is a text document identifier, and the 'value' field is a binary packed list of vector elements in a form (index, TF-IDF). A binary format of the 'value' field is compatible with an accelerator input stream form, so no processing is necessary while computing. The CPU sends 'value' straight to the FPGA accelerator for calculations. The task of the MapReduce application is to calculate the similarity between the reference vector (user document) and the documents that are stored in the database. The output results should be sorted according to the similarity score. It calculates the similarity between the reference document and database document in FPGA, and outputs ('similarity value', 'document identifier') pair. The MapReduce output is a list of database documents that are ordered by the similarity with the user vector.

In the FPGA-enabled node, the hardware accelerator is integrated with the CPU by Xillybus [12] interface. The CPU loads reference and database data into the node's memory and sends data to programmable logic using system queues provided by Xillybus. Thus, computation of document similarity takes place in the FPGA. The CPU takes calculation results using another queue. The system is designed to store a set of reference vectors in the internal accelerator's memory. This approach allows for a concurrent calculation of separate cosine similarities of the different reference vectors and vectors from the database. In our tests, hardware was set to store eight reference vectors simultaneously. In the software processing scheme, which is provided for reference, cosine similarity metric was calculated sequentially. The database is split among available CPU cores. That way each core calculates metrics for all reference vectors and the fraction of the database.

4   Experiment Description

4.1   Test Platforms

To widen the perspective, we prepared two substantially different hardware platforms, on which we performed all experiments. The first computing platform is the low-power ZedBoard development kit [13]. It is based on the XC7Z020-CLG484-1 device from Xilinx Zynq-7000 All Programmable SoC family. It combines the embedded ARM processor and FPGA on a single silicon chip. The 'Xillinux', which is an 'Ubuntu' related Linux OS for Zynq, was installed on the ZedBoard node. The 'Xillybus' was used as the CPU-FPGA interface.

The second platform is a typical server-class node. It is built on ASUS P6X58E-D motherboard, with Intel Core-i7 950 running at 3,066 MHz, 12 GB DDR3, and 512 GB HDD. Additionally, the server hosted the Xilinx Virtex-7 FPGA VC707 Evaluation Kit [14] connected with PCIe Gen2x8. The VC707 features Xilinx VX485T device.

4.2   Experiment Outline

We computed cosine similarity metric for each pair of the reference vector and the database vector in the experiments. We used pre-computed data of TF-IDF vectors for the similarity calculation. This kind of processing allowed us to search for the most similar pairs of vectors (documents). We used random articles downloaded from the Internet as the database. The database contained 12,500 different articles with an average number of 214 different terms each. The document similarity results were satisfactory, but they are not discussed in this paper. We were not concerned about the meaning of similarity results, so to mimic a bigger database, we copied downloaded articles multiple times in presented experiment. Afterward, we performed various search tasks for the different number of provided input reference vectors. As the processing time may slightly vary depending on the size of reference vectors, all used reference vectors originated from the same text document, which contained 947 different terms.

5   Results

5.1   Performance

We performed a series of experiments in a single node acceleration scheme. We set database size to 200,000 documents first. Then, we measured the time that hardware needed to process the different amount of reference vectors. Figure 1 presents the processing time for ZedBoard. The figure distinguishes runtimes of the CPU (labeled as ARM) and CPU-FPGA couple (labeled as ARM + FPGA). It is clear that runtime increases linearly for CPU; it is because each reference vector requires an additional database search. The use of unique hardware architecture, which is specially adjusted to the similarity algorithm and implemented in the FPGA, eliminates that effect, as reference vectors are already loaded into the internal accelerator's memory, and a single database run is sufficient to perform many comparison tasks in parallel. For eight reference vectors that are put into eight internal channels of the FPGA accelerator, the CPU-FPGA configuration reached the speedup of 11.7 times, comparing to the software application running on two processor cores. It is worth to note that the speedup reaches 1.5 when a single reference vector is processed.

Similar behavior can be observed for the server platform (Fig. 1). The CPU-FPGA couple (labeled as i7 + FPGA in the case of the server platform) allows for 1.4 times faster calculations for a single reference vector, and up to 10.5 for eight vectors. The software program was divided into eight threads in the case the CPU only experiment (labeled as i7), and each thread processed separate part of the database with the same reference vector. Eight number of threads had been found to be optimal for our processor.

For the experiment that is presented in Fig. 2, all platforms are tested with the higher amount of reference vectors. We can observe interesting effect when the number of reference vectors exceeds internal channel number of the hardware accelerator. Its processing time is about constant below that number, and then

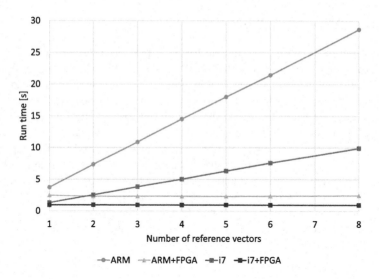

Fig. 1. The processing time of 200,000 database vectors. Up to eight reference vectors.

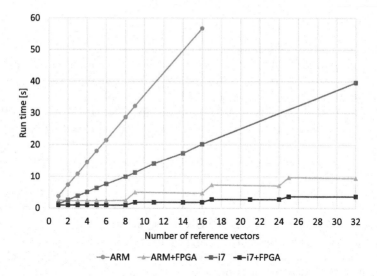

Fig. 2. The processing time of 200,000 database vectors. A high number of reference vectors.

it rises. The architecture of the hardware causes that non-linearity. The runtime depends mostly on how fast the database is transferred from memory to FPGA. There are eight internal channels in the tested hardware setup, so processing of the ninth vector requires an extra transfer of the whole database. The increase of the channels number on accelerated platforms can further amplify the speedup level.

Table 1. FPGA resource utilization.

| Resource type | Zynq PL | | | Virtex | | |
	Acc	Acc + Bus	Available	Acc	Acc + Bus	Available
Slice LUT	2,648 (4.98 %)	8,362 (15.71 %)	53,200	2,722 (0.9 %)	10,977 (3.62 %)	303,600
Slice register	3,967 (3.73 %)	9,366 (8.8 %)	106,400	3,975 (0.65 %)	11,653 (1.91 %)	607,200
Block RAM	123.5 (88.21 %)	127 (90.71 %)	140	123.5 (11.99 %)	132.5 (12.86 %)	1,030
DSP48	8 (3.63 %)	8 (3.63 %)	220	8 (0.29 %)	8 (0.29 %)	2,800

Table 1 presents the utilization of internal FPGA resources for the Zynq and Virtex devices. The table concerns resources of the accelerator alone (Acc) and the accelerator plus interface to host (Acc + Bus). Operating frequency of internal logic for Zynq was 100 MHz. The Zynq is not a high-performance processing device, but there is enough logic and BRAM resources to implement a total number of 32 channels for the reference vectors; that could quadruple the performance. Much higher gain can be achieved with the Virtex device which internal clock frequency was 250 MHz. Although the processing system implemented in programmable logic is identical, the resource utilization of Virtex is slightly higher than Zynq's. The difference in the interface to host causes that. Virtex is connected to the host with the PCIe bus, which is much more complex and, therefore, requires more logic than the AXI bus that is used in Zynq. Available Virtex's resources allow a designer to implement up to 256 internal channels. That could significantly rise the performance. Although, higher logic utilization increases overall power consumption, energy that is required for single operation is lower at the same time. All solutions scale well with the growth of the database size maintaining linear time rise.

5.2 Energy Consumption

The energy efficiency is another very important issue regarded in this paper. We measured the power that is consumed by the computing nodes during documents processing. It was important to establish a unified measure to compare the different platforms and settings directly. We chose the average energy value that is required to process a pair of vectors for that purpose. Our measure is derived from the total energy consumed during processing which is divided by the number of similarity calculations. For example, if platform required 1 Joule to compare two reference vectors with ten database vectors, it gives $1 J/(2 \cdot 10) = 0.05 J$ of an average energy per operation. We used the external power meter to measure the power consumption of a whole processing node, what is most similar to the real data center scenario.

Figures 3 and 4 present the results. The number of reference vectors does not influence the *energy per operation* consumed by the processor, therefore a plot is a flat line. However, the measure falls hyperbolically for FPGAs. It is because most energy consumed by FPGA does not rise with the number of used internal channels. ZedBoard platform used 3.99 W when only ARM processed documents. Accordingly, it needed 4.35 W when hardware accelerator was

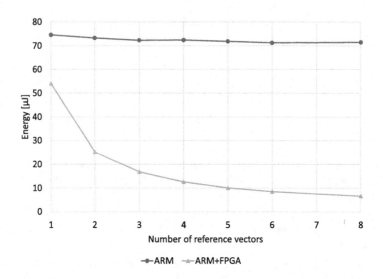

Fig. 3. The normalized energy consumption of Zedboard.

additionally used. The Vivado Power Estimator tool reported that total Zynq power is 2.1 W (including ARM processing system), so other Zedboard's components consume the rest of measured power *e.g.* DDR3 and Ethernet interface. Moreover, the hardware accelerator system containing eight channels is reported to consume total 0.175 W, *i.e.* 0.021 W for a single channel. Therefore, a number of channels have a slight impact on overall power consumption.

The average power consumption is 220 W for the class server node when it uses only the CPU processor (Fig. 4). It is 180 W while hardware accelerator is used additionally. The Vivado Power Estimator reports that total Virtex power is 3.24 W, where GTX Transceivers used for PCIe interface consume 2.2 W. The hardware accelerator consumes 0.43 W and the single channel takes 0.051 W, which also have minimal impact on total power usage. The Virtex's power is around 2.5 times higher than Zynq's because Virtex runs with 2.5 higher clock frequency.

On ZedBoard, the power that is consumed by the accelerated processing system was higher than for the non-accelerated one, because both ARM cores were used together with an FPGA device. On server class node, the power dissipation profile is different because a hardware accelerated computing utilizes only three out of eight available logical processor cores while in the processor-only processing scheme all eight cores are utilized. However, the energy required for the single operation was reduced for both platforms. Thanks to the hardware accelerator, it was reduced 10.8 times using Zynq, and 12.9 times using Virtex. Virtex energy consumption is close to the low-power platform for eight channels, as Table 2 shows.

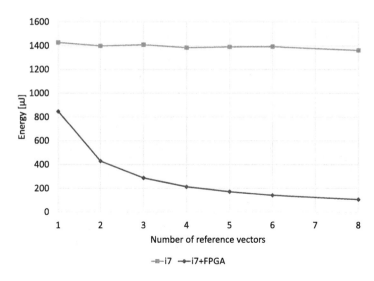

Fig. 4. The normalized energy consumption of the server.

Table 2. An average energy consumption per single comparison operation [μJ].

Platform configuration	Number of reference vectors	
	1	8
ARM	74	71
ARM + FPGA	54	6
i7	1,427	1,362
i7 + FPGA	846	105

6   Conclusions

The presented experiments clearly indicate that the use of FPGAs to accelerate computations significantly boosts both the performance and energy efficiency. The speedup varies from 11.7 for the low-power platform to 10.5 for the high-performance platform. The energy consumption was lowered 10.8 and 12.9 times from initial respectively. It is possible to implement a total of 32 channels what still should quadruple the performance on the ZedBoard platform. However, it is possible to implement 256 channels on Virtex what could probably lower the energy required for single operation below $10\,\mu J$.

Acknowledgments. This research was supported by the European Regional Development Program no. POIG.02.03.00-12-137/13 PL-Grid Core.

References

1. Wielgosz, M., Janiszewski, M., Russek, P., Pietroń, M., Jamro, E., Wiatr, K.: Implementation of a system for fast text search and document comparison. In: Bembenik, R., et al. (eds.) Intelligent Tools for Building a Scientific Information Platform: From Research to Implementation, pp. 173–186. Springer, Switzerland (2014)
2. Perera, D.G., Li, K.F.: Parallel computation of similarity measures using an FPGA-based processor array. In: 2008 22nd International Conference on Advanced Information Networking and Applications, AINA 2008, pp. 955–962, March 2008
3. Russek, P., Wiatr, K.: The regular expression matching algorithm for the energy efficient reconfigurable SoC. In: Wyrzykowski, R., Dongarra, J., Karczewski, K., Waśniewski, J. (eds.) PPAM 2013, Part I. LNCS, vol. 8384, pp. 545–556. Springer, Heidelberg (2014)
4. Wang, W., Guo, K., Gu, M., Ma, Y., Wang, Y.: A universal FPGA-based floating-point matrix processor for mobile systems. In: 2014 International Conference on Field-Programmable Technology (FPT), pp. 139–146, December 2014
5. Dorrance, R., Ren, F., Marković, D.: A scalable sparse matrix-vector multiplication kernel for energy-efficient sparse-BLAS on FPGAs. In: ACM/SIGDA International Symposium on FPGA, pp. 161–170, February 2014
6. Jamro, E., Pabiś, T., Russek, P., Wiatr, K.: The algorithms for FPGA implementation of sparse matrices multiplication. Comput. Inform. **33**(3), 667–684 (2015)
7. Salton, G., Wong, A., Yang, C.S.: A vector space model for automatic indexing. Commun. ACM **18**(11), 613–620 (1975)
8. Kiela, D., Clark, S.: A systematic study of semantic vector space model parameters. In: Proceedings of the 2nd Workshop on Continuous Vector Space Models and Their Compositionality (CVSC) at EACL, pp. 21–30, April 2014
9. Karwatowski, M., Koryciak, S., Wiatr, K.: Cosine similarity metric calculation on low power heterogeneous computing platform. In: Proceedings of the KU KDM 2015: Eighth ACC Cyfronet AGH users' Conference, Zakopane, 11–13 March 2015, pp. 111–112 (2015)
10. Russek, P., Karwatowski, M., Wielgosz, M., Frączek, R., Wiatr, K.: Documents similarity calculation in the low-power cluster. In: Proceedings of the KU KDM 2015: Eighth ACC Cyfronet AGH Users' Conference, Zakopane, 11–13 March 2015, pp. 37–38 (2015)
11. Karwatowski, M., Wielgosz, M., Russek, P., Wiatr, K.: FPGA-based low-energy cluster for acceleration of the document similarity analysis. In: Proceedings of the Cracow Grid Workshop, CGW 2014, 27–29 October 2014, Krakow, Poland, pp. 57–58 (2014)
12. Xillybus project site. http://xillybus.com
13. Zedboard community site. http://zedboard.org
14. Xilinx Virtex-7 FPGA VC707 Evaluation Kit. http://www.xilinx.com

Numerical Algorithms and Parallel
Scientific Computing

A Bucket Sort Algorithm for the Particle-In-Cell Method on Manycore Architectures

Andreas Jocksch[1]([✉]), Farah Hariri[2], Trach-Minh Tran[2], Stephan Brunner[2], Claudio Gheller[1], and Laurent Villard[2]

[1] CSCS, Swiss National Supercomputing Centre,
Via Trevano 131, 6900 Lugano, Switzerland
{jocksch,gheller}@cscs.ch

[2] Swiss Plasma Center, Ecole Polytechnique Fédérale
de Lausanne (EPFL), 1015 Lausanne, Switzerland
{farah.hariri,trach-minh.tran,stephan.brunner,laurent.villard}@epfl.ch

Abstract. The Particle-In-Cell (PIC) method is effectively used in many scientific simulation codes. In order to optimize the performance of the PIC approach, data locality is required. This relies on efficient sorting algorithms. We present a bucket sort algorithm with small memory footprint for the PIC method targeting Graphics Processing Units (GPUs). Our sorting algorithm shows an increased performance with the amount of storage provided and with the orderliness of the particles. For our application where particles are presorted it performs better and requires less memory than other sorting algorithms in the literature. The overall PIC algorithm performs at its best if the sorting is applied.

Keywords: Particle-In-Cell · GPU · Bucket sort

1 Introduction

New computer architectures develop based on massive parallelism, which is obtained by utilizing many nodes and many arithmetic units per node. With hundreds of cores and fine-grained multi-threading, Graphics Processing Units (GPUs) offer massive amounts of on-chip parallelism. Using general-purpose GPUs (GPGPUs) enables the development of high-performance scientific applications. It helps accelerating codes by offloading compute intensive portions of the application to the GPU, while the remainder of the code still runs on the CPU. However, designing algorithms coping with these new architectural features becomes challenging. A key consideration is to account for many parallel threads while providing data locality for achieving higher performance.

In Particle-In-Cell (PIC) simulations [5] particles are moved by a force field resulting from a source distribution. The latter results from particle charge and current contributions. In simulations which store the particles in a global array, adjacent particles are located randomly with respect to the mesh. Consequently, the memory reads and writes of field quantities in the routines where particles

© Springer International Publishing Switzerland 2016
R. Wyrzykowski et al. (Eds.): PPAM 2015, Part I, LNCS 9573, pp. 43–52, 2016.
DOI: 10.1007/978-3-319-32149-3_5

interact with the force field do produce many cache misses, hence hampering simulation performance. For these algorithms, sorting the particles, plays an essential role in providing data locality.

Various sorting algorithms have been implemented on the GPU. Sintorn and Assarsson [12] introduced a complete GPU-based sorting algorithm where bucket sort (bin sort) is used for a first coarse sorting step. Satish et al. [11] implemented a radix sort and a merge sort. Merrill and Grimshaw [8] developed a fast radix sort. A serial (CPU) in-place implementation of bucket sort has been introduced by Burnetas et al. [2].

In the context of the PIC method (on the GPU), sorting algorithms have been employed before. Decyk and Singh [4] sorted particles in a two-dimensional arrangement of bins (buckets) assuming that a particle can move at most to the neighboring bins every time step. Their computations were performed in single precision. Stantchev et al. [13] used an efficient algorithm also in order to exchange particles between neighboring cells. They suggest a quicksort like approach in order to account for fast particles. Chen et al. [3] implemented a particle in cell method with an implicit timestep, i. e., particles move between all cells within one timestep and they are sorted accordingly. Shared memory was used for the particle to cell interpolation step. A particle reordering method with a bucket sort algorithm has been developed by Mertmann et al. [9]. The particle in cell method based on a triangular mesh was considered by Joseph et al. [7]. A bucket sort algorithm was applied to the particles which makes use of the shared memory on the GPU. Rozen et al. [10] implemented a bucket sort algorithm using linked lists.

We consider a PIC application where particles might move fast with respect to a cartesian grid. It is considered as a testbed of complex plasma physics simulations as performed by Jolliet et al. [6]. For our application, bucket sort is advantageous since only a limited data locality is required and the algorithm is linear in time. In this work, we apply a new variant of the bucket sort algorithm with small memory footprint to the PIC method. In Sect. 2, we introduce the details of the algorithm. In Sect. 3, we show its overall performance when applied to the PIC method. The summary and conclusions are presented in Sect. 4.

2 Parallel Bucket Sort

High data locality is needed for good performance. That is why we sort the particle data of the PIC algorithm. The data is stored in a two-dimensional array $aparticles[numpart, natt]$ where $numpart$ and $natt$ are the number of particles and the number of attributes belonging to one entry, respectively (see Figs. 1 and 2). In our case we have six attributes: three spatial coordinates and three velocity components. The array is sorted according to three-dimensional splits of the domain in the x, y and z directions, where a bucket sort algorithm is applied. Depending on the values of the particle data a key which presents the target bin is assigned to each particle. The sorting is done based on the keys.

The reference sorting algorithm considered here is the bucket sort algorithm introduced by Sintorn and Assarsson [12] (SIAS) as part of their complete sorting

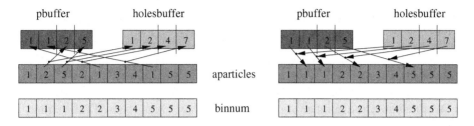

Fig. 1. Bucket sort MIME, numbers are keys except in holesbuffer where they are offsets

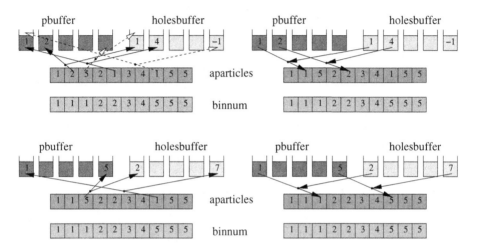

Fig. 2. Bucket sort FIME, numbers are keys except in holesbuffer where they are offsets

algorithm. It is an out-of-place sorting algorithm, if the particles are required in the source array a copy back operation is added. For our problem this algorithm is throughout faster than the radix sort algorithm provided by the Thrust and CUDPP libraries [8]. If the number of bins is smaller than a threadblock (not considered here) the bucket sort algorithm of Joseph et al. [7] should be more efficient than SIAS.

In our bucket sort algorithm only particles which change their bin are moved. Thus we exploit the fact that in the PIC application the array can be considered as presorted. Furthermore a small amount of buffer space is required not only for the presorted case but also for the general one with a tradeoff between performance and memory provided.

A particle buffer array *pbuffer* with intermediate bins is used where the particles are stored with respect to their target bins. Furthermore a buffer array of holes *holesbuffer* is used where the indices of the particles that have to be moved are stored. In addition arrays of counters *counter_pbuffer* and *counter_holes* are used which present the number of entries in *pbuffer* and *holesbuffer*.

We follow two different strategies: The first one is to allocate as much memory as it is required to fit all moving particles and all holes. For this $pbuffer[bufsize, natts]$ and $holesbuffer[bufsize]$ are chosen to be two- and one-dimensional arrays, respectively, since they provide contiguous storage and thus require the least amount of memory. We denote this algorithm minimum memory MIME (see Fig. 1 for the algorithm).

As a first step the number of particles per target bin is counted (histogram). Second the prefixsum is computed and a temporary array $binnum$ is created which contains the sorted keys as reference. Afterwards the number of particles which are not in their target bins is counted (second histogram) and with a prefixsum the overall size of $pbuffer$ and $holesbuffer$ and the offsets for the individual bins are determined. The counters for every bin are reset to zero. After the allocation of the buffer arrays all particles which are not in their target bin are moved to $pbuffer$ in their target bin, the holes left in $aparticles$ are marked in $holesbuffer$, the counters which determine the indices within the buffers are incremented with **atomic** operations. Then the particles are moved from $pbuffer$ to $aparticles$ to the positions indicated in $holesbuffer$.

The overall memory requirement of this algorithm is:

$$reqmem_{MIME} = numpart \cdot sizeof(integer) + numpartmoved \cdot [natt$$
$$\cdot sizeof(double) + sizeof(integer)] + 3 \cdot nbins \cdot sizeof(integer) \quad (1)$$

and the overall runtime is linear with the number of particles $numpart$ and with the number of particles moved $numpartmoved$ ($runtime = C1 \cdot numpart + C2 \cdot numpartmoved$) if conflicts resolved by **atomic** operations are neglected (the prefixsum is also neglected).

In the second strategy we provide a fixed amount of memory for the buffers possibly smaller than total space required for all particles moving. The buffers are repeatedly used. For this second approach – fixed memory FIME – we choose three- and two-dimensional arrays for the buffers $pbuffer[maxpbuffer, nbins, natts]$ and $holesbuffer[maxholes, nbins]$, respectively, where the size of the particle buffer $maxpbuffer$ and holes buffer $maxholes$ can be chosen independently, see Fig. 2 for a sketch of the algorithm.

The algorithm starts as for the first strategy: The particles are counted, $binnum$ is established and the counters are set to zero. Afterwards follows the main loop: A $flag$ is unset. The indices of particles which are not in the right bin are saved in the holes buffer $holesbuffer$ and the particle data is copied to their target bins in the particle buffer $pbuffer$. The counters of the particle buffer and the holes buffer are updated with **atomic** operations. In order to avoid an overflow of the particle buffer or of the holes buffer the counters are only incremented if allowed. The particles counter is incremented after the holes counter where if the increment of the particles counter is not successful the value of the allocated hole is set to a sentinel value (-1). In case of a successful increment of both counters the particle is moved into $pbuffer$ and the position of the hole is stored in $holesbuffer$ otherwise the $flag$ is set. A compaction of $holesbuffer$ is done by taking the sentinel values out. In the next step the particles from the buffer are moved to their target bins. The particle position is given with the array of holes. Then the main loop is repeated or it finishes if the buffers had a sufficient size, i. e., if the counters could be incremented successfully, the *flag* has not been set. In the last execution of the holes filling procedure the number of particles matches exactly the number of holes.

The sizes of *pbuffer* and *holesbuffer* determine the number of executions of the main loop and thus the performance.

The fact that *holesbuffer* might contain dummy entries might lead to an infinite loop if very small buffers are provided only. In this exceptional case, according to a criterion – here if no particles could be moved in one pass – the algorithm switches to a serial execution. The serial algorithm has one difference to the parallel one: the sentinel values are not required. This is guaranteed to converge to the sorted particles. The serial sorting is done on the host.

The FIME algorithm has the memory requirement:

$$reqmem_{FIME} = numpart \cdot sizeof(integer) + nbins \cdot [maxpbuffer \cdot natt$$
$$\cdot sizeof(double) + maxholes \cdot sizeof(integer) + 2 \cdot sizeof(integer)] \quad (2)$$

If the algorithm is executed within one loop iteration its runtime is linear, as is MIME.

For the counting of particles – the first histogram – shared memory is used if its available amount permits [1]. If the number of bins is too large for the available shared memory the histogram is computed directly in global memory as the second histogram of MIME. In both cases **atomic** operations are used. The prefixsum is computed in parallel with a call of the Thrust library.

3 Application to the Particle-In-Cell Method

3.1 Generic Problem

We use a simplified 3D PIC simulation code for the benchmarking in this work. A regular Cartesian mesh (x, y, z) with periodic boundary conditions in all directions is employed. The PIC computational cycle is summarized by the following series of operations:

- init_part: initialize new particles with positions **x** and velocities **v**
    !– Start time loop
    - setrho: compute charge density ρ (*Particle-to-Grid*)
    - Compute Electric Potential ϕ and Electric Field **E**
    - accel: $\mathbf{v} = \mathbf{v} + \mathbf{E}\,\Delta t$ (*Grid-to-Particle*)
    - push: $\mathbf{x} = \mathbf{x} + \mathbf{v}\,\Delta t$
    !– End time loop

The routines accel(), push() and setrho() are called consecutively in the timeloop. The particle acceleration called accel is done with an explicit method using fields linearly interpolated from the mesh. The particle charge deposition, setrho, is done using linear weighting.

The code is written in Fortran using OpenACC (Cray®compiler 8.3.10), only the histogram and part of setrho are written in CUDA C (6.5.14). The implementation on the host using OpenMP serves as a benchmark testbed to compare the GPU results with an implementation on a single host CPU. All our tests are in double precision. We use the NVIDIA®Tesla®K20X card with an Intel®Xeon®E5-2670 processor on a Cray®XC30 system.

We consider two problems with 10^6 particles and $16 \cdot 10^6$ particles on a 2D Cartesian grid of size $512 \times 256 \times 1$ as a realistic range for the target physical cases. Thus, we use the bucket sort effectively in x and y direction, in z-direction we apply only one layer of bins. This choice of parameters is inspired from production runs using the application code ORB5, which solves the problem of gyrokinetic turbulence in magnetically confined plasmas [6]. Our 2D mesh represents one layer (node/GPU) of the 3D decomposed mesh of the production run. The initial particle distribution is uniform in space and velocity with $v_{max} = 1$. In a typical PIC application, the time step is executed a very large number of times, with a wall clock time per timestep which is usually varying only a little from timestep to timestep. Also, there are usually many more particles than spatial grid points. That is the reason why we give all timings in this paper as wall clock time/ timestep/ particle, averaging over 10 timesteps, and ignoring the initialization part.

The routines setrho(), accel() and push() are parallelized for the GPU with threads on the particles. In setrho() the updates of the mesh values lead to conflicts which are resolved with **atomic** operations. The threads on the particles approach has the advantage of ideal load balance independent from the distribution of the particles on the grid.

For the performance of the routines accel() and setrho() data locality is essential. In order to increase the data locality, we sort the array of particles in a routine psort() according to the algorithm described in Sect. 2 called after push() and in init_part(). In accel() the cache saves the grid data for different particles in a bin and thus exploits the data locality. For the sorted array, the setrho() operations are local. Thus the shared memory of the card can be used where the **atomic add** operations are first done to local memory and the result is put also with **atomic add** to the global memory. For performance reasons the array *binnum* for psort() is computed in push().

3.2 Overall Performance

Figure 3 shows the timings for the different PIC components in dependency of the number of bins for the two different testcases (a) with 10^6 particles and (b) with $16 \cdot 10^6$ particles, the MIME algorithm is applied. While a low number of bins produces a coarse sorting a high number of bins is equivalent to a fine grain sorting with high data locality. For the sorting a resolution of 64×32 subdomains shows the best timings for both cases. The sorting psort() tends to become more expensive for many subdomains (with exceptions of the first datapoints in Fig. 3a and b and the last datapoint of Fig. 3b). The timing of the function push() is not affected by the sorting, involving no particle to/from grid

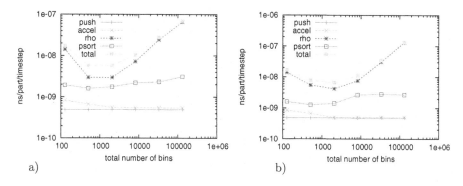

Fig. 3. Timings for the different components in dependency of the number of bins. (a) 10^6 particles (b) $16 \cdot 10^6$ particles

operations. The function `accel()` becomes faster with higher data locality while `setrho()` becomes also faster up to a certain granularity of sorting. Beyond this level it becomes slower again. This is explained by the fact that with increasing data locality the **atomic** operations in `setrho()` have to resolve more conflicts. Furthermore for high data locality the problem of little occupancy of the GPU occurs (for the 10^6 particle case).

In the following, we consider the generic case of 64×32 bins for which the code shows the optimum performance. For our 10^6 particle case every bin fills approximately one threadblock (we use as default 128 threads per block).

Figure 4 shows the cost for the sorting routines MIME and SIAS in dependency of the number of particles moved, as a result of using different timesteps Δt. SIAS is presented with and without the costs for copy back. SIAS becomes slightly more efficient with an increasing number of particles moved. This effect is related to the particular pattern of particle movement. For the testcases (a) and (b) MIME shows throughout linear growth of time with the number of particles moved.

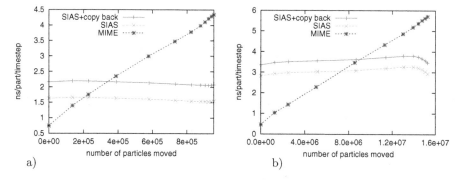

Fig. 4. Comparison of different sorting routines in dependency of the number of particles moved. (a) 10^6 particles (b) $16 \cdot 10^6$ particles.

For our application where only a small portion of particles has to be resorted – $\Delta t = 1$ corresponds to an average of 230000 and 2490000 particles out of 10^6 and $16 \cdot 10^6$ particles, respectively – our algorithm is faster than SIAS including the copy back and it is more memory efficient. While SIAS requires memory 1.083 times the particle data size, MIME only requires memory 0.332 and 0.252 times the particle data size for 10^6 and $16 \cdot 10^6$ particles, respectively.

Note that, not only the number of particles moved but the particle movement pattern matters, if the particles are not redistributed randomly but mostly move to certain buckets – in our case the neighboring ones – our algorithm shows better performance. This is due to the fact that the threads of a warp write to the same counters.

Figure 5 shows timings in dependency of the number of particles moved for different amounts of storage space provided (FIME). It is clearly visible that with an increasing amount of memory the sorting becomes faster. For the $16 \cdot 10^6$ particle case, when many particles move the switch to serial execution is necessary, visible in the kinks of the curves at $11.32 \cdot 10^6$ and $12.98 \cdot 10^6$ particles for 15 % and 22 % memory with respect to the particle data size, respectively.

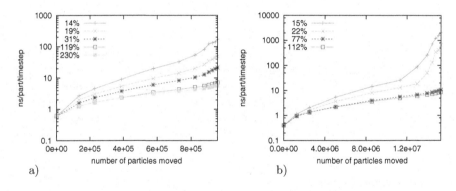

Fig. 5. Timings for different numbers of particles moved (a) 10^6 particles (b) $16 \cdot 10^6$ particles and varying storage space specified in percent of the particle data size.

In Table 1 the timings of the different subroutines are compared for the cases with and without sorting.

On the GPU the PIC algorithm with sorting is 2.2 (1.9) times faster than the algorithm without sorting for 10^6 ($16 \cdot 10^6$) particles. The OpenACC version on one GPU is faster than the OpenMP version on 8 CPU cores. For the 10^6 ($16 \cdot 10^6$) particle case the sorting brings a speedup of factor 3.4 (2.2).

4   Summary and Conclusions

We have developed an in-place bucket sort algorithm in the context of plasma PIC simulations. The algorithm has two versions. The first one uses a buffer

Table 1. Comparison of the main components of the PIC algorithm using 8 OMP threads and the GPU (with and without bins).

	Non-sorted OMP		Non-sorted GPU		Sorted 64 × 32 bins GPU	
Number of particles	10^6	$16 \cdot 10^6$	10^6	$16 \cdot 10^6$	10^6	$16 \cdot 10^6$
push() (ns/part/timestep)	2.291	2.152	0.4946	0.4847	0.4965	0.4847
accel() (ns/part/timestep)	7.585	7.156	3.851	3.757	0.5631	0.5059
rho() (ns/part/timestep)	9.364	5.161	8.015	8.031	2.922	4.146
psort() (ns/part/timestep)					1.656	1.345
Total (ns/part/timestep)	19.24	14.47	12.36	12.27	5.639	6.482

array of a size chosen by the algorithm to fit all moving particle data into the buffer whereas in the second one the buffer has a fixed size and is used repeatedly. Particles which change their bin are copied from their source bin to the buffer and from there to the target bin. The counters are updated using atomic operations which are an essential part of the algorithm. If only a part of the particles change their bin – in our case up to 33 % out of 10^6 particles and 50 % out of $16 \cdot 10^6$ particles –, the method is faster than a standard bucket sort followed by a copy back. Furthermore, the method is more memory efficient than a standard bucket sort.

On the GPU, we show that the PIC method reaches a maximum performance with a sorting of the particles of medium grainsize (64 × 32 bins). As the grain size increases the sorting becomes (in most cases) slightly more expensive. While the push() routine is not affected by the sorting, accel() becomes faster with increasing data locality. The charge assignment setrho() becomes also faster up to a certain grainsize equal to (64 × 32 bins) then it slows down again. The performance optimum of setrho() is a trade-off between high data locality for fast memory access and low data locality for little collisions.

Finally, the newly proposed bucket sort algorithm opens a promising way for efficient production PIC codes on the GPU.

Acknowledgments. The authors wish to thank Peter Messmer and Jakob Progsch from NVIDIA for helpful discussions.

References

1. http://devblogs.nvidia.com/parallelforall/gpu-pro-tip-fast-histograms-using-shared-atomics-maxwell/
2. Burnetas, A., Solow, D., Agarwal, R.: An analysis and implementation of an efficient in-place bucket sort. Acta Informatica **34**, 687–700 (1997)
3. Chen, G., Chacón, L., Barnes, D.C.: An efficient mixed-precision, hybrid CPU-GPU implementation of a nonlinearly implicit one-dimensional particle-in-cell algorithm. J. Comput. Phys. **231**, 5374–5388 (2012)
4. Decyk, V.K., Singh, T.V.: Particle-in-cell algorithms for emerging computer architectures. Comput. Phys. Commun. **185**(3), 708–719 (2014)

5. Hockney, R.W., Eastwood, J.W.: Computer Simulation Using Particles. Hilger, Bristol (1988)
6. Jolliet, S., Bottino, A., Angelino, P., Hatzky, R., Tran, T.M., Mcmillan, B.F., Sauter, O., Appert, K., Idomura, Y., Villard, L.: A global collisionless PIC code in magnetic coordinates. Comput. Phys. Commun. **177**, 409–425 (2007)
7. Joseph, R.G., Ravunnitkutty, G., Ranka, S., D'Azevedo, E., Klasky, S.: Efficient GPU implementation for particle in cell algorithm. In: 25th IEEE International Parallel and Distributed Processing Symposium (IPDPS), Anchorage (Alaska), May 2011
8. Merrill, D., Grimshaw, A.: High performance and scalable radix sorting: a case study of implementing dynamic parallelism for GPU computing. Parallel Process. Lett. **21**(2), 245–272 (2011)
9. Mertmann, P., Eremin, D., Mussenbrock, T., Brinkmann, R.P., Awakowicz, P.: Fine-sorting one-dimensional particle-in-cell algorithm with Monte-Carlo collisions on a graphics processing unit. Comput. Phys. Commun. **182**, 2161–2167 (2011)
10. Rozen, T., Boryczko, K., Alda, W.: GPU bucket sort algorithm with applications to nearest-neighbour search. J. WSCG **16**, 161–167 (2008)
11. Satish, N., Kim, C., Chhugani, J., Nguyen, A.D., Lee, V.W., Kim, D., Dubey, P.: Fast sort on CPUs and GPUs: a case for bandwidth oblivious SIMD sort. In: SIGMOD 2010, Indinapolis (Indiana), June 2010
12. Sintorn, E., Assarsson, U.: Fast parallel GPU-sorting using a hybrid algorithm. J. Parallel Distrib. Comput. **68**, 1381–1388 (2008)
13. Stantchev, G., Dorland, W., Gumerov, N.: Fast parallel particle-to-grid interpolation for plasma PIC simulations on the GPU. J. Parallel Distrib. Comput. **68**, 1339–1349 (2008)

Experience on Vectorizing Lattice Boltzmann Kernels for Multi- and Many-Core Architectures

Enrico Calore[1,2], Nicola Demo[1], Sebastiano Fabio Schifano[1,2][✉],
and Raffaele Tripiccione[1,2]

[1] Università di Ferrara, Ferrara, Italy
[2] INFN Ferrara, Ferrara, Italy
schifano@fe.infn.it

Abstract. Current development trends of fast processors calls for an increasing number of cores, each core featuring wide vector processing units. Applications must then exploit both directions of parallelism to run efficiently. In this work we focus on the efficient use of vector instructions. These process several data-elements in parallel, and memory data layout plays an important role to make this efficient. An optimal memory-layout depends in principle on the access patterns of the algorithm but also on the architectural features of the processor. However, different parts of the application may have different requirements, and then the choice of the most efficient data-structure for vectorization has to be carefully assessed. We address these problems for a Lattice Boltzmann (LB) code, widely used in computational fluid-dynamics. We consider a state-of-the-art two-dimensional LB model, that accurately reproduces the thermo-hydrodynamics of a 2D-fluid. We write our codes in C and expose vector parallelism using directive-based programming approach. We consider different data layouts and analyze the corresponding performance. Our results show that, if an appropriate data layout is selected, it is possible to write a code for this class of applications that is automatically vectorized and performance portable on several architectures. We end up with a single code that runs efficiently onto traditional multi-core processors as well as on recent many-core systems such as the Xeon-Phi.

Keywords: Directive based compilation · Memory data layout · Vectorization · Accelerator processors

1 Introduction

Lattice Boltzmann (LB) methods are widely used in computational fluid dynamics. This class of applications – discrete in time and momenta and living on a discrete and regular grid of points, see later for details – offers a large amount of easily identified parallelism, making LB an ideal target for modern HPC systems [1–4]. However, exploiting available parallelism is becoming more and more difficult on recent processor architectures, exhibiting a large number of cores, each core being in turn able to execute SIMD instructions; both levels of parallelism have to be used.

© Springer International Publishing Switzerland 2016
R. Wyrzykowski et al. (Eds.): PPAM 2015, Part I, LNCS 9573, pp. 53–62, 2016.
DOI: 10.1007/978-3-319-32149-3_6

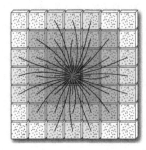 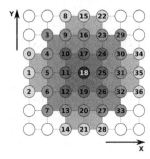

Fig. 1. Left: velocity vectors for the LB populations in the D2Q37 model. Right: populations are identified by an arbitrary label, identifying the lattice hop that they perform in the *propagate* phase.

For regular LB applications, it is easy to apply core parallelism assigning tiles of the physical lattice to different cores. However, exploiting vectorization requires additional care, and in particular two aspects are relevant: how to introduce and expose vector instructions in the code, and memory data-layout to enable efficient vector processing. Vector instructions can be explicitly introduced defining vector variables and processing them with specific functions – called *intrinsics* – which are mapped by the compiler onto the corresponding assembly instruction. However, even if potentially efficient, this approach prevents compiler to make all possible optimizations, and codes are not portable. Moreover, unskilled use can make the code less efficient than plain C code. In this work, we use the *directive* based approach provided by OpenMP 4.0 and supported by the Intel compiler, which allows to annotate standard C-codes with *pragma* directives to specify regions of the code to vectorize; this approach leaves to the compiler all optimization steps specific for the target architecture, and makes the code portable. Finding the best data structure layout to enable vector processing is relevant to ensure that data operated upon by SIMD operations are allocated on contiguous memory addresses so read (and write back) of data is fast enough not to starve the processing engine. This involves to find the best compromise between conflicting requirements between different parts of the code. Currently this is difficult to achieve automatically by programming tools, even if some experimental stencil compilers, such as PLUTO [5] ad POCHOIR [6] are promising solutions.

In this work, we use the directive approach for programming, and experiment with several memory data layouts. We assess the corresponding performance results, and we end up with just one implementation of our LB application that can be automatically and efficiently vectorized onto traditional multi-core processors and many-core processors such as the Xeon-Phi accelerator. Analyses of optimal data layouts for LB have been made in [7–9]. However, [7] focuses only on the *propagate* step, one of the two key kernels in LB codes, while [8] does not take into account vectorization; in [9] vectorization is considered using *intrinsics* functions only. None of these papers consider accelerators. We extend these results in several ways: first, we take into account both *propagate* and *collision* steps used in LB simulations. Then we use a high level approach based on compiler directives, and we take into account also accelerators.

This paper is structured as follows: Sect. 2 gives an overview of LB methods, while Sect. 3 describes in details our implementations. Finally, Sect. 4 analyzes our performance results and compare them with those of earlier codes for the same LB application we have written in CUDA for GPU, and using *intrinsics* for traditional multi-core CPUs and the Xeon-Phi processor.

2 Lattice Boltzmann Methods

In this section, we sketchily introduce the computational method that we adopt, based on an advanced Lattice Boltzmann (LB) scheme. LB methods (see, e.g. [10] for an introduction) are discrete in position and momentum spaces; they are based on the synthetic dynamics of *populations* sitting at the sites of a discrete lattice. At each time step, populations hop from lattice-site to lattice-site and then they *collide*, mixing and changing their values accordingly.

Over the years, several LB models have been developed, describing flows in 2 or 3 dimensions, and using sets of populations of different size (a model in x dimensions based on y populations is labeled as $DxQy$). Populations $(f_l(x,t))$, each having a given lattice velocity c_l, are defined at the sites of a discrete and regular grid; they evolve in (discrete) time according to the Bhatnagar-Gross-Krook (BGK) equation:

$$f_l(\boldsymbol{x}, t + \Delta t) = f_l(\boldsymbol{x} - \boldsymbol{c}_l \Delta t, t) - \frac{\Delta t}{\tau} \left(f_l(\boldsymbol{x} - \boldsymbol{c}_l \Delta t, t) - f_l^{(eq)} \right) \qquad (1)$$

The macroscopic physics variables, density ρ, velocity $\boldsymbol{u}$ and temperature T are defined in terms of the $f_l(x,t)$ and c_ls:

$$\rho = \sum_l f_l, \qquad \rho \boldsymbol{u} = \sum_l \boldsymbol{c}_l f_l, \qquad D\rho T = \sum_l |\boldsymbol{c}_l - \boldsymbol{u}| \, 2 f_l; \qquad (2)$$

the equilibrium distributions $(f_l^{(eq)})$ are themselves functions of these macroscopic quantities [10]. With an appropriate choice of the set of lattice velocities c_l and of the equilibrium distributions $f_l^{(eq)}$, one shows that, performing an expansion in Δt and renormalizing the values of the physical velocity and temperature fields, the evolution of the macroscopic variables obeys the thermo-hydrodynamical equations of motion and the continuity equation:

$$\partial_t \rho + \rho \partial_i u_i = 0 \qquad (3)$$

$$\rho D_t u_i = -\partial_i p - \rho g \delta_{i,2} + \nu \partial_{jj} u_i, \qquad \rho c_v D_t T + p \partial_i u_i = k \partial_{ii} T; \qquad (4)$$

where $D_t = \partial_t + u_j \partial_j$ is the material derivative and we neglect viscous heating; c_v is the specific heat at constant volume for an ideal gas, $p = \rho T$, and ν and k are the transport coefficients; g is the acceleration of gravity, acting in the vertical direction. Summation of repeated indexes is implied.

In our case we study a 2-dimensional system ($D = 2$ in the following), and the set of populations has 37 elements (hence the D2Q37 acronym) corresponding to (pseudo-)particles moving up to three lattice points away, as shown in Fig. 1. This LB model, that automatically enforces the equation of state of a perfect gas ($p = \rho T$), was recently developed in [11,12];. Our optimization efforts have made it possible to perform large scale simulations of convective turbulence in several physics regimes (see e.g., [13,14]);

An LB code starts with an initial assignment of the populations, corresponding to a given initial condition at $t = 0$ on some spatial domain, and iterates Eq. 1 for each population and lattice site and for as many time steps as needed; boundary conditions are enforced at the edges of the domain after each time step by appropriately modifying population values at and close to the boundary.

From the computational point of view, the LB approach offers a huge degree of easily identified available parallelism. Inspecting Eq. (1) one easily identifies the overall structure of the computation that evolves the system by one time step Δt: for each point x in the discrete grid the code: (i) gathers from neighboring sites the values of the fields f_l corresponding to populations drifting towards x with velocity c_l (propagate step) and then, (ii) performs all mathematical operations associated to the r.h.s. of Eq. (1) (collide step). One quickly sees that there is no correlation between different lattice points, so both steps can be performed in parallel on all grid points according to any convenient schedule, with the only constraint that step 1 precedes step 2. All other steps of a complete simulations have a negligible computational cost.

As already remarked, our D2Q37 model correctly and consistently describes the thermo-hydrodynamical equations of motion and the equation of state of a perfect gas; this translates into a more complex implementation than earlier 2D LB models as well as in demanding hardware requirements for memory bandwidth and floating-point throughput. Indeed, propagate implies accessing 37 neighbor cells to gather all populations, while collide executes ≈7000 double-precision floating point operations per lattice point.

3   Implementation and Optimization of LB Kernels

As already remarked, data organization plays a key role; popular layouts for LB methods are arrays of structures (AoS) or structure of arrays (SoA). In the AoS layout, population data for each lattice site are stored one after the other at successive memory locations, while in SoA, for each population of index i, all sites are stored one after the other. AoS enjoys locality for all data of each site; on the other hand, same-index populations of different sites are stored at non-unit strided addresses. We store the lattice in column-major order (Y direction) and instantiate two copies, that are alternatively read and written. Although this solution allocates more memory than having a single copy, it is required to process all lattice sites in parallel.

```
typedef struct {
  double *p[NPOP];
} pop_soa_t;

  // snippet of propagate code to move population index 0
  for ( xx = XMIN; xx < stopx; XMAX++ ) {
    #pragma vector nontemporal
    for( yy = YMIN; yy < YMAX; yy++ ) {
      idx = IDX(xx,yy);
      (nxt->p[0])[idx] = (prv->p[0])[idx+OXM3YP1];
    }
  }
```

Fig. 2. Snippet of sample code for **propagate**, moving f_0 according to Fig. 1, right. OXM3YP1 is the memory address offset associated to the population hop. The lattice is stored using the SoA layout.

3.1   Optimization of propagate

The **propagate** kernel moves populations for each lattice site according to the pattern of Fig. 1 (left) involving accesses at lattice-cells at distance up to 3 in the physical grid. Earlier works – e.g. [7] – has considered two implementation schemes: *push* and *pull*. The *push* scheme moves all populations of one site to appropriate neighbor sites, while *pull* gathers to one site populations belonging to neighbor sites. However, *push* performs aligned-reads and misaligned-writes, while *pull* does the opposite, and this results more efficient on modern architectures since aligned writes can bypass the cache hierarchy.

Vectorization of this kernel using SIMD instructions apply each move shown in Fig. 1 to several lattice sites in parallel depending on the size of vectors supported by the target processor, e.g. 4 and 8 for those we have considered. This simple vectorization is not possible if one uses an AoS layout, as populations of different sites are stored at non-contiguous memory addresses, requiring multiple memory accesses. In contrast, an SoA layout has unit stride for populations of fixed index i belonging to different sites, so data can be fetched in parallel from memory using vector instructions. This is the main reason for selecting SoA rather than AoS memory arrangement.

Figure 2 shows the C-code moving population of index 0 which – see Fig. 1 (right) – comes from sites three steps left and one up in the physical lattice. All other populations are moved in the same way. The code sweeps all lattice with two loops in X and Y; the inner loop is on Y, as elements are stored in column-major order. The **#pragma vector** directive notifies the compiler that the next loop can be vectorized; this is an Intel-specific directive, corresponding to **#pragma omp simd** in the OpenMP standard. This directive vectorize the inner loop in chunks of 4 or 8 words for the Haswell CPU and the Xeon-Phi accelerator respectively. We add the **nontemporal** attribute specifying that data is not used again by this kernel, so *non-temporal* stores can be used. The latter bypass the cache hierarchy, and reduce memory traffic by 1/3, with a corresponding save in time. Unfortunately OpenMP currently does not support this directive.

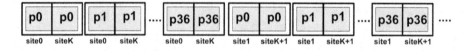

```
typedef struct { double c[VL]; } vdata_t;   // cluster

typedef struct { vdata_t p[NPOP]; } caosoa_t; // CAoSoA type definition

// snippet of propagate code to move population index 0
for ( xx = startx; xx < stopx; xx++ ) {
  for ( yy = 0; yy < SIZEYOVL; yy++, idx++ ) {
    idx =  IDX(xx,yy);
    #pragma vector aligned nontemporal
    for(tt = 0; tt < VL; tt++) {
      nxt[idx].p[0].c[tt] = prv[idx+OPOVL0].p[0].c[tt];
    }
  }
}

// snippet of part of collide code to compute density rho
vdata_t rho;
for (ii =0; ii < NPOP; ii++)
  #pragma vector aligned
  for (tt=0; tt < VL; tt++)
    rho.c[tt] = rho.c[tt] + prv[idx].p[ii].c[tt];
```

Fig. 3. Top: data arrangement for the CAoSoA layout; for illustration purposes, we take VL = 2. Bottom: snippet of sample codes for `propagate` and `collide` using this layout.

3.2 Optimization of `collide`

The `collide` kernel updates populations at each lattice site at the next time step, performing all mathematical operations associated to Eq. 1 (this is called *collision* in LB jargon). Input data are the populations gathered by the previous `propagate` phase. For each lattice site, this floating point intensive kernel uses only data belonging to the site on which it operates – previously gathered by `propagate` – so the available parallelism equals the size of the full lattice.

The SoA layout allows in principle to vectorize the code: to do so, the machine code should load in sequence blocks of words for all populations, each block as long as the vector size, and then perform all operations in SIMD mode using these blocks as operands. In practice, this implies reading relatively short data chunks from scattered memory locations, that the memory controller is not able to do efficiently. We therefore anticipate a limited performance as the compute-unit in the processor becomes data starved.

We then define a new data layout, offering to compilers a different handle to vectorization. We divide each population array in VL parts along Y (VL is the vector size), pack the populations at each site of each part into an array of VL elements (we call this block a *cluster*) and store clusters for all 37 populations one after the other, see Fig. 3. We call this layout a *Clustered Array of Structure of Array* (CAoSoA). This layout obviously allows vectorization of inner structures (clusters) of size VL, and at the same time improve loacality of population

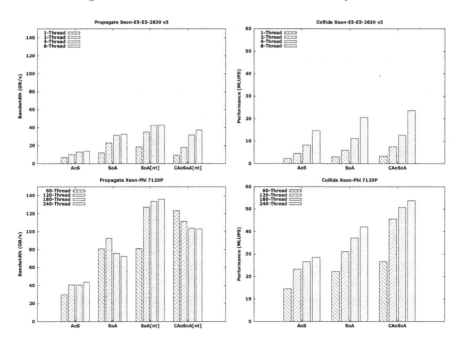

Fig. 4. Benchmark results for `propagate` and `collide` on a lattice of 2160×8192 points, using different data layouts. Top: Xeon E5-2630-v3 CPU; bottom: Xeon-Phi 7120P accelerator.

keeping at contiguous and aligned addresses all data items needed to handle each lattice point in the kernel. Figure 3 shows the definition of the `vdata_t` data type, corresponding to a *cluster*, and a snippet of representative small sections of the code. Cluster variables are processed iterating on all elements of the cluster through a loop over VL; `pragma vector aligned` instructs the compiler to fully vectorize the loop since all iterations are independent and memory accesses are aligned. This layout is probably less efficient for *propagate*: in the next section we discuss the corresponding tradeoffs.

Table 1. Performance comparison of propagate and collide kernels using the CaAoSoA scheme with intrinsic functions and pragma directives. The rightmost column lists performance figures for an NVIDIA K40 GPU. Data refer to a lattice of 2160×8192 points.

	E5-2630 v3 CAoSoA		Intel Xeon 7120 CAoSoA		NVIDIA K40 SoA
	Intrinsics	Directives	Intrinsics	Directives	CUDA
Propagate (GB/s)	36	37	86	103	187
Collide (MLUPS)	14	24	55	54	108
Collide (GFs)	94	154	365	351	703
Global (MLUPS)	12	17	40	41	81

4 Results and Conclusions

We have tested our code on an 8-core Xeon E5-2630-v3 CPU, based on the Haswell micro-architecture running at 2.4 GHz, and on a 61-core Xeon-Phi 7120P accelerator based on the *Many Integrated Core* architecture.

Figure 4 shows performance results for all data layouts that we have considered. For the *propagate* kernel we measure the effective bandwidth, while for *collide* we use *Millions Lattice Updates per Second* (MLUPS) figures, a standard performance metric for this application.

We first consider CPU results. Both kernels have been run using 1 thread per core since use hyperthreading does not improve performances. For *propagate*, the bandwidth increases with the number of threads used, as using only one thread does not saturate the bandwidth made available by all memory channels. The AoS scheme has poor performance since it prevents vectorization and use of non-temporal stores, while the SoA scheme reaches the best results with 4 and 8 threads and non-temporal stores enabled; in this case the bandwidth is ≈43 GB/s, that is ≈70 % of theoretical peak (59 GB/s) and ≈86 % of bandwidth measured with STREAM benchmark [15] (≈50 GB/s). As expected, the CAoSoA layout reduces performance by ≈15 % (≈37 GB/s). The performance of *collide* grows with the number of threads. The best performance value we measure is 24 MLUPS using 8 cores and the CAoSoA layout; this is 1.2X faster than SoA and 1.6X faster than AoS.

On the Xeon-Phi the situation is somewhat different. In this case we have used hyperthreading since this is required to hide the huge latency in accessing external memory. For *propagate*, performance depends weakly on the number of threads per core, reflecting a complex interaction between the complexity of the memory system and the role of multi-threading to mitigate latency effects. Using the SoA[nt] layout the peak value of bandwidth is 136 GB/s. Using the CAoSoA[nt] layout, the bandwidth decreases to 103 GB/s, after prefetching hints have been added to the code. These figures are 30 % or less of the theoretical peak (≈350 GB/s), but an encouraging ≈62 % of the STREAM benchmark [15], a widely acknowledged reference test for the Xeon-Phi, that measures an effective bandwidth of ≈165 GB/s. In much the same way as for the CPU case, *collide* scales in performance with the number of threads up to 240 threads for all storage layouts. The best value is obtained using the CAoSoA layout, approximately 1.3X faster than SoA, and 1.8X faster than the AoS.

Table 1 summarizes our results, comparing performances for different data layouts on different processors; it considers implementations using intrinsics instructions [16–18] and directive-based ones (this work). We also show global MLUPS figures, a reasonably accurate performance metric of a complete code assuming that propagate and collide kernels are executed and applied to the lattice one after the other; moreover we also quote previous results for GPUs using the SoA layout [19–21].

In summary, we verify that propagate and collide have conflicting requirements for data layout. The hybrid CAoSoA scheme combines the benefits of SoA since clusters can be easily vectorized, with those of AoS where population locality boosts the performance of the `collide` kernel. All in all, considering the overall performance the CAoSoA layout is the preferred one for both strands of processor architectures. It is interesting to note that our directive-based code is as fast or even faster than those using intrinsic functions, as in the former case the compiler is able to apply a larger set of optimization strategies.

Coming now to a comparison of performance across different processors, we see that for this class of codes – typical of a large class of HPC applications – Xeon-Phi accelerators offer roughly 2X better performance than state-of-the-art CPUs, while a NVIDIA K40 GPU is approximately 4X better. We also remark that GPUs have a more efficient memory interface, so outstanding performance is obtained without a careful adjustment of the SoA data layout.

In conclusion, an important result of this work is that it is possible to write directive-based programs that are: (i) as efficient or even more efficient than handcrafted codes and, (ii) reasonably performance-portable across different present available (and hopefully future) architectures. This requires to invest a significant amount of effort in the definition of an appropriate data layout, choosing a non-trivial optimal (or almost optimal) layout among conflicting requirements. This step is clearly out of the scope of current programming environment; considering a longer time horizon, we see here a potential large space for data definitions that do not prescribe a fixed memory layout, leaving the compiler free to make appropriate choices during the compilation process.

Acknowledgements. This work was done in the framework of the COKA, COSA and SUMA projects of INFN. We would like to thank CINECA (Italy) for access to their HPC systems.

References

1. Williams, S., et al.: Lattice Boltzmann simulation optimization on leading multicore platforms. In: IEEE International Symposium on Parallel and Distributed Processing (2008). doi:10.1109/IPDPS.2008.4536295
2. Williams, S., et al.: Optimization of a Lattice Boltzmann computation on state-of-the-art multicore platforms. J. Parallel Distrib. Comput. **69**, 762–777 (2009). doi:10.1016/j.jpdc.2009.04.002
3. Bernaschi, M., et al.: A flexible high-performance Lattice Boltzmann GPU code for the simulations of fluid flows in complex geometries. Concurrency Comput. Pract. Experience **22**(1) (2010). doi:10.1002/cpe.1466
4. Ye, Z.: Lattice Boltzmann based PDE solver on the GPU. Vis. J. **24**(5), 323–333 (2008). doi:10.1007/s00371-007-0191-y
5. Bondhugula, U., et al.: A practical and automatic polyhedral program optimization system. In: Proceedings of the ACM SIGPLAN 2008 Conference on Programming Language Design and Implementation (2008). doi:10.1145/1375581.1375595
6. Tang, Y., et al.: The pochoir stencil compiler. In: Proceedings of the 23rd ACM Symposium on Parallelism in Algorithms and Architectures (2011). doi:10.1145/1989493.1989508
7. Wittmann, M., et al.: Comparison of different Propagation Steps for the Lattice Boltzmann Method, 3 November 2011. arXiv:1111.0922vI

8. Shet, A.G., et al.: Data structure and movement for lattice-based simulations. Phys. Rev. E **88**, 013314 (2013). doi:10.1103/PhysRevE.88.013314

9. Shet, A.G., et al.: On vectorization for lattice based simulations. Int. J. Mod. Phys. C **24**, 1340011 (2013). doi:10.1142/S0129183113400111

10. Succi, S.: The Lattice-Boltzmann Equation. Oxford University Press, Oxford (2001)

11. Sbragaglia, M., et al.: Lattice Boltzmann method with self-consistent thermo-hydrodynamic equilibria. J. Fluid Mech. **628**, 299–309 (2009). doi:10.1017/S002211200900665X

12. Scagliarini, A., et al.: Lattice Boltzmann methods for thermal flows: Continuum limit and applications to compressible Rayleigh-Taylor systems. Phys. Fluids **22**(5), 055101 (2010). doi:10.1063/1.3392774

13. Biferale, L., et al.: Second-order closure in stratified turbulence: simulations and modeling of bulk and entrainment regions. Phys. Rev. E **84**(1), 016305 (2011). doi:10.1103/PhysRevE.84.016305

14. Biferale, L., et al.: Reactive Rayleigh-Taylor systems: front propagation and non-stationarity. EPL (Europhys. Lett.) **94**(5), 54004 (2011). doi:10.1209/0295-5075/94/54004

15. McCalpin, J.: The STREAM Benchmark: Computer Memory Bandwidth. http://www.streambench.org/

16. Mantovani, F., et al.: Exploiting parallelism in many-core architectures: Lattice Boltzmann models as a test case. J. Phys. Conf. Ser. **454**, 012015 (2013). doi:10.1088/1742-6596/454/1/012015

17. Mantovani, F., et al.: Performance issues on many-core processors: a D2Q37 Lattice Boltzmann scheme as a test-case. Comp. Fluids **88** (2013). doi:10.1016/j.compfluid.2013.05.014

18. Crimi, G., et al.: Early experience on porting and running a Lattice Boltzmann code on the Xeon-phi co-processor. Proc. Comput. Sci. **18**, 551–560 (2013). doi:10.1016/j.procs.2013.05.219

19. Biferale, L., et al.: An optimized D2Q37 Lattice Boltzmann code on GP-GPUs. Comput. Fluids **80** (2013). doi:10.1016/j.compfluid.2012.06.003

20. Biferale, L., et al.: A multi-GPU implementation of a D2Q37 Lattice Boltzmann code. In: Wyrzykowski, R., Dongarra, J., Karczewski, K., Waśniewski, J. (eds.) PPAM 2011, Part I. LNCS, vol. 7203, pp. 640–650. Springer, Heidelberg (2012)

21. Kraus, J., et al.: Benchmarking GPUs with a parallel Lattice-Boltzmann code. In: Proceedings of Computer Architecture and High Performance Computing (SBAC-PAD), pp. 160–167 (2013). doi:10.1109/SBAC-PAD.2013.37

Performance Analysis of the Kahan-Enhanced Scalar Product on Current Multicore Processors

Johannes Hofmann[1]([✉]), Dietmar Fey[1], Michael Riedmann[2], Jan Eitzinger[3],
Georg Hager[3], and Gerhard Wellein[3]

[1] Chair for Computer Architecture, University Erlangen-Nuremberg,
Erlangen, Germany
johannes.hofmann@fau.de
[2] AREVA GmbH, Erlangen, Germany
[3] Erlangen Regional Computing Center (RRZE), University Erlangen-Nuremberg,
Erlangen, Germany

Abstract. We investigate the performance characteristics of a numerically enhanced scalar product (dot) kernel loop that uses the Kahan algorithm to compensate for numerical errors, and describe efficient SIMD-vectorized implementations on recent Intel processors. Using low-level instruction analysis and the execution-cache-memory (ECM) performance model we pinpoint the relevant performance bottlenecks for single-core and thread-parallel execution, and predict performance and saturation behavior. We show that the Kahan-enhanced scalar product comes at almost no additional cost compared to the naive (non-Kahan) scalar product if appropriate low-level optimizations, notably SIMD vectorization and unrolling, are applied. We also investigate the impact of architectural changes across four generations of Intel Xeon processors.

Keywords: Scalar product · Kahan algorithm · SIMD · Performance model · Multicore

1 Introduction and Related Work

Accumulating finite-precision floating-point numbers in a scalar variable is a common operation in computational science and engineering. The consequences in terms of accuracy are inherent to the number representation and have been well known and studied for a long time [1]. There is a number of summation algorithms that enhance accuracy while maintaining an acceptable throughput [2,3], of which Kahan [4] is probably the most popular one. However, the topic is still subject to active research [5–8]. A straightforward solution to the inherent accuracy problems is arbitrary-precision floating point arithmetic, which comes at a significant performance penalty. Naive summation and arbitrary precision arithmetic are at opposite ends of a broad spectrum of options, and balancing performance vs. accuracy is a key concern when selecting a specific solution.

Naive summation, which simply adds each successive number in sequence to an accumulator, requires appropriate unrolling for Single Instruction Multiple Data

© Springer International Publishing Switzerland 2016
R. Wyrzykowski et al. (Eds.): PPAM 2015, Part I, LNCS 9573, pp. 63–73, 2016.
DOI: 10.1007/978-3-319-32149-3_7

(SIMD) vectorization and pipelining. The necessary code transformations are performed automatically by modern compilers, which results in optimal in-core performance. Such a code quickly saturates the memory bandwidth of modern multi-core CPUs when the data is in memory.

This paper investigates implementations of the scalar product, a kernel which is relevant in many numerical algorithms. Starting from an optimal naive implementation it considers scalar and SIMD-vectorized versions of the Kahan algorithm using various SIMD instruction set extensions on a range of current Intel processors. Using an analytic performance model we point out the conditions under which Kahan comes for free, and we predict the single core performance in all memory hierarchy levels as well as the scaling behavior across the cores of a chip.

2   Performance Modeling on the Core and Chip Level

The ECM model [9–11] is an extension of the well-known Roofline model [12]. It estimates the number of CPU cycles required to execute a number of iterations n_{it} of a loop on a single core of a multicore chip. It considers the time for executing the iterations with data coming from the L1 cache as well as the time for moving the required cache lines (CLs) through the cache hierarchy. In the following we will assume fully inclusive caches, which is appropriate for current Intel architectures. We give a brief overview of the model here; details can be found in [11].

The ECM model considers the time to execute the instructions of a loop kernel on the processor core, assuming that there are no cache misses, and the time to transfer data between its initial location and the L1 cache. The in-core execution time T_{core} is determined by the unit that takes the most cycles to execute the instructions. Since data transfers in the memory hierarchy occur in units of cache lines (CLs), we always consider one cache line's "worth of work." E.g., with a loop kernel that handles single-precision floating-point arrays with unit stride, one unit of work is $n_{it} = 16$ iterations.

The time needed for all data transfers required to execute one work unit is the "transfer time." We neglect all latency effects, so the cost for one CL transfer is set by the maximum bandwidth. E.g., on the Intel IvyBridge architecture, one CL transfer takes two cycles between adjacent cache levels. Getting a 64-byte CL from memory to L3 or back takes 64 bytes· f/b_S cycles, where f is the CPU clock speed and b_S is the memory bandwidth. Note that in practice we encounter the problem that the model is too optimistic for in-memory data sets on some processors. This can be corrected by introducing a latency penalty. See Sect. 3 for details.

The in-core execution and transfer times must be put together to arrive at a prediction of single-thread execution time. If T_{data} is the transfer time, T_{OL} is the part of the core execution that overlaps with the transfer time, and T_{nOL} is the part that does not, then

$$T_{core} = \max(T_{nOL}, T_{OL}) \quad \text{and} \quad T_{ECM} = \max(T_{nOL} + T_{data}, T_{OL}) . \quad (1)$$

The model assumes that (i) core cycles in which loads are retired do not overlap with any other data transfer in the memory hierarchy, but all other in-core cycles (including pipeline bubbles) do, and (ii) the transfer times up to the L1 cache are mutually non-overlapping. A shorthand notation is used to summarize the relevant information about the cycle times that comprise the model for a loop: We write the model as $\{T_{\mathrm{OL}} \,\|\, T_{\mathrm{nOL}} \,|\, T_{\mathrm{L1L2}} \,|\, T_{\mathrm{L2L3}} \,|\, T_{\mathrm{L3Mem}}\}$, where T_{nOL} and T_{OL} are as defined above, and the other quantities are the data transfer times between adjacent memory hierarchy levels. Cycle predictions for data sets fitting into any given memory level can be calculated from this by adding up the appropriate contributions from T_{data} and T_{nOL} and applying (1). For instance, if the ECM model reads $\{2 \,\|\, 4 \,|\, 4 \,|\, 4 \,|\, 9\}$ cy, the prediction for L2 cache will be $\max(2, 4 + 4)$ cy $= 8$ cy. As a shorthand notation for predictions we use a similar format but with "⌉" as the delimiter. For the above example this would read as $T_{\mathrm{ECM}} = \{4 \,\rceil\, 8 \,\rceil\, 12 \,\rceil\, 21\}$ cy. Converting from time (cycles) to performance is done by dividing the work W (e.g., flops) by the runtime: $P = W/T_{\mathrm{ECM}}$. If T_{ECM} is given in clock cycles but the desired unit of performance is flops/s, we have to multiply by the clock speed.

We assume that the single-core performance scales linearly until a bottleneck is hit. On modern Intel processors the only bottleneck is the memory bandwidth, which means that an upper performance limit is given by the Roofline prediction for memory-bound execution: $P_{\mathrm{BW}} = I \cdot b_{\mathrm{S}}$, where I is the computational intensity of the loop code. The performance scaling for n cores is thus described by $P(n) = \min(nP_{\mathrm{ECM}}^{\mathrm{mem}}, I \cdot b_{\mathrm{S}})$ if $P_{\mathrm{ECM}}^{\mathrm{mem}}$ is the ECM model prediction for data in main memory. The performance will saturate at $n_{\mathrm{S}} = \lceil T_{\mathrm{ECM}}^{\mathrm{mem}}/T_{\mathrm{L3Mem}} \rceil$ cores. In the following section we will use the ECM model to describe performance properties of different dot implementations.

3   Optimal Implementations and Performance Models for Dot

Table 1 gives an overview of the relevant architectural details of the four generations of Intel Xeon processors used in this work. The CPUs were released in successive years between 2012 and 2015. Intel Haswell-EP marks the big microarchitectural change, with a new SIMD instruction set extension (AVX2) and several fused multiply-add instructions (FMA3). There are also notable improvements in the memory hierarchy: The access path width of load/store units was widened from 16 bytes to 32 bytes, and the bus width between the L2 and the L1 cache was enlarged from 32 bytes to 64 bytes. Note that we run the Haswell chip in "Cluster on Die" (CoD) mode, which is a configuration option that splits the 14-core chip logically into two separate ccNUMA domains with seven cores and half the L3 cache size each. This leads to a slightly better efficiency of the memory hierarchy. We also disable "Uncore frequency scaling," an energy-saving feature that can reduce the clock speed of the Uncore part of the chip (L3 cache, memory interface) if only one or two cores are active. The Broadwell chip is a very recent power-efficient "Xeon D" variant. All results for Broadwell are preliminary since we only had access to a pre-release version of the chip.

Table 1. Test machine specifications and micro-architectural features (one socket). The cache line length is 64 bytes in all cases. The SIMD register width is 16 bytes for SSE and 32 bytes for AVX. The Haswell-EP chip was run in "Cluster on Die" mode.

Microarchitecture	SandyBridge-EP	IvyBridge-EP	Haswell-EP	Broadwell-D
Shorthand	SNB	IVB	HSW	BDW
Xeon model	E5-2680	E5-2660 v2	E5-2695 v3	D-1540
Year	03/2012	09/2013	09/2014	03/2015
Clock speed (fixed)	2.7 GHz	2.2 GHz	2.3 GHz	1.8 GHz
Cores/threads	8/16	10/20	$2 \times 7/2 \times 14$	8/16
Load/store throughput per cycle				
AVX(2)	1 LD & 1/2 ST	1 LD & 1/2 ST	2 LD & 1 ST	2 LD & 1 ST
SSE/scalar	2 LD ‖ 1 LD & 1 ST	2 LD ‖ 1 LD & 1 ST	2 LD & 1 ST	2 LD & 1 ST
L1 port width	$2 \times 16{+}1 \times 16$ B	$2 \times 16{+}1 \times 16$ B	$2 \times 32{+}1 \times 32$ B	$2 \times 32{+}1 \times 32$ B
ADD throughput	1/cy	1/cy	1/cy	1/cy
MUL throughput	1/cy	1/cy	2/cy	2/cy
FMA throughput	n/a	n/a	2/cy	2/cy
L2-L1 data bus	32 B	32 B	64 B	64 B
L3-L2 data bus	32 B	32 B	32 B	32 B
LLC size	20 MiB	25 MiB	2×17.5 MiB	12 MiB
Main memory	$4 \times$ DDR3-1600	$4 \times$ DDR3-1600	$4 \times$ DDR4-2133	$2 \times$ DDR4-2133
Peak memory BW	51.2 GB/s	51.2 GB/s	2×34.1 GB/s	34.1 GB/s
Load-only BW	43.6 GB/s (85 %)	46.1 GB/s (90 %)	32.4 GB/s (95 %)	33 GB/s (97 %)
T_{L3Mem} per CL	3.96 cy	3.05 cy	2.43 cy	3.49 cy

We first discuss variants for dot in single precision (SP) for the Intel IvyBridge microarchitecture. The differences to double precision (DP) and the impact of architectural changes are covered in Sect. 3. To eliminate variations introduced by compilers we implemented all kernels directly in assembly language using the `likwid-bench` microbenchmarking framework [13].

Naive Scalar Product. The naive scalar product in single precision serves as the baseline (see Fig. 1a). Sufficient unrolling must be applied to hide the ADD pipeline latency for the recursive update on the accumulation register and to apply SIMD vectorization. Both optimizations introduce partial sums and are therefore not compatible with the C standard as the order of non-associative operations is changed. With higher optimization levels the current Intel compiler (version 15.0.2) is able to generate optimal code. Note that partial sums usually improve the accuracy of the result [8].

This kernel is limited by the throughput of the LOAD unit on the IVB architecture (see Table 1). Two AVX loads per vector (a and b) are required to cover one unit of work (16 scalar loop iterations), resulting in $T_{nOL} = 4$ cy. The overlapping part is $T_{OL} = 2$ cy since two MULT and two ADD instructions must be executed. Data transfers between cache levels require two cycles per CL, so that $T_{L1L2} = T_{L2L3} = 4$ cy.

(a)

```
float sum = 0.0;

for (int i=0; i<N; i++) {
    sum = sum + a[i] * b[i]
}
```

(b)

```
float sum = 0.0;
float c = 0.0;
for (int i=0; i<N; ++i) {
    float prod = a[i]*b[i];
    float y = prod-c;
    float t = sum+y;
    c = (t-sum)-y;
    sum = t;
}
```

Fig. 1. (a) Naive scalar product code in single precision. (b) Kahan-compensated scalar product code.

For T_{L3Mem} we calculate the number of cycles per CL transfer from the maximum memory bandwidth and the clock speed (last row in Table 1) and arrive at $T_{L3Mem} = 6.1$ cy. The full ECM model thus reads $\{2 \,\|\, 4\,|\,4\,|\,4\,|\,6.1\}$ cy. On newer Intel chips (notably IVB and HSW) unknown peculiarities in the design of the Uncore lead to extra latency penalties per cache line from memory. We take these deviations into account by introducing a penalty parameter that is fixed empirically. This parameter is an additive contribution to T_{L3Mem}, so that the final model is $\{2 \,\|\, 4\,|\,4\,|\,4\,|\,6.1 + 2.9\}$ cy, leading to a runtime prediction of $\{4\,\rceil\,8\,\rceil\,12\,\rceil\,18.1 + 2.9\}$ cy. At a clock speed of 2.2 GHz the expected serial performance is thus

$$P = \frac{16 \, \text{updates} \cdot 2.2 \, \text{Gcy/s}}{\{4\,\rceil\,8\,\rceil\,12\,\rceil\,18.1 + 2.9\} \, \text{cy}} = \{8.80\,\rceil\,4.40\,\rceil\,2.93\,\rceil\,1.68\} \, \text{GUP/s} . \qquad (2)$$

We choose an "update" (two flops) as the basic unit of work to make performance results for different implementations comparable. The resulting unit is Giga updates per second: GUP/s. The predicted saturation point is at $n_S = \lceil (18.1 + 2.9)/6.1 \rceil = 4$ cores. Note that the maximum memory bandwidth has to be taken into account for the saturation point, so we divide by 6.1 cy. The Roofline "light speed," i.e., the memory bandwidth-limited saturated performance, can be calculated from the computational intensity of one update per eight bytes: $P_{BW} = (1 \, \text{update}/8 \, \text{B}) \cdot b_S = 5.76 \, \text{GUP/s}$.

All versions of the enhanced scalar product described in the next section will be compared to the optimal naive implementation.

Kahan-Enhanced Scalar Product on IvyBridge. Figure 1b shows the implementation of the Kahan algorithm for dot. Compilers have problems with this loop code for two reasons: First, the compiler detects (correctly) a loop-carried dependency on c, which prohibits SIMD vectorization and modulo unrolling. Second, the compiler may recognize that, arithmetically, c is always equal to zero. With high optimization levels it may thus reduce the code to the naive scalar product, defeating the purpose of the Kahan algorithm. This is the reason why we use hand-coded assembly throughout this work.

One iteration comprises one multiplication, four additions or subtractions, and two loads. The bottleneck on the IVB core level is thus the ADD unit (ADD and SUB are handled by the same pipeline). In the following we construct the ECM model for scalar, SSE, and AVX versions of the Kahan loop. Independent of vectorization we always establish proper modulo unrolling for best pipeline utilization.

Scalar implementation. In scalar mode, one unit of work amounts to $16 \times 4 = 64$ instructions in the ADD unit, resulting in $T_{\mathrm{OL}} = 64$ cy. Since two scalar loads can be executed per cycle on the IVB core, the 32 loads lead to $T_{\mathrm{nOL}} = 16$ cy. The contributions from in-cache and memory transfers are the same as for the naive variant above, so the complete ECM model is $\{64 \,\|\, 16\,|\,4\,|\,4\,|\,6.1 + 2.9\}$ cy, and the runtime prediction is $\{64 \,\rceil\, 64 \,\rceil\, 64 \,\rceil\, 64\}$ cy. According to the model the scalar variant should not be able to saturate the memory bandwidth using all cores on the ten-core chip, since $n_S = \lceil 64/6.1 \rceil = 11$ cores. The analysis shows that the scalar variant of Kahan is limited by the instruction throughput, specifically on the ADD pipeline, regardless of where the data resides. We thus expect the same performance $P = 16 \cdot 2.2/64\,\mathrm{GUP/s} = 0.55\,\mathrm{GUP/s}$ in all memory hierarchy levels for single-threaded execution, and close to perfect scalability across the cores of the chip.

SSE implementation. SSE uses 16-byte wide registers, and all instructions required for the Kahan algorithm exist in SSE variants, so the overall number of instructions is reduced by a factor of four compared to the scalar version, but the same throughput limits apply for the ADD and the LOAD unit. This leads to an ECM model of $\{16 \,\|\, 4\,|\,4\,|\,4\,|\,6.1 + 2.9\}$ cy and a prediction of $\{16 \,\rceil\, 16 \,\rceil\, 16 \,\rceil\, 18.1 + 2.9\}$ cy, which yields $P = \{2.20 \,\rceil\, 2.20 \,\rceil\, 2.20 \,\rceil\, 1.68\}$ GUP/s. The SSE code is limited by the instruction throughput up to the L3 cache since all data transfer contributions can be overlapped with the ADD instructions. The optimal $4\times$ speed-up of SSE is thus observed in this case. For data in main memory the speed-up is just about $64/21 \approx 3\times$, and the single-core performance and saturation behavior are identical to the naive scalar product.

AVX implementation. AVX further reduces the runtime for the ADD operations by a factor of two, so $T_{\mathrm{OL}} = 8$ cy. Although the number of LOAD instructions is also cut in half, the non-overlapping time T_{nOL} does not change, because the two LOAD ports of the L1 cache are only 16 bytes wide. Therefore only one LOAD instruction can be retired per cycle. The complete ECM model is $\{8 \,\|\, 4\,|\,4\,|\,4\,|\,6.1 + 2.9\}$ cy, the runtime prediction is $\{8 \,\rceil\, 8 \,\rceil\, 12 \,\rceil\, 18.1 + 2.9\}$ cy (leading to $P = \{4.40 \,\rceil\, 4.40 \,\rceil\, 2.93 \,\rceil\, 1.68\}$ GUP/s), and the saturation behavior is the same as for the SSE variant of Kahan and the naive scalar product. The AVX code is limited by the instruction throughput up to the L2 cache, and the full $2\times$ advantage versus SSE can be observed in this case. Starting from L3 there is a slight impact on runtime by data transfers, leading to a reduced speed-up of $1.3\times$ in L3 and none at all in main memory. Again the saturation behavior is expected between three and four cores.

The conclusion from this analysis is that there is no expected performance difference for in-memory working sets between the naive scalar product and the

Kahan version if any kind of vectorization is applied to Kahan. With AVX, Kahan comes for free even in the L3 or the L2 cache. Only for in-L1 data we expect a 2× slowdown for Kahan versus the naive version even with the best possible code.

Influence of Processor Architecture. In this section we compare the model-based analysis across four generations of Intel CPUs: SandyBridge-EP (SNB), IvyBridge-EP (IVB), Haswell-EP (HSW), and Broadwell (BDW, in a power-efficient "Xeon D" variant). This covers four Intel Xeon microarchitectures over a time of three years and involves one major architectural step (from IVB to HSW). We always consider the optimal AVX code for the comparisons. There is no major change expected between SNB and IVB, since no dot-relevant hardware features were added. All observed performance differences are thus rooted in the clock speed and memory bandwidth (first row in Table 2). Note that despite the lower memory bandwidth of the SNB test system compared to IVB, the in-memory performance is higher due to the faster clock speed of SNB. The HSW microarchitecture has new features which influence dot performance: It can sustain two AVX loads and one AVX store per cycle, effectively doubling LOAD/STORE throughput. In addition, the L1-L2 bus width was doubled, allowing for a full CL transfer per cycle. These changes result in $T_{nOL} = 2\,\text{cy}$ and $T_{L1L2} = 2\,\text{cy}$ (third row in Table 2). It is interesting that BDW only requires half a cycle of latency penalty per CL in memory, so the uncorrected ECM model works very well already. BDW performance is insensitive to data transfers up to the L3 cache.[1]

Table 2. Comparison of the ECM model for optimal AVX implementations across the multicore Xeon CPUs in the testbed (see Table 1). The consequences of relevant architectural changes to the preceding generation are highlighted.

	ECM model [cy]	Prediction [cy/CL]	Pred. performance [GUP/s]
SNB	$\{8 \parallel 4 \mid 4 \mid 4 \mid 7.9 + 5.1\}$	$\{8 \rceil 8 \rceil 12 \rceil 19.9 + 5.1\}$	$\{5.40 \rceil 5.40 \rceil 3.60 \rceil 1.73\}$
IVB	$\{8 \parallel 4 \mid 4 \mid 4 \mid 6.1 + 2.9\}$	$\{8 \rceil 8 \rceil 12 \rceil 18.1 + 2.9\}$	$\{4.40 \rceil 4.40 \rceil 2.93 \rceil 1.68\}$
HSW	$\{8 \parallel \mathbf{2} \mid \mathbf{2} \mid 4 \mid 8.6 + 2.4\}$	$\{8 \rceil 8 \rceil 8 \rceil 16.6 + 2.4\}$	$\{4.60 \rceil 4.60 \rceil 4.60 \rceil 2.11\}$
BDW	$\{8 \parallel \mathbf{2} \mid \mathbf{2} \mid 4 \mid 7 + \mathbf{1}\}$	$\{8 \rceil 8 \rceil 8 \rceil 15 + 1\}$	$\{3.60 \rceil 3.60 \rceil 3.60 \rceil 1.8\}$

Double vs. Single Precision. The model prediction in terms of cycles per CL does not change for the SIMD variants of Kahan when going from SP to DP, but one CL update represents twice as much useful work (scalar iterations) in the SP case. However, the penalty for going from SIMD to scalar is only half as big as for SP, since the scalar register width is eight bytes instead of four. The ECM model for the DP scalar version of the Kahan dot on IVB is $\{32 \parallel 8 \mid 4 \mid 4 \mid 6.1 + 2.9\}$ cy and

---

[1] Note that our test system was a pre-release Xeon D; production systems and mainstream Xeon Broadwell chips may show a different behavior.

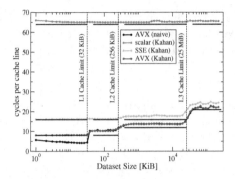

Fig. 2. Single-core cycles per CL vs. data set size for various implementations of the Kahan scalar product and the AVX version of the naive scalar product in SP on IVB. The horizontal lines represent the ECM model predictions for scalar (top), SSE (middle), and AVX (bottom) Kahan variants.

the according runtime prediction is $\{32 \rceil 32 \rceil 32 \rceil 32\}$ cy, with $P = 0.55$ GUP/s. The reduced cycle count (32 instead of 64) for DP leads to saturation at a smaller number of cores for in-memory working sets: $n_S = \lceil 32/6.1 \rceil = 6$. Hence, even the scalar DP variant of Kahan exerts sufficient pressure on the memory interface to reach saturation. The saturated DP performance according to the Roofline model is $P_{BW} = (1 \text{ update}/16 \text{ B}) \cdot b_S = 2.88$ GUP/s.

4   Performance Results and Model Validation

Single-core benchmarking results for single precision on IVB are shown in Fig. 2. The model predicts the overall behavior very well. The naive and the AVX Kahan version show identical performance in L2 cache and beyond. As predicted there is no performance drop for the SSE Kahan version from L1 to L2. Both AVX Kahan and the compiler-generated naive version fall slightly short of the prediction in L2. This is a general observation with many loop kernels, and we interpret it as a consequence of the L2-L1 hardware prefetcher doing a better job in latency hiding for SSE than for AVX due to the more relaxed timings in the SSE case. Since the details of prefetching are undisclosed, we have no way to prove or refute this hypothesis. Finally, the constant performance of the scalar Kahan variant across all memory levels is perfectly predicted by the model.

In-memory scaling results on the chip level are shown in Fig. 3a. The dashed lines are the model predictions (for clarity we only show models for scalar and AVX). As anticipated via the ECM model, the scalar version cannot saturate the memory bandwidth even if all cores are used. Since any code that is able to saturate the bandwidth is "perfect," any kind of vectorization will make the Kahan algorithm as fast as the naive scalar product. Note, however, that on a CPU with a faster clock speed or more cores saturation will be easily achieved even with scalar code. This effect illustrates the general observation that more parallelism can "heal" low single-core performance. For comparison we also show

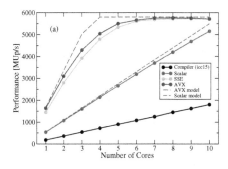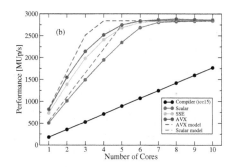

Fig. 3. In-memory scaling (10 GB working set size) for different implementations of the Kahan scalar product on IVB for (a) single precision and (b) double precision. Model predictions are shown with dashed lines for the scalar and AVX versions. One update is equivalent to five flops (one MULT, four ADDs).

the compiler-generated variant of Kahan. As described earlier, the code is devastatingly slow since the compiler cannot resolve the loop-carried dependency.

The only relevant difference between DP and SP is the smaller performance penalty of the scalar variant for DP (see Fig. 3b), which leads to strong saturation at about six cores as predicted by the model.

In order to compare different architectures, we show single-core data for the AVX-vectorized Kahan scalar product in Fig. 4a. Since we report runtime in cycles per CL, the influence of different clock speeds and memory bandwidths is only visible for the in-memory case. In the L1 cache all processors show the same runtime, because the architectural improvements with HSW and BDW do not address the bottleneck of the algorithm at hand (the ADD throughput). In L2 and L3, HSW and BDW show higher performance than the previous generations due to their doubled L2-L1 bandwidth and LOAD throughput. The step from L2 to L3 is important, since it marks the transition to the Uncore, which is a shared resource across the cores. Although a notable improvement is seen in L3 with each new architecture, we observe efficiency issues in the Uncore on IVB and HSW that prevent those CPUs from attaining the expected performance. In memory, HSW falls behind BDW due to a larger latency penalty. BDW seems to have corrected those issues, but we must stress again that these observations were made on an eight-core single-socket "Xeon D" chip, and it is unclear if the to be released multi-socket variants with larger core counts can live up to the expectations raised here. It is also worth emphasizing that, as already mentioned above, in practice any code that can saturate the memory bandwidth is "good enough." Fig. 4b shows the in-memory performance scaling for all four architectures with the AVX Kahan scalar product: The differences in saturated performance indeed reflect the differences in saturated memory bandwidth. Again, and certainly as expected from the model, vectorization makes the Kahan algorithm come for free.

There is one additional optimization on HSW and BDW that we have not mentioned yet. The two FMA units can theoretically increase the ADD throughput

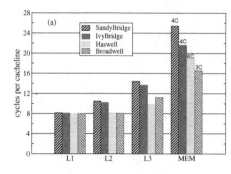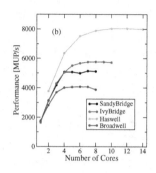

Fig. 4. Comparison between four Intel Xeon multi-core architectures using the single-precision AVX Kahan scalar product: (a) Measured single-core runtime in cycles per CL in different memory hierarchy levels. The saturation point n_S is indicated above the bars for the memory-bound case. (b) Measured performance scaling with in-memory working set. The scaling data for Haswell was obtained in "Cluster on Die" mode with threads scattered across the two ccNUMA domains. Hence, only even core counts are available.

by a factor of two. Both units can execute FMA and MULT instructions, but only one of them can handle stand-alone ADDs. This is not a problem with hand-crafted assembly since one can endow an FMA instruction with a unit multiplicand to act like an ADD. The downside is that the FMA instruction has a higher latency of five cycles (ADD only has three) and therefore requires deeper unrolling to hide the pipeline latency. Both architectures hence run out of registers and only achieve a 20 % speed-up from FMA with data in L1, and no noticeable improvement beyond L1.

5   Conclusion

We have investigated the performance of naive and Kahan-enhanced variants of the scalar product on a range of recent Intel multicore chips. Using the ECM model the single-core performance in all memory hierarchy levels and the multi-core scaling for in-memory data were accurately described. The most important result is that even the single-threaded Kahan algorithm comes with no performance penalties on all standard multicore architectures under investigation in the L2 cache, the L3 cache, and in memory if implemented optimally. Depending on the particular architecture and whether single or double precision is used, even scalar code may achieve bandwidth saturation in memory when using multiple threads. Performance improvements between successive generations of Intel CPUs could be attributed to specific architectural advancements, such as increased LOAD throughput on Haswell or a more efficient Uncore on Broadwell.

We emphasize that the approach and insights described here for the special case of the Kahan scalar product can serve as a blueprint for other load-dominated streaming kernels.

Acknowledgement. We thank Intel Germany for providing an early access Broadwell test system. This work was partially funded by BMBF under grant 01IH13009A (project FEPA), and by the Competence Network for Scientific High Performance Computing in Bavaria (KONWIHR).

References

1. Goldberg, D.: What every computer scientist should know about floating-point arithmetic. ACM Comput. Surv. **23**(1), 5–48 (1991)
2. Linz, P.: Accurate floating-point summation. Commun. ACM **13**(6), 361–362 (1970)
3. Gregory, J.: A comparison of floating point summation methods. Commun. ACM **15**(9), 838 (1972)
4. Kahan, W.: Pracniques: further remarks on reducing truncation errors. Commun. ACM **8**(1), 40 (1965)
5. Rump, S.M., Ogita, T., Oishi, S.: Accurate floating-point summation Part I: faithful rounding. SIAM J. Sci. Comput. **31**(1), 189–224 (2008)
6. Zhu, Y.K., Hayes, W.B.: Algorithm 908: online exact summation of floating-point streams. ACM Trans. Math. Softw. **37**(3), 1–13 (2010)
7. Demmel, J., Nguyen, H.D.: Fast reproducible floating-point summation. In: 21st IEEE Symposium on Computer Arithmetic, pp. 163–172, April 2013
8. Dalton, B., Wang, A., Blainey, B.: SIMDizing pairwise sums: a summation algorithm balancing accuracy with throughput. In: Proceedings of the 2014 Workshop on Programming Models for SIMD/Vector Processing, WPMVP 2014, pp. 65–70. ACM, New York (2014)
9. Treibig, J., Hager, G.: Introducing a performance model for bandwidth-limited loop kernels. In: Wyrzykowski, R., Dongarra, J., Karczewski, K., Wasniewski, J. (eds.) PPAM 2009, Part I. LNCS, vol. 6067, pp. 615–624. Springer, Heidelberg (2010)
10. Hager, G., Treibig, J., Habich, J., Wellein, G.: Exploring performance and power properties of modern multicore chips via simple machine models. Concurrency Comput.: Pract. Exper. (2013). doi:10.1002/cpe.3180
11. Stengel, H., Treibig, J., Hager, G., Wellein, G.: Quantifying performance bottlenecks of stencil computations using the execution-cache-memory model. In: Proceedings of the 29th ACM International Conference on Supercomputing, ICS 2015. ACM, New York (2015)
12. Williams, S., Waterman, A., Patterson, D.: Roofline: an insightful visual performance model for multicore architectures. Commun. ACM **52**(4), 65–76 (2009)
13. Treibig, J., Hager, G., Wellein, G.: likwid-bench: An extensible microbenchmarking platform for x86 multicore compute nodes. In: Brunst, H., et al. (eds.) Tools for High Performance Computing 2011, pp. 27–36. Springer, Heidelberg (2012)

Performance Analysis of the Chebyshev Basis Conjugate Gradient Method on the K Computer

Yosuke Kumagai[1(✉)], Akihiro Fujii[1], Teruo Tanaka[1], Yusuke Hirota[2,3],
Takeshi Fukaya[2,3,4], Toshiyuki Imamura[2,3], and Reiji Suda[5]

[1] Kogakuin University, Tokyo, Japan
`em14006@ns.kogakuin.ac.jp`
[2] RIKEN Advanced Institute for Computational Science, Kobe, Japan
[3] JST CREST, Tokyo, Japan
[4] Hokkaido University, Hokkaido, Japan
[5] The University of Tokyo, Tokyo, Japan

Abstract. The conjugate gradient (CG) method is useful for solving large and sparse linear systems. It has been pointed out that collective communication needed for calculating inner products becomes serious performance bottleneck when executing the CG method on massively parallel systems. Recently, the Chebyshev basis CG (CBCG) method, a communication avoiding variant of the CG method, has been proposed, and theoretical studies have shown promising results, particularly for upcoming exascale supercomputers. In this paper, we evaluate the CBCG method on an actual system, namely the K computer, to examine the potential of the CBCG method. We first construct a realistic performance model that reflects the computation on the K computer, and the model indicates that the CBCG method is faster than CG method if the number of cores is sufficient large. We then measure the execution time of both methods on the K computer, and obtained results agree with our estimation.

Keywords: Communication avoiding · Conjugate gradient method · Linear solver

1 Introduction

Supercomputers are used to perform large-scale scientific computing. According to the TOP 500 List [1], the performance of supercomputers improves as the number of cores increases. However, parallel processing requires communication (i.e. data transfer among processes), and its cost is usually proportional to the number of cores. Therefore, reducing communication costs when designing algorithms is required.

The conjugate gradient (CG) method [2] has been widely used for solving a system of linear equations whose coefficient matrix is symmetric positive definite. The CG method performs one sparse matrix vector product (SpMV) operation

© Springer International Publishing Switzerland 2016
R. Wyrzykowski et al. (Eds.): PPAM 2015, Part I, LNCS 9573, pp. 74–85, 2016.
DOI: 10.1007/978-3-319-32149-3_8

and two inner product operations per iteration. When the CG method is parallelized in a distributed system, assuming that the matrix is split by block rows, each process possesses only a part of vector data. Therefore, inner product operations require collective communication, namely MPI_AllReduce, and introduce a serious performance bottleneck in massively parallel computing.

There are many studies that aim at reducing the cost of collective communication in the CG method. Ghysels et al. proposed the pipelined CG method [3], in which the number of collective communication calls is reduced to half that in the CG method. The s-step CG method (by Chronopoulos et al.) [4] and the Krylov basis CG method (by Toledo) [5] make a further reduction possible. These methods are mathematically equivalent to the CG method, but in finite precision arithmetic, the residual history tends to be less stable than the CG method because of the effects of round-off error.

Recently, Hoemmen has proposed the communication avoiding CG (CA-CG) method [6], in which a certain polynomial basis of the Krylov subspace is employed to improve the numerical stability. The Chebyshev basis CG (CBCG) method [7] is one of its variants, in which the Krylov subspace is generated based on the Chebyshev polynomial. It has been reported that the CBCG and other CA-CG type methods reach convergence with the same number of iterations as the CG method, and they have been regarded as numerically stable. Therefore, these methods attract the interest of many researchers nowadays.

However, to the best of our knowledge, many studies on the CA-CG method focus on the theoretical aspects, and there are few studies that report the performance results on actual large-scale supercomputers. For example, in the study by Carson et al., which investigates the deflation techniques for the CA-CG method, the execution time of the method is evaluated only by a simple performance model [8]. Besides, the model parameters in their model seem to represent a future system rather than an actual exiting system. Such studies are of course important, but it is also crucial to examine the effectiveness of the CA-CG method on existing supercomputers because the cost for collective communication in the CG method is already a serious problem.

Considering these situations, in this paper, we aim for evaluating the CA-CG method, particularly the CBCG method, on the K computer, which is one of the top large-scale supercomputers. We first construct a performance model that reflects the parallel computation on the K computer, and our model indicates that the CBCG method is faster than the CG method when the number of cores is sufficient large. We then measure the execution time of both methods by using the K computer (up to 98,304 cores) and find that the CBCG method is faster than the CG method when the number of cores is sufficient large. These experimental results agree well with the prediction by the model. Although our test problem is simple (i.e. a coefficient matrix derived from the 7-point stencil structure), we confirm that the CBCG method is indeed effective on the K computer.

The rest of the paper is organized as follows: in Sect. 2, we briefly explain the CBCG method. In Sect. 3, we construct a performance model and predict the performance of the CG and CBCG methods by using the model. We then present the results of experiments by using the K computer in Sect. 4. Finally, we give our concluding remarks in Sect. 5.

2   The Chebyshev Basis Conjugate Gradient Method

The CG method generates the Krylov subspace from the matrix A and the initial residual vector $r_0 = b - Ax_0$ and updates the solution vector x through expanding the Krylov subspace by one dimension per iteration. The CBCG method, whose algorithm is shown in Algorithm 1, does the same but expands it by k dimensions per iteration through performing SpMV k times (lines 6 through 12). In the CBCG method, the 3-term recurrence formula of the Chebyshev polynomials is employed to improve the numerical stability. Note that λ_{min} and λ_{max} (lines 1 and 2) are estimated values of the minimum and maximum eigenvalues of the coefficient matrix, respectively. We also mention that B_{n+1} (line 17) and a_{n+1} (line 21) are calculated through the QR decomposition of $(Q^T A Q)$ because this matrix could be singular. For further details of the CBCG method, see [6,7].

Let us consider parallelizing the CBCG method; we assume that Q and S are split by block rows and that each process has the partial data. In this case, MPI_AllReduce is required three times per iteration: for calculating $Q_n^T (AS_{n+1})$ (line 17), $Q_{n+1}^T (AQ_{n+1})$ and $Q_{n+1}^T r_{nk}$ (line 21), and $\| r_{nk} \|_2$ (line 25). It is worth noting that the multiplication of A is not necessary here because both AS_{n+1} and AQ_{n+1} are already obtained (lines 12 and 15/19). Since one iteration of the CBCG method is mathematically equivalent to k iterations in the CG method, it follows that the CBCG method reduces the number of MPI_AllReduce calls to approximately $1/k$ that the CG method.

3   Performance Prediction Using a Performance Model

3.1   Parallel Execution Model

In this study, we model parallel execution time basically following Carson et al. [8]. Let F be the total number of floating-point operations (i.e. additions and multiplications), S be the total number of messages, and W be the total amount of data. We then assume that the parallel execution time T is expressed as

$$T = \gamma \cdot F + \alpha \cdot S + \beta \cdot W, \tag{1}$$

where γ is the floating-point throughput (i.e. the cost for one floating point operation), α is the setup cost (i.e. latency) of one communication operation, and β is the inverse of the network bandwidth. Note that these three model parameters need to be set depending on a target system of interest.

Algorithm 1. The CBCG method.

1: $\eta \leftarrow 2/(\lambda_{max} - \lambda_{min})$
2: $\zeta \leftarrow (\lambda_{max} + \lambda_{min})/(\lambda_{max} - \lambda_{min})$
3: $\boldsymbol{r}_0 \leftarrow \boldsymbol{b} - A\boldsymbol{x}_0$
4: $n \leftarrow 0$
5: **repeat**
6:    $\boldsymbol{s}_0 \leftarrow \boldsymbol{r}_{ik}$
7:    $\boldsymbol{s}_1 \leftarrow \eta A\boldsymbol{s}_0 - \zeta\boldsymbol{s}_0$
8:    **for** $i = 2$ to k **do**
9:       $\boldsymbol{s}_i \leftarrow 2\eta A\boldsymbol{s}_{i-1} - 2\zeta\boldsymbol{s}_{i-1} - \boldsymbol{s}_{i-2}$
10:   **end for**
11:   $S_{n+1} \leftarrow (\boldsymbol{s}_0, \boldsymbol{s}_1, \cdots, \boldsymbol{s}_{k-1})$
12:   $AS_{n+1} \leftarrow (A\boldsymbol{s}_0, A\boldsymbol{s}_1, \cdots, A\boldsymbol{s}_{k-1})$
13:   **if** $n = 0$ **then**
14:      $Q_{n+1} \leftarrow S_{n+1}$
15:      $AQ_{n+1} \leftarrow AS_{n+1}$
16:   **else**
17:      $B_{n+1} \leftarrow \left(Q_n^T AQ_n\right)^{-1} Q_n^T AS_{n+1}$
18:      $Q_{n+1} \leftarrow S_{n+1} - Q_n B_{n+1}$
19:      $AQ_{n+1} \leftarrow AS_{n+1} - AQ_n B_{n+1}$
20:   **end if**
21:   $\boldsymbol{a}_{n+1} \leftarrow \left(Q_{n+1}^T AQ_{n+1}\right)^{-1} Q_{n+1}^T \boldsymbol{r}_{nk}$
22:   $\boldsymbol{x}_{(n+1)k} \leftarrow \boldsymbol{x}_{nk} + Q_{n+1}\boldsymbol{a}_{n+1}$
23:   $\boldsymbol{r}_{(n+1)k} \leftarrow \boldsymbol{r}_{nk} - AQ_{n+1}\boldsymbol{a}_{n+1}$
24:   $n \leftarrow n + 1$
25: **until** $\| \boldsymbol{r}_{nk} \|_2 / \| \boldsymbol{b} \|_2 < \epsilon$

3.2 Performance Modeling of the CG and CBCG Methods

We now consider modeling the performance of the CG and CBCG methods; for each method, we count the amount of F, S, and W along the critical path. We assume that the load balance is uniform and that there is no overlap between computation and communication. In addition, we deal with MPI_AllReduce as multiple calls of point-to-point communication (MPI_Send/Recv); we assume that one MPI_AllReduce operation is done by $2\log_2 P$ times point-to-point communication calls (i.e. reduce and scatter along a binary tree), where P is the number of processes.

In the CG and CBCG methods, SpMV, vector-vector (e.g. inner product), and dense matrix-matrix (only in CBCG) operations are performed. It has been reported that distinguishing computational kernels makes a performance model more realistic [9]. For this reason, we sophisticate the model as

$$T = \gamma_{\text{spmv}} \cdot F_{\text{spmv}} + \gamma_{\text{vec}} \cdot F_{\text{vec}} + \gamma_{\text{mat}} \cdot F_{\text{mat}} + \alpha \cdot S + \beta \cdot W, \qquad (2)$$

where spmv, vec, and mat correspond to SpMV, vector–vector, and matrix-matrix operations, respectively.

It is very difficult to model the performance of a general SpMV operation; its computation and communication generally depend on the size, structure, and distribution of the matrix. In this study, we thus consider a simple case where the matrix is obtained from the 7-point central difference of the Poisson equation defined on a cubic domain; in other words, the matrix is derived from the 7-point stencil structure. We assume that the domain is discretized into $\sqrt[3]{N} \times \sqrt[3]{N} \times \sqrt[3]{N}$ points, so that the dimension of the matrix is N. We also assume that each process uniformly has $\sqrt[3]{N/P} \times \sqrt[3]{N/P} \times \sqrt[3]{N/P}$ points (i.e. a partial cubic domain of the original one).

In this situation, it is easy to verify that F_{spmv} is $14N/P$; each element of the resulting vector of SpMV is calculated with 7 elements of the input vector. Since the matrix is derived from the 7-point stencil structure, communication is required only for calculating the elements on the surface of each local cubic domain. This means that each process communicates with six neighbors and that the amount of data transferred is $(\sqrt[3]{N/P})^2$ per neighbor. Assuming that each process can send and receive data at the same time, S and W in SpMV are 6 and $6(\sqrt[3]{N/P})^2$, respectively.

We then show the overall results in Table 1, in which we compare the costs for one iteration of the CBCG method with those for k iterations of the CG method. Here, we ignore the cost for calculating QR decompositions in the CBCG method. This is because it depends only on k, which is usually sufficiently small (e.g. $k = 10$ or 20).

Table 1. The computation and communication costs for the CG and CBCG methods: N is dimension of the coefficient matrix, and P is the number of processes in parallel computing.

	CG (k iterations)	CBCG (one iteration)
F_{spmv}	$(14N/P)k$	$(14N/P)k$
F_{vec}	$(10N/P + 2\log_2 P)k$	$(8k - 2)N/P + \log_2 P$
F_{mat}	0	$(8k^2 + 2k + 2)N/P + (2k^3 + 2k)\log_2 P$
S	$(6 + 4log_2 P)k$	$6k + 6\log_2 P$
W	$\{6(\sqrt[3]{N/P})^2 + 4\log_2 P\}k$	$6k(\sqrt[3]{N/P})^2 + (4k^2 + 4k + 2)\log_2 P$

3.3   Estimation by Performance Model

We predict the execution time of the CG and CBCG methods based on our performance model and compare them. Through preliminary executions on the K computer, we obtained the following parameters: $\gamma_{\mathrm{spmv}} = 2.7 \times 10^{-9}$, $\gamma_{\mathrm{vec}} = 6.6 \times 10^{-14}$, $\gamma_{\mathrm{mat}} = 3.4 \times 10^{-10}$, $\alpha = 2.7 \times 10^{-6}$, and $\beta = 3.3 \times 10^{-8}$. Here the parameters related to floating-point operation are calculated from the results measured with 6,144 cores, and those related to communication are obtained from results of the benchmark of point-to-point communication (i.e. ping-pong benchmark).

Setting $N = 27,000,000$ (i.e. the grid is $300 \times 300 \times 300$) and assuming FlatMPI execution (i.e. each core has one MPI process), we predict the execution time of the CG and CBCG methods. We show the results in Fig. 1, where we plot the time per iteration of the CG method and $1/k$ of the time per iteration of the CBCG method. This graph clearly illustrates that the CBCG method is slower than the CG method when the number of cores is small, which is due to the additional floating-point operations required in the CBCG method, but that the CBCG method is faster when the number of cores is sufficient large. The graph also shows that the CBCG method becomes more efficient than the CG method as the number of cores increases. This result indicates that the CBCG method can be effective on the K computer.

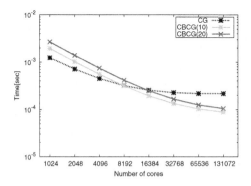

Fig. 1. Prediction of the parallel execution time of the CG and CBCG methods by the performance mode: assuming a coefficient matrix derived from the 7-point stencil structure ($N = 27,000,000$), the time for one iteration in the CG method and $1/k$ of the time for one iteration in the CBCG method are plotted.

4   Experimental Results

4.1   Experimental Environments and Problem Setting

We conduct environments using the K computer [10,11]. In the K computer, each node is equipped with one SPARC64 VIIIfx processor (2.0GHz, 8cores) and 16GB memory (DDR3 SDRAM, 64GB/s), and is connected by the 6D mesh/tours network (5GB/s/link, bidirectional). We implemented the program of the CG and CBCG methods in the C programming language with the BLAS (Basic Linear Algebra Subprograms) and MPI libraries; vector and dense matrix operations are done by BLAS routines. We used the "`mpifccpx`" compiler with the options "`-Kfast,ipo,restp=all -lm -SSL2`". We linked the BLAS and MPI libraries provided by Fujitsu. The version of the language environment was "`K-1.2.0-15`".

Launching one process per core (FlatMPI model), we executed our programs by using up to 98,304 cores (12,288 nodes) and measured the execution time. We set the convergence criterion of the CG and CBCG methods to be the relative residual (2-norm) below 10^{-12}. In the CBCG method, estimates for the maximum and minimum eigenvalues are required, and we determined the former by using the power method and set the latter to 0. Note that the execution time of the CBCG method does not include the time for the power method.

In the experiments, we solved linear systems whose coefficient matrix is obtained from the Poisson equation arising in a groundwater flow problem in 3-dimensional heterogeneous porous media [12]. Note that the water conductivity values were determined by the sequential Gaussian algorithm [13], and that the minimum and maximum water conductivity values are 10^{-5} and 10^5, respectively, with the average value of 1.0. The problem domain is discretized into $300 \times 300 \times 300$ points, so that the number of unknowns in linear systems (i.e. the dimension of the coefficient matrix) is 27,000,000. We divided the original domain into rectangular parallelepiped domains and allocated each of them to each process. More precisely, we factorized P into $P_X \times P_Y \times P_Z$ as evenly as possible and divided 300 into P_X, P_Y, and P_Z processes in X, Y, and Z directions, respectively. We list how P was factorized in the experiments in Table 2 along with the shape of the local domain that each process owned. Note that each process stored the data of the coefficient matrix explicitly in the compressed row (CRS) format although it can be stored implicitly because the matrix derived from the stencil structure.

Table 2. The shape of the process grid and the local domain on each process in the experiments.

N	P	P_X	P_Y	P_Z	The shape of local domain per process
27,000,000	6,144	16	16	24	$19 \times 19 \times 13$
	12,288	16	32	24	$19 \times 10 \times 13$
	24,576	16	32	48	$19 \times 10 \times 7$
	49,152	32	32	48	$10 \times 10 \times 7$
	98,304	32	64	48	$10 \times 5 \times 7$

4.2   Results

First, we report the number of iterations required for convergence. The CG method converged in 7,449 iterations, the CBCG ($k = 10$) method in 745 iterations (equivalent to 7,450 iterations of the CG method), and the CBCG ($k = 20$) method in 373 iterations (equivalent to 7,460 iterations of the CG method). Taking into account that 744×10 and 372×20 are less than 7,449, this result is reasonable and reflects the mathematical equivalence between the CG and CBCG methods.

We then show the measured execution time required for convergence in Fig. 2, where the problem size is fixed and the number of cores increases (i.e. strong scaling). From Fig. 2, it is clear that the runtime of the CG method rarely decreases in the region where the number of cores is greater than 12,288. In contrast, the runtime of the CBCG method decreases as the number of cores increases, although the CBCG(10) method stagnates a little when the number of cores is 98,304. It is also observed that the CBCG method becomes faster than the CG method when the number of cores is greater than 24,576. In the case of using 98,304 cores, the CBCG(20) method is approximately twice faster than the CG method.

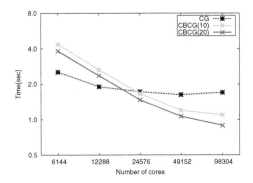

Fig. 2. Measured execution time of the CG and CBCG methods on the K computer: the coefficient matrix is derived from the 7-point stencil structure, and the size of the matrix is 27,000,000. The strong scaling is evaluated.

In addition to the above results, we present some detailed results in Fig. 3, where we show the detailed breakdown of the execution time. The left graph in Fig. 3 clearly shows that the time for communication accounts for almost all the portion in the CG method. By contrast, the dominant cost in the CBCG method is that for computation, and it decreases significantly as the number of cores increases, which is conductive to the decrease of the total time of the CBCG method. Actually, we can observe that the time for communication in the CG method is nearly twice that in the CBCG method.

We also mention the absolute performance. When using 98,304 cores, the obtained performance of the CG method is 2,860 GFLOPS, and that of the CBCG(20) method is 41,360 GFLOPS. The former is 0.18 % of the theoretical performance (1,572,864 GFLOPS), and latter is 2.63 %. Here, using 98,304 cores for solving the problem with 27,000,000 unknowns sometimes seems to be impractical; the number of cores might be too large for the problem size. However, due to the recent increase of large-scale supercomputers, there is a requirement for solving a small size problem as fast as possible by using huge computational resource, even if the effective performance is not high. In such

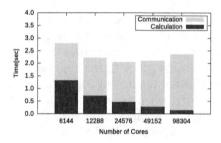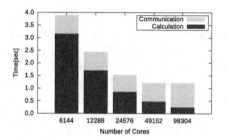

Fig. 3. The detailed breakdown of the measured execution time (left: CG, right: CBCG(20)): the target matrix is the same as in Fig. 2.

situation, reducing communication and synchronization are vital as in our comparison between the CG and CBCG methods.

4.3   Comparison with the Prediction by the Performance Model

We compare the measured results with the predicted results by our performance model discussed in the previous section. In order to evaluate the accuracy of the prediction, we show both results in Fig. 4, where we plot the time for one iteration of the CG method and the equivalent time in the CBCG method. Figure 4 shows that the predicted results are in good agreement with the measured results, and therefore our model can be considered to be reasonable at least for this test problem.

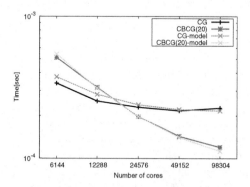

Fig. 4. Comparison between the measured and predicted results for the coefficient matrix derived from the 7-point stencil structure: the measured/predicted time for one iteration in the CG method and that for the equivalent in the CBCG(20) method is plotted.

Although the cost for SpMV strongly depends on the coefficient matrix, the number of SpMV operations required in the CG method equals to that in the

CBCG method assuming their convergence histories are same. The essential differences between the CG and CBCG methods reside in some part other than SpMV. Thus, the difference of the runtime between the CG and CBCG methods is likely predicted by the model without the generalization of SpMV.

We conducted an experiment using another coefficient matrix derived by applying the finite element method with voxel mesh to the same Poisson equation described in Sect. 4.1; the discretized equation has the 27-point stencil structure. Note that the dimension of the matrix is $8,000,000$. We measured the runtime on the K computer and show the results along with the prediction by the model in Fig. 5. Here, we plot the runtime required for one iteration excepting a SpMV operation in the CG method and the equivalent in the CBCG method.

Although there is a gap between the measured and predicted results, Fig. 5 illustrates that the model captures the behavior of the runtime; the crossover point in the predicted results is very close to that in the measured results. It is also observed that the gap becomes smaller as the number of cores increases. From this observation, we surmise that the model predicts the time for communication more accurately than that for computation because the former becomes more dominant than the latter as the number of cores increases. These results indicate that modeling only the parts excepting SpMV can predict the faster algorithm regardless of the structure of the coefficient matrix.

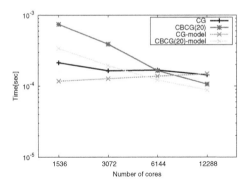

Fig. 5. Comparison between the measured and predicted results excepting SpMV for the coefficient matrix derived from the 27-point stencil structure: the measured/predicted time for one iteration in the CG method excepting the SpMV operation and that for the equivalent in the CBCG(20) method is plotted.

5   Conclusion

In this paper, we evaluated the performance of the CG and CBCG methods on the K computer. We first constructed a performance model to predict the parallel execution time of both methods on the K computer. Our method indicated that the CBCG method becomes faster than the CG method if the number of cores is sufficiently large. Then, we actually implemented both methods and measured the runtime by using the K computer. In our experiments, the CBCG method was faster than the CG method when the number of cores was greater than 24,576, which was predicted by our performance model. We also observed that the CBCG method was about twice faster than the CG method when using 98,304 cores. The detailed breakdown of the measured execution time showed the impact of reducing the communication cost in the CBCG method. Although we evaluated the CBCG method with a simple problem, the matrix derived from the 7-point stencil structure, it was shown that the CBCG method has the potential to be efficient not only on upcoming exascale systems but also on current petascale systems such as the K computer.

Let us conclude with remarks on future directions. In order to strengthen the effectiveness of the CBCG method on an actual system, evaluations with other problems and on other systems are required. From the viewpoint of practical applications, an investigation into the case of combining a preconditioner is necessary. It is also of importance to implement and evaluate the matrix powers kernel [14,15], which will contribute to reducing the communication time in the CBCG method. Besides, our results presented in this paper indicates the possibility of predicting the region in which the CBCG method is faster than the CG method, and research on performance modeling will be thus useful to practical use of the CBCG method.

Acknowledgments. The authors would like to thank the anonymous referees for their valuable comments. This research used the results of the "RIKEN AICS HPC computational science internship program 2014". This research also used the computational resources of the K computer provided by the RIKEN Advanced Institute for Computational Science(Project ID: ra000005). This work was partially supported by the Japan Society for the Promotion of Science KAKENHI (grant numbers 25330144, 15H02708, and 15K16000).

References

1. TOP500 Supercomputer Sites. http://www.top500.org/
2. Hestenes, M.R., Stiefek, E.: Method of conjugate gradient for solving linear systems. J. Res. Natl. Bur. Stan. **49**, 408–436 (1952)
3. Ghysels, P., Vanrose, P.: Hiding synchronization latency in the preconditioned conjugate gradient algorithm. Parallel Comput. **40**, 224–238 (2014)
4. Chronopoulos, A., Gear, C.: S-step iterative methods for symmetric linear systems. J. Comput. Appl. Math. **25**, 153–168 (1989)
5. Toledo, S.A.: Quantitative performance modeling of scientific computations and creating locality in numerical algorithms. Ph.D. thesis, Massachusetts Institute of Technology (1995)
6. Hoemmen, M.: Communication-avoiding Krylov subspace methods. Ph.D. thesis, University of California Berkeley (2010)
7. Suda, R., Motoya, T.: Chebyshev basis conjugate gradient method. In: IPSJ SIG High Performance Computing Symposium, p. 72 (2013)

8. Carson, E., Knight, N., Demmel, J.: An efficient deflation technique for the communication-avoiding conjugate gradient method. Electron. Trans. Numer. Anal. **43**, 125–141 (2014)
9. Fukaya, T., Imamura, T., Yamamoto, Y.: Performance analysis of the householder-type parallel tall-skinny QR factorizations toward automatic algorithm selection. In: Daydé, M., Marques, O., Nakajima, K. (eds.) VECPAR 201. LNCS, vol. 8969, pp. 269–283. Springer, Heidelberg (2015)
10. RIKEN Advanced Institute for Computational Science. http://www.aics.riken.jp/en/
11. K computer - Fujitsu Global. http://www.fujitsu.com/global/about/businesspolicy/tech/k/
12. Nakajima, K.: OpenMP/MPI hybrid parallel multigrid method on Fujitsu FX10 supercomputer system. In: IEEE International Conference on Cluster Computing Workshops, pp. 199–206 (2012)
13. Deutsch, C.V., Journel, A.G.: GSLIB Geostatistical Software Library and User's Guide, 2nd edn. Oxford University Press, Oxford (1998)
14. Demmel, J., Hoemmen, M., Mohiyuddin, M., Yelick, K.: Avoiding communication in sparse matrix computations. In: IEEE International Parallel and Distributed Processing Symposium, pp. 1–12 (2008)
15. Demmel, J., Hoemmen, M., Mohiyuddin, M., Yelick, K.: Minimizing communication in sparse matrix solvers. In: Proceedings of the ACM/IEEE Conference on Supercomputing (2009)

Dense Symmetric Indefinite Factorization on GPU Accelerated Architectures

Marc Baboulin[1]([⊠]), Jack Dongarra[2], Adrien Rémy[1], Stanimire Tomov[2], and Ichitaro Yamazaki[2]

[1] University of Paris-Sud and Inria, Orsay, France
{baboulin,aremy}@lri.fr
[2] University of Tennessee, Knoxville, USA
{dongarra,tomov,iyamazaki}@eecs.utk.edu

Abstract. We study the performance of dense symmetric indefinite factorizations (Bunch-Kaufman and Aasen's algorithms) on multicore CPUs with a Graphics Processing Unit (GPU). Though such algorithms are needed in many scientific and engineering simulations, obtaining high performance of the factorization on the GPU is difficult because the pivoting that is required to ensure the numerical stability of the factorization leads to frequent synchronizations and irregular data accesses. As a result, until recently, there has not been any implementation of these algorithms on hybrid CPU/GPU architectures. To improve their performance on the hybrid architecture, we explore different techniques to reduce the expensive communication and synchronization between the CPU and GPU, or on the GPU. We also study the performance of an LDL^T factorization with no pivoting combined with the preprocessing technique based on Random Butterfly Transformations. Though such transformations only have probabilistic results on the numerical stability, they avoid the pivoting and obtain a great performance on the GPU.

Keywords: Dense symmetric indefinite factorization · Communication-avoiding · Randomization · GPU computation

1 Introduction

A symmetric matrix A is called indefinite when its quadratic form $x^T A x$ can take both positive and negative values. Dense linear systems of equations with symmetric indefinite matrices appear in many studies of physics, including physics of structures, acoustics, and electromagnetism. For instance, such systems arise in the linear least-squares problem for solving an augmented system [15, p. 77], or in the electromagnetism where the discretization by the Boundary Element Method results in linear systems with dense complex symmetric (non Hermitian) matrices [21]. The efficient solution of these linear systems demands a high performance implementation of a dense symmetric indefinite solver that can efficiently use the current hardware architecture. In particular, the use of accelerators has

© Springer International Publishing Switzerland 2016
R. Wyrzykowski et al. (Eds.): PPAM 2015, Part I, LNCS 9573, pp. 86–95, 2016.
DOI: 10.1007/978-3-319-32149-3_9

become pervasive in scientific computing due to their high-performance capabilities. To achieve the performance, however, the algorithms must be designed for high parallelism, high flops to data ratio, and be architecture-aware. A dense symmetric indefinite solver which can efficiently use the GPU's high computing power could lead to new discoveries in the field of physics. The use of the GPU is also motivated by its low energy consumption. For example, a single K40 NVIDIA GPU has a double precision peak of 1,689 Gflop/s for a thermal design power (TDP) of 235 W. Optimized large dense matrix computations, e.g., matrix-matrix multiplications, reach 1,200 Gflop/s for a power draw of about 200 W, i.e., ≈ 6 Gflop/W. In contrast, two Sandy Bridge E5-2670 CPUs have about the same TDP ($2 \times 115 = 230$ W) as the K40 but for a peak of 333 Gflop/s, which translates to only 1.4 Gflop/W.

To solve a symmetric indefinite linear system of equations, $Ax = b$, a classical method decomposes the matrix A into an LDL^T factorization,

$$PAP^T = LDL^T, \tag{1}$$

where L is unit lower triangular, D is block diagonal with either 1-by-1 or 2-by-2 diagonal blocks, and P is a permutation matrix to ensure the numerical stability of the factorization. Then the solution x is computed by successively solving the triangular and block-diagonal systems. The pivoting strategies to compute the permutation matrix P for the LDL^T factorization include complete pivoting (Bunch-Parlett algorithm) [11], partial pivoting (Bunch-Kaufman algorithm) [12], rook pivoting (bounded Bunch-Kaufman) [4, p. 523], and fast Bunch-Parlett [4, p. 525]. In particular, the Bunch-Kaufman and rook pivoting are implemented in LAPACK [2], a set of dense linear algebra routines on multicore CPUs that are used extensively in many scientific and engineering simulations. The routines implemented in LAPACK are based on block algorithms that can exploit the memory hierarchy on modern architectures, using BLAS-3 matrix operations.

Another promising method for solving a symmetric indefinite linear system is the Aasen's method [1], which computes the LTL^T factorization of the matrix A,

$$PAP^T = LTL^T, \tag{2}$$

where T is now a symmetric tridiagonal matrix. The left-looking formulation of the algorithm requires about the same number of floating point operations (flops) that are required to compute the LDL^T factorization. A block algorithm for computing the LTL^T factorization was also proposed [23]. Though the block implementation performs slightly more flops, it can exploit a modern's computer memory hierarchy and obtain performance similar to the Bunch-Kaufman algorithm implemented in LAPACK.

To maintain numerical stability, the pivoting techniques mentioned above involve between $\mathcal{O}(n^2)$ and $\mathcal{O}(n^3)$ comparisons to search for pivots and possible interchanges of selected columns and rows. This leads to synchronization and data movement at each step of the factorization, which have become significantly more expensive compared to the arithmetic operations on modern computers.

Furthermore, due to the symmetric storage used to store A, the symmetric pivoting requires irregular data access. This increases dramatically the cost of the data movement, making it difficult to obtain the higher performance of the symmetric indefinite factorization. Recently, a communication-avoiding variant of the Aasen's algorithm was proposed [9], which can compute the factorization with a minimum amount of communication. However, the pivoting must still be applied symmetrically, leading to irregular data access. Due to these performance challenges, ScaLAPACK [10], which is the extension of LAPACK for distributed-memory machines, does not support the symmetric indefinite factorization, and until recently, there were no implementations of the algorithm, that could exploit a GPU.[1] This motivated our efforts to review the different factorization algorithms, develop their efficient implementations on multicores with a GPU to address their current limitations, and show the new state-of-the-art outlook for this important problem. Another technique studied in this paper is a symmetric version of Random Butterfly Transformations (RBT) [22] on the GPU. RBT can be combined with an LDL^T factorization to probabilistically improve the stability of the factorization without pivoting. The performance of RBT has been studied on multicore systems [8] and distributed-memory systems [6], but its performance has not been investigated on a GPU.

This paper is organized as follows. Section 2 describes three methods for solving dense symmetric indefinite systems (Bunch-Kaufman and Aasen's algorithms, random butterfly transformations) and their implementation for hybrid CPU/GPU architectures. Section 3 presents performance results for the factorizations using the above algorithms and how they compare with the LU factorization. Section 4 contains concluding remarks.

2   Symmetric Indefinite Factorizations with a GPU

2.1   Bunch-Kaufman Algorithm

The most widely used algorithm for solving a symmetric indefinite linear system is based on the block LDL^T factorization with the Bunch-Kaufman algorithm [12], which is also implemented in LAPACK [2]. The pseudo-code of the algorithm is shown in Fig. 1a. To select the pivot at each step of the factorization, it scans two columns of the trailing submatrix, and depending on the numerical values of the scanned matrix entries, it uses either a 1-by-1 or a 2-by-2 pivot. This algorithm is backward stable, subject to the growth factor [20, p. 219]. Then a variant of the Bunch-Kaufman algorithm, also called "rook pivoting", was proposed in [4] that provides a better accuracy by bounding the triangular factors. However, depending on the matrix, the rook pivoting method could perform $O(n^3)$ comparisons, each of which requires expensive synchronization. Hence, in this paper, we focus on the Bunch-Kaufman algorithm as a baseline for our performance comparison.

---

[1] A Bunch-Kaufman implementation became recently available in the cuSolver library as part of the CUDA Toolkit v7.5 from NVIDIA.

$$\alpha = (1 + \sqrt{17})/8, \quad k = 1$$
while $k < n$ **do**
$\quad \omega_1 = \max_{i>k} |a_{ik}| := |a_{rk}|$
$\quad$ **if** $\omega_1 > 0$ **then**
$\quad\quad$ **if** $|a_{kk}| \geq \alpha\omega_1$ **then**
$\quad\quad\quad s = 1$
$\quad\quad\quad$ Use a_{kk} as a 1×1 pivot.
$\quad\quad$ **else**
$\quad\quad\quad \omega_r = \max_{i \geq k; i \neq r} |a_{ir}|$
$\quad\quad\quad$ **if** $|a_{kk}|\omega_r \geq \alpha\omega_1^2$ **then**
$\quad\quad\quad\quad s = 1$
$\quad\quad\quad\quad$ Use a_{kk} as a 1×1 pivot.
$\quad\quad\quad$ **else**
$\quad\quad\quad\quad$ **if** $|a_{rr}| \geq \alpha\omega_r$ **then**
$\quad\quad\quad\quad\quad s = 1$
$\quad\quad\quad\quad\quad$ Swap rows/columns (k, r)
$\quad\quad\quad\quad\quad$ Use a_{rr} as a 1×1 pivot.
$\quad\quad\quad\quad$ **else**
$\quad\quad\quad\quad\quad s = 2$
$\quad\quad\quad\quad\quad$ Swap rows/columns $(k+1, r)$
$\quad\quad\quad\quad\quad$ Use $\begin{pmatrix} a_{kk} & a_{rk} \\ a_{rk} & a_{rr} \end{pmatrix}$ as 2×2 pivot.
$\quad\quad\quad\quad$ **end if**
$\quad\quad\quad$ **end if**
$\quad\quad$ **end if**
$\quad$ **else**
$\quad\quad s = 1$
$\quad$ **end if**
$\quad k = k + s$
end while

(a) Bunch-Kaufman.

for $j = 1, 2, \ldots, \frac{n}{n_b}$ **do**
$\quad$ **for** $i = 2, 3, \ldots, j - 1$ **do**
$\quad\quad X = T_{i,i-1} L_{j,i-1}^T$
$\quad\quad Y = T_{i,i} L_{j,i}^T$
$\quad\quad Z = T_{i,i+1} L_{j,i+1}^T$
$\quad\quad W_{i,j} = 0.5Y + Z$
$\quad\quad H_{i,j} = X + Y + Z$
$\quad$ **end for**

$\quad C = A_{j,j} - L_{j,2:j-1} W_{2:j-1,j}$
$\quad\quad\quad - W_{2:j-1,j}^T L_{j,2:j-1}^T$
$\quad T_{j,j} = L_{j,j}^{-1} C L_{j,j}^{-T}$
$\quad$ **if** $j < n$ **then**
$\quad\quad$ **if** $j > 1$ **then**
$\quad\quad\quad H_{j,j} = T_{j,j-1} L_{j,j-1}^T + T_{j,j} L_{j,j}^T$
$\quad\quad$ **end if**

$\quad\quad E = A_{j+1:n,j} - L_{j+1:n,2:j} H_{2:j,j}$
$\quad\quad [L_{j+1:n,j+1}, H_{j+1,j}, P^{(j)}] = \mathrm{LU}(E)$

$\quad\quad T_{j+1,j} = H_{j+1,j} L_{j,j}^{-T}$

$\quad\quad L_{j+1:n,2:j} = P^{(j)} L_{j+1:n,2:j}$
$\quad\quad A_{j+1:n,j+1:n} \qquad\qquad =$
$\quad\quad P^{(j)} A_{j+1:n,j+1:n} P^{(j)T}$
$\quad\quad P_{j+1:n,1:n} = P^{(j)} P_{j+1:n,1:n}$
$\quad$ **end if**
end for

(b) Communication-avoiding Aasen's.

Fig. 1. Symmetric indefinite factorization algorithm.

Our first implementation of the Bunch-Kaufman algorithm is based on a hybrid CPU/GPU programming paradigm where the block column (commonly referred to as the *panel*) is factorized on the CPU (e.g., using the multithreaded MKL library [3]), while the trailing submatrix is updated on the GPU. This is often an effective programming paradigm for many of the LAPACK subroutines because the panel factorization is based on BLAS-1 or BLAS-2, which can be efficiently implemented on the CPU, while BLAS-3 is used for the submatrix updates, which exhibits high data parallelism and can be efficiently implemented on the GPU [24]. Unfortunately, at each step of the panel factorization, the Bunch-Kaufman algorithm may select the pivot from the trailing submatrix. Hence, though copying the panel from the GPU to the CPU can be overlapped with the update of the rest of the trailing submatrix on the GPU, the *look-ahead* – a standard optimization technique to overlap the panel factorization on the CPU with the trailing submatrix update on the GPU – is prohibited. In addition, when the pivot column is on the GPU, this leads to an expensive data transfer between the GPU and the CPU at each step of the factorization. To avoid this expensive data transfer, our second implementation performs the entire factorization on the GPU. Though the CPU may be more efficient at performing the BLAS-1 and BLAS-2 based panel factorization, this implementation often obtains higher performance by avoiding the expensive data transfer.

When the entire factorization is implemented on the GPU, up to two columns of the trailing submatrix must be scanned to select a pivot at each step of the Bunch-Kaufman algorithm – the current column and the column with index corresponding to the row index of the element with the maximum modulus in the first column. This not only leads to the expensive global reduce on the GPU, but also to irregular data accesses since only the lower-triangular part of the submatrix is stored. This makes it difficult to obtain high performance on the GPU. In the next two sections, we describe two other algorithms (i.e., communication-avoiding and randomization algorithms) that aim at reducing this bottleneck.

2.2 Aasen's Algorithm

To solve a symmetric indefinite linear system, Aasen's Algorithm [1] factorizes A into an LTL^T decomposition. The left-looking algorithm takes advantage of the symmetry of A and performs $\frac{1}{3}n^3 + O(n^2)$ flops, which is the same flop count as that of the Bunch-Kaufman algorithm. In addition, like the Bunch-Kaufman algorithm, it is backward stable subject to a growth factor. To maintain the stability, at each step of the factorization, it uses the largest element of the current column being factorized as the pivot, leading to more regular data access compared to the Bunch-Kaufman algorithm. To exploit the memory hierarchy of modern computers, a blocked version of the algorithm was developed [23], which is based on a left-looking panel factorization, followed by a right-looking trailing submatrix update using BLAS-3 routines. Compared to the column-wise algorithm, this blocked algorithm performs slightly more flops, requiring $\frac{1}{3}(1 + \frac{1}{n_b})n^3 + O(n^2 n_b)$ flops with a block size n_b, but BLAS-3 can be used to perform most of these flops. However, the panel factorization is still based on BLAS-1 and BLAS-2, which often obtains only a small fraction of the peak performance. To improve the performance of the panel factorization, another variant of the algorithm was proposed [9]. This other variant computes an LTL^T factorization of A, where T is a banded matrix with its half-bandwidth equal to the block size n_b, and then uses a banded matrix solver to compute the solution. This algorithm factorizes each panel using an existing LU factorization algorithm, such as recursive LU [16,19,25] or communication-avoiding LU (TSLU, for the panel) [17,18]. In comparison with the panel factorization algorithm used in the block Aasen's algorithm, these LU factorization algorithms reduce communication, and are likely to speed up the whole factorization process. This is referred to as a communication-avoiding (CA) variant of the Aasen's algorithm, and its pseudocode is shown in Fig. 1b.

The GPU has a greater memory bandwidth than the CPU, but the memory accesses are still expensive compared to the arithmetic operations. Hence, our implementation is based on the CA Aasen's algorithm. Though this algorithm performs most of the flops using BLAS-3 (e.g., xGEMM), most of the operations are on the submatrices of the block size n_b. In order to exploit parallelism between the small BLAS calls, we use GPU streams extensively. In addition, for an efficient application of the symmetric pivots after the panel factorization, we apply the pivots in two steps. The first step copies all the columns of the trailing submatrix, which need to be swapped, into an n-by-$2n_b$ workspace. Here, because of the symmetry, the k-th block column consists of the blocks in the k-th block row and those in the k-th block column. Then, in the second step, we copy the columns of the workspace back to a block column of the submatrix after the column pivoting is applied. The same pivoting strategy is used to exploit the parallelism on multicore CPU [13]. We tested using the LU factorization with partial pivoting as the panel factorization, using either the multithreaded MKL library on the CPU or using its native GPU implementation in MAGMA on the GPU. Though the BLAS-1 and BLAS-2 based panel factorization may be more efficient on the CPU, the second approach avoids the expensive data transfer required to copy the panel from the GPU to the CPU.

2.3 Random Butterfly Transformations

Random Butterfly Transformation (RBT) is a randomization technique initially described by Parker [22] and recently revisited for dense linear systems, either general [5] or symmetric indefinite [6]. It has also been applied recently to a sparse direct solver in a preliminary paper [7]. The procedure to solve $Ax = b$, where A is a symmetric indefinite matrix, using a random transformation and the LDL^T factorization is summarized in Algorithm 1. The random matrix U is chosen among a particular class of matrices called *recursive butterfly matrices*. A *butterfly matrix* is an $n \times n$ matrix of the form

$$B^{<n>} = \frac{1}{\sqrt{2}} \begin{bmatrix} R_0 & R_1 \\ R_0 & -R_1 \end{bmatrix}$$

where R_0 and R_1 are random diagonal $\frac{n}{2} \times \frac{n}{2}$ matrices. A *recursive butterfly matrix* of size n and depth d is defined recursively as

$$W^{<n,d>} = \begin{bmatrix} B_1^{<n/2^{d-1}>} & & \\ & \ddots & \\ & & B_{2^{d-1}}^{<n/2^{d-1}>} \end{bmatrix} \cdot W^{<n,d-1>}, \text{ with } W^{<n,1>} = B^{<n>}$$

where the $B_i^{<n/2^{d-1}>}$ are butterflies of size $n/2^{d-1}$, and $B^{<n>}$ is a butterfly of size n. The application of RBT to symmetric indefinite problems was studied in [14] where it is shown that in practice, $d = 1$ or 2 gives satisfactory results (possibly using a few steps of iterative refinement). It is also shown that random butterfly matrices are cheap to store and apply ($O(nd)$ and $O(dn^2)$ respectively). An implementation for the multicore library PLASMA was described in [8].

For the GPU implementation, we use a recursive butterfly matrix U of depth $d = 2$. Only the diagonal values of the blocks are stored into a vector of size $2 \times N$ as described in [5]. Applying the depth 2 recursive butterfly matrix U consists of multiple applications of depth 1 butterfly matrices on different parts of the matrix A. The application of a depth 1 butterfly matrix is performed using

---

Algorithm 1. Random Butterfly Transformation Algorithm

Generate recursive butterfly matrix U

Apply randomization to update the matrix A and compute the matrix $A_r = U^T A U$

Factorize the randomized matrix using LDL^T factorization with no pivoting

Compute right-hand side $U^T b$, solve $A_r y = U^T b$, then $x = Uy$

---

a CUDA kernel where the computed part of the matrix A is split into blocks. For each of these blocks, the corresponding part of the matrix U is stored in the shared memory to improve the memory access performance. Matrix U is small enough to fit into the shared memory due to its packed storage.

To compute the LDL^T factorization of A_r without pivoting, we implemented a block factorization algorithm on multicore CPUs with a GPU. In our implementation, the matrix is first copied to the GPU, then the CPU is used to compute the LDL^T factorization of the diagonal block. Once the resulting LDL^T factors of the diagonal block are copied back to the GPU, the corresponding off-diagonal blocks of the L-factor are computed by the triangular solve on the GPU. Finally, we update each block column of the trailing submatrix calling a matrix-matrix multiply on the GPU.

3   Experimental Results

Figure 2 compares the performance of the symmetric indefinite factorizations on multicores with a GPU, where the "Gflop/s" is computed as the ratio of the number of flops required for the LDL^T factorization (i.e., $n^3/3$) over time (in seconds) for the particular dimension of the matrix, n. Note that, for normalization of the graph, we also consider the same flop count for LU, even though it performs twice more flops. The experiments were conducted on two eight-core Intel SandyBridge CPUs with an NVIDIA K40c GPU. The code is compiled using the GNU `gcc` version 4.4.7 and the `nvcc` version 7.0 with the optimization flag `-O3` and linked with Intel's Math Kernel Library (MKL) version xe_2013_sp1.2.144. First, when the matrix size is large enough (i.e., $n > 10,000$), the performance of the Bunch-Kaufman algorithm can be improved using the GPU over the multithreaded MKL implementation (routine `dsytrf`) on the 16 cores of two Sandy Bridge CPUs. In addition, performing the panel factorization on the GPU avoids the expensive data transfer between the CPU and GPU, and may improve the performance of the hybrid CPU/GPU implementation. Next, the communication-avoiding variant of the Aasen's algorithm further improves the performance of the Bunch-Kaufman by reducing the synchronization and communication costs required for selecting the pivots. The RBT approach outperforms the Bunch-Kaufman and Aasen factorizations but, as mentioned in [8], it may not be numerically stable for some matrices. However, the performance of all the symmetric factorizations with provable stability was lower than that of the LU factorization, demonstrating the cost of the irregular data access associated with the symmetric storage. In addition, though our current implementations

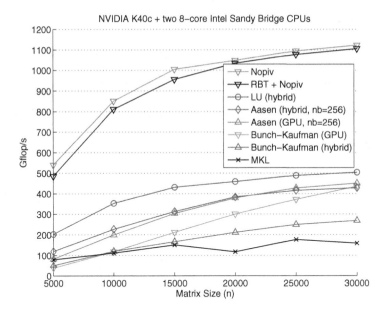

Fig. 2. Performance of symmetric factorization (double precision) (Color figure online).

of the Bunch-Kaufman and Aasen's algorithms were slower than the LU factorization, they preserve the symmetry which can reduce the runtime or memory requirement for the rest of the software (e.g., sparse symmetric factorization, or any simulation code).

In some physical applications involving dense symmetric complex non hermitian systems, it is not necessary to pivot in the LDL^T factorization (see e.g., [20, p. 209] for more information on this class of matrices). These systems are classically solved using an LU factorization since ScaLAPACK does not provide symmetric factorization for this type of matrix. Here we consider test matrices (in single complex precision) discretized by the boundary element method, used to approximate the solution of harmonic acoustic problems. Tables 1 and 2 present numerical results for the solution based on our LDL^T factorization with no pivoting on the GPU (see Sect. 2.3), applied to two sample matrices with comparison to LU factorization. Due to the smaller number of flops, our LDL^T factorization enables us to accelerate the calculation by about 48 %, while keeping a similar accuracy, expressed here by computing the scaled residual $||b - Ax||_\infty/(N||A||_\infty \times ||x||_\infty)$.

Table 1. Human head (matrix size is 10,424 in single complex precision).

	Time (sec)	Scaled residual
LU	1.34	1.44e-10
LDL^T NoPiv	0.69	1.37e-10

Table 2. Car motor (matrix size is 15,135 in single complex precision).

	Time (sec)	Scaled residual
LU	3.74	7.46e-11
LDL^T NoPiv	1.93	9.28e-11

4   Conclusion

We presented the performance of symmetric indefinite factorizations on multi-core CPUs accelerated by a GPU. The symmetric pivoting required to maintain the numerical stability of the factorization leads to frequent synchronizations and exhibits irregular memory accesses which are difficult to optimize on a GPU. As a result, until recently, there were no implementations of the algorithms that can utilize the GPU. To enhance performance, we investigated several techniques to reduce the expensive communication required for pivoting (e.g., native GPU and communication-avoiding implementations). Unfortunately, the overhead associated with the symmetric pivoting can still be significant. However, these algorithms preserve the symmetry, which is required in several physical applications, and reduces the runtime and memory requirement for the rest of the application software. Though it only has a probabilistic error bound, randomization using RBT followed by an LDL^T factorization without pivoting outperforms other algorithms, and is about twice as fast as the LU factorization. Our current implementations are based on standard BLAS/LAPACK routines, and we are improving the performance of factorization by developing specialized GPU kernels. These implementations will be integrated in a new release of MAGMA.

Acknowledgments. The authors would like to thank the NSF grant #ACI-1339822, NVIDIA, and MathWorks for supporting this research effort. The authors are also grateful to Nicolas Zerbib (ESI Group) for his help in using test matrices from acoustics.

References

1. Aasen, J.: On the reduction of a symmetric matrix to tridiagonal form. BIT **11**, 233–242 (1971)
2. Anderson, E., Bai, Z., Dongarra, J.J., Greenbaum, A., McKenney, A., Du Croz, J., Hammarling, S., Demmel, J.W., Bischof, C., Sorensen, D.: LAPACK: a portable linear algebra library for high-performance computers. In: Proceedings of the ACM/IEEE Conference on Supercomputing (1990)
3. Intel, Math Kernel Library (MKL). http://www.intel.com/software/products/mkl/
4. Ashcraft, C., Grimes, R.G., Lewis, J.G.: Accurate symmetric indefinite linear equation solvers. SIAM J. Matrix Anal. Appl. **20**(2), 513–561 (1998)
5. Baboulin, M., Dongarra, J.J., Hermann, J., Tomov, S.: Accelerating linear system solutions using randomization techniques. ACM Trans. Math. Softw. **39**(2), 8 (2013)
6. Baboulin, M., Becker, D., Bosilca, G., Danalis, A., Dongarra, J.J.: An efficient distributed randomized algorithm for solving large dense symmetric indefinite linear systems. Parallel Comput. **40**(7), 213–223 (2014)
7. Baboulin, M., Li, X.S., Rouet, F.-H.: Using random butterfly transformations to avoid pivoting in sparse direct methods. In: Proceedings of International Conference on Vector and Parallel Processing (VecPar 2014), Eugene (OR), USA
8. Baboulin, M., Becker, D., Dongarra, J.J.: A parallel tiled solver for dense symmetric indefinite systems on multicore architectures. In: Parallel Distributed Processing Symposium (IPDPS) (2012)

9. Ballard, G., Becker, D., Demmel, J., Dongarra, J., Druinsky, A., Peled, I., Schwartz, O., Toledo, S., Yamazaki, I.: Communication-avoiding symmetric-indefinite factorization. SIAM J. Matrix Anal. Appl. **35**, 1364–1460 (2014)
10. Blackford, L., Choi, J., Cleary, A., D'Azevedo, E., Demmel, J.W., Dhillon, I., Dongarra, J.J., Hammarling, S., Henry, G., Petitet, A., Stanley, K., Walker, D., Whaley, R.: ScaLAPACK Users Guide. SIAM, Philadelphia (1997)
11. Bunch, J.R., Parlett, B.N.: Direct methods for solving symmetric indefinite systems of linear equations. SIAM J. Numer. Anal. **8**, 639–655 (1971)
12. Bunch, J.R., Kaufman, L.: Some stable methods for calculating inertia and solving symmetric linear systems. Math. Comput. **31**, 163–179 (1977)
13. Ballard, G., Becker, D., Demmel, J., Dongarra, J., Druinsky, A., Peled, I., Schwartz, O., Toledo, S., Yamazaki, I.: Implementing a blocked Aasen's algorithm with a dynamic scheduler on multicore architectures. In: Proceedings of the 27th International Symposium on Parallel and Distributed Processing, pp. 895–907 (2013)
14. Becker, D., Baboulin, M., Dongarra, J.: Reducing the amount of pivoting in symmetric indefinite systems. In: Wyrzykowski, R., Dongarra, J., Karczewski, K., Waśniewski, J. (eds.) PPAM 2011, Part I. LNCS, vol. 7203, pp. 133–142. Springer, Heidelberg (2012)
15. Björck, Å.: Numerical Methods for Least Squares Problems. SIAM, Philadelphia (1996)
16. Castaldo, A., Whaley, R.: Scaling LAPACK panel operations using parallel cache assignment. In: Proceedings of the 15th AGM SIGPLAN Symposium on Principle and Practice of Parallel Programming, pp. 223–232 (2010)
17. Demmel, J., Grigori, L., Hoemmen, M., Langou, J.: Communication-optimal parallel and sequential QR and LU factorizations. SIAM J. Sci. Comput. **34**, A206–A239 (2012). Technical report (UCB/EECS-2008-89), EECS Department, University of California, Berkeley
18. Grigori, L., Demmel, J., Xiang, H.: CALU: a communication optimal LU factorization algorithm. SIAM. J. Matrix Anal. Appl. **32**(4), 1317–1350 (2011)
19. Gustavson, F.: Recursive leads to automatic variable blocking for dense linear-algebra algorithms. IBM J. Res. Dev. **41**, 737–755 (1997)
20. Higham, N.J.: Accuracy and Stability of Numerical Algorithms. SIAM, Philadelphia (2002)
21. Nédélec, J.-C.: Acoustic and Electromagnetic Equations. Integral Representations for Harmonic Problems. Applied Mathematical Sciences, vol. 144. Springer, New York (2001)
22. Parker, D.S.: Random butterfly transformations with applications in computational linear algebra. Technical report CSD-950023, UCLA Computer Science Department (1995)
23. Rozložník, M., Shklarski, G., Toledo, S.: Partitioned triangular tridiagonalization. ACM Trans. Math. Softw. **37**(4), 1–16 (2011)
24. Tomov, S., Dongarra, J., Baboulin, M.: Towards dense linear algebra for hybrid GPU accelerated manycore systems. Parallel Comput. **36**(5&6), 232–240 (2010)
25. Toledo, S.: Locality of reference in LU decomposition with partial pivoting. SIAM J. Matrix Anal. Appl. **18**(4), 1065–1081 (1997)
26. University of Tennessee: PLASMA Users' Guide, Parallel Linear Algebra Software for Multicore Architectures, Version 2.3 (2010)

A Parallel Multi-threaded Solver for Symmetric Positive Definite Bordered-Band Linear Systems

Peter Benner[1], Pablo Ezzatti[2], Enrique S. Quintana-Ortí[3],
and Alfredo Remón[1($\boxtimes$)]

[1] Max Planck Institute for Dynamics of Complex Technical Systems,
39106 Magdeburg, Germany
{benner,remon}@mpi-magdeburg.mpg.de
[2] Instituto de Computación, Universidad de la República,
11300 Montevideo, Uruguay
pezzatti@fing.edu.uy
[3] Dep. de Ingeniería y Ciencia de la Computación,
Universidad Jaime I, 12701 Castellón, Spain
quintana@icc.uji.es

Abstract. We present a multi-threaded solver for symmetric positive definite linear systems where the coefficient matrix of the problem features a bordered-band non-zero pattern. The algorithms that implement this approach heavily rely on a compact storage format, tailored for this type of matrices, that reduces the memory requirements, produces a regular data access pattern, and allows to cast the bulk of the computations in terms of efficient kernels from the Level-3 and Level-2 BLAS. The efficiency of our approach is illustrated by numerical experiments.

Keywords: Bordered-band matrices · Symmetric positive definite linear systems · Cholesky factorization · Multicore architectures

1   Introduction

The solution of linear systems $Ax = b$, where $A \in \mathbb{R}^{n \times n}$, $b, x \in \mathbb{R}^n$, and x contains the sought-after solution, is a fundamental problem in many scientific and engineering applications [6]. In a number of these problems, the coefficient matrix A is sparse (i.e., a large fraction of its elements is null), and storing/operating exclusively with the non-zero entries generally yields important savings in the memory/computational requirements.

Symmetric bordered-band matrices are a particular case of (structured) sparse matrices with all the non-zero elements symmetrically placed in a reduced number of rows/columns of the bottom/rightmost side of the matrix as well as some diagonals next to its main diagonal [13]. Specialized instances of these matrices appear in certain symmetric inverse eigenvalue and inverse Sturm-Liouville problems for particle physics, geology, and vibration analysis; numerical integration methods based on Gaussian quadrature rules [8]; eigenvalue problems

© Springer International Publishing Switzerland 2016
R. Wyrzykowski et al. (Eds.): PPAM 2015, Part I, LNCS 9573, pp. 96–105, 2016.
DOI: 10.1007/978-3-319-32149-3_10

of large sparse matrices via Lanczos method [5]; updating eigenvalue problems; generating orthogonal polynomials; and in polynomial least squares approximation; see [13] and the references therein. These matrices can also be found in finite-element applications, when the underlying finite-element domain is partitioned into non-overlapping subdomains [2,7] and in alignment algorithms for complex tracking detectors [4,11].

In this paper we present an efficient solver for symmetric positive definite bordered-band (SPBB) linear systems, with borders of size p and lower/upper bandwidth of size b in the coefficient matrix, making the following contributions:

- Our solver introduces a compact data structure to keep the non-zero entries of a symmetric bordered-band matrix, similar to that in LAPACK [1] for symmetric band matrices and close to the layout proposed in [4,11]. The proposal delivers relative low storage overhead when $p, b \ll n$ (specifically, $pb + p^2/2 + b^2/2$ over $pn + bn$ numbers).
- The solver relies on the Cholesky factorization to exploit the symmetric definite positivity property of the problem, yielding a computational cost of $\mathcal{O}(nb^2 + np^2 + n^2p)$ floating-point arithmetic operations (flops) for the factorization, and $\mathcal{O}(nb + np)$ flops for each one of the subsequent triangular bordered-band (TBB) systems.
- Our implementations of the factorization stage and triangular solves decompose these operations into a collection of standard fine-grain numerical kernels that can be efficiently executed via a highly-tuned multi-threaded version of BLAS.
- The software is written in C following the "LAPACK"-style and guidelines for the interface, with separate routines for the factorization and triangular solve, dpbbtrf and dtbbtrs, respectively. We evaluate the new SPBB solver on a server equipped with two Intel Xeon E5520 processors, with four cores each, comparing its performance against the sparse direct solver in Intel MKL PARDISO [9]. The new solver is designed for the solution of SPBB linear systems with "moderate" levels of density in both the band structure and borders. We recognize that, for a SPBB system, with sparser blocks, PARDISO may be a more efficient option.

The rest of the paper is structured as follows. In Sect. 2 we introduce the new data storage format for SPBB and TBB matrices, and in Sect. 3 we describe the algorithms for the solution of linear systems involving this type of matrices. The experimental evaluation of these solvers follows in Sect. 4, and we close the paper with a few concluding remarks in Sect. 5.

2   Storage Layout for Symmetric Bordered Band Matrices

The benefits of adopting a structured/sparse matrix storage scheme over a dense layout depend on the amount of null elements of the matrix. In general, sparse formats reduce the storage requirements as well as the volume of data transferred between the memory subsystem and the processor floating-point units (FPUs).

However, they are often more cumbersome to deal with, especially when the sparsity pattern is irregular.

We next introduce a compact storage scheme specific for symmetric (positive definite or not) bordered-band matrices that basically maintains only half of the elements inside of the band and border "areas" of the matrix. The storage layout is illustrated in Fig. 1. Our solution partially mimics that of LAPACK for symmetric band matrices, storing only half of the entries of the lower (or upper) band area into a $b \times n$ block B. Here, each row contains the elements of one of the (non-zero) diagonals, with the first entry of each diagonal aligned with the first column of B. Thus, the diagonals are stored, from the main diagonal down, into consecutive rows of B, from top to bottom. In addition, the bottom (or upper) border area of the symmetric bordered-band matrix is kept in a $p \times n$ block P, and both blocks are aligned with B on top of P, as shown in the figure. With this solution, the number of entries that store zeros (overhead) amounts to $bp + p^2/2 + b^2/2$. This represents a small to moderate increment in the memory requirements (usually $b, p \ll n$) but, compared with a storage scheme for irregular sparse matrices, facilitates the access to the data as there exists a one-to-one (direct) relation between the coordinates of a matrix element and its address in memory. Furthermore, if b, p are very small, the memory requirements of this scheme can be lower than those of a standard sparse storage, like e.g., the compressed row storage (CSR) or coordinate (COO) formats [12], as no extra space is necessary to maintain the coordinates of the non-zero entries.

A more efficient storage layout can be obtained by storing B and P as two separate arrays [4,11] since, in such case, the overhead due to the $b \times p$ rectangle disappears. However, this implies that the codes (routines) would have to employ two separate parameters to identify the matrix contents. Given that we expect the overhead to be small in practice, we adopt the first solution, with a single compact memory space recording the complete matrix, as illustrated in Fig. 1.

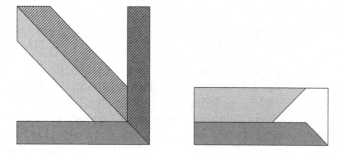

Fig. 1. Symmetric bordered-band matrix (left) and the proposed storage in memory (right). The entries in the shadow areas are not explicitly stored.

3   Solution of SPBB Linear Systems

The solution of a symmetric positive definite linear system of equations $Ax = b$, via the Cholesky factorization, commences by decomposing the coefficient matrix into the product of two triangular factors: $A = LL^T$, with $L \in \mathbb{R}^{n \times n}$ lower triangular. (Alternatively, A can be decomposed as $A = U^T U$, with $U \in \mathbb{R}^{n \times n}$ upper triangular.) This is followed by the lower triangular solve $Ly = b$ for $y \in \mathbb{R}^n$, and finally the upper triangular solve $L^T x = y$ for x.

When the coefficient matrix A is SPBB, parameterized by b and p, the Cholesky factorization yields a lower TBB factor L with its non-zero entries lying into the b subdiagonals closest to its main diagonal and a border consisting of p rows at the bottom of the matrix. This basically corresponds to the non-shadowed areas in the right-hand side plot of Fig. 1, which can then be stored overwriting the original entries of the matrix in compact form.

In the next two subsections, we describe efficient algorithms for the factorization and triangular solves involving SPBB and TBB matrices, respectively.

3.1   Cholesky Factorization of SPBB Linear Systems

Figure 2 introduces a blocked algorithm to compute the Cholesky factorization of a SPBB matrix using the FLAME notation [3]. The algorithm performs only the flops that are strictly necessary. Furthermore, it is rich in Level-3 BLAS kernels such as matrix-matrix products (GEMM), triangular system solves with multiple right-hand sides (TRSM) and symmetric rank-k updates (SYRK), which turns it very appealing for modern hardware architectures. The loop body of the algorithm identifies the corresponding BLAS kernel next to each operation. All the blocks except A_{11} are updated via a Level-3 BLAS level operation, while the update of A_{11} requires the Cholesky factorization of a square dense block, an operation that is supported by routine POTF2 from LAPACK.

The adoption of the matrix storage scheme described in the previous section allows to cast the operations underlying algorithm PBBTRF$_{BLK}$ in terms of routines in the BLAS and LAPACK specifications. In our actual implementation, all these operations are performed by invoking the corresponding routines of a multi-threaded high performance implementation of BLAS and LAPACK, in particular the Intel MKL library. As a result we can expect high performance combined with low storage requirements by adopting this storage format. Figure 3 shows how the blocks that are updated during a particular iteration of algorithm PBB-TRF$_{BLK}$ are mapped to the storage scheme. The layout of these blocks is conformal with standard storage solutions for dense or triangular matrices, depending on the shape of the block, enabling the use of kernels from BLAS and LAPACK via a correct selection of the leading dimension lda. Concretely, for those blocks located in the border area of the matrix (A_{51}, A_{52}, A_{53} and A_{54}), lda has to be $b + p$, while for the remaining blocks it is $b + p - 1$.

The routine which implements algorithm PBBTRF$_{BLK}$ incorporates an additional modification to improve performance. Specifically, the algorithm progresses along the matrix, from the top-left corner to the bottom-right one,

Algorithm: $[A] := \text{PBBTRF}_{BLK}(A, k, l)$

Partition $A \rightarrow \left(\begin{array}{c|c|c} A_{TL} & A_{TM} & A_{TR} \\ \hline A_{ML} & A_{MM} & A_{MR} \\ \hline A_{BL} & A_{BM} & A_{BR} \end{array} \right)$

where A_{TL} is 0×0, A_{MM} is $k \times k$

while $m(A_{TL}) < m(A)$ **do**

  Determine block size nb

  Repartition

$\left(\begin{array}{c|c|c} A_{TL} & \star & \star \\ \hline A_{ML} & A_{MM} & \star \\ \hline A_{BL} & A_{BM} & A_{BR} \end{array} \right) \rightarrow \left(\begin{array}{c|c|c|c|c|c} A_{00} & \star & \star & & & \star \\ \hline A_{10} & A_{11} & \star & \star & & \star \\ \hline A_{20} & A_{21} & A_{22} & \star & \star & \star \\ \hline & A_{31} & A_{32} & A_{33} & \star & \star \\ \hline & & A_{42} & A_{43} & A_{44} & \star \\ \hline A_{50} & A_{51} & A_{52} & A_{53} & A_{54} & A_{55} \end{array} \right)$

  where A_{11}, A_{33} are $nb \times nb$, A_{22} is $p \times p$, **and** A_{55} is $l \times l$,
  with $p := \min(k - nb, m(A) - m(A_{TL}) - nb)$

$A_{11} := \text{CHOL}(A_{11})$	(POTF2)
$A_{21} := A_{21}\,\text{TRIL}(A_{11})^{-T}$	(TRSM)
$A_{31} := A_{31}\,\text{TRIL}(A_{11})^{-T}$	(TRSM)
$A_{22} := A_{22} - A_{21}A_{21}^T$	(SYRK)
$A_{32} := A_{32} - A_{31}A_{21}^T$	(TRMM)
$A_{33} := A_{33} - A_{31}A_{31}^T$	(SYRK)
$A_{51} := A_{51}\,\text{TRIL}(A_{11})^{-T}$	(TRSM)
$A_{52} := A_{52} - A_{51}A_{21}^T$	(GEMM)
$A_{53} := A_{53} - A_{51}A_{31}^T$	(TRMM)
$A_{55} := A_{55} - A_{51}A_{51}^T$	(SYRK)

Continue with

$\left(\begin{array}{c|c|c} A_{TL} & \star & \star \\ \hline A_{ML} & A_{MM} & \star \\ \hline A_{BL} & A_{BM} & A_{BR} \end{array} \right) \leftarrow \left(\begin{array}{c|c|c|c|c|c} A_{00} & \star & \star & & & \star \\ \hline A_{10} & A_{11} & \star & \star & & \star \\ \hline A_{20} & A_{21} & A_{22} & \star & \star & \star \\ \hline & A_{31} & A_{32} & A_{33} & \star & \star \\ \hline & & A_{42} & A_{43} & A_{44} & \star \\ \hline A_{50} & A_{51} & A_{52} & A_{53} & A_{54} & A_{55} \end{array} \right)$

endwhile

Fig. 2. Algorithm PBBTRF_{BLK} for the factorization of an SPBB matrix $A = LL^T$. In this notation, $\text{POTF2}(\cdot)$ is an unblocked version of the Cholesky algorithm for dense matrices, and $\text{TRIL}(\cdot)$ returns the lower triangular part of a matrix.

updating 10 blocks at each iteration. However, when all the elements in the first $(n - p)$ columns of L have been computed, the remaining submatrix does no longer present the arrowhead structure but, instead, it is a regular dense matrix. At this point, the implementation exits the iterative process and computes the Cholesky factorization of A_{55}.

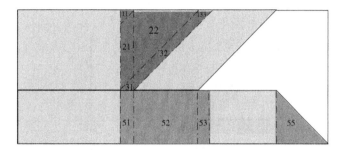

Fig. 3. Symmetric bordered-band layout and storage of blocks updated during an iteration of algorithm PBBTRF$_{BLK}$. Pink and blue colors are employed to highlight updated blocks from the band and border areas of the matrix, respectively. The grey color identifies zones that are not modified.

3.2 Triangular Solver for TBB Linear Systems

We next focus on the (first) triangular solve $Ly = b$, though a similar procedure to that described next applies to the "transposed" case $L^T x = y$.

The adoption of the compact symmetric/triangular bordered-band storage scheme enables us to rely, once again, on kernels from the BLAS library for this operation. This turns the implementation of the triangular system solve simple. In particular, the lower TBB matrix is partitioned into the three blocks, as shown in Fig. 4, and the solve is then decomposed into the following three steps:

1. Solve the triangular band linear system $L_{00}y_0 = b_0$ for y_0.
2. Update the right-hand side vector: $b_1 := b_1 - L_{10} \cdot y_0$.
3. Solve the (dense) triangular linear system $L_{11}y_1 = b_1$ for y_1.

BLAS provides the necessary support for all three steps. Concretely, routine TBSV performs the first solve and its application is straight-forward, since L_{00} is maintained conformally with the storage for band matrices adopted by BLAS. Routine GEMV can be employed to perform the matrix-vector product that updates b_1. Finally, the triangular solver involved in the last step is implemented as routine TRSV. These three kernels belong to the Level-2 BLAS specification. To conclude this section, we note that lda must be set to $b+p-1$ for block L_{00}, while lda is $b+p$ for blocks L_{10} and L_{11}.

4   Experimental Evaluation

We next perform an experimental evaluation of the SPBB linear system solver. The target platform is furnished with two Intel Xeon 5520 processors, with a total of 8 cores running at 2.27 GHz, and 24 Gbytes of DDR3 RAM. The codes were compiled using Intel icc v.12.1.3, with the -O3 optimization flag enabled, and linked to Intel MKL v9.293 [10] for the BLAS and LAPACK kernels. Parallelism was exploited by invoking multi-threaded routines in the Intel MKL library.

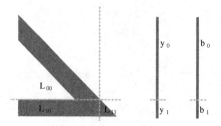

Fig. 4. Partitioning applied to the lower TBB matrix L and vectors y, b during the triangular solve.

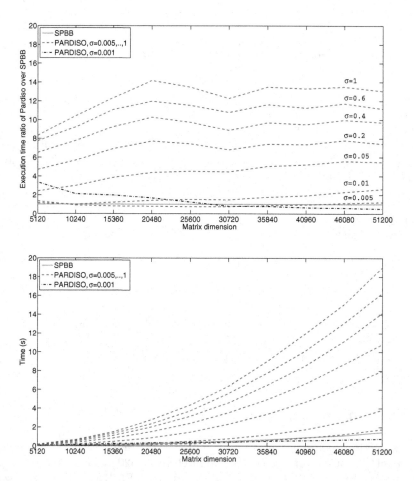

Fig. 5. Execution time ratio of PARDISO over SPBB (top) and execution time (bottom) of the solvers for SPBB linear systems of varying dimension n, with b, p both equal to 1 % of the matrix dimension.

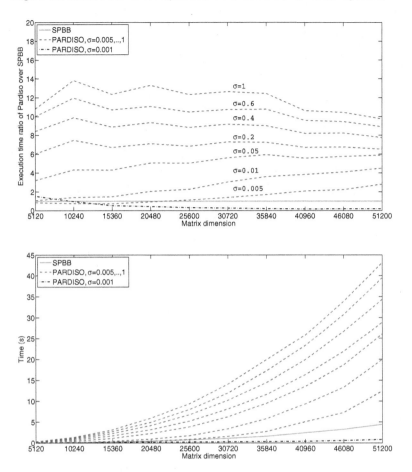

Fig. 6. Execution time ratio of PARDISO over SPBB (top) and execution time (bottom) of the solvers for SPBB linear systems of varying dimension n, with b, p both equal to 2 % of the matrix dimension.

The performance of the SPBB solver is compared against Intel MKL PARDISO [9], a direct solver based on the Cholesky factorization for sparse matrices. Our executions with PARDISO apply the optimizations enabled in this solver by the following sequence of calls:

```
dss_chk( dss_define_structure( handle, sym, ... ) );
dss_reorder( handle, opt, 0);
```

We evaluate both solvers using linear systems of dimension $5,120 \leq n \leq 51,200$. The bandwidth for each coefficient matrix depends on its dimension, with two instances being tested: $b = 1$ % and 2 % of n. The number of rows in the border area is $p = b$. Additionally, we initialized the bordered–band regions with several degrees of density (i.e., non-zero elements), ranging from $\sigma = 0.1$ %, 0.5 %, . . . , to

100 %. This parameter does not influence the SPBB solver because it does not take into account the zeros inside these parts, but it is important for PARDISO.

Figures 5 and 6 respectively report the results obtained by the solvers when $b = p = 1\%$ and 2% of n. Two plots are included for each instance, one for the performance ratio between PARDISO and the SPBB linear system solver (top), and the other with the execution time (bottom). The performance of SPBB compared with PARDISO depends on the values of b, p and the density inside of the structure σ. Thus, when $\sigma = 100$ (i.e., all entries in the bordered–band area are non-zero), SPBB is 13× faster than Intel MKL PARDISO but, when the density is decreased, the performance gap between both solvers gets narrower. In the end, if $\sigma = 0.5\%$, both solvers present similar results, and PARDISO outperforms our solver for smaller values of σ.

We obtain a similar behavior when $b = p = 2\%$ of n (Fig. 6). In this case, SPBB outperforms Intel MKL PARDISO even when $\sigma = 0.5\%$, while PARDISO is faster when $\sigma = 0.1\%$. From the trends in Figs. 5 and 6, we can infer that larger values of b and p will benefit the SPBB solver, turning it the best option in systems even with lower values of σ.

From the perspective of storage requirements, when the sparsity inside of the bordered-band structure is considerable, PARDISO may present larger memory savings. In contrast, if the density inside of the bordered-band area is high, then the SPBB solver outperforms the direct solver from MKL in both memory savings and execution time.

5   Concluding Remarks

We have presented a solver for symmetric positive definite bordered-band linear systems. The algorithms for the factorization and triangular system solves rely on a novel compact storage layout with a number of advantages. First, the compact layout reduces the memory requirements with respect to alternative sparse storage schemes, because the memory address of each element can be directly inferred from its coordinates. In addition, the storage format yields a regular memory access pattern, providing a more efficient usage of the processor memory hierarchy than that often exhibited in (irregular) sparse solvers. Finally, the compact format allows to decompose the operation into a collection of highly-efficient parallel BLAS kernels with conformal access patterns.

Our experimental evaluation on an Intel Xeon-based system demonstrates the performance of the new solver, with significant advantages over the direct sparse solver in the MKL library, even when the density of non-zero elements inside of the structure is low (down to $\sigma \geq 0.5\%$ in the experiments), with acceleration factors up to 13× with respect to the MKL solver. However, with lower densities Intel MKL PARDISO is superior in both, performances and storage requirements.

Acknowledgments. The researcher from *Universidad Jaime I* was supported by the CICYT project TIN2011-23283 of the *Ministerio de Economía y Competitividad* and FEDER. We would like to thank Matthias Bollhöfer, from TU Braunschweig, for his help during the evaluation of these codes.

References

1. Anderson, E., Bai, Z., Bischof, C., Blackford, L.S., Demmel, J., Dongarra, J.J., Du Croz, J., Hammarling, S., Greenbaum, A., McKenney, A., Sorensen, D.: LAPACK Users' Guide, 3rd edn. Society for Industrial and Applied Mathematics, Philadelphia (1999)
2. Bollhöfer, M., Kressner, D., Mehl C., Stykel, T., Benner, P. (eds.) Numerical Algebra, Matrix Theory, Differential-Algebraic Equations and Control Theory. Festschrift in Honor of Volker Mehrmann. Springer, Berlin (2015)
3. Bientinesi, P., Gunnels, J.A., Myers, M.E., Quintana-Ortí, E.S., van de Geijn, R.A.: The science of deriving dense linear algebra algorithms. ACM Trans. Math. Softw. **31**(1), 1–26 (2005)
4. Blobel, V., Kleinwort, C., Meier, F.: Fast alignment of a complex tracking detector using advanced track models. Comput. Phys. Commun. **182**(9), 1760–1763 (2011)
5. de Boor, C., Golub, G.: The numerically stable reconstruction of a Jacobi matrix from spectral data. Linear Algebra Appl. **21**(3), 245–260 (1978)
6. Demmel, J.W.: Applied Numerical Linear Algebra. Society for Industrial and Applied Mathematics, Philadelphia (1997)
7. Duff, I.S., Scott, J.A.: Stabilized bordered block diagonal forms for parallel sparse solvers. Parallel Comput. **31**(3–4), 275–289 (2005)
8. Golub, G., Welsch, J.H.: Calculation of Gauss quadrature rules. Math. Comput. **23**(106), 221–230 (1969)
9. Intel Corporation. https://software.intel.com/en-us/node/470282
10. Intel Corporation. http://www.intel.com/
11. Kleinwort, C.: General broken lines as advanced track fitting method. Nucl. Instrum. Methods Phys. Res. Sect. A Accelerators Spectrometers Detectors Assoc. Equip. **673**, 107–110 (2012)
12. Saak, J.: Effiziente numerische Lösung eines Optimalsteuerungsproblems für die Abkühlung von stahlprofilen. Diplomarbeit, Fachbereich 3/Mathematik und Informatik, Universität Bremen, D-28334 Bremen, September 2003
13. Van Huffel, S., Park, H.: Efficient reduction algorithms for bordered band matrices. Numer. Linear Algebra Appl. **2**(2), 95–113 (1995)

Parallel Algorithm for Quasi-Band
Matrix-Matrix Multiplication

Dharma Teja Vooturi$^{(\boxtimes)}$ and Kishore Kothapalli

International Institute of Information Technology, Hyderabad,
Gachibowli, Hyderabad 500032, India
dharmateja.vooturi@research.iiit.ac.in, kkishore@iiit.ac.in

Abstract. Sparse matrices arise in many practical scenarios. As a result, support for efficient operations such as multiplication of sparse matrices (spmm) is considered to be an important research area. Often, sparse matrices also exhibit particular characteristics that can be used towards better parallel algorithmics. In this paper, we focus on quasi-band sparse matrices that have a large majority of the non-zeros along the diagonals. We design and implement an efficient algorithm for multiplying two such matrices on a many-core architecture such as a GPU.

Our implementation outperforms the corresponding library implementation by a factor of 2x on average over a wide variety of quasi-band matrices from standard datasets. We analyze our performance over synthetic quasi-band matrices.

Keywords: Band · Quasi-band · spmm · Real-world · Synthetic

1 Introduction

Multiplying two sparse matrices, denoted spmm, is an important and challenging problem in parallel computing with applications to a wide variety of disciplines including climate modeling, computational fluid dynamics, and molecular dynamics [10]. Due to the importance of spmm, a number of works aimed at efficient algorithms and their implementations on a variety of architectures are reported in the literature [2,6,7].

A current trend in parallel algorithm engineering is to focus on customizing algorithms based on input characteristics. Such a customization allows the algorithms to benefit from the properties of the input. Recent examples include finding the strongly connected components of real-world graphs by Hong et al. [4], mapping graph traversals to a CPU+GPU heterogeneous platform by Gharibieh et al. [3], sparse matrix-vector multiplication and matrix-matrix multiplication of scale free matrices by Indarapu et al. [5] and Ramamoorthy et al. [7].

In this context, we note that applications such as aerodynamics and computational fluid dynamics [10] produce sparse matrices called as *quasi-band* matrices that exhibit a near-diagonal nature of sparsity. As noted by Yang et al. [11], many sparse matrices also can be reordered or divided into a near-diagonal form.

© Springer International Publishing Switzerland 2016
R. Wyrzykowski et al. (Eds.): PPAM 2015, Part I, LNCS 9573, pp. 106–115, 2016.
DOI: 10.1007/978-3-319-32149-3_11

In this paper, we focus on quasi-band matrices and design a GPU algorithm to multiply two such matrices. Our work extends the work of Yang et al. [11] who design a GPU algorithm for multiplying a sparse quasi-band matrix with a dense vector. Our algorithm starts by separating the input quasi-band sparse matrices into a diagonal part and the rest as a sparse part. Once such a separation is achieved, we introduce specific optimizations to perform the four multiplications: the diagonal/sparse part with the diagonal/sparse part.

Our main technical contributions can be summarized as follows.

- We propose an algorithm (see Sect. 3) for multiplying two sparse quasi-band matrices. Our algorithm identifies, to a reasonable extent, the indices of the output matrix that will have nonzero entries and uses this information to manage the space required and an estimate of the work required.
- An implementation of our algorithm on an Nvidia K40 GPU achieves a speedup of 5x and 2x on average over a collection of band and quasi-band matrices, respectively, taken from the University of Florida dataset [10]. (See Sect. 4).
- We also perform experiments on synthetic quasi-band matrices to understand the effect of the nature of the matrix on the speedup. (See Sect. 4).

2   Preliminaries

We start with a few definitions. If all the nonzero elements in a matrix are in a single diagonal then that matrix is called a *uni-diagonal* matrix. If there exists a diagonal index pair (i, j) such that $i \leq j$ and all the non-zero elements of a matrix are present between diagonals d_i (leftOffset), d_j (rightOffset) then such a matrix is said to be a *uni-band* matrix with a bandwidth of $(j - i + 1)$. If multiple such disjoint pairs of indices exist then it is called a *multi-band matrix*. Uni-band and multi-band matrices are commonly referred to as *band* matrices. A band matrix along with some non-zero elements in the non-band diagonals is called a *quasi-band* matrix. We also use the terms defined in Table 1 that indicate some of the properties of quasi-band matrices.

Table 1. Glossary of terms

Term	Description
nnz	Number of nonzero elements in a matrix
nnzPercentage	Percentage of NNZ to all elements in matrix
bandPercentage	Percentage of NNZ in band part to NNZ in matrix
bandOccupancy	Percentage of NNZ in band part to all elements in band part
bandCount	Number of bands present in the matrix
diagonalCount	Total number of diagonals across all the bands in the matrix

2.1 Matrix Representation

In our work, we make use of several representations to store sparse matrices. These are described below in brief.

DIA Format: Each diagonal in the matrix with at least one nonzero is stored as a column array of length equal to the number of rows in the matrix. Diagonals are numbered starting with zero for the principal diagonal and -1 and $+1$ for the diagonals to the left and right of the principal diagonal, and so on. These numbers, called as *diagonal offsets*, are stored in a separate array called offsetArray.

COOSR (COOSC) Format: These can be thought of as a mix of the COO and the CSR (CSC) formats reported in [1]. Each non-zero element in a matrix can be mapped to a triplet (value, row, column). In this format we store three arrays *data, rowIndex and colIndex* each of length nnz(number of non-zero elements in the matrix). Elements are ordered row (`resp.` column) wise in the arrays. An array rowPtr (colPtr) of length $[rows + 1]([columns + 1])$ is also stored. An index i in rowPtr (colPtr) array maps to the index value of first element of row (column) i in *value* array.

3 Our Algorithm

One of the prime difficulties of sparse matrix multiplication include the possible lack of any relation between the nature and degree of sparsity of the input matrices and their product matrix. On multi- and many-core architectures, other difficulties such as load imbalance imply that efficient sparse matrix multiplication is often challenging. One way of addressing this difficulty is to look for particular properties of the input matrices and their impact on the product matrix. To this end, focusing on quasi-band matrices, we start with the following observations.

Observation 1. *Multiplying two uni-diagonal matrices A and B with diagonal offsets a_o and b_o respectively results in another uni-diagonal matrix with diagonal offset $(a_o + b_o)$.*

Observation 2. *When a matrix A having a single non-zero element a_{ij} is multiplied with a uni-band matrix B defined by diagonal offsets $(left_o, right_o)$, in the product matrix $C = A \times B$, only elements in row i with column indices between $[max(0, j + left_o) : min(B_{cols} - 1, j + right_o)]$ are effected.*

To make use of the above observations, in the matrix product $C = A \times B$, we start by partitioning quasi-band matrices A and B into their band and non-band components denoted A_{bands} and A_{sparse}, B_{bands} and B_{sparse} respectively. We then compute four matrix products $C_{\text{bb}} = A_{\text{bands}} \times B_{\text{bands}}$, $C_{\text{sb}} = A_{\text{sparse}} \times B_{\text{bands}}$, $C_{\text{bs}} = A_{\text{bands}} \times B_{\text{sparse}}$ and $C_{\text{ss}} = A_{\text{sparse}} \times B_{\text{sparse}}$. The matrix C is the result of adding the matrices $C_{\text{bb}}, C_{\text{sb}}, C_{\text{bs}}$, and C_{ss}. An outline of the above approach is shown in Algorithm 1.

The computation of each of the four matrix products along with how to achieve the required partitioning of A and B is described in Subsects. 3.1–3.3. For computing C_{ss} we use the spmm kernel from NVIDIA's cusparse library [8].

Algorithm 1. Algorithm for multiplying quasi-band matrices A and B into C.

1: $A = A_{bands} + A_{sparse}$ (A_{bands} is $A_{band}^1 + A_{band}^2 + \ldots + A_{band}^m$)
2: $B = B_{bands} + B_{sparse}$ (B_{bands} is $B_{band}^1 + B_{band}^2 + \ldots + B_{band}^n$)
3: $C_{bb} = A_{bands} \times B_{bands}$
4: $C_{sb} = A_{sparse} \times B_{bands}$
5: $C_{bs} = A_{bands} \times B_{sparse}$
6: $C_{ss} = A_{sparse} \times B_{sparse}$
7: $C = MERGE(C_{bb}, C_{sb}, C_{bs}, C_{ss})$

3.1 Matrix Partition

We first calculate diaOccupancy for each diagonal in the matrix. DiaOccupancy for a diagonal is defined as the percentage of non-zero elements in that diagonal to the rows of the matrix. We filter all the diagonals which have diaOccupancy greater than a threshold $diaOcc$ to a set S. Diagonals in set S are potential pivots of bands. We start by identifying the band surrounding the diagonal with the largest diaOccupancy in S. This is done by including this diagonal and the diagonals to its left and right so long as their bandOccupancy is more than a threshold $bandOcc$. These diagonals are then recognized as one band and the diagonals from S which intersect with the band are removed from S. The process is repeated to locate other bands of diagonals until S is empty. The diagonals that are chosen as part of some band are arranged in the DIA format. The left over elements are arranged in the COOSR format for A_{sparse} and COOSC format for B_{sparse}. As threshold $diaOcc$ is used for finding pivot diagonals of bands, we keep the value at 50 %. We also set $bandOcc$ to be 40 %, to ensure that bands are nearly half dense.

3.2 Multiplying A_{bands} with B_{bands}

In this section we present optimizations that can be applied when we multiply two band matrices. We perform the multiplication in two steps: a preprocessing step and a computation step.

In the preprocessing step, we use Observation 1 to allocate the correct space for the matrix C_{bb} which we store in the DIA format. Each diagonal from A_{bands} having offset a_o can then be multiplied with each diagonal in B_{bands} having offset b_o in parallel. Furthermore each row element r in the resultant diagonal can be updated atomically in parallel using equation $C_{bb}(a_o + b_o, r) += A_{bands}(a_o, r) \times B_{bands}(b_o, r + a_o)$ provided diagonal offset $a_o + b_o$ and row index $r + a_o$ are valid. (For a matrix M represented in the DIA format, $M(o, r)$ refers to the element with diagonal offset o and row number r.). An outline of computation step is shown in Algorithm 2.

Algorithm 2. Multiplying A_{bands} with B_{bands}.

1: **for** a_o in $A_{bands}.offsetArray$ **do**
2:    **for** b_o in $B_{bands}.offsetArray$ **do**
3:       **for** r in $[0 : A_{rows})$ **do**
4:          **if** $(-A_{rows} < a_o + b_o < B_{cols})$ and $(0 <= r + a_o < A_{rows})$ **then**
5:             $C_{bb}(a_o + b_o, r) = A_{bands}(a_o, r) * B_{bands}(b_o, r + a_o)$
6:          **end if**
7:       **end for**
8:    **end for**
9: **end for**

3.3 Multiplying A_{sparse} with B_{bands}

We now use Observation 2 in this computation. We start with a preprocessing step where we find all the column segments effected by $A_{\text{sparse}} \times B_{bands}^{i}$ for any i. These segments may overlap with each other. We proceed by merging these segments so that we can then allocate necessary storage for the matrix C_{sb} in the COOSR format.

In the actual computation step, we note from Observation 2 that an element e from A_{sparse} at row r and column c can be multiplied with each diagonal in B_{bands} having offset b_o in parallel. Such a computation effects at most one element in C_{sb} with row r and column $(c + b_o)$. As C_{sb} is stored in the COOSR format, this element needs to be introduced in to row r of C_{sb} by doing binary search on column indices of row r. An outline of computation step is shown in Algorithm 3. Using the COOSR format instead of CSR format increases the efficiency of our algorithm as we have simultaneous access to the row and the column indices of elements. This can be observed in steps 3 & 4 in Algorithm 3.

The multiplication of A_{bands} with B_{sparse} is similar to that of $A_{\text{sparse}} \times B_{\text{bands}}$. In this case row segments will be effected, instead of column segments. Another change to note is that we store the matrix C_{bs} in the COOSC format.

Algorithm 3. Multiplying A_{sparse} with B_{sparse}.

1: **for** i in $[0 : A_{sparse}.NNZ)$ **do**
2:    **for** b_o in $B_{bands}.offsetArray$ **do**
3:       $r = A_{sparse}.rowIndex[i]$
4:       $c = A_{sparse}.colIndex[i] + b_o$
5:       **if** $0 \leq c < B.cols$ **then**
6:          index $= SEARCH(c, C_{sb}.colIndex[C_{sb}.rowPtr[r]:C_{sb}.rowPtr[r+1]-1])$
7:          $C_{sb}.data[index] += A_{sparse}.data[i] * B_{bands}(b_o, c)$
8:       **end if**
9:    **end for**
10: **end for**

3.4 Merging

In the merging step, an element e at row r and column c in the matrix C has to be accumulated from corresponding elements in the four sub-products C_{bb}, C_{sb}, C_{bs} and C_{ss}. This is done in four steps. In the first step, for every element with row index r and column index c in the matrix C_{bs} we check if there is a corresponding element in any of the matrices C_{sb}, C_{ss}, and C_{bb}. If it exists, we consolidate the contribution of the element (r, c) in C_{bs} with one of the other three matrices. In the second step, we consolidate overlapping elements in C_{ss} with elements in matrices C_{sb} and C_{bb}. In the third step, we consolidate overlapping elements in C_{sb} with elements in matrix C_{bb}. Having easy access to both the row and column indices via the COOSR/COOSC formats makes the merge process efficient compared to using CSR/CSC formats.

We now have all the four subproducts that do not have any overlapping elements, but they all are not in the same format. In the fourth step, we convert matrices C_{bs} and C_{bb} into the COOSR format leaving us with four non-overlapping matrices in the COOSR format. It is now easy to combine these four matrices to a single matrix C in the COOSR format. (See also [7].)

Table 2. Properties of band matrices. Letter K stands for a thousand, and letter M stands for a million

Matrix	Rows	NNZ	Band-Count	Diagonal-Count	Band-Occupancy	Band-Spread
Bai/af23560	24K	484K	5	33	62.5	12.9
Boeing/crystm03	25K	583K	27	39	88.8	689.21
Castrillon/denormal	89K	1156K	5	13	99.75	6.67
Averous/epb1	15K	95K	3	11	58.96	3.77
Norris/fv1	10K	87K	3	9	99.32	4.08
Nasa/nasa2146	2K	72K	3	45	76.63	13.79
Boeing/pcrystk03	25K	1751K	9	99	72.66	76.53
Oberwolfach/windscreen	22K	1482K	9	99	79.36	909.68
Nemeth/nemeth21	10K	1173K	1	169	73.39	0

4 Experimental Results and Analysis

4.1 Datasets

We experiment with real-world band and quasi-band matrices from the University of Florida sparse matrix collection [10]. The matrices we use and some of the properties are shown in Tables 2 and 3. We also experiment with a variety of synthetic datasets that help us understand the impact of the nature of the matrix on our algorithm. These are described in Sect. 4.4.

Table 3. Properties of Quasi-band matrices.

Matrix	Rows	NNZ	Nonband-Elements	Band-Count	Diagonal-Count	Band-Percentage	Band-Occupancy
Schenk_IBMSDS/matrix_9	103K	2M	9.5K	5	31	99.55	66.73
Fluorem/PR02R	161K	8.2M	61K	5	108	99.25	47
Boeing/pwtk	218K	11M	2.1M	3	71	81.46	61.31
Simon/raefsky3	21K	1.5M	34K	3	93	97.71	75.48
Schenk_AFE/af_shell9	505K	18M	10M	1	25	42.93	59.83
Norris/heart1	3557	1.4M	0.95M	1	239	31.51	52.32
DNVS/trdheim	22K	1.9M	1.4M	1	35	25.92	64.88
HB/cegb2802	2802	277K	156K	1	67	43.73	65
MathWorks/Sieber	2290	15K	8K	1	5	46.18	60.02
Muite/Chebyshev3	4101	37K	16K	1	9	55.59	44.43

4.2   Experimental Platform

We use the K40 GPU from the NVidia Tesla series in our experiments. The host on which K40 is mounted is an Intel i7-4790K CPU with 32 GB of global memory. To program the GPU we used CUDA API Version 6.5.

4.3   Results on Real-World Datasets

We show results as speedup when compared to spmm kernel(cusparseScsrgemm) in NVIDIA's cusparse library [8] and spmm kernel(mkl_scsrmultcsr) from theIntel MKL library [9]. The computation is done in single precision.

Band Matrices: Figure 1a shows the speedup achieved by our algorithm on band matrices from Table 2. In our algorithm, only steps partitioning and computing $A_{bands} \times B_{bands}$ need to be executed as the input matrices are band matrices. Speedup variations can be partly explained as follows.

Firstly, notice that for matrices with more bandOccupancy, the number of unproductive computations in our algorithm reduce. Hence, for our algorithm, a high bandOccupancy is helpful. On the other hand, given that the input and the output matrices are band matrices, the number of nonzeros in the output matrix depends on how bands are spread in the input matrices. We capture the above via parameter bandSpread which is calculated as follows. Take the middle diagonal in each band to be a pivot diagonal. Calculate the distance (absolute difference of the diagonal offset values) between all pairs of pivot diagonals. Divide sum by dimension and multiply it by 100 to arrive at bandSpread of the matrix. Table 2 shows the bandSpread value for all matrices considered. It can now be noticed that the performance of the spmm routine from cusparse degrades as bandSpread increases. Hence, we expect that a high bandSpread usually results in a bigger speedup keeping other parameters fixed.

Following the above, consider matrices af23560, crystm03, pcrystk03, and windscreen that have near equal size. Matrix af23560 has less bandOccupancy and less bandSpread when compared to matrix crystm03. Hence the former has less speedup. Matrix windscreen has high bandOccupancy and high bandSpread compared to that of matrix pcrystk03 resulting in a better speedup. Matrices epb1, fv1, and nasa2146 have very small size and NNZ compared to others. To offset the cost of partitioning and pre-processing, a reasonable size and NNZ are desirable. Hence these matrices show a lesser speedup. Among matrices epb1 and fv1, epb1 has low bandOccupancy and low bandSpread resulting in low speedup.

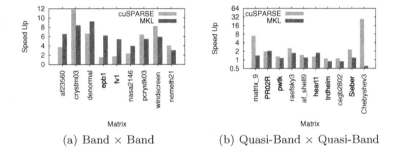

(a) Band × Band   (b) Quasi-Band × Quasi-Band

Fig. 1. Results on real world datasets from [10].

Quasi-band Matrices: Figure 1b shows the speedup achieved by our algorithm on quasi-band matrices from Table 3. For quasi-band matrices, the performance of our algorithm and also that of the spmm library routine from cusparse depends on various factors such as nnzPercentage, bandPercentage, bandOccupancy, bandCount, diagonalCount, bandSpread, and the spatial distribution of non-band elements. Because of various dependencies among these factors, it is in general not possible to explain the speedup achieved.

On matrix_9, the high speedup achieved is due to its high bandPercentage. On the other hand, though the matrix PR02R has a bandPercentage comparable to that of matrix_9, the speedup is not as high due to poor bandOccupancy. For matrix trdheim, the low speedup can be attributed to the fact that all the nonzero elements are present in a small band of diagonal indices (-531 to 531). This ensures that also the cusparse library routine performs well. The matrix Chebyshev3 has its nonzero elements outside of the bands cohesively within the first four rows. Such a structure is likely to create load imbalance in the cusparse library routine for spmm which results in a high speedup for our algorithm. (See also Experiment 3 in Sect. 4.4).

Profile: In Fig. 2, we show the time taken by various steps of our algorithm on matrices from Table 3 as a percentage of the overall time. As we use cusparse library for computing C_{ss}, this suggests that indeed multiplying sparse matrices is always difficult and focusing on the nature of sparsity is usually beneficial.

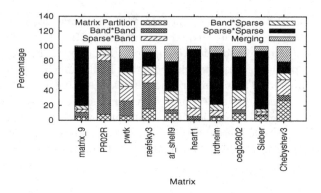

Fig. 2. Percentage of time taken across the various steps of our algorithm on quasi-band matrices from Table 3.

4.4 Synthetic Datasets and Results

We conduct three experiments with synthetic quasi-band matrices to understand the factors that influence the speedup and runtime of our algorithm. We consider synthetic matrices with 16,000 rows, an nnzPercentage of 0.5, a bandOccupancy of 90 and a bandPercentage of 95. The number of bands and the position of the bands are chosen uniformly at random.

In Experiment 1, we study the impact of bandPercentage and bandOccupancy. The results of this study, shown in Fig. 3a, indicate that a high bandPercentage is favorable to our algorithm compared to bandOccupancy. In Experiment 2 we study the impact of nnzPercentage. The results of this study, shown in Fig. 3b, indicate that a high nnzPercentage improves the performance of our algorithm. This can be understood from the highly parallel strucutre of the algorithm.

In Experiment 3, we consider three data layouts i.e., RANDOM,ROW-BLOCK and COLUMN-BLOCK for the non-band elements in matrix. In case of RANDOM, the non-band elements are spread uniformly at random in non-band part. In case of ROW-BLOCK and COLUMN-BLOCK, the non-band elements are filled in contiguous rows and columns respectively. Figure 3c shows the speedup as we vary the bandPercentage for a given data layout. As can be observed, our algorithm too suffers in the RANDOM distribution model but is faster compared to the library routine from **cusparse** once the bandPercentage crosses 85 %. Further, the speedup on COLUMN-BLOCK distribution is lower compared to that of ROW-BLOCK. The reason for this can be attributed to the fact that that the library routine from **cusparse** does 5X times better on COLUMN-BLOCK compared to ROW-BLOCK as there is lesser chance of load imbalance in the COLUMN-BLOCK.

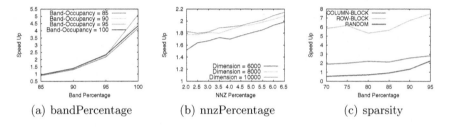

(a) bandPercentage   (b) nnzPercentage   (c) sparsity

Fig. 3. Studies on quasi-band matrices with varying bandPercentage, nnzPercentage and spatial distribution of non-band elements.

5   Conclusions and Future Work

In this paper, we demonstrated that paying attention to the nature of sparsity will be beneficial when performing operations on sparse matrices on architectures such as the GPU. In future, we would like to focus on other operations such as linear solver, matrix inversion while focusing on quasi-band matrices.

References

1. Bell, N., Garland, M.: Implementing sparse matrix-vector multiplication on throughput-oriented processors. In: Proceeding SuperComputing (SC), pp. 1–11 (2009)
2. Buluc, A., Gilbert, J.R.: Challenges and advances in parallel sparse matrix-matrix multiplication. In: Proceeding International Conference on Parallel Processing, pp. 503–510 (2008)
3. Gharaibeh, A., Costa, B., Santos-Neto, E., Ripeanu, M.: On Graphs, GPUs, and Blind Dating: a workload to processor matchmaking quest. In: Proceeding International Parallel & Distributed Processing Symposium (IPDPS), pp. 851–862 (2013)
4. Hong, S., Rodia, N.C., Olukotun, K.: On fast parallel detection of strongly connected components in small-world graphs. In: Proceedings of the SC (2013). Article No. 92
5. Indarapu, S., Maramreddy, M., Kothapalli, K.: Architecture- and workload-aware algorithms for spare matrix- vector multiplication. In: Proceeding of ACM India Computing Conference (2014). Article No. 3
6. Liu, W., Vinter, B.: An efficient GPU general sparse matrix-matrix multiplication for irregular data. In: Proceeding of IPDPS, pp. 370–381 (2014)
7. Ramamoorthy, K.R., Banerjee, D.S., Srinathan, K., Kothapalli, K.: A novel heterogeneous algorithm for multiplying scale-free sparse matrices. In: Proceeding of IPDPS Workshops, pp. 637–646 (2015)
8. Nvidia sparse matrix library (cuSPARSE). http://developer.nvidia.com/cusparse
9. Intel Math Kernel Library. https://software.intel.com/en-us/articles/intel-mkl/
10. University of Florida UF sparse matrix collection (2011). http://www.cise.ufl.edu/research/sparse/matrices/groups.html
11. Yang, W., Li, K., Liu, Y., Shi, L., Wan, L.: Optimization of quasi-diagonal matrix-vector multiplication on GPU. Int. J. High Perform. Comput. Appl. **28**(2), 183–195 (2014)

Comparative Performance Analysis of Coarse Solvers for Algebraic Multigrid on Multicore and Manycore Architectures

Alex Druinsky[1(✉)], Pieter Ghysels[1], Xiaoye S. Li[1], Osni Marques[1],
Samuel Williams[1], Andrew Barker[2], Delyan Kalchev[2],
and Panayot Vassilevski[2]

[1] Lawrence Berkeley National Laboratory, Berkeley, USA
adruinsky@lbl.gov
[2] Lawrence Livermore National Laboratory, Livermore, USA

Abstract. We study the performance of a two-level algebraic-multigrid
algorithm, with a focus on the impact of the coarse-grid solver on performance. We consider two algorithms for solving the coarse-space systems:
the preconditioned conjugate gradient method and a new robust HSS-
embedded low-rank sparse-factorization algorithm. Our test data comes
from the SPE Comparative Solution Project for oil-reservoir simulations.
We contrast the performance of our code on one 12-core socket of a Cray
XC30 machine with performance on a 60-core Intel Xeon Phi coprocessor.
To obtain top performance, we optimized the code to take full advantage
of fine-grained parallelism and made it thread-friendly for high thread
count. We also developed a bounds-and-bottlenecks performance model
of the solver which we used to guide us through the optimization effort,
and also carried out performance tuning in the solver's large parameter
space. As a result, significant speedups were obtained on both machines.

Keywords: Algebraic multigrid · HSS matrices · Manycore machines

1 Introduction

We study the performance of a novel algebraic multigrid algorithm that was
recently introduced by Brezina and Vassilevski [4] for solving difficult elliptic
PDEs with a variable coefficient that can be resolved only using a fine-grained
discretization. We use the two-level variant of the method, implemented in the
serial C++ code SAAMGe [11,12], and our focus is on the impact of the coarse-
grid solver on the performance of the algorithm. The coarse grid represents the

This material is based upon work supported by the US Department of Energy
(DOE), Office of Science, Office of Advanced Scientific Computing Research (ASCR),
Applied Mathematics program under contract number DE-AC02-05CH11231. This
work was performed under the auspices of the DOE under Contract DE-AC52-
07NA27344, and used resources of the National Energy Research Scientific Computing Center, which is supported by ASCR under contract DE-AC02-05CH11231.

© Springer International Publishing Switzerland 2016
R. Wyrzykowski et al. (Eds.): PPAM 2015, Part I, LNCS 9573, pp. 116–127, 2016.
DOI: 10.1007/978-3-319-32149-3_12

parallel bottleneck in the code, and therefore solving the corresponding systems efficiently is crucial for the solver's performance. Outside of the coarse grid, most of the work in SAAMGe is formulated in terms of sparse matrix-vector multiplications (SpMVs) that involve large matrices with regular sparsity structure. Optimizing such SpMVs is a well-studied problem and optimized implementations are available for many architectures.

We consider two coarse-grid solvers. One is the preconditioned conjugate gradient method (PCG), which we precondition using a single step of the Jacobi iteration. Although careful optimization of PCG can make it a powerful coarse-grid solver, using it when convergence is slow can be expensive due to its low arithmetic intensity. As an alternative, we use STRUMPACK, a new HSS-embedded low-rank sparse-factorization algorithm [10]. It has a higher arithmetic intensity, and it is robust, meaning that it can be used as a direct solver.[1] The use of low-rank structure in the factorization reduces the amount of work, memory space, and memory bandwidth that we expend, making the solver potentially more efficient than earlier sparse-factorization algorithms.

There exist thorough studies of parallel-performance optimization and analysis of algebraic multigrid (AMG) [2,3,8,9]. Our paper differs from these studies by focusing on the architecture level and performing detailed performance-bound modeling, taking into account the arithmetic intensity of the different algorithmic components. We are the first to apply this methodology to AMG, and this allows us to bridge the gap in the understanding of the limits of AMG performance on individual multicore nodes with large core counts.

Our main contributions are the following. We make a comprehensive study of the impact of the coarse-grid solver on the performance of a two-level AMG solver. We perform extensive optimization on two multicore architectures, one of which is the challenging Xeon Phi, characterized by having a large number of relatively slow cores and a high memory latency. We incorporate a novel randomized HSS-embedded sparse-factorization algorithm as the coarse-grid solver. Finally, we develop and validate a bounds-and-bottlenecks Roofline model which helps us to identify performance optimization opportunities on our target architectures and to characterize the limitations of our code.

2 Background

2.1 Test Problems and Machines

We use the oil-reservoir simulation benchmark SPE10 (model 2) from the SPE Comparative Solution Project [6]. Here, fluid flow in porous media is described by the Darcy equation (in primal form):

$$- \nabla \cdot (\kappa(x) \nabla p(x)) = f(x) \tag{1}$$

---

[1] The STRUMPACK library can use the factorization either to solve the system directly, or to precondition the flexible GMRES iteration [16]. In our study, we found that performance is best if we tune STRUMPACK's parameters so that GMRES is not required. We expect this effect to be problem dependent. For details, see Sect. 4.

where $p(x)$ is the pressure field and $\kappa(x)$ is the permeability of the medium. The model is described on a regular Cartesian grid of $60 \times 220 \times 85$ cells. The coefficient $\kappa(x)$ admits a wide range of variation between distinct horizontal layers of the medium, which makes this a challenging problem to solve. Finite-element discretization of the problem produces a fine-grid matrix of order 1.2 million with a regular sparsity structure and an average of 26.4 nonzero elements in each row. From this matrix, the SAAMGe algorithm produces a coarse-grid matrix whose dimensions and sparsity pattern depend on the algorithm's parameters, as we explain below.

We carried out our study on two machines at the National Energy Research Scientific Computing Center in Oakland, CA. One machine is Edison, a Cray XC30 machine of 5,600 Ivy Bridge EP nodes. The other is Babbage, a commodity Intel cluster in which each node contains two Xeon Phi Knights Corner coprocessors. In the following, we refer to these machines as IVB-EP and KNC, respectively. In this paper, we focus on one CPU socket of an IVB-EP node and one KNC coprocessor. Each IVB-EP socket consists of 12 cores with a theoretical 230.4 peak gflop/s rate. A KNC coprocessor has 60 cores with a theoretical 1,010.9 peak gflop/s rate. Although KNC has more computational power and memory bandwidth, it has slower cores, a wider SIMD architecture and more primitive hardware stream prefetchers. As a result, SpMV-like operations are much more difficult to optimize on KNC. The cache hierarchies of both machines are coherent and have the same net capacity—30 MB. However, whereas IVB-EP provides a 30 MB unified L3 cache, KNC maintains 60 caches of 512 KB each. This results in superfluous data movement and coherency transactions that can impede the effective bandwidth on KNC. Furthermore, ineffective software prefetching can highlight the fact that KNC's memory latency can be an order of magnitude higher than that of IVB-EP [13]. As a result, although KNC has 5× the nominal bandwidth of IVB-EP, it can often be underutilized or squandered given complex memory access patterns endemic in sparse methods.

2.2 Algebraic Multigrid

The SA-ρAMGe method that we study [4] works by forming a coarse-grid matrix A_c and solving the problem iteratively by repeating the following steps:

1. Pre-smoothing: $y_k \leftarrow x_k + M^{-1}(b - Ax_k)$
2. Coarse-grid correction: $z_k \leftarrow y_k + PA_c^{-1}P^T(b - Ay_k)$
3. Post-smoothing: $x_{k+1} \leftarrow z_k + M^{-1}(b - Az_k)$

Here, M^{-1} is a polynomial smoother and P is the so-called prolongation operator. The operator P is formed by representing A as the sum of local stiffness matrices (which requires knowledge of the finite-element discretization) and computing the eigenvectors that correspond to the smallest eigenvalues of each such matrix. We form the *tentative prolongator* $\bar{P}$, which is a rectangular block-diagonal matrix whose diagonal blocks correspond to the local stiffness matrices and consist of the eigenvectors that we computed. The ultimate prolongator has the form $P = S\bar{P}$, where S is a matrix-polynomial smoother. The coarse-grid matrix is obtained by forming the sparse-matrix product $A_c = P^T A P$.

2.3   HSS Sparse Solver

PCG is relatively easy to implement and parallelize, but its convergence can be slow for numerically difficult problems. Sparse-factorization methods can serve as powerful alternatives. Here, we consider an HSS-embedded sparse-factorization method that has asymptotically lower complexity than traditional factorizations. The algorithm has a shared-memory parallel implementation in the package STRUMPACK [10], which uses *nested-dissection* ordering and *multifrontal* factorization. The sparse solver consists of the following steps: preprocessing (e.g., sparsity-preserving ordering by a graph-partitioning algorithm such as METIS, and symbolic factorization), numerical factorization and solution.

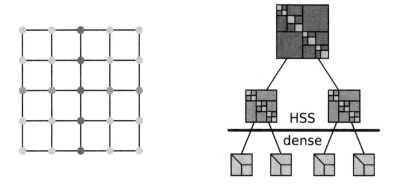

Fig. 1. A regular two-dimensional grid, partitioned using nested dissection (left) and the corresponding separator tree (right). The nodes of the separator tree represent frontal matrices. Only the nodes of the top levels are compressed using HSS. (Color figure online)

The novelty is to represent the dense frontal matrix corresponding to a node of the separator tree as a hierarchically-semiseparable (HSS) matrix [10,14,18]. The HSS structure exploits the *data-sparsity* of the dense matrix using low-rank compression. Furthermore, the *hierarchical partitioning* of the matrix blocks and the use of *nested bases* lead to factorization and solve algorithms that are asymptotically faster than the classical ones and use less memory.

Figure 1 illustrates the nested-dissection procedure on a small regular mesh. The blue points denote the root separator, splitting the domain in two unconnected parts. The nested-dissection procedure is then repeated on both parts recursively. Each of the separators corresponds to one frontal matrix, placed in a data structure called the separator or elimination tree, illustrated in Fig. 1 (right). The largest frontal matrices, corresponding to the top ℓ_s levels of the elimination tree, are approximated as HSS matrices. The other smaller frontal matrices are stored as full-rank dense matrices. The factorization is computed by a bottom-up traversal of the elimination tree, computing a partial factorization of each front and passing the Schur complement from child to parent. The structure of an HSS matrix is also illustrated in Fig. 1 (right). In an HSS matrix, diagonal blocks are recursively partitioned until at the finest level diagonal blocks are stored as full-rank dense matrices. Off-diagonal blocks are represented as low-rank products $A_{ij} = U_i B_{ij} V_j^* + \mathcal{O}(\epsilon)$.

The STRUMPACK solver can be used as a preconditioner, where the quality of the preconditioner is controlled by the accuracy of the low-rank approximations in the HSS structure and by the number of HSS levels ℓ_s. The HSS approximation accuracy can be controlled by a user specified tolerance ϵ. A single solve with the STRUMPACK preconditioner is more expensive than a single PCG iteration, but it is more effective for numerically challenging PDEs.

3   Code Optimizations

Obtaining top performance on a traditional multicore architecture such as the IVB-EP is a well-studied problem, and therefore we focus in this section on the performance optimizations that we conducted on KNC. The challenges on this platform are due to a large core count, wide SIMD FPU and distributed coherent L2 caches.

3.1   PCG Thread-Friendly Optimizations

In contrast with IVB-EP, a large proportion of the time on KNC is spent on solving coarse-grid systems, and this proportion is increasing as we use more threads. Ultimately, using 180 threads, coarse-grid PCG accounts for more than 50 % of the total time on KNC. Furthermore, our study in Sect. 4 showed that AMG performs best when the coarse-grid system is small, and so the optimizations that we require are different from those that are typically done in large-scale implementations.

The initial version of PCG in our code was implemented as a serial code that launched parallel OpenMP kernels such as dot product, vector linear combination and SpMV. Arranging the computation in this way incurs an overhead of entering and exiting an OpenMP `parallel` region for each computational kernel. For this reason, we introduced an alternative implementation, **omp-for-all**, in which the whole iteration is nested inside a single OpenMP `parallel` region. This yielded a speedup of 1.55 (using 180 threads).

We accomplished a further 1.69 speedup by introducing the **omp-for-spmv** variant, in which all kernels are serial, except for SpMV, which is parallel. Our explanation for the speedup in this case is that the coarse grid is represented by a small matrix and therefore the overhead of parallelizing the dot-product and vector-linear-combination kernels outweighs the benefits of such parallelization.

Finally, we also considered the **omp-parallel-spmv** variant, which we obtained from the original code by replacing all parallel kernels with serial ones, except for SpMV. Similarly to **omp-for-spmv**, in this version all kernels except SpMV are serial and there is only one `parallel` region. However, in contrast with **omp-for-spmv**, the parallel region here is inside the main loop and therefore incurs an overhead in each iteration. This version is slightly slower than **omp-for-spmv**, reaching 88 % of **omp-for-spmv**'s performance using 180 threads. Nevertheless, **omp-parallel-spmv** is competitive with **omp-for-spmv** and it allows us to use an external library that implements the SpMV kernel, such as the one proposed in [13].

3.2   HSS Thread-Friendly Optimizations

We now consider the use of STRUMPACK for solving coarse-grid systems. Table 1 and Fig. 2 show the running time of the algorithm.

Table 1. Runtime (seconds) of the HSS solver, broken down into the time dedicated to ordering, symbolic factorization, numeric factorization and solve. The coarse-grid matrix is of order 53,709 and has 24.8 million nonzeros, and was generated from a fine-grid matrix of order 2.4 million. The HSS compression level is 1 and tolerance is 10^{-4}, which corresponds to an HSS rank of 217.

Machine	Factorization (s)			Solve (s)	Threads
	Ordering	Symbolic	Numeric		
IVB-EP	0.44	0.34	5.6	0.23	12
KNC	3.4	0.83	19.1	0.56	60

The solve time on both machines is more than an order of magnitude faster than factorization, which is for the better, because factorization is required only once, whereas solve is required in each AMG iteration. On both machines, the computation scales well, with the exception of ordering and symbolic factorization on KNC, where performance stagnates early, at about 10 threads. These steps involve purely combinatorial algorithms which are hard to parallelize on a machine architecture optimized for a high flop rate. Parallel scaling on IVB-EP is better than on KNC, with a speedup of 12 threads over one thread of 6.5× and 3.3× for factorization and solve, respectively, compared to speedups of 12.8× and 7.6× using 60 threads on KNC.

The solver is implemented using OpenMP task parallelism, so that the numeric-factorization phase is represented by a single parallel region, and therefore the only barriers in the code correspond to dependencies between the tasks. We took the following additional steps to improve performance on KNC. We replaced the default memory allocator by the more scalable TBB [1] allocator. Tasks are generated by recursive functions. We tuned the total number of tasks, i.e., the task granularity, specifically for each machine. We replaced the SCOTCH graph partitioner, which we were using for ordering the matrix, with

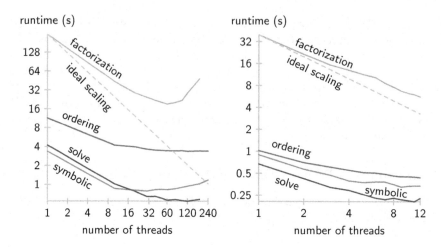

Fig. 2. Wall-clock time for solving the coarse-grid system on the Xeon Phi (left) and the Ivy Bridge EP (right). We used the optimized version of STRUMPACK as described in Sect. 3.2, and solved using three iterations of HSS-preconditioned GMRES.

METIS, and thereby gained a 10× time savings in the ordering step on KNC.[2,3] We disabled a preprocessing step that performs permutation and scaling using the MC64 code [7]. This step is not required because our matrix is symmetric positive-definite. Finally, we disabled code for counting the number of executed flops, which we found was causing a 3× slowdown in the solve phase on KNC.

4   Performance Comparison

To complement our optimization effort, we conducted a comprehensive performance assessment of the solvers from the architectural viewpoint (two machines) as well as the algorithmic one (varying the algorithm's parameters). In particular, we considered five parameters of SAAMGe, and explored the effect of changing these parameters within a five-dimensional space. For each parameter we consider a range of values, as described in Table 2.[4] There are altogether 216 configurations explored on each machine. We used 60 cores on KNC and 12 cores on IVB-EP.

The following summarizes our findings. First, choosing the proper algorithm parameters is crucial to achieve good performance. The difference in runtime can

---

[2] Based on this experience, we changed the STRUMPACK default to METIS.

[3] The ordering phase consists of running METIS, applying the computed permutation to the matrix and sorting the column indices within each row of the permuted matrix. METIS runs serially, but the rest of the work is done in parallel.

[4] We also used the following parameters to control the accuracy of the computed solution. For the HSS algorithm, we used four levels of compression with compression tolerance 10^{-4} and zero GMRES iterations. For PCG, we used relative tolerance 10^{-4}. These were chosen so as to maximize performance without sacrificing accuracy.

Table 2. The parameters of the algorithm and the corresponding values that we explored. The number of elements per agglomerate determines the size of the local stiffness matrices; ν_P and ν_{M-1} are respectively the polynomial degrees of the interpolator smoother S and the relaxation smoother M^{-1}; and θ is the spectral tolerance, which determines how many eigenvectors of each local stiffness matrix represent that matrix in the tentative prolongator.

Parameter	Values
Coarse solver	HSS, PCG
Elements-per-agglomerate	64, 128, 256, 512
ν_P	0, 1, 2
ν_{M-1}	1, 3, 5
θ	0.001, 0.001 $\times$ 10$^{0.5}$, 0.01

be as large as an order of magnitude among the 216 configurations. Taking PCG on IVB-EP as an example, the fastest configuration took 9.6 s, while the slowest took 168.1 s — more than a 17× difference. Second, on the same machine, the HSS coarse-grid solver always won over PCG. For the best configurations, HSS is 1.54× faster than PCG on IVB-EP and 1.34× on KNC. Finally, with the best configurations, IVB-EP is 1.7× and 1.49× faster than KNC using HSS and PCG, respectively.

5   Roofline Performance Model

We developed a bounds-and-bottlenecks Roofline model to drive the performance optimization of our OpenMP code [17].[5] The goal is to gain insight about the machine's performance bottlenecks and terminating performance optimization. Here, we focus on the AMG solution cycle; modeling AMG setup is future work.

The model consists of formulas, one for each component of the algorithm, expressing the number of bytes that we move between the levels of the memory hierarchy and the number of flops that we carry out. To obtain runtime estimates from this model, we divide the total memory traffic by the machine bandwidth, and also divide the total number of flops by the machine flops rate. This yields two lower bounds on the runtime: one that corresponds to memory bandwidth being the bottleneck, and the other to the floating-point units.

Concurrent with traditional Roofline analysis, the inputs to our model are: (1) The machine peak flop rate and its sustainable memory bandwidth, measured using a modified STREAM benchmark [15]; (2) The dimensions of A and A_c, denoted by n and n_c, respectively; (3) The number of nonzeros in A, A_c and P, denoted by **nza**, **nzc** and **nzp**, respectively; (4) The number of AMG cycles; and (5) Parameters that are specific to the coarse solver: the average number of PCG iterations per AMG cycle when we use PCG, and the memory size of the HSS factors when we use HSS.

---

[5] See also [5] for earlier work on such models.

5.1   The Model for the Combination of AMG with PCG

The model that we obtain for the version of the solver in which we use PCG to solve coarse-grid systems is shown in Table 3. We used the following combination of parameters: elements-per-agglomerate is set to 400, $\nu_{M^{-1}} = 3$ and $\theta = 0.001$. The corresponding runtime bounds on IVB-EP are shown in Fig. 3.[6]

Table 3. The costs associated with each AMG cycle.

Stage	Bytes	Flops
Pre- and post-smooth	$(3\nu + 1)(12\mathbf{nza} + 3 \cdot 8n)$	$2(3\nu + 1)(\mathbf{nza} + 2n)$
Restriction	$12\mathbf{nza} + 12\mathbf{nzp} + 3 \cdot 8n$	$2(\mathbf{nza} + \mathbf{nzp})$
One coarse solve (PCG/J)		
Multiply by A_c	$12\mathbf{nzc}$	$2\mathbf{nzc}$
Preconditioner	$2 \cdot 8n\_c$	n_c
Vector operations	$5 \cdot 8n\_c$	$2 \cdot 5n\_c$
Interpolation	$12\mathbf{nzp} + 8n$	$2\mathbf{nzp}$
Stopping criterion	$12\mathbf{nza} + 4 \cdot 8n$	$2(\mathbf{nza} + n)$

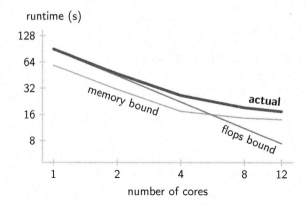

Fig. 3. Runtime bounds and actual runtime (seconds) of the AMG iteration on the Ivy Bridge EP.

When 1 or 2 cores are used, our flops-based bound is within 1 % and 7 % of actual runtime respectively. As the number of cores is increased, memory bandwidth becomes the bottleneck. For 4, 8 and 12 cores, our memory-bandwidth-based bound is within 23 % of the actual runtime. We attribute the difference

---

[6] Roofline models often use a corrected machine gflop/s rate that accounts for an imbalanced mix of multiply and add operations in the computation. We do not do this here, because in our computation, multiplies and adds are almost perfectly balanced. The only exception is multiplications by a diagonal matrix in the polynomial smoother and the Jacobi preconditioner, but these multiplications correspond to a small fraction of the work.

to the extremely different memory access patterns in AMG compared to the STREAM benchmark.

5.2   The Model for the Combination of AMG with HSS

Following the same practice as in Sect. 5.1, we conduct a performance-bound analysis when HSS is used as the coarse solver. Comparing to Table 3, the costs are the same for smoothing, restriction, interpolation, and termination. The difference is in the coarse solve, where the code needs to stream through the factored matrix. For our test cases, the factors are larger than the largest cache of the machines. Therefore, we assume that the factors are read from DRAM. Table 4 shows the performance upper bound based on the DRAM sustained bandwidth. On IVB-EP, the best configuration is: elements-per-agglomerate = 256, $\nu_{M-1} = 1$ and $\theta = 0.01$. On KNC, the best configuration is: elements-per-agglomerate = 256, $\nu_{M-1} = 3$ and $\theta = 0.01$. The bandwidth-based performance bound is quite accurate on IVB-EP, yielding an estimate within a gap of 31 % of the actual time. Among all the stages, the best match between model and reality is the smoothing step—about an 18 % gap. The worst gap corresponds to the coarse solve—about 55 %.

On the other hand, the estimated time on KNC is far less than the actual time, implying that the memory bandwidth is severely underutilized. Attributing this significant performance difference to either architecture or model is an area of continued investigation.

Table 4. Runtime bounds and actual runtime (seconds) of the AMG iteration. HSS is used as coarse-grid solver. The column "R/I" represents the combined restriction and interpolation steps. "Stopping" refers to the evaluation of the stopping criterion.

Machine		Memory bandwidth model			Stopping	Total
		Smoothing	R/I	HSS		
IVB-EP	Model	3.2	1.5	0.9	0.41	6.0
(12-core)	Actual	3.8	2.0	1.4	0.88	7.9
KNC	Model	1.8	0.22	0.13	0.09	2.25
(60-core)	Actual	7.4	0.86	1.9	1.0	10.7

6   Conclusion

We proposed a series of optimizations to improve the performance of the coarse-grid solver. The optimizations aim to expose fine-grained parallelism, exploit high memory bandwidth, and reduce OpenMP overheads. These led to a 2.6× reduction of the AMG solve time on a 60-core Xeon Phi KNC machine. We expect these optimizations to be effective on other manycore architectures as well. We also compared the performance of PCG with STRUMPACK when the two algorithms are used as coarse-grid solvers. We found that PCG is at a disadvantage because of its slow convergence. HSS usually leads to a faster AMG cycle, up to 2× faster than PCG. We expect the relative performance of PCG and HSS to depend on the problem. If the problem is such that the AMG parameters can be tuned so as to produce a well-conditioned coarse-grid matrix, then PCG could outperform HSS. Additionally, we explored the parameter space of our AMG algorithm and found high variation in performance. This makes the algorithm a good candidate for an autotuning approach. Our roofline model yields a bound that is within 23 % (for PCG) and 31 % (for HSS) of the actual performance on the Ivy Bridge EP. The gap is much more significant on the Xeon Phi KNC, which indicates that the bottleneck on that machine is not the memory bandwidth or the FPU but rather the high memory latency. More effective prefetching could hide this latency, but achieving this is challenging because of the relatively primitive hardware prefetchers on that machine, and because of the irregular memory access pattern in our coarse-grid solvers. We are exploring this optimization.

Finally, the only aspect of performance that we considered in our study was time to solution. This is the objective that users care most about. Nevertheless, in future work it would also be valuable to consider other parameters, such as the financial cost of the hardware and its energy efficiency.

Acknowledgments. We thank the anonymous referees for their many comments that greatly helped to improve the paper.

References

1. Intel threading building blocks. https://www.threadingbuildingblocks.org
2. Baker, A.H., Schulz, M., Yang, U.M.: On the performance of an algebraic multigrid solver on multicore clusters. In: Palma, J.M.L.M., Daydé, M., Marques, O., Lopes, J.C. (eds.) VECPAR 2010. LNCS, vol. 6449, pp. 102–115. Springer, Heidelberg (2011)
3. Bolz, J., Farmer, I., Grinspun, E., Schröoder, P.: Sparse matrix solvers on the GPU: conjugate gradients and multigrid. ACM Trans. Graph. **22**(3), 917–924 (2003)
4. Brezina, M., Vassilevski, P.S.: Smoothed aggregation spectral element agglomeration AMG: SA-ρAMGe. In: Lirkov, I., Margenov, S., Waśniewski, J. (eds.) LSSC 2011. LNCS, vol. 7116, pp. 3–15. Springer, Heidelberg (2012)
5. Callahan, D., Cocke, J., Kennedy, K.: Estimating interlock and improving balance for pipelined architectures. J. Parallel Distrib. Comput. **5**(4), 334–358 (1988)
6. Christie, M.A., Blunt, M.J.: Tenth SPE comparative solution project: Comparison of upscaling techniques. SPE Reserv. Eval. Eng. **4**(4), 308–317 (2001)
7. Duff, I.S., Koster, J.: The design and use of algorithms for permuting large entries to the diagonal of sparse matrices. SIAM J. Matrix Anal. Appl. **20**, 889–901 (1999)
8. Gahvari, H., Baker, A.H., Schulz, M., Yang, U.M., Jordan, K.E., Gropp, W.: Modeling the performance of an algebraic multigrid cycle on HPC platforms. In: Proceedings of ICS, pp. 172–181 (2011)
9. Gahvari, H., Gropp, W., Jordan, K.E., Schulz, M., Yang, U.M.: Modeling the performance of an algebraic multigrid cycle using hybrid MPI/OpenMP. In: Proceedings of ICPP, pp. 128–137 (2012)

10. Ghysels, P., Li, X.S., Rouet, F.H., Williams, S., Napov, A.: An efficient multi-core implementation of a novel HSS-structured multifrontal solver using randomized sampling. SIAM J. Sci. Comput. (2014) preprint
11. Kalchev, D., Ketelsen, C., Vassilevski, P.S.: Two-level adaptive algebraic multigrid for a sequence of problems with slowly varying random coefficients. SIAM J. Sci. Comput. **35**(6), B1215–B1234 (2013)
12. Kalchev, D.: Adaptive Algebraic Multigrid for Finite Element Elliptic Equations with Random Coefficients. Master's thesis, Sofia University, Bulgaria (2012)
13. Liu, X., Smelyanskiy, M., Chow, E., Dubey, P.: Efficient sparse matrix-vector multiplication on x86-based many-core processors. In: Proceedings of ICS, pp. 273–282 (2013)
14. Martinsson, P.: A fast randomized algorithm for computing a hierarchically semiseparable representation of a matrix. SIAM J. Matrix Anal. Appl. **32**(4), 1251–1274 (2011)
15. McCalpin, J.D.: Memory bandwidth and machine balance in current high performance computers. In: IEEE TCCA Newsletter, pp. 19–25 (1995)
16. Saad, Y.: A flexible inner-outer preconditioned GMRES algorithm. SIAM J. Sci. Comput. **14**(2), 461–469 (1993)
17. Williams, S., Waterman, A., Patterson, D.: Roofline: An insightful visual performance model for multicore architectures. Commun. ACM **52**(4), 65–76 (2009)
18. Xia, J., Chandrasekaran, S., Gu, M., Li, X.S.: Fast algorithms for hierarchically semiseparable matrices. Numer. Linear Algebra Appl. **17**(6), 953–976 (2010)

LU Preconditioning for Overdetermined Sparse Least Squares Problems

Gary W. Howell[1]([✉]) and Marc Baboulin[2]

[1] North Carolina State University, Raleigh, USA
gwhowell@ncsu.edu
[2] Université Paris-Sud and Inria, Orsay, France
marc.baboulin@lri.fr

Abstract. We investigate how to use an LU factorization with the classical `lsqr` routine for solving overdetermined sparse least squares problems. Usually L is much better conditioned than A and iterating with L instead of A results in faster convergence. When a runtime test indicates that L is not sufficiently well-conditioned, a partial orthogonalization of L accelerates the convergence. Numerical experiments illustrate the good behavior of our algorithm in terms of storage and convergence.

Keywords: Sparse linear least squares · Iterative methods · Preconditioning · Conjugate gradient algorithm · `lsqr` algorithm

1 Introduction to LU Preconditioning for Least Squares

Linear least squares (LLS) problems arise when the number of linear equations is not equal to the number of unknown parameters. For example LLS problems occur in many parameter estimation and constrained optimization problems [2, 22]. Commonly, nonlinear least squares problems are solved via algorithms which solve sparse linear least squares problems at each step [14]. Levenberg-Marquardt algorithms [21] are one example.

Here we consider the overdetermined full rank LLS problem

$$\min_{x \in \mathbb{R}^n} \|Ax - b\|_2, \tag{1}$$

with $A \in \mathbb{R}^{m \times n}, m \geq n$ and $b \in \mathbb{R}^m$. When A is sparse, direct methods based on QR factorization or the normal equations are not always suitable because the R factor or $A^T A$ can be dense. A common iterative method to find the least squares solution x is to solve the normal equations

$$A^T A x = A^T b, \tag{2}$$

by applying the conjugate gradient (CG) algorithm to $A^T A$. In this case the matrix $A^T A$ does not need to be explicitly formed, avoiding possible fill-in in the formation of $A^T A$. As with other sparse linear systems, preconditioning

© Springer International Publishing Switzerland 2016
R. Wyrzykowski et al. (Eds.): PPAM 2015, Part I, LNCS 9573, pp. 128–137, 2016.
DOI: 10.1007/978-3-319-32149-3_13

techniques based on incomplete factorizations can improve convergence. One method to precondition the normal equations (2) is to perform an incomplete Cholesky decomposition of $A^T A$ (e.g., RIF preconditioner [5]).

When $A^T A$ and its Cholesky factorization are denser than A, it is natural to wonder if the LU factorization of A can be used in solving the least squares problem. In this paper we use an LU factorization of the rectangular matrix $A = \begin{pmatrix} A_1 \\ A_2 \end{pmatrix}$ where L is unit lower trapezoidal and U is upper triangular. Following [11, p. 102], such a factorization exists when A_1 is non singular, which for A full rank, can be obtained by permuting rows of A. For the nonpivoting case, the normal equations (2) become

$$L^T L y = c, \qquad (3)$$

with $c = L^T b, U x = y$, and we can apply CG iterations on (3).

Least squares solution using LU factorization has been explored by several authors. Peters and Wilkinson [24] and Björck and Duff [7] give direct methods. This work follows Björck and Yuan [8] using conjugate gradient methods based on LU factorization, an approach worth revisiting because of the recent progress in sparse LU factorization. In contrast to the SP1, SP2, and SP3 algorithms of [8], the lsqrLU algorithm presented here uses a lower trapezoidal L returned from a direct solver package, easing implementation. Where Saunders, cited in [8], iterated with $U^{-1}A$, we iterate directly with L from $LU = PA$, an approach amenable to parallel implementation and to further preconditioning, if necessary. Here we use the Matlab and Octave fast sparse LU factorizations built on Davis' UMFPACK package [10]. Because other direct solver packages (e.g., [1,19,20,26]) offer scalable sparse LU factorizations, it appears likely that the algorithm used here can also be used to solve larger problems.

The rate of linear convergence for CG iterations on the normal equations is (see [6, p. 289])

$$K = \frac{\kappa - 1}{\kappa + 1},$$

where $\kappa = \sqrt{\text{cond}(A^T A)} = \text{cond}(A)$ and $\text{cond}(A)$ denotes the 2-norm condition number of A (ratio of largest and smallest singular values of A). CG methods work acceptably well when $\text{cond}(A) = O(100)$, but converge very slowly, if at all, when $\text{cond}(A) > O(1000)$.

In our experiments, L is often much better conditioned than A, so convergence of the CG method is relatively rapid. Moreover, the total number of nonzeros in L and U is usually less than in the sparse Cholesky factorization of $A^T A$. Successful iteration with L depends on the good conditioning of L, often requiring partial pivoting in the factorization. For parallel computations, pivoting for the LU decomposition may be expensive or not available, so we may also need to further precondition the problem. In the test problems here, partial orthogonalization of L (described in the next paragraph) is an effective strategy.

If a condition estimate indicates that L is not sufficiently well conditioned for fast convergence, it is natural to use a drop tolerance on L to get L_{drop}. If R denotes the R-factor in the QR decomposition $QR = L_{drop}$, then we expect LR^{-1}

to be better conditioned than L, allowing faster convergence for the conjugate gradient iteration. A partial orthogonalization using a drop tolerance for A was proposed in [17] and developed as a multilevel algorithm by Li and Saad [18].

For the test set of 51 full rank matrices described in the next section, iteration with L was sufficient to get convergence (using at most n iterations, relative error less than 10^{-6}) in all but three cases. In all 51 cases, partial orthogonalization of L gave convergence, typically in fewer iterations.

An inexpensive estimate of the conditioning [13,15] of square triangular matrices allows the condition estimate to be used at runtime to determine whether partial orthogonalization is needed. If partial orthogonalization is chosen, the condition estimate is used to control the drop tolerance. Computing a drop tolerance for L from the estimated $cond(L(1:n,1:n))$ gives a more reliable algorithm than computing a drop tolerance for A from the estimated $cond(A(1:n,1:n))$ (which also requires an LU factorization). Numerical experiments show the robustness of the least squares algorithm combining LU factorization and possibly (depending on a runtime estimate of L conditioning) partial orthogonalization of L.

The remainder of this paper is organized as follows. Section 2 describes a set of test matrices, the LU factorization used, the observed conditioning of L, and "fill" for L and U. Section 3 presents an algorithm to precondition least squares using the LU factorization of A, and shows convergence results on the test matrices, comparing the lsqr algorithm to the lsqr algorithm preconditioned by LU. Section 4 shows a way to use partial orthogonalization of L to improve convergence and presents results of numeric experiments on the same set of matrices. Section 5 has concluding remarks.

2 LU Decomposition on Rectangular Matrices and a Set of Test Matrices

In this paper, the LU decomposition is computed using Octave [4] and Matlab. Both frameworks call UMFPACK [10]. While both partial (row) pivoting $L_p U_p = PA$ and row and column pivoting algorithms $L_q U_q = PAQ$ are implemented, row pivoting is more commonly available in parallel packages, [12,19] and thus a logical choice for test problems. L was computed with maximal element row pivoting, so that L is unit lower trapezoidal.

As a numerical test bed, we considered a set of 59 matrices adapted from the University of Florida collection [9]. We took most unsymmetric matrices larger than 500 and of maximal dimension at most 5000, small enough that we could explicitly compute the singular values, and could use a QR algorithm to compare solutions of the LLS problem. We wanted rectangular matrices with more rows than columns, so transposed if necessary. The "more column than row" matrices are mainly from linear programming problems, for which the transposed (dual) problem requires a least squares solution. So for these matrices the least squares problem has practical interest. For square matrices, we randomly selected 10 % of rows, duplicating them and appending them to the end of the matrix, randomly perturbing each nonzero entry by at most ten per cent.

The 51 full rank matrices were derived from add20, bwm2000, cage9, cavity11, cavity12, cavity13, cavity14, cavity15, Chebyshev3, circuit_2, crew1, ex24, ex26, ex27, ex28, ex29, ex31, heart1, heart3, lhr02, olm5000, orsreg_1, piston, poli, psmigr_2, psmigr_3, raefsky1, raefsky2, rajat02, rajat04, rajat05, rajat11, rajat12, rajat14, rajat19, rbsa480, rdb2048, rdb2048_noL, rdb5000, rdist1, rdist2, sherman5, shermanACa, swang1,swang2, thermal, tols4000, utm3060, viscoplastic1, wang1, and wang2. By limiting matrix size to around 5000, we were able to compute singular values and thus the condition number as the ratio of largest to smallest singular values. Eight matrices, derived from adder_dcop, extr1, Kohonen, lhr04, lns_3937, raefsky6, SciMet, and sherman3, were rank deficient, with computed condition numbers greater than $1.e16$. For the eight rank deficient matrices, `Matlab` QR could not compute least squares solutions. Dropping the rank deficient matrices from the set of 59 resulted in the set of 51 matrices used as a test collection for comparing convergence.

For the collection of 59 matrices (see Fig. 1), we computed the 2-norm condition number of L_p from $PA = L_pU_p$ as the ratio of largest to smallest singular value. By rounding larger condition numbers down to $1.e16$, and computing the multiplicative mean by averaging the logarithms of $cond(L_p)/cond(A)$, we observed that the condition numbers L_p are on average 4000 times smaller than the condition numbers of A. In Fig. 1, the better conditioning of L_p is indicated by the closer clustering of maximal and minimal singular values to 1.

The results for the $PAQ = L_qU_q$ case (row and column pivoting with multipliers bounded by 10) were similar, L_q in most cases being better conditioned than L_p. For 38 of the 59 matrices, the total number of nonzeros in L_p and U was less than the number of nonzeros in $chol(A)$. For most of the 59 matrices, L_q is sparser than L_p.

3   Iterative Algorithms with L and Experiments on Convergence

CGNE, CGNR, (see for example [25]) and `lsqr` [23] are all methods of solving the normal equations by a conjugate gradient algorithm. CGNE is appropriate for minimizing the solution x for an underdetermined system. CGNR and `lsqr` minimize $\|r\|_2^2 = \|Ax - b\|_2^2$. We found that the decline of $\|r\|_2$ is more monotonic for `lsqr` than for CGNR and that when many iterations are required, `lsqr` convergence is more likely. The `lsqrLU` algorithm is given in Algorithm 3.1 (see [23] for more details).

If L is denser than U, multiplications by $U^{-1}A$ in `lsqr` could replace multiplications by L. Since U is upper triangular, a solve $Uw = y$ efficiently replaces $w = U^{-1}y$ (and similarly for multiplications in `lsqr` by $L^T = U^{-T}A^T$). If storage is comparable, multiplication by L and L^T is likely to execute more efficiently, particularly for parallel computations.

Figure 2 compares convergence results obtained by iterating with `lsqr` on L_p and A, where L_p is the L factor obtained using the LU factorization of A with partial pivoting ($PA = L_pU_p$). The graph plots the relative error taking

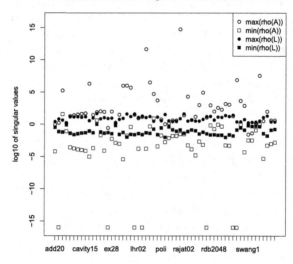

Fig. 1. Maximal and minimal singular values of L_p are closer to one than for A: L_p is better conditioned than A.

Algorithm 3.1. lsqrLU

---

function $[x, its] = $ lsqrLU$(A, b, tol, maxit)$
% A is an input matrix of m rows and n columns
% b is an input m-vector
% tol is an input scalar larger than zero, convergence tolerance.
% $maxit$ is the maximal number of iterations
% output x is an n vector to minimize $\|Ax - b\|_2$
% its is the actual number of iterations performed
    [L,U,prow] $\leftarrow$ lu$(A, $"vector"$)$;
    r $\leftarrow$ permute$(b$,prow$)$;
    $[x, its] \leftarrow$ lsqr(L,r,$tol, maxit$);
    $x \leftarrow$ U $\setminus x$;
    if $(its \geq maxit)$, "maxits exceeded, did not converge"
end

---

the exact solution as that obtained using a QR factorization algorithm. There is a maximum of $2n$ lsqr iterations (or convergence with a tolerance of $1.e$-10). The relative errors for lsqrLU are plotted at convergence with tolerance $1.e$-10 (or after at most n iterations). The set of 51 matrices was described in Sect. 2. The vector x minimizes $\|Ax - b\|_2$ where b is randomly generated, so that the typical problem is overdetermined, with nonzero residual.

    Though we allow twice as many lsqr iterations compared to lsqrLU, many more matrices converge with iterations on L than with iterations on A. For lsqrLU, only three matrices (those that had not converged after $n = $ maxit

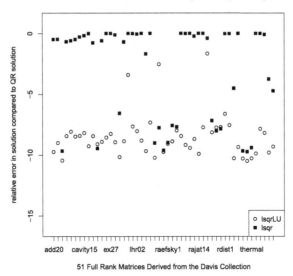

Fig. 2. Relative difference from QR solution, `lsqr` and `lsqrLU`. `lsqrLU` has better convergence because $cond(L_p) << cond(A)$

iterations) had relative error larger than $1.e\text{-}6$. For this $tol = 1.e\text{-}10$, the relative error obtained from iterating with L is smaller than that from iterating on A. For more data on this numeric experiment see the R data frame [16].

4   Partial Orthogonalization

Iteration with L from partial pivoting is often satisfactory. For larger problems, partial pivoting may not be available (e.g., for distributed memory SuperLU [20]) or may be much slower than an LU factorization without pivoting (as can happen in MUMPS [1]). Since the speed of convergence depends on the condition of L, a natural idea is to improve the conditioning of L. Similarly, Jennings and Ajiz [17] improved the conditioning of A. When the estimated condition of L_p is larger than 10^2, we construct $QR = L_{drop}$ and the `lsqr` iteration matrix is taken as with $L_p R^{-1}$. To avoid having R dense, we would like to drop many entries of L_p. L_{drop} is obtained from L_p by zeroing all column entries with absolute value less than $colmax/condest(L_p)^\alpha$, where $\alpha = 0.25$ and $colmax$ is the maximal column entry for each column. For example, if $condest(L_p) = 10^4$, the drop tolerance is $colmax/10$, and for $condest(L_p) = 10^8$, the drop tolerance is $colmax/100$. Here, L_p was obtained by partial pivoting with $colmax = 1$. Using a larger α would give a lower drop tolerance, better conditioning and faster convergence, but also more nonzero entries in R.

We denoted `lsqr` using L from LU decomposition as `lsqrLU` (see Sect. 3). When we use partial orthogonalization of L, then we obtain the `lsqrLUQR` algorithm described in Algorithm 4.1.

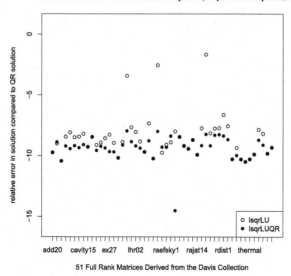

Fig. 3. Relative difference from QR solution, at convergence or max of n iterations (n is the number of matrix columns). lsqr using L and preconditioning with partial orthogonalization lsqrLUQR converges for all members of this set of matrices.

Algorithm 4.1. lsqrLUQR: LU preconditioning with partial orthogonalization

function $[x, its]$ = lsqrLUQR($A, b, tol, maxit$)
  % $A, b, tol, maxit, x, its$ same as for lsqrLU
    [L,U,prow] $\leftarrow$ lu(A,'vector');
    r $\leftarrow$ permute(b,prow);
    C_{max} $\leftarrow$ 100
    $\alpha \leftarrow .25$
    C_{est} $\leftarrow$ condest(L(1:n,1:n))
    if $C_{est} > C_{max}$,
        $\beta \leftarrow 1/C_{est}^{\alpha}$
        L_{drop} $\leftarrow$ drop(L,β)
          % if $|l_{i,j}| < \beta$, $l_{drop,i,j} \leftarrow 0$ else $l_{drop,i,j} \leftarrow l_{i,j}$
        R_{drop} $\leftarrow$ qr(L_{drop}) % Q_{drop} R_{drop} = L_{drop}, Q_{drop} not stored
        $[x, its]$ $\leftarrow$ lsqr(LR_{drop}^{-1},r,$tol,maxit$);
          % L R_{drop}^{-1} not computed,
          % lsqr solves with R_{drop}, R_{drop}^T, multiplies by L, L^T.
    else
        $[x, its]$ $\leftarrow$ lsqr(L,r,$tol,maxit$);
    end
    x $\leftarrow$ U $\setminus$ x;
    if ($its \geq maxit$),"maxits exceeded, did not converge"
  end

For Fig. 3, the lsqrLUQR data points take $C_{max} = 100$ and $\alpha = .25$, i.e. used lsqrLU for $condest(L(1:n, 1:n)) < 100$, else also use partial orthogonalization with $\alpha = .25$. As the Matlab or Octave function $condest$ [13,15] gets a fast estimate of the 1-norm $cond(L)$ (from the square submatrix $L(1:n, 1:n)$ of L), it can be used for a runtime algorithm decision.

For a given convergence tolerance, choosing C_{max} larger means that lsqrLU is more likely to be used, entailing less storage and more iterations. Decreasing α (i.e. making the drop tolerance larger) also decreases storage and increases the number of required iterations.

Figure 3 represents the convergence on L preconditioned by partial orthogonalization. The figure plots the quantity

$$\log_{10} \frac{\|x_{QR} - x_{alg}\|_2}{\|x_{QR}\|_2},$$

to compare the solutions x_{alg} computed with the algorithms lsqr with LU preconditioning lsqrLU as in Fig. 1 (open circles), and lsqrLUQR as described above (solid circles) with the solution x_{QR} obtained using a sparse QR factorization.

In Fig. 3, lsqr with L and partial orthogonalization converges for each member of this set of matrices. lsqrLUQR had relative error less than 1.e-6 for all 51 matrices. If $R = Q^T L_{drop}$ is computed, then the total storage is on average about the same as required for $chol(A^T A)$. If only L and U are needed, storage requirements are usually less than for $chol(A^T A)$. Data (R data frames) for the assertions of this section and additional plots can be downloaded from [16]. Data includes required storage, arithmetic operations, and iterations for each of five solution methods for each matrix in the data set.

For comparison, we also applied partial orthogonalization directly to the sparse matrix. Using the first n rows of A to estimate condition number and then dropping entries a_{kj} such that

$$|a_{kj}| < \frac{max_i |a_{ij}|}{condest(A)^{.25}},$$

we computed $QR = A_{drop}$ and iterated with lsqr on AR^{-1}. Similarly to the procedure with L, we estimated the condition of A by $condest(A(1:n, 1:n))$.

Then only 11 of the 51 matrices gave relative error (compared to QR solution) less than 1.e-6. A problem may be that the condition estimate based on the first n rows is less useful for A than it is for the triangular matrices L_p. An acceptable partial orthogonalization AR^{-1} would require fewer A entries to be dropped and hence increased storage.

5   Conclusions and Future Work

The lsqrLU and lsqrLUQR preconditioned versions of lsqr were reliable in our numerical experiments. lsqrLU requires on average less storage than computing the Cholesky factorization of $A^T A$. If an additional partial preconditioning step is considered, the total storage is on average about the same as for the Cholesky factorization.

Condition estimates allow a runtime decision on whether to use partial pivoting, and what drop tolerance can be used. To address larger problems, it would be interesting to investigate how to use fast statistical methods to estimate the condition number of L_p, similarly to [3].

Since LU factorization with partial pivoting is available in distributed packages for sparse direct solvers [1,12], we expect lsqrLU to be scalable to larger problems. Scalable versions of lsqrLUQR may use MIQR [18] for partial orthogonalization.

References

1. Amestoy, P.R., Guermouche, A., L'Excellent, J.-Y., Pralet, S.: MUMPS: a MUltifrontal Parallel sparse direct Solver (2014). http://mumps.enseeiht.fr/index.php?page=home
2. Baboulin, M., Giraud, L., Gratton, S., Langou, J.: Parallel tools for solving incremental dense least squares problems. Application to space geodesy. J. Algorithms Comput. Technol. **3**(1), 117–133 (2009)
3. Baboulin, M., Gratton, S., Lacroix, R., Laub, A.J.: Statistical estimates for the conditioning of linear least squares problems. In: Wyrzykowski, R., Dongarra, J., Karczewski, K., Waśniewski, J. (eds.) PPAM 2013, Part I. LNCS, vol. 8384, pp. 124–133. Springer, Heidelberg (2014)
4. Bateman, D., Adler, A.: Sparse matrix implementation in octave (2006). arxiv.org/pdf/cs/0604006.pdf
5. Benzi, M., Tuma, M.: A robust incomplete factorization preconditioner for positive definite matrices. Numer. Linear Algebra Appl. **10**, 385–400 (2003)
6. Björck, Å.: Numerical Methods for Least Squares Problems. SIAM, Philadelphia (1996)
7. Björck, A., Duff, I.S.: A direct method for the solution of sparse linear least squares problems. Linear Algebra Appl. **34**, 43–67 (1980)
8. Björck, A., Yuan, J.Y.: Preconditioners for least squares problems by LU factorization. Electron. Trans. Numer. Anal. **8**, 26–35 (1999)
9. Davis, T.: The University of Florida sparse matrix collection. http://www.cise.ufl.edu/research/sparse/matrices/
10. Davis, T.: Direct Methods for Sparse Linear Systems. SIAM, Philadelphia (2006)
11. Golub, G.H., Van Loan, C.F.: Matrix Computations, 3rd edn. The Johns Hopkins University Press, Baltimore (1996)
12. Gupta, A.: Improved symbolic and numerical factorization algorithms for unsymmetric sparse matrices. SIAM J. Matrix Anal. Appl. **24**, 529–552 (2002)
13. Hager, W.W.: Condition estimates. SIAM J. Sci. Stat. Comput. **5**, 311–316 (1984)
14. Heath, M.T.: Numerical methods for large sparse linear squares problems. SIAM J. Sci. Stat. Comput. **5**(4), 497–513 (1984)
15. Higham, N.J., Tisseur, F.: A block algorithm for matrix 1-norm estimation with an application to 1-norm pseudospectra. SIAM J. Matrix Anal. Appl. **21**, 1185–1201 (2000)
16. Howell, G., Baboulin, M.: Data and plots for lsqr, lsqrLU and lsqrLUQR (2015). http://ncsu.edu/hpc/Documets/Publications/gary_howell/contents.html

17. Jennings, A., Ajiz, M.A.: Incomplete methods for solving $A^T Ax = b$. SIAM J. Sci. Stat. Comput. **5**(4), 978–987 (1984)
18. Li, N., Saad, Y.: MIQR: a multilevel incomplete QR preconditioner for large sparse least-squares problems. SIAM J. Matrix Anal. Appl. **28**(2), 524–550 (2006)
19. Li, X.S.: An overview of SuperLU: algorithms, implementation, and user interface. ACM Trans. Math. Softw. **31**(3), 302–325 (2005)
20. Li, X.S., Demmel, J.W.: SuperLU_DIST: a scalable distributed-memory sparse direct solver for unsymmetric linear systems. ACM Trans. Math. Softw. **29**(9), 110–140 (2003)
21. Lourakis, M.: Sparse non-linear least squares optimization for geometric vision. Eur. Conf. Comput. Vis. **2**, 43–56 (2010)
22. Nocedal, J., Wright, S.J.: Numerical Optimization. Springer, New York (1999)
23. Paige, C., Saunders, M.: An algorithm for sparse linear equations and sparse least squares. ACM Trans. Math. Softw. **8**(1), 43–71 (1982)
24. Peters, G., Wilkinson, J.H.: The least squares problem and pseudo-inverses. Comput. J. **13**, 309–316 (1970)
25. Saad, Y.: Iterative Methods for Sparse Linear Systems, 2nd edn. SIAM, Philadelphia (2000)
26. Schenk, O., Gärtner, K.: PARDISO User Guide (2014). http://www.pardiso-project.org/manual/manual.pdf

Experimental Optimization of Parallel 3D Overlapping Domain Decomposition Schemes

Sofia Guzzetti$^{(\boxtimes)}$, Alessandro Veneziani, and Vaidy Sunderam

Department of Mathematics and Computer Science,
Emory University, Atlanta, GA, USA
{sofia.guzzetti,avenez2,vss}@emory.edu

Abstract. Overlapping domain decomposition is a possible convenient technique for solving complex problems described by Partial Differential Equations in a parallel framework. The performance of this approach strongly depends on the size and the position of the overlap since the overlapping has a positive impact on the number of iterations required by the numerical scheme and the relatively flexible and judicious choice of the interface may lead to a reduction of the communication time. In this paper we test the overlapping domain decomposition method on the finite element discretization of a diffusion reaction problem in both idealized and real 3D geometries. Results confirm that the detection of the optimal overlapping in real cases is not trivial but has the potential to significantly reduce the computational costs of the entire solution process.

Keywords: Applied computing in medicine · Computational fluid dynamics · Parallel overlapping domain decomposition · Performance analysis

1 Introduction

Advanced applications in several fields of engineering based on partial differential equations (PDEs) challenge high performance and parallel computing to reduce computational costs within timelines of practical relevance. Among others, we mention applications related to computational fluid dynamics for vascular diseases and their applications in clinical trials (Computer Aided Clinical Trials), where complex simulations need to be performed on a large number of patients and possibly on computational facilities remotely located one to the other. In this context, *domain decomposition techniques* (DD) provide an important framework to associate mathematical formalism to the parallel solution of a complex PDEs system - see e.g. [7,13]. With this technique, the problem over a region of interest Ω is decomposed in subproblems to be iteratively solved by single processors or clusters up to the fulfilment of a convergence criterion stating that the solution found is equivalent to the one of the unsplit problem. Several types

Research supported in part by National Science Foundation grant OCI-1124418.

R. Wyrzykowski et al. (Eds.): PPAM 2015, Part I, LNCS 9573, pp. 138–149, 2016.
DOI: 10.1007/978-3-319-32149-3_14

of DD approaches have been investigated from the viewpoint of mathematical and numerical performances. Their relevance is increasing with the adoption of heterogeneous computational resources and the detection of optimal splitting is of paramount importance for the global performance of the parallel solver. In fact, splitting is important not only from the mathematical point of view (for the numerical efficiency of the iterative-by-subdomain solver), but also from the computational viewpoint, since the subdivision drives the communications among the different processors and therefore it has a major impact on the global computational time.

In [11] the author investigates the performance of a non-overlapping DD solver for problems in fluid dynamics of blood in cerebral arteries when using heterogeneous architectures. *Non overlapping* means that the different subdomains do not share portions of Ω. The information among subdomains is exchanged iteratively at the interfaces. In this work, we focus on a different approach, based on the so called *overlapping* DD. These methods were introduced by Schwarz as a tool for solving analytically equations on domains that could be decomposed into geometrical primitives where it was possible to find an analytical solution [2]. Generally speaking, in the context of numerical iterative schemes, the solution of overlapping methods requires fewer iterations (in the extreme case of 100 % overlapping it requires one) than a non overlapping one (for a 1D exemplification with convergence analysis see [5]) and it has more flexibility in the selection of the interface conditions - as these conditions in non overlapping methods must guarantee consistence between split and unsplit problems. On the other hand, with overlapping DD the PDE problem is solved multiple times on the overlapping regions, with a computational overhead due to the duplication. The interplay of (i) additional numerical costs due to the overlap, (ii) efficiency advantages induced in the specific iterative methods and (iii) flexibility of the selection of domain interfaces (and the associated conditions) for the communication time, is not trivial in problems of practical interest. Numerical testing in 2D for second order purely diffusive problems (Poisson equation) has been carried out in [4]. Here we want to perform a similar extensive analysis of the performances of overlapping DD in more general problems (reaction-diffusion) in both idealized and realistic 3D domains. The specific goal is to verify the performances of these schemes in problems where the geometry of the domain of interest plays a major role. For our specific interest in Computer Aided Clinical Trials for cardiovascular diseases, we aim at simulating vascular pipe-like geometries such cerebral arteries - ultimately for running simulations on large data sets of patients. Specifically, the realistic geometries here are retrieved from a data base of cerebral arteries associated with the so-called Aneurisk project [1].

Our preliminary results conducted in a virtual parallel setting with two subdomains show that an appropriate selection of overlap and interfaces introduces significant computational benefits, consistent with the finding of [4]. In fact, for a 3D cylindrical geometry, we found an optimal overlapping of around 20 % of the entire domain, which is consistent with the results of that paper for rectangular domains. Simulations in nontrivial geometries point out however that the morphology of the domain plays a major role and that the definition of a general rule for the optimal overlapping deserves further investigations. The outline of the paper is as follows: we recall basics of Overlapping DD in Sect. 2. Detailed description of the software used for numerical tests is provided in Sect. 3. Section 4 describes the test cases solved, while results are commented in Sect. 5. Conclusions and perspectives are drawn in Sect. 6.

2   Overlapping DD in a Nutshell

Let us consider the following differential *Diffusion-Reaction* (DR) problem

$$-\sum_{i=1}^{3} \frac{\partial}{\partial x_i} \left(\mu \frac{\partial u}{\partial x_i} \right) + \sigma u = f \tag{1}$$

for $(x_1, x_2, x_3) \in \Omega \subset \mathbb{R}^3$ with $\mu > 0$ and σ coefficients for simplicity assumed to be constant. The forcing term f is a given function of x_1, x_2, x_3. Hereafter it will be set to 0. To the equation, we associate the boundary conditions $u(\Gamma_D) = g(x_1, x_2, x_3)$, $\frac{\partial u}{\partial n}(\Gamma_N) = 0$, where Γ_D and Γ_N are two disjoint portions of the boundary of Ω such that $\Gamma_D \cup \Gamma_N = \partial \Omega$. Here the unknown u may represent the density of a species in a region where it diffuses with diffusivity μ and it undergoes to a chemical reaction with rate σ. As for our application, this species could be either Oxygen or a drug injected in the system. To take advantage of domain decomposition, we first split the domain Ω into two overlapping subdomains Ω_1 and Ω_2, such that $\Omega_1 \cap \Omega_2 = \Omega_o$ and $\Omega_1 \cup \Omega_2 = \Omega$. Let us denote by Γ_j the interfaces between the two subdomains $(j = 1, 2)$, that is the portion of the boundary of Ω_j that is not also boundary of Ω, in short $\Gamma_j \equiv \partial \Omega_j \setminus (\partial \Omega_j \cap \partial \Omega)$. The solution of the problem in each subdomain will be denoted by $u_j(x_1, x_2, x_3)$. We reformulate the original problem in an iterative fashion. Given an initial guess $u_j^{(0)}$ (typically $= 0$), we solve on each subdomain for $k = 1, 2, \ldots$

$$-\sum_{i=1}^{3} \frac{\partial}{\partial x_i} \left(\mu \frac{\partial u_j^{(k)}}{\partial x_i} \right) + \sigma u_j^{(k)} = f \qquad \text{in } \Omega_j, j = 1, 2 \tag{2}$$

with boundary conditions

$$u_j^{(k)}(\Gamma_D \cap \partial \Omega_j) = g(x_1, x_2, x_3), \quad \frac{\partial u}{\partial n}(\Gamma_N \cap \partial \Omega_j) = 0, \quad u_j^{(k)}(\Gamma_j) = u_{\hat{j}}^{(k-1)}(\Gamma_j), \tag{3}$$

(where $\hat{j} = 2$ for $j = 1$ and $\hat{j} = 1$ for $j = 2$) up to the fulfillment of an appropriate convergence condition. In our case this condition checks that the solution in the overlapping region is not changing significantly along the iterations (as we will detail later on). Notice that at each iteration we solve two independent problems in each subdomain, while the communication by subdomain occurs in the latter of boundary conditions (3).

The convergence of the iterative scheme depends on the size of the overlapping region. Should the overlapping be 100 % of Ω (clearly a case of no interest for the parallel computing), convergence is trivially guaranteed as at the first iteration (2)–(3) we are solving the unsplit problem. On the other hand, if the overlapping reduces to a volume-zero region, convergence is not guaranteed, as in general the juxtaposition of the two problems does not coincide with the original problem (this occurs only if the interface conditions are chosen properly) - see e.g. [7].

The one presented here is the so called *additive* formulation of the overlapping DD method, where the two subdomain problems can be solved simultaneously - as opposed to the *multiplicative* version, where one subdomain can be solved only when the problem on the other subdomain is completed. In the multiplicative formulation a faster convergence is guaranteed in terms of number of iterations (about one half of the additive scheme), but the advantage of the parallel setting is limited by the sequential structure of the algorithm. From now on, we refer only to the additive algorithm. A simplified analysis of the convergence rate of the different methods in 1D problems can be found in [5].

The numerical solution of the differential subdomain problems (2)–(3) can be performed for instance by the so called Finite Element Method. This is done by introducing a tessellation or *mesh* of each subdomain in a (generally large) number of tetrahedra where the solution is locally assumed to be polynomial. The number of the tetrahedra on each mesh is denoted by N. The more tetrahedra we have, the more accurate the numerical approximation is, but the more computationally expensive it will be since we resort to solving a larger system. For the same reason, the larger the computational domain, the higher the computational costs. For this reason, to optimize the parallel performance by minimizing the latency time it is crucial that subdomain splitting guarantees a uniform load balancing (so that the number of elements is similar in each subdomain). However, the selection of the interfaces Γ_j should also minimize the number of degrees of freedom where information is exchanged by subdomains, so as to minimize the communication time. The latter may become a bottleneck in particular for remotely located computing resources. In this respect, overlapping DD guarantees a better flexibility in determining the interfaces. Here we want to investigate the optimal selection as the result of the trade-off between the computational cost of each subproblem and the reduction of the communications between processors. It is worth noting that the problem of combining the optimization of the load and communication volume was earlier investigated in [3] in a non overlapping context for partitioning sparse matrices.

3   Software Tools

We list hereafter the steps needed to obtain the numerical solution once a mathematical representation of Ω is available. (a) The tessellation or mesh is performed by a public domain software called NetGen. A mesh for complex geometries like the ones we consider in this paper in general cannot rely on particular assumptions on the shape of Ω and it is called *unstructured*. A text file storing the coordinates of the vertexes of the reticulation and their connectivity is the result of this step. The maximal size of elements is denoted by h, which is generally inversely proportional to N. (b) The mesh is split by subdomains using a specific library based on the package ParMETIS. This is a MPI-based parallel library that implements a variety of algorithms for partitioning large unstructured meshes. ParMETIS extends the functionality of the serial package METIS [6] for large-scale parallel numerical simulations. In particular it aims at (i) reducing the communication time by minimizing the number of edges between non-overlapping subdomains; (ii) balancing the number of elements for subdomain so to create a nearly uniform load distribution or to optimize the communication volume, according to the different possible settings. The original package does not manage overlapping. For this reason, ad hoc MATLAB scripts were prepared to extend two pre-computed non-overlapping partitions symmetrically with respect to the original splitting, yielding a symmetric region of intersection with respect to the original nonoverlapping subdivision. The extension of the subdomains is performed by subsequently adding a desired number of layers of elements to the interface. At the end of this additional step, we have two meshes and the map of the nodes of each interface Γ_j in the corresponding domain $\Omega_{\hat{j}}$. In the follow-up of the present work, we plan to perform this step with the *replicated bipartitioning* approach introduced in the multilevel tool PaToH [10]. At this preliminary stage we followed a manual definition of the overlapping region to test several options in particular in nontrivial geometries, where the number of degrees of freedom at the interface may change significantly, as in cerebral aneurysms. (c) On each subdomain, after a proper labeling of the vertexes to identify the portions of the boundary/interface nodes associated with different boundary conditions (3), we solve the problem (2) by the linear Finite Elements, so that on each element tetrahedron the solution is assumed to be linear-in-space. Equations (2)–(3) can be then conveniently reformulated as a linear algebraic system. The assembly of these matrices has been performed through the MATLAB package `fast_fem_assembly` [8]. The system is then solved with a built-in MATLAB function that implements the GMRes iterative method [9].

4   Benchmarking

In our experiments to measure computational time, we consider only the iterative-by-subdomain solution. In fact, meshing, partitioning and matrix assembly are not included in this analysis, since they are off-line costs that do not depend on the specific solution procedure. The time $T_{it}^{(k)}$ of each iteration (k) is computed as the maximum of the two parallel subdomain solution times $T_j^{(k)}$,

$$T_{it}^{(k)} = \max_{j=1,2} T_j^{(k)}.$$

The single processor time is given by the time for solving the linear system added by the communication time to read from the other processor the last of

(3), $T_j^{(k)} = T_{j,sol}^{(k)} + T_{j,com}^{(k)}$. For this particular problem, we speculate that the computational cost per iteration is constant (denoted by $\overline{T_{sol} + T_{com}}$), so we get

$$T = \sum_{k=1}^{N_{it}} T_{it}^{(k)} \approx N_{it}(\overline{T_{sol} + T_{com}}).$$

If we denote by p the percentage of overlap in the domain splitting (i.e. the ratio of the volume of the intersection of the domains to the total volume of the geometry), we expect in general that N_{it} *decreases* with p [5], T_{sol} increases with p while T_{com} depends on the position of the interfaces (precisely on the number of vertexes of the mesh on the interface), so it changes with p depending on where the interfaces are placed. The flexibility granted by overlapping DD in placing the interfaces comes to play here. Therefore, p has a major impact on the solver performances depending on the different geometries. We test this hypothesis in the next Section. To this aim, we consider both idealized and real geometries from computational hemodynamics (described hereafter). It is worth noting that the total cost is a function of the mesh size N as $T_j^{(k)} = T_j^{(k)}(N)$. In this case, both factors N_{it} and $\overline{T_{sol} + T_{com}}$ get larger with h. This is a price to pay to the improvement of the accuracy of the approximated solution achieved in this way. The sensitivity of our results to N will be tested too.

Idealized Geometries. *Test 1: Cylinder.* We consider a cylinder of length $L = 6$ cm and radius $R = 0.5$ cm. We use five meshes with different values of N, for each of the sizes of the overlap. The numerical solution shown in Fig. 1a, representing a longitudinal cut of the 3D solution, displays the overlapping sub-domains (bottom panels) and the associated solution matching at the interfaces.

Test 2: Idealized Aneurysm. In this case (shown in Fig. 1b) we consider an idealized representation of a cerebral aneurysm (a pathology of the cerebral circulation) where a curved cylinder (namely a torus with radius $r = 2$ cm) representing an artery is merged with a sphere of radius $R = 0.5$ cm, representing the abnormal sac of the aneurysm. This test intends to emphasize the role of communication time. In fact a splitting with an interface intersecting the sac has more vertices than with interfaces involving only the artery. Overlapping DD allows to manage the location of the interfaces so to avoid many vertices on the interface (and correspondingly lowering the communication time) yet preserving workload balance between the subdomains.

Real Geometries. A web repository of cerebral aneurysm geometries is freely available [1]. We selected two cases with significant features for the purpose of the paper. As we pointed out, computational hemodynamics is our ultimate application of interest, however, investigations performed here have more general applicability. In particular, any engineering problem of internal fluid dynamics (typically occurring in pipe-like domains) is informed by this analysis.

Test 3: Patient 1. This geometry is approximately a cylindrical shape with a terminal spherical aneurysm located at the bifurcation (see Fig. 2a). In spite

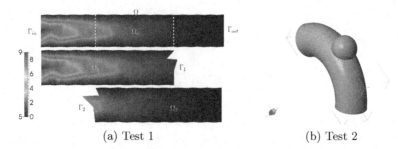

(a) Test 1   (b) Test 2

Fig. 1. (a) Overlapping DD solution in a cylinder (longitudinal section): in the aggregated domain Ω (top) and by subdomains (bottom). (b) Idealized geometry of an aneurysm, made of a sphere merging with a portion of a torus.

of the simplicity, the collateral ramifications make the *a priori* prediction of performances quite problematic.

Test 4: Patient 2. This geometry is more complicated for the curvature and the torsion of the main vessel (see Fig. 2b). The size of the aneurysm is smaller and a major bifurcation occurs along the main vessel (the carotid siphon).

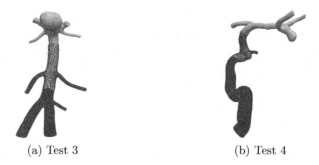

(a) Test 3   (b) Test 4

Fig. 2. Geometries of patient-specific cerebral aneurysms from the AneuriskWeb Repository, used for Test 3 and 4 respectively.

5   Results

Test 1. Figure 3a shows the parallel running time as a function of p. The varying dependence of number of iterations and cost per iteration on p results in a convex behavior of the computational time. This behavior is expected, since for small p the high number of iterations dominates the cost, while beyond a certain value it does not decrease any longer (Fig. 3a), while the cost per processor increases. The size of the meshes for all the tests is $N \simeq 3,000$ (very coarse), $N \simeq 6,000$ (coarse), $N \simeq 12,000$ (medium), $N \simeq 24,000$ (fine), $N \simeq 300,000$ (very fine). In addition, the value of the optimal p changes with N (Fig. 3a). For the very coarse

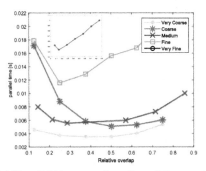 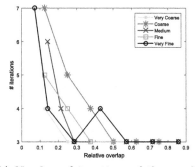

(a) Parallel time as a function of the overlap. (b) Number of iterations of the parallel solver.

Fig. 3. Parallel time performed by a diffusion-reaction solver as a function of the relative overlap for five levels of refinement of the mesh on a cylinder (a). Corresponding number of iterations for very coarse ($\bullet$), coarse ($*$), medium ($\times$), fine ($\square$) and very fine ($\bigcirc$) meshes (b).

and coarse meshes, the optimal overlap happens between 37.5 % and 50 %. The medium and fine meshes perform better with 20–37.5 % overlap, while the very fine grid reaches the minimum at 14 %. The optimal overlap shifts leftward as the mesh size decreases. In fact, for coarse meshes, the cost-per-iteration is small and a large overlap is totally beneficial. For fine meshes, a larger overlap results in a significant burden for the solution on the subdomains. Note that the number of nodes that lie on the inner interfaces - directly proportional to the communication time - is nearly constant with p since we have a simplified geometry with a constant numbers of vertexes along the transversal sections. However, it gets larger when N increases. Notice that the optimal range for the medium-refined meshes agrees with the results in [4] on a 2D rectangular domain.

Test 2. The numerical results in Fig. 4 show that medium and fine grids perform better with an overlapping around 45 %. Interestingly, the curve related to the very fine mesh features a minimum at a fraction of overlap of ~ 30 %. This corresponds to the minimal overlap including the sphere entirely, so that the interfaces do not cross the bulb. In this way, the number of vertices is contained (as opposed to the case with an interface through the aneurysm) with a benefit for the communication time. This confirms that the overlap allows the minimization of communication time thanks to a flexible positioning of the interfaces. For coarser meshes, the extra computational cost associated with an increase of p does not have a great impact on the cost of an iteration and a wider intersection is allowed. In addition, Fig. 4b shows that the minimum fraction of volume at which the minimum number of iterations is attained grows when N increases. In particular, the very fine mesh (marked by $\bigcirc$ symbols in the figure) requires the greatest value of p to minimize N_{it} to 3 iterations, as expected from the theory of DD.

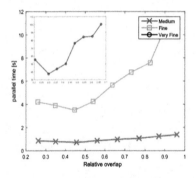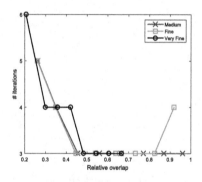

Fig. 4. Left: Parallel time performed as a function of p for three levels of refinement of the mesh on an idealized aneurysm (a). Right: Corresponding number of iterations for medium ($\times$), fine ($\square$) and very fine ($\bigcirc$) meshes (b).

Test 3. In the real case, the complexity of the geometry makes the analysis of the results more difficult. We comment only the case we ran on the very fine mesh, since similar considerations hold for the other cases, but this is the case of practical interest. For this geometry the optimal value of overlap is localized around 7 % (Fig. 5). This happens when the number of subdomain iterations is minimized to the value of 3 and a further increment of the overlap (in the range considered here) does not bring any further reduction of this number. In particular Table 1 points out that the optimal partitioning occurs when a large number of interface vertices is corresponded by a well balanced workload of the subdomains. This partition is located at the branching of collateral vessels of the artery. A shift of the interfaces would include the entire set of those vessels in one of the two subdomains, with an additional burden to the solution in that subdomain. In fact, in this case the complexity of the problem weights the cost per subdomain more than the communication time.

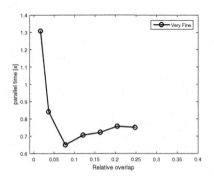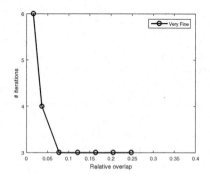

Fig. 5. Parallel time performed by a diffusion-reaction solver as a function of the relative overlap (left) for a very fine mesh of Patient 1. Corresponding number of iterations (right).

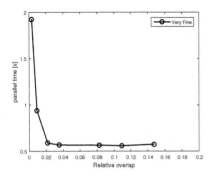 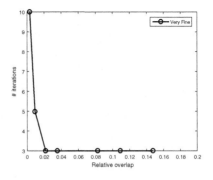

Fig. 6. Same of the previous Figure for Patient 2.

Table 1. Number of nodes of each partition (DoF1, DoF0), total number of nodes on the interfaces (InterNodes) and parallel times (in s) for different levels of overlap on a very fine mesh for Case C0095.

% overlap	DoF1	DoF0	InterNodes	Parallel Time [s]
1.60	25,469	28,017	446	1.30770
3.63	25,906	28,273	549	0.84232
7.74	27,420	28,810	869	0.65015
12.13	29,105	19,194	601	0.70613
16.34	30,320	29,618	510	0.72133
20.51	31,201	30,127	534	0.75693
24.73	32,132	30,620	489	0.75093

Test 4. For Patient 2 we find similar results to Patient 1 (Fig. 6). The optimal partition penalizes the size of the local problem more than the communication time induced by the number of vertices at the interface. The minimal time corresponds therefore to interfaces cutting collateral vessels so to balance the workload between subdomains. The tortuosity of the vessel and the size of the aneurysm do not play any significant role in this respect. We speculate this to be related to the particular problem we are solving, where the species undergoes to "isotropic" dynamics (diffusion and reaction) as opposed to more directional phenomena like the presence of drift. This will be subject of further investigations.

6 Conclusions and Perspectives

This work presents preliminary results on finding the optimal partition in an overlapping DD approach to minimize the computational time a for parallel solution. Different aspects have competitive dynamics resulting in a nontrivial optimization. As a matter of fact, the dependence of the number of iterations on the iterative-by-subdomain method generally decreases with the overlap, while the cost of the solution on each subdomain increases. The communication time depends on the location of the interface. Overlapping DD has some advantages vs nonoverlapping approaches, in particular in the positioning of the interfaces. The definition of the optimal partitioning depends on many aspects related to both the mathematical algorithm and the computer architecture. The nature of the PDE problem to be solved has generally an impact too. For this reason it is not easy to draw general conclusions. However, our results in realistic geometries point out the efficacy of an appropriate selection of overlapping to reduce costs in a parallel computing setting. In general, a small amount of overlap results in a good trade-off of all the competitive mechanisms affecting the total computational time. This is just the first of several experiments we intend to pursue, in view of massive use of scientific computing in Computer Aided Clinical Trials. There are several limitations in the present study. (i) First of all, we are solving a simplified diffusion-reaction linear problem and we will consider later incompressible fluid dynamics (Navier-Stokes equations) as mathematical benchmark (as done in [12] for nonoverlapping partitions). (ii) In addition, we limited our analysis to just two domains for simplicity, with a manual setting of the overlapping partition. We intend to move to a multidomain analysis with automatic tools for partitioning [10]. This most likely will require the introduction of a *coarse solver* to control the number of iterations [2]. (iii) We used Dirichlet interface conditions (3), however we plan to perform a similar analysis with optimized conditions, according to the *Optimized Schwarz approach*, that are supposed to impact the number of iterations needed for convergence. Related to this, we will consider flexible strategies, when the accuracy pursued to solve each subdomain is reduced to accelerate the procedure, possibly performing more iterations (see [4]) but with an overall reduction of the computational cost. (iv) Finally, we will perform this in truly parallel settings with heterogeneous and distributed architectures, to emphasize the importance of communication time reduction in real scenarios. In spite of these limitations, the tests presented here on nontrivial geometries - sufficiently diverse for our applications - point out the interesting potential of overlapping DD methods in problems of real interests and justify further investigations where the different limitations will be removed.

References

1. Antiga, L., Passerini, T., Piccinelli, M., Veneziani, A.: Aneurisk web. http://ecm2. mathcs.emory.edu/aneuriskweb/index
2. Cai, X.: Overlapping domain decomposition methods. In: Langtangen, H.P., Tveito, A. (eds.) Advanced Topics in Computational Partial Differential Equations, pp. 57–95. Springer, Heidelberg (2003)
3. Catalyurek, U.V., Aykanat, C.: Hypergraph-partitioning-based decomposition for parallel sparse-matrix vector multiplication. IEEE Trans. Parallel Distrib. Syst. **10**(7), 673–693 (1999)
4. Darjany, D., Englert, B., Kim, E.H.: Implementing overlapping domain decomposition methods on a virtual parallel machine. In: Min, G., Di Martino, B., Yang, L.T., Guo, M., Rünger, G. (eds.) ISPA Workshops 2006. LNCS, vol. 4331, pp. 717–727. Springer, Heidelberg (2006)

5. Formaggia, L., Saleri, F., Veneziani, A.: Solving Numerical PDEs: Problems, Applications, Exercises: Problems, Applications, Exercises. Springer Science & Business Media, New York (2012)
6. Lab, K.: Metis. http://glaros.dtc.umn.edu/gkhome/views/metis
7. Quarteroni, A., Valli, A.: Domain decomposition methods for partial differential equations. Technical report. Oxford University Press (1999)
8. Rahman, T., Valdman, J.: Fast MATLAB assembly of FEM matrices in 2D and 3D: nodal elements. Appl. Math. Comput. **219**(13), 7151–7158 (2013)
9. Saad, Y.: Iterative Methods for Sparse Linear Systems. Siam, Philadelphia (2003)
10. Selvitopi, R.O., Turk, A., Aykanat, C.: Replicated partitioning for undirected hypergraphs. J. Parallel Distrib. Comput. **72**(4), 547–563 (2012)
11. Slawinski, J.: Adaptive Approaches to Utility Computing for Scientific Applications. Ph.D. thesis, Emory University (2014)
12. Slawinski, J., Passerini, T., Villa, U., Veneziani, A., Sunderam, V.: Experiences with target-platform heterogeneity in clouds, grids, and on-premises resources. In: 2012 IEEE 26th International Parallel and Distributed Processing Symposium Workshops & PhD Forum (IPDPSW), pp. 41–52 (2012)
13. Toselli, A., Widlund, O.: Domain Decomposition Methods: Algorithms and Theory. Springer Series in Computational Mathematics, vol. 34. Springer, Heidelberg (2005)

Parallel Implementation of the FETI DDM Constraint Matrix on Top of PETSc for the PermonFLLOP Package

Alena Vasatova[1,2], Martin Cermak[1], and Vaclav Hapla[1,2(✉)]

[1] IT4Innovations National Supercomputing Center, Ostrava, Czech Republic
[2] Department of Applied Mathematics, FEI, VSB, Technical University of Ostrava, Ostrava, Czech Republic
vaclav.hapla@vsb.cz

Abstract. This paper deals with implementation of the FETI non-overlapping domain decomposition method within our new software toolbox PERMON, combining quadratic programming algorithms and domain decomposition methods. It is built on top of the PETSc framework for numerical computations. Particularly, we focus on parallel implementation of the matrix which manages connectivity between subdomains within the FETI method. We present a basic idea of our approach based on processing local and global numberings of the degrees of freedom on subdomain interfaces.

Keywords: PERMON · PermonFLLOP · PETSc · FETI · DDM · Constraint matrix · Star forest

1 Introduction

Many real world problems may be described by partial differential equations (PDEs). To be solved with computers, they have to be discretized, e.g. with the popular Finite Element Method (FEM). This discretization leads to large sparse linear systems of equations.

Huge problems not solvable on usual personal computers can be solved only in parallel on supercomputers. Suitable numerical methods are needed for that such as domain decomposition methods (DDM) or multigrid. DDM solve the original problem by splitting it into smaller subdomain problems that are independent, allowing natural parallelization. Finite Element Tearing and Interconnecting (FETI) methods [2,4,5,12] form a successful subclass of DDM. They belong to non-overlapping methods and combine iterative and direct solvers [11]. The FETI methods allow highly accurate computations scaling up to tens of thousands of processors and billions of unknowns.

In FETI, subdomain stiffness matrices are assembled, factorized and solved independently whereas continuity of the solution across subdomain interfaces is enforced by separate linear equality constraints. In our specific FETI subclass

© Springer International Publishing Switzerland 2016
R. Wyrzykowski et al. (Eds.): PPAM 2015, Part I, LNCS 9573, pp. 150–159, 2016.
DOI: 10.1007/978-3-319-32149-3_15

called Total FETI (TFETI) [2], Dirichlet boundary conditions are enforced in the same way, too. The goal of this paper is to focus on the parallel implementation of the FETI constraint matrix. This assembly process is not described in detail elsewhere. Our approach does not need information about neighbouring subdomains. It just needs local and global numberings of the degrees of freedom (DOFs) on the subdomain interfaces. This approach was implemented as a part of the PermonFLLOP package, being part of our set of libraries called PERMON (Parallel, Efficient, Robust, Modular, Object-oriented, Numerical) [10].

The rest of the paper reads as follows. The TFETI method is briefly described in Sects. 2 and 4 introduces the PERMON toolbox, and Sect. 5 deals with the implementation of the gluing itself. Finally, Sect. 6 shows the performance of the proposed approach.

2   TFETI Overview

FETI-1 [4,6] is a non-overlapping domain decomposition method which is based on decomposing the original spatial domain into non-overlapping subdomains. They are "glued together" by Lagrange multipliers which have to satisfy certain equality constraints which will be discussed later. The original FETI-1 method assumes that the Dirichlet conditions are embedded in the usual way into the linear system arising from the FEM discretization. This means physically that subdomains whose interfaces intersect the Dirichlet boundary are fixed while others are kept floating; in the linear algebra speech, the corresponding subdomain stiffness matrices are non-singular and singular, respectively.

The basic idea of the Total-FETI (TFETI) method [2,11,13,15] is to keep all subdomains floating and enforce the Dirichlet boundary conditions by means of the constraint matrix and Lagrange multipliers, similarly to the gluing conditions. This simplifies implementation of the stiffness matrix pseudoinverse. The key point is that kernels of subdomain stiffness matrices are known a priori, have the same dimension and can be formed without any computation from the mesh data. Furthermore, each local stiffness matrix can be regularized cheaply, and the inverse of the resulting nonsingular matrix is at the same time a pseudoinverse of the original singular one [3].

Let N_p, N_d, N_n, N_c denote the primal dimension, the dual dimension, the null space dimension and the number of processes available for our computation. The primal dimension means the number of all DOFs including those resulting from duplication of interface DOFs due to the non-overlapping domain decomposition. The dual dimension is the total number of constraints. Let us consider a partitioning of the global domain Ω into N_S subdomains $\Omega^s, s = 1, \ldots, N_S$ ($N_S \geq N_c$). To each subdomain Ω^s corresponds the subdomain stiffness matrix $\mathbf{K}^s$, the subdomain nodal load vector $\mathbf{f}^s$, the matrix $\mathbf{R}^s$ whose columns span the nullspace (kernel) of $\mathbf{K}^s$, and the signed boolean matrix $\mathbf{B}^s$ defining connectivity of the subdomain s with all its neighbouring subdomains. In case of TFETI,

$\mathbf{B}^s$ also enforces Dirichlet boundary conditions. This special matrix is described more in detail in Sect. 5. The local objects $\mathbf{K}^s$, $\mathbf{f}^s$, $\mathbf{R}^s$ and $\mathbf{B}^s$ constitute global objects

$$\mathbf{K} = \mathrm{diag}(\mathbf{K}^1, \ldots, \mathbf{K}^{N_S}) \in \mathbb{R}^{N_p \times N_p},$$
$$\mathbf{R} = \mathrm{diag}(\mathbf{R}^1, \ldots, \mathbf{R}^{N_S}) \in \mathbb{R}^{N_p \times N_n},$$
$$\mathbf{B} = [\mathbf{B}^1, \ldots, \mathbf{B}^{N_S}] \in \mathbb{R}^{N_d \times N_p},$$
$$\mathbf{f} = [(\mathbf{f}^1)^T, \ldots, (\mathbf{f}^{N_S})^T]^T \in \mathbb{R}^{N_p \times 1},$$

where diag means a block-diagonal matrix consisting of the diagonal blocks in parentheses. Note that columns of $\mathbf{R}$ span the kernel of $\mathbf{K}$ just as $\mathbf{R}^s$ do for $\mathbf{K}^s$.

Let us apply the convex QP duality theory to the primal decomposed problem

$$\min \frac{1}{2} \mathbf{u}^T \mathbf{K} \mathbf{u} - \mathbf{u}^T \mathbf{f} \quad \text{s.t.} \quad \mathbf{B} \mathbf{u} = \mathbf{o}, \tag{1}$$

and let us establish the following notation

$$\mathbf{F} = \mathbf{B} \mathbf{K}^\dagger \mathbf{B}^T, \quad \mathbf{G} = \mathbf{R}^T \mathbf{B}^T, \quad \mathbf{d} = \mathbf{B} \mathbf{K}^\dagger \mathbf{f}, \quad \mathbf{e} = \mathbf{R}^T \mathbf{f},$$

where $\mathbf{K}^\dagger$ denotes a pseudoinverse of $\mathbf{K}$, satisfying $\mathbf{K} \mathbf{K}^\dagger \mathbf{K} = \mathbf{K}$. We obtain a new QP

$$\min \frac{1}{2} \boldsymbol{\lambda}^T \mathbf{F} \boldsymbol{\lambda} - \boldsymbol{\lambda}^T \mathbf{d} \quad \text{s.t.} \quad \mathbf{G} \boldsymbol{\lambda} = \mathbf{e}. \tag{2}$$

In order to solve the problem (2) efficiently, several further reformulations are carried out, not mentioned here for sake of space limitations, see [2,3,9].

3   Constraint Matrix Structure

The FETI method is a non–overlapping DDM and thus the submeshes of the global mesh are handled completely separately; there are no overlapping cells, no ghost layer. The DOFs on submesh interfaces are duplicated into each intersecting submesh, i.e. each submesh is "complete" and "self-contained". It is possible to do that just with subdomain surface meshes. Volume meshing and subsequent FEM matrix assembly can then be done completely separately for each submesh. Let us introduce several DOF numberings, useful for further discussions:

1. *global* – a unique global DOF numbering before the DOF duplication connected with the non–overlapping domain decomposition,
2. *local* – after the decomposition and DOF duplication, DOFs of each subdomain are numbered starting from 0 independently of other subdomains,
3. *interface* – similar to local, but only the DOFs residing on subdomain interfaces are numbered.

The matrix $\mathbf{B}$ mentioned in Sect. 2 can be split into two parts, the first implementing the Dirichlet conditions and the second implementing gluing between subdomains,

$$\mathbf{B} = \begin{bmatrix} \mathbf{B}_d \\ \mathbf{B}_g \end{bmatrix} = \begin{bmatrix} \mathbf{B}_d^1 & \cdots & \mathbf{B}_d^{N_S} \\ \mathbf{B}_g^1 & \cdots & \mathbf{B}_g^{N_S} \end{bmatrix}.$$

To express connectivity between subdomains, we use operators described in [7]. First one is the "local trace" Boolean operator $\mathbf{T}^s$ which selects from all DOFs of subdomain Ω^s only those that intersect with the interface. By contrast, $(\mathbf{T}^s)^T$ prolongs the interface data to the whole subdomain, setting values corresponding to the internal DOFs to zero.

Data lying on a subdomain interface have to be exchanged with its neighbouring subdomains, leading to the global "assembly" operator $\mathbf{A}$, constructed in the following way. Let two subdomains Ω^i and Ω^j share a common interface, $i < j$. Let the subdomain Ω^i own a DOF d_i and the subdomain Ω^j own a DOF d_j, both in the *interface* numbering, while d_i and d_j represent the same DOF d in the *global* numbering. Then a row r is added into $\mathbf{A}$ with all zeros except 1 at position d_i in the block $\mathbf{A}^i$ and -1 at position d_j in $\mathbf{A}^j$.

The matrix $\mathbf{B}_g^s$ can then be represented as a "composed operator", $\mathbf{B}_g^s = \mathbf{A}^s \mathbf{T}^s$. Implementation of $\mathbf{B}_d^s$ can be done in a similar way except only one side of interface is taken into account.

4   PERMON Toolbox

PERMON [10] is our newly emerging set of tools which combines advanced quadratic programming algorithms and domain decomposition methods. It incorporates our own codes, and makes use of renowned open source libraries, especially PETSc [1,14]. We focus so far mainly on linear elasticity and contact problems, but investigate also applications in medical imaging, ice-sheet melting modelling, statistical methods and others.

The core of PERMON depends on PETSc and uses its coding style. It consists of the PermonQP and PermonFLLOP modules. PermonQP provides a base for solution of linear systems and quadratic programming (QP) p roblems. It includes data structures, transformations, algorithms, and supporting functions for QP. It supports any combination of linear equality, box and general linear inequality constraints, just like quadprog function in MATLAB Optimization Toolbox. PermonFLLOP is an extension of PermonQP providing support for DDM of the FETI type. This combination of DDM and QP algorithms is what makes PERMON unique. PermonQP and PermonFLLOP are licensed under the BSD 2-Clause license and we currently prepare them for publishing.

Other PERMON modules include application-specific ones such as PermonPlasticity or PermonMultiBody, discretization tools such as PermonCube, interfaces with external discretization software, and support tools. PermonCube can be described as a library for parallel generation of simple finite element meshes and their FEM processing, and serves as provider of testing data for a massively parallel DDM solver such as PermonFLLOP.

5   Gluing Matrix Implementation

5.1   PETSc Distributed Matrices and Their Transposition

Let us mention some PETSc features concerning distributed matrices. Elements of vectors and matrices are distributed among processors; each process owns only its local part. The local part consists of a contiguous range of rows. See also [14].

Concerning the matrix $\mathbf{B}^T$ (see Sect. 3), $(\mathbf{B}^s)^T$ is its local part in the above-mentioned sense. This leads us to store the matrix $\mathbf{B}$ as $(\mathbf{B}^{T_E})^{T_I}$, where T_E means an explicit transposition and T_I is an implicit transposition. The explicit transposition is implemented in PETSc with the `MatTranspose` routine or by direct assembling, whereas the implicit one is implemented by the `MatCreateTranspose` function. This way, column distribution can be mimicked while using physical row distribution. `MatCreateTranspose` returns a new envelope matrix of the `MATTRANSPOSE` type, wrapping the original matrix and swapping the meanings of its `MatMult` and `MatMultTranspose` methods.

We have implemented a new convenience function `PermonMatTranpose`. The demanded type of transpose is specified by the additional enum argument having one of `EXPLICIT`, `IMPLICIT`, `CHEAPEST` values. The last one stands for the variant which is the computationally cheapest for the current matrix. In case of `MATTRANSPOSE` the inner wrapped matrix is returned, while in other cases `MATTRANSPOSE` is created wrapping the current matrix. This function allows transparent handling of all transposes and easy switching of the transpose type.

5.2   Custom Gluing Matrix Assembly

Previous approaches to assembling the $\mathbf{B}_g$ operator are based on the knowledge of the subdomain adjacency [8]. Our latest approach described below needs only the $i2g$ (interface to global) and $i2l$ (interface to local) mappings. One of them can be replaced by the $l2g$ (local to global) mapping, because the replaced one is easily computed.

We physically assemble the matrix $\mathbf{B}_g^{T_E}$ whereas $\mathbf{B}_g = (\mathbf{B}_g^{T_E})^{T_I}$ in the sense of Subsect. 5.1. The matrix $\mathbf{B}_g$ is in fact composed as $\mathbf{B}_g = \mathbf{AT} = ((\mathbf{T})^{T_E}(\mathbf{A})^{T_E})^{T_I}$, see Sect. 3. $\mathbf{T}^s$ is implemented using the PETSc `MATSCATTER` implicit matrix type.

`PetscSF` is a PETSc class for setting up and managing the communication of certain entries of arrays and vectors between MPI processes. It uses star forest graphs to indicate and determine the communication patterns concisely and efficiently. A star is a graph consisting of one root vertex with zero or more leaves. A union of disjoint stars is called a star forest. In the PETSc implementation, all operations are split into matching begin and end phases, which allows interleaving communication by computation. The following list introduces functions implemented in PermonFLLOP.

`QPFetiAssembleGluing` returns the matrix $\mathbf{B}_g$ by calling all functions mentioned below. For illustration, we consider an elementary geometry with decomposition into 4 subdomains (Fig. 1) and only one DOF per node for the sake of simplicity.

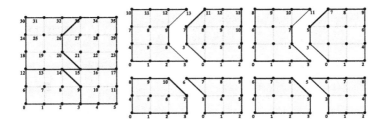

Fig. 1. Decomposition, and global, local and interface interface DOF numbering.

QPFetiGetI2LMapping, QPFetiGetI2GMapping form *i2l* or *i2g* mapping if missing.
QPFetiGetAtSF constructs $\mathbf{A}^{T_E}$ using the *i2g* mapping and PetscSF object, the
    procedure is described bellow.
QPFetiConvertAtToBgt forms the matrix $\mathbf{T}^{T_E}$ from the *i2l* mapping. The matrix
    $\mathbf{A}^{T_E}$ is then implicitly pre-multiplied by $\mathbf{T}^{T_E}$ (MATCOMPOSITE can be used for
    that). The resulting product $\mathbf{B}_g^{T_E}$ acts as having zero rows corresponding to
    non-gluing DOFs.

Let us now describe the QPFetiGetAtSF function with illustrative figures
which correspond to domain decomposition and numberings in Fig. 1 and non-
redundant connections.

1. Make first PetscSF from *i2g* mapping (PetscSFSetGraphLayout). Each proces-
    sor has a local part of roots and leaves.

2. Compute root degrees (PetscSFComputeDegreeBegin/End) and broadcast it to
    leaves (PetscSFBcastBegin/End), count connections.

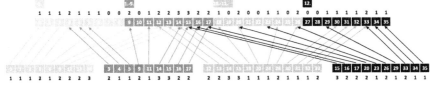

3. Remove non-gluing leaves (light), multiply appropriate leaves with multiple
    connections (dark).

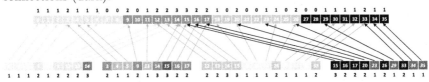

4. Create second `PetscSF` (with non-contiguous local indices).

5. Scatter indices of connections (`PetscSFScatterBegin/End`).

6. Make third `PetscSF` from connection indices.

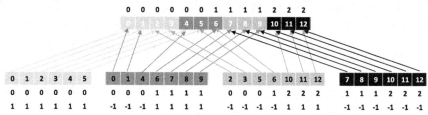

7. Send the lowest rank number of each leaf to roots (`PetscSFReduceBegin/End`), then send it back from roots to leaves (`PetscSFBcastBegin/End`). Set value to 1 if the broadcasted number is same as the rank of the leaf, and to -1 otherwise, for each leaf.

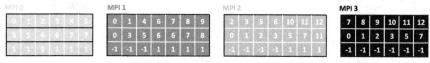

8. End up with all information needed to assemble $(\mathbf{A}^s)^{T_E}$ - indices of columns (from step 5) and rows (from step 4), and values to input (from step 7).

5.3   MATNEST, MATBLOCKDIAG and MatMatMultByColumns

The resulting constraint $\mathbf{B}_g = \mathbf{o}$ is then passed to the PermonQP solver module using its `QPAddEq` method, separately from $\mathbf{B}_D$. The global matrix $\mathbf{B} = \begin{bmatrix} \mathbf{B}_d \\ \mathbf{B}_g \end{bmatrix}$ is implemented internally in PermonQP using the composite matrix type `MATNEST`. It provides implicit nesting of matrices; the nested matrices stay stored separately but the matrix-vector multiply function (`MatMult`) of the nesting matrix behaves like they were stored interleaved in both directions by processes. This design allows decoupled assembly and storage of constraint matrices related to different type of constraints, in our case $\mathbf{B}_d$ and $\mathbf{B}_g$.

We had to cope with the fact that MATNEST does not support matrix-matrix multiplication. However, the only place where it is actually needed, is the multiplication

$$\mathbf{G} = \mathbf{R}^T \mathbf{B}^T = (\mathbf{BR})^T,$$

where $\mathbf{R} = \mathrm{diag}(\mathbf{R}^1, \ldots, \mathbf{R}^{N_S})$ is a block-diagonal matrix and its diagonal blocks $\mathbf{R}^s$, $s = 1, \ldots, N_S$, are dense with just few columns, e.g. 1 for Laplace, 3 for 2D elasticity, and 6 for 3D elasticity. Thus this matrix-matrix multiplication can be implemented using matrix-vector product efficiently. The block-diagonal matrix is implemented using the new composite matrix type MATBLOCKDIAG.

New functions PMat/PMatTransposeMatMultByColumns were implemented for this reason. To carry out the matrix-matrix product $\mathbf{Z} = \mathbf{XY}$, they firstly allocate $\mathbf{Z}$ as a dense matrix, use newly implemented type-specific functions PMatGetColumnsVectors and PMatRestoreColumnsVectors to extract the column vectors of $\mathbf{Y}$ and $\mathbf{Z}$, and iterate over the columns of $\mathbf{Y}$. For the i-th column $\mathbf{Y}(:, i)$, the matrix-vector product $\mathbf{X} * \mathbf{Y}(:, i) = \mathbf{Z}(:, i)$ is performed (MATLAB notation). For dense matrices, PMat{Get,Restore}ColumnsVectors creates cheaply vectors sharing the inner array with the original matrix.

In our case, $\mathbf{X} = \mathbf{B}$, $\mathbf{Y} = \mathbf{R}$, and $\mathbf{Z} = \mathbf{G}^T$. Thus we have to deal with $\mathbf{X}$ being MATNEST and $\mathbf{Y}$ being MATBLOCKDIAG. If the type of $\mathbf{Y}$ is MATBLOCKDIAG, then PMatMatMultByColumns proceeds as follows. It firstly gets the explicit transpose $\mathbf{X}^{T_E}$; this is cheap provided $\mathbf{X}$ is stored as $(\mathbf{X}^{T_E})^{T_I}$ which is actually our case. PMatGetLocalMat is then called to obtain local parts $(\mathbf{X}^s)^{T_E}$ owned by each process $s = 1, \ldots, N_S$. Furthermore, PMatTransposeMatMultByColumns is used to carry out $\mathbf{Z}^s = ((\mathbf{X}^s)^{T_E})^{T_I} \mathbf{Y}^s$ independently on each process. Then local matrices $(\mathbf{Z}^s)^{T_E}$ are computed, and concatenated with no communication into the global matrix $\mathbf{Z}^{T_E}$ using the PMatConcat function. The final result is $\mathbf{Z} = (\mathbf{Z}^{T_E})^{T_I}$.

We implemented all the aforementioned functions also for the MATNEST type; these type-specific implementations just recursively delegate the operations to the nested blocks. The result of multiplying MATNEST $\mathbf{B}$ with MATBLOCKDIAG $\mathbf{R}$ is then a MATNEST matrix $\mathbf{G}^{T_I} = \begin{bmatrix} \mathbf{G}_d^{T_I} \\ \mathbf{G}_g^{T_I} \end{bmatrix}$.

6   Numerical Experiments

Parallel numerical experiments were performed using the Salomon cluster operated by IT4Innovations National Supercomputing Center, Czech Republic. This cluster consists of 1008 compute nodes, giving a total of 24192 compute cores with 129 TB RAM and over 2 Pflop/s theoretical peak performance. Each node is a x86–64 computer with two Intel Xeon E5-2680v3 12-core processors (24 cores per node) and at least 128 GB RAM. Nodes are interconnected by 7D Enhanced hypercube Infiniband network. Salomon consists of 576 nodes without accelerators and 432 nodes equipped with Intel Xeon Phi MIC accelerators.

As a benchmark for parallel tests numerical model of an elastic cube was used. The primal data **K**, **f**, **R** as well as $i2l$ and $l2g$ mappings were generated by the PermonCube package. Timings of the individual operations introduced in Sect. 5.2 are shown in Table 1. There are also timings for the old approach using information about neighbouring subdomains (`QPFetiAssembleGluingNeigh`). As you can see the timings are comparable, thus we can achieve similar times without the neighbouring information.

Table 1. Timings of the individual operations performed on Salomon for different number of subdomains with constant number of 8000 elements and 27783 DOFs per subdomain and $N_s = N_c$.

N_S	216	512	1 000	2 197	4 096	8 000
N_p	6 001 128	14 224 896	27 783 000	61 039 251	113 799 168	222 264 000
N_d	820 248	2 015 076	4 019 328	9 001 071	16 980 948	33 505 068
QPFetiAssembleGluingNeigh	3.42E–01	3.46E–01	4.03E–01	6.28E–01	4.75E–01	1.17E+00
QPFetiAssembleGluing	3.33E–01	6.20E–01	7.76E–01	1.20E+00	1.42E+00	2.27E+00
QPFetiGetI2Lmapping	2.44E–03	2.13E–02	4.83E–03	2.35E–02	1.72E–02	6.86E–02
QPFetiGetAtSF	3.29E–01	5.94E–01	7.64E–01	1.16E+00	1.36E+00	2.10E+00
QPFetiConvertAtToBgt	2.46E–03	4.44E–03	9.04E–03	1.74E-02	5.30E–02	1.19E–01

7   Conclusion

We have presented results related to parallel implementation of the FETI constraint matrix within our PERMON software toolbox, particularly its "gluing" part responsible for subdomain connectivity. We have briefly reviewed the TFETI method and the PERMON toolbox modules. We have shown and evaluated our new approach needing only local and global numberings of subdomain interface DOFs, and several needed implementation features. The results show the current implementation scales at least up to thousands of cores.

Acknowledgements. This work was supported by the European Regional Development Fund in the IT4Innovations Centre of Excellence project (CZ.1.05/1.1.00/02.0070); Project of major infrastructures for research, development and innovation of Ministry of Education, Youth and Sports with reg. num. LM2011033; by the EXA2CT project funded from the EUs Seventh Framework Programme (FP7/2007–2013) under grant agreement no. 610741; by the internal student grant competition project SP2015/186 "PERMON toolbox development"; and by the Grant Agency of the Czech Republic (GACR) project no. 15-18274S.

References

1. Balay, S., Gropp, W.D., McInnes, L.C., Smith, B.F.: Efficient management of parallelism in object oriented numerical software libraries. In: Arge, E., Bruaset, A.M., Langtangen, H.P. (eds.) Modern Software Tools in Scientific Computing, pp. 163–202. Birkhäuser Press (1997)

2. Dostál, Z., Horák, D., Kučera, R.: Total FETI - an easier implementable variant of the FETI method for numerical solution of elliptic PDE. Commun. Numer. Methods Eng. **22**(12), 1155–1162 (2006)
3. Dostál, Z., Kozubek, T., Markopoulos, A., Menšík, M.: Cholesky decomposition of a positive semidefinite matrix with known kernel. Appl. Math. Comput. **217**(13), 6067–6077 (2011)
4. Farhat, C., Mandel, J., Roux, F.X.: Optimal convergence properties of the FETI domain decomposition method. Comput. Methods Appl. Mech. Eng. **115**, 365–385 (1994)
5. Farhat, C., Roux, F.X.: A method of finite element tearing and interconnecting and its parallel solution algorithm. Int. J. Numer. Methods Eng. **32**(6), 1205–1227 (1991)
6. Farhat, C., Roux, F.X.: An unconventional domain decomposition method for an efficient parallel solution of large-scale finite element systems. SIAM J. Sci. Stat. Comput. **13**, 379–396 (1992)
7. Gosselet, P., Rey, C.: Non-overlapping domain decomposition methods in structural mechanics. Arch. Comput. Methods Eng. **13**(4), 515–572 (2006)
8. Hapla, V., Cermak, M., Markopoulos, A., Horak, D.: FLLOP: a massively parallel solver combining FETI domain decomposition method and quadratic programming. In: 2014 IEEE International Conference on High Performance Computing and Communications (HPCC 2014), pp. 320–327 (2014)
9. Hapla, V., Horak, D.: TFETI coarse space projectors parallelization strategies. In: Wyrzykowski, R., Dongarra, J., Karczewski, K., Waśniewski, J. (eds.) PPAM 2011, Part I. LNCS, vol. 7203, pp. 152–162. Springer, Heidelberg (2012)
10. Hapla, V., et al.: PERMON (Parallel, Efficient, Robust, Modular, Object-oriented, Numerical) web pages (2015). http://industry.it4i.cz/en/products/permon/
11. Hapla, V., Horák, D., Merta, M.: Use of direct solvers in TFETI massively parallel implementation. In: Manninen, P., Öster, P. (eds.) PARA. LNCS, vol. 7782, pp. 192–205. Springer, Heidelberg (2013)
12. Klawonn, A., Widlund, O.B.: FETI and Neumann-Neumann iterative substructuring methods: connections and new results. Commun. Pure Appl. Math. **54**(1), 57–90 (2001)
13. Merta, M., Vašatová, A., Hapla, V., Horák, D.: Parallel implementation of Total-FETI DDM with application to medical image registration. In: Erhel, J., Gander, M.J., Halpern, L., Pichot, G., Sassi, T., Widlund, O. (eds.) Domain Decomposition Methods in Science and Engineering XXI. Lecture Notes in Computational Science and Engineering, vol. 98, pp. 917–925. Springer, Switzerland (2014)
14. Smith, B.F., et al.: PETSc users manual. Technical report. ANL-95/11 - Revision 3.5, Argonne National Laboratory (2014). http://www.mcs.anl.gov/petsc
15. Čermák, M., Hapla, V., Horák, D., Merta, M., Markopoulos, A.: Total-FETI domain decomposition method for solution of elasto-plastic problems. Adv. Eng. Softw. **84**, 48–54 (2015)

Accelerating Sparse Arithmetic in the Context of Newton's Method for Small Molecules with Bond Constraints

Carl Christian Kjelgaard Mikkelsen[1]([✉]), Jesús Alastruey-Benedé[2],
Pablo Ibáñez-Marín[2], and Pablo García Risueño[3,4,5]

[1] Department of Computing Science and HPC2N, Umeå University, Umeå, Sweden
spock@cs.umu.se
[2] Instituto Universitario de Investigación en Ingeniería de Aragón (I3A),
Universidad de Zaragoza, Zaragoza, Spain
{jalastru,imarin}@unizar.es
[3] Institut für Physik, Humboldt Universität zu Berlin, Berlin, Germany
risueno@physik.hu-berlin.de
[4] Fritz-Haber Institut (MPG), Berlin, Germany
[5] Instituto de Biocomputación y Física de Sistemas Complejos, Zaragoza, Spain

Abstract. Molecular dynamics is used to study the time evolution of
systems of atoms. It is common to constrain bond lengths in order to
increase the time step of the simulation. Here we accelerate Newton's
method for solving the constraint equations for a system consisting of
many identical small molecules. Starting with a modular and generic
base code using a sequential data layout, we apply three different optimization techniques. The compiled code approach is used to generate
subroutines equivalent to a single step of Newton's method for a user
specified molecule. Differing from the generic subroutines, these specific
routines contain no loops and no indirect addressing. Interleaving the
data describing different molecules generates vectorizable loops. Finally,
we apply task fusion. The simultaneous application of all three techniques increases the speed of the base code by a factor of 15 for single
precision calculations.

Keywords: Newton's method · Non-linear equations · Molecular
dynamics · Constraints · SHAKE · RATTLE · LINCS · Compiled code
approach · Vector level parallelism · Vectorizing compiler · SIMD

1 Introduction

Molecular dynamics (MD) of bio-molecules and organic compounds is at present
an extremely important tool for bio-medical purposes and in the chemical industry [1,2]. It is central for understanding phenomena within the human body and
for the design of novel drugs. For instance, MD simulations of proteins have
been instrumental in the design of HIV therapies [3]. In industry, MD simulations enable the detailed analysis of a wide range of phenomena such as catalysis
and adsorption [4–6].

© Springer International Publishing Switzerland 2016
R. Wyrzykowski et al. (Eds.): PPAM 2015, Part I, LNCS 9573, pp. 160–171, 2016.
DOI: 10.1007/978-3-319-32149-3_16

In the context of MD, it is common to *constrain* internal degrees of freedom (usually bond lengths and bond angles), i.e. to keep their values constant throughout the simulation. Constraining fast degrees of freedom allows for an increase in the time step of the MD simulation, so that larger systems and intervals of real time can be simulated [7].

The imposition of constraints requires the solution of nonlinear equations. The most widely used methods are SHAKE, RATTLE and LINCS which all converge linearly [8–10]. Solving the constraint equations to the limits of machine precision is normally out of the question, unless the constraint block is allowed to consume a significant fraction of the total computational time, defeating the purpose of imposing constraints in the first place. For reasons of efficiency, accuracy and stability, it is desirable to develop a constraint algorithm which can satisfy the constraints within the limits imposed by machine precision and perform the corresponding calculations in a very efficient manner. Several authors have already sought to solve the constraint equations using Newton's method, which is locally second order convergent, together with a direct or an iterative method for the linear systems. However, their proposals were generally not satisfactory due to efficiency [11] or generality [12] reasons. In this paper we apply three different optimization techniques to Newton's method together with a direct linear solver and we solve the bond constraint equations for several solvents each consisting of many identical molecules.

1. **Compiled code approach**: We construct a code generator which reads a description of a molecule and writes loop free subroutines with direct addressing, which are then compiled and used to process all molecules of the given type.
2. **Data layout transformations**: We enable vector-level parallelism for functions with irregular patterns of memory access and computation by interleaving the data describing different molecules and linear systems of the same type.
3. **Task fusion**: We fuse distinct stages of Newton's method in order to facilitate data reuse and reduce the number of memory operations.

The combined effect of our three optimization techniques is a 15-fold increase in the single precision computational speed as demonstrated by our experiments with several commonly used organic solvents.

2   Newton's Method for Molecules with Bond Constraints

In MD the most commonly constrained degrees of freedom are general bond lengths and the hydrogen bond angles. For the sake of simplicity, in this paper we only tackle bond length constraints. However, note that hydrogen bond angles can be constrained by imposing a bond length-like constraint between two atoms that are not actually covalently bonded, so our treatment is rather general.

Consider a molecule with m atoms and let $\mathbf{r}_i = (x_i, y_i, z_i)^T \in \mathbb{R}^3$ denote the coordinates of the ith atom. A bond length constraint is an equation of the form

$$\|\mathbf{r}_{a_k} - \mathbf{r}_{b_k}\|_2^2 - \sigma_k^2 = 0, \tag{1}$$

where $\sigma_k > 0$ is the length of the bond between atoms a_k and b_k, and $\| \cdot \|_2$ denotes the Euclidean norm. Physical bond lengths are on the order of 100 pm, where 1 pm $= 10^{-12}$ m. Let n denote the number of constrained bonds lengths and let

$$\mathbf{g} : \mathbb{R}^{3m} \to \mathbb{R}^n, \quad \mathbf{g}(\mathbf{r}) = (g_1(\mathbf{r}), g_2(\mathbf{r}), \dots, g_n(\mathbf{r}))^T \tag{2}$$

denote the constraint function where

$$g_k(\mathbf{r}) = \frac{1}{2} \left(\| \mathbf{r}_{a_k} - \mathbf{r}_{b_k} \|_2^2 - \sigma_k^2 \right) \tag{3}$$

represents the bond between atoms a_k and b_k. The celebrated SHAKE and RATTLE algorithms are contingent upon the solution of nonlinear equations of the form $\mathbf{f}(\mathbf{r}) = \mathbf{g}(\phi(\mathbf{r})) = 0$, where $\phi \in \mathbb{R}^{3m} \to \mathbb{R}^{3m}$ is a function which depends on the algorithm. For the sake of clarity, this presentation is limited to the simplest case of $\phi(\mathbf{r}) = \mathbf{r}$, which corresponds to finding initial coordinates for each of the m atoms such that the n bond length constraints are satisfied. If we assume that the Jacobian $\mathbf{Dg}(\mathbf{r}) \in \mathbb{R}^{n \times 3m}$ has full row rank, then Newton's method for the constraint equation $\mathbf{g}(\mathbf{r}) = 0$ takes the form

$$\mathbf{r} := \mathbf{r} - \mathbf{Dg}(\mathbf{r})^T (\mathbf{Dg}(\mathbf{r})\mathbf{Dg}(\mathbf{r})^T)^{-1} \mathbf{g}(\mathbf{r}). \tag{4}$$

The problem of evaluating each Newton iteration (4) can be split into the following tasks:

1. Compute the n components of $\mathbf{g}(\mathbf{r})$.
2. Build a sparse representation of the symmetric matrix $\mathbf{A} = \mathbf{A}(\mathbf{r})$ given by

$$\mathbf{A}(\mathbf{r}) = \mathbf{Dg}(\mathbf{r})\mathbf{Dg}(\mathbf{r})^T. \tag{5}$$

3. Expand the representation of $\mathbf{A}$ into extended arrays which can accept any fill-in during the factorization. Simultaneously, overwrite any fill-in from previous factorizations.
4. Compute a sparse Cholesky factorization of $\mathbf{A} = \mathbf{LL}^T$.
5. Solve the linear system $\mathbf{Ly} = \mathbf{g}(\mathbf{r})$ using forward substitution.
6. Solve the linear system $\mathbf{L}^T \mathbf{z} = \mathbf{y}$ using backward substitution.
7. Do the linear update $\mathbf{r} := \mathbf{r} - \mathbf{Dg}(\mathbf{r})^T \mathbf{z}$.

Normally, one can only monitor the components of the residual, i.e. $\mathbf{g}(\mathbf{r})$, but in our case we can estimate the relative constraint violation. Specifically, if $\| \mathbf{r}_{a_k} - \mathbf{r}_{b_k} \|_2 \approx \sigma_k$, then

$$\frac{g_k(\mathbf{r})}{\sigma_k^2} = \frac{1}{2} \frac{\| \mathbf{r}_{a_k} - \mathbf{r}_{b_k} \|_2^2 - \sigma_k^2}{\sigma_k^2} \approx \frac{\| \mathbf{r}_{a_k} - \mathbf{r}_{b_k} \|_2 - \sigma_k}{\sigma_k}, \tag{6}$$

which allows us to terminate the iteration when the constraints are satisfied to a specific tolerance. The factorization (Task 4) need not be the Cholesky factorization and there are several variants to choose from including a right-looking, a left-looking, and a multi-frontal factorization [15].

3   Compiled Code Optimization

The compiled code approach has been applied to the solution of linear systems in [13] and is rediscovered periodically in this context according to [14]. However, to the best of our knowledge, it has not been applied to all aspects of a complete Newton step.

3.1   Transforming a Numerical Routine into a Code Generator

Consider a single step of Newton's method with a sparse direct solver based on Cholesky's decomposition. As there is no need to pivot for the sake of stability, the order of the instructions depends only on the chemical structure of the molecule and not on the actual values of the spatial coordinates of the atoms.

Starting from an implementation of Newton's method, which uses a sparse direct solver and indirect addressing, we developed a generator, which writes loop-free subroutines that use direct addressing and are equivalent to a complete Newton step for a molecule of a specific type.

The process of developing the generator can be illustrated in terms of Task 5, the solution of a lower triangular linear system $\mathbf{Lx} = \mathbf{b}$. Figure 1 contains a C implementation of forward substitution for matrices in compressed sparse column (CSC) format. Here n is the dimension of the system, adj[k] is the row index, val[k] is the value of the kth nonzero of the matrix $\mathbf{L}$, and xadj[i] marks the start of the ith column inside arrays adj[] and val[].

```
void forward(int n, int *xadj, int *adj,
             float * restrict L, float * restrict b) {
  for (int i=0; i<n; i++) {
    b[i]=b[i]/L[xadj[i]];
    for (int j=xadj[i]+1; j<xadj[i+1]; j++) {
      b[adj[j]]=b[adj[j]]-L[j]*b[i]; // inner loop body
    }
  }
}
```

Fig. 1. A C subroutine for solving a non-singular lower triangular linear system $\mathbf{Lx} = \mathbf{b}$ in CSC format (xadj, adj, val) using forward substitution.

We obtained our subroutine generator by systematically replacing every computation involving floating point numbers with an instruction writing direct addressing statements equivalent to the original computation to a file. This transformation has been applied to the subroutine displayed in Fig. 1 producing the generator given in Fig. 2.

We eliminated Task 3 (matrix expansion) by mapping the nonzero entries of $\mathbf{A}$ directly to their correct location inside an extended array, which is exactly large enough to absorb the fill-in during the numerical factorization. Moreover, our generator tracks the status of all entries during the symbolic factorization and issues instructions which treat an entry as zero until it fills in.

```
void generate_forward(FILE *fp, int n, int *xadj, int *adj) {
  fprintf(fp,"void forward(float *L, float *b)\n{\n");
  for (int i=0; i<n; i++) {
    fprintf(fp,"  b[%d]=b[%d]/L[%d];\n",i,i,xadj[i]);
    for (int j=xadj[i]+1; j<xadj[i+1]; j++) {
      fprintf(fp,"  b[%d]=b[%d]-L[%d]*b[%d];\n",adj[j],adj[j],j,i);
    }
  }
  fprintf(fp,"}\n\n");
}
```

Fig. 2. Given the adjacency graph of a lower triangular matrix **L**, this code generates a C subroutine `forward` with the linear list of instructions necessary for solving $\mathbf{Lx} = \mathbf{b}$ using forward substitution.

Since the required number of subroutine generators is low, they were all developed manually. If necessary, this task can be done automatically using a parser that transforms computation statements into `fprintf()` instructions that resolve indirect addressing.

3.2   Generated Code

Figure 3 shows the forward substitution subroutine written by the code generator in the case of the chloroform solvent. When comparing the new subroutine with the generic subroutine in Fig. 1 we observe that:

1. The code is fully unrolled, i.e. it is loop-free.
2. There is no indirect addressing.

Replacing indirect addressing with direct addressing reduces the number of memory operations. For instance, the read access of `b[1]` in the subroutine `forward_chloroform()` is performed by a single assembly instruction and requires just one memory access to the array `b[]`. The corresponding access in the generic `forward()` (`b[adj[xadj[i]+1]]`) is a memory read with double indirection (three memory accesses) which requires many instructions.

4   Vectorization Through Data Transformations

It is hard to exploit vector-level parallelism in subroutines such as those in Figs. 1 and 3 due to their irregular patterns of memory access and computation. Nevertheless, since the solvent is composed of identical molecules, we can generate vectorizable loops that process a group of molecules/linear systems simultaneously by interleaving their data, instead of storing the data for each item in a single contiguous block.

Figure 4 shows a generic subroutine for solving p lower triangular linear systems. The subroutine assumes that the kth nonzero entries from each of the p systems are stored contiguously in memory. The two loops over t can be vectorized by a compiler like GCC or ICC if the iteration count p is sufficiently large.

```
void forward_chloroform(float * restrict L, float * restrict b) {
    b[0]=b[0]/L[0];
    b[1]=b[1]-L[1]*b[0]; // inner loop body
    b[2]=b[2]-L[2]*b[0]; // inner loop body
    b[3]=b[3]-L[3]*b[0]; // inner loop body
    b[1]=b[1]/L[4];
    b[2]=b[2]-L[5]*b[1]; // inner loop body
    b[3]=b[3]-L[6]*b[1]; // inner loop body
    b[2]=b[2]/L[7];
    b[3]=b[3]-L[8]*b[2]; // inner loop body
    b[3]=b[3]/L[9];
}
```

Fig. 3. Generated code for solving $Lx = b$ using forward substitution for the chloroform solvent.

```
void forward_interleaved(int n, int *xadj, int *adj,
                         float * restrict L, float * restrict *b) {
    int i, j, t;
    const int p=LOOP_LENGTH;
    for (i=0; i<n; i++) {
        for (t=0; t<p; t++) b[p*i+t]=b[p*i+t]/L[p*xadj[i]+t];
        for (j=xadj[i]+1; j<xadj[i+1]; j++) {
            #pragma GCC ivdep // b[p*adj[j]+t] and b[p*i+t] do not overlap
            for (t=0; t<p; t++) b[p*adj[j]+t]=b[p*adj[j]+t]-L[p*j+t]*b[p*i+t];
        }
    }
}
```

Fig. 4. A C code for solving a group of p non-singular lower triangular linear systems $L_i x_i = b_i$ in CSC format and all with the same sparsity pattern. The parameter LOOP_LENGTH should be set at compile time.

In our experiments with GCC and the AVX2 instruction set (256 bits, 8 single precision floating point numbers), the optimal value of p depends on the level of optimization. Generic codes peak at higher values of the loop length, compared with loop free codes, see Subsect. 6.2.

The generator produces vectorizable code by generalizing a statement such as b[1]=b[1]-L[1]*b[0] into a simple loop with no dependencies over p systems:

$$\text{for } (t = 0;\ t < p;\ t + +)\ b[p * 1 + t] = b[p * 1 + t] - L[p * 1 + t] * b[p * 0 + t];$$

Vectorial performance can be improved by ensuring that the arrays are properly aligned at a suitable boundary. It can also be increased by issuing the directive #pragma gcc ivdep which asserts that there is no aliasing in the loop. This avoids loop versioning and runtime checks.

5   Task Fusion

Modular programming is a key concept in software development and reflects how we think mathematically: a large problem is broken into smaller pieces which are solved separately. Modular programming facilitates code reuse, testing, debugging, maintenance, and future development, but it can increase the number of memory operations as intermediate results have to be stored and retrieved.

It is possible to fuse the construction of the right hand side (Task 1) and the matrix (Task 2) with the factorization (Task 4) and the forward sweep (Task 5). This hinges on the fact that pivoting is not necessary for systems which are symmetric positive definite, but a left looking, rather than a right looking factorization is required. In our case we merely fused the factorization of the matrix with the forward sweep. It is not necessary to compute all components of z (Task 6) before initiating the linear update (Task 7). Specifically, since

$$\mathbf{r} := \mathbf{r} - \mathbf{Dg}(\mathbf{r})^T \mathbf{z} = \mathbf{r} - \sum_{k=1}^{n} \mathbf{v}_k z_k \tag{7}$$

where $\mathbf{v}_k$ is the kth column of $\mathbf{Dg}(\mathbf{r})^T$, we can define $\mathbf{r}^{(n+1)} = \mathbf{r}$ and compute $\mathbf{r}^{(k)} := \mathbf{r}^{(k+1)} - \mathbf{v}_k z_k$ as soon as the backward substitution sweep has produced z_k. The vector $\mathbf{r}^{(1)}$ will then contain the result of the linear update (7). In this manner we fused the backward sweep and the linear update.

6   Numerical Experiments

In this paper we do not discuss how to integrate Newton's method into existing libraries for molecular dynamics. This allows us to concentrate on the effect of the three different optimization techniques and we avoid the discussion of a number of questions which are application or even library specific.

6.1   Methodology

In order to demonstrate the effect of the three different optimization techniques we wrote $8 = 2^3$ different implementations with the generic name `newtonXYZ`. The coding is as follows:

1. Sequential data layout (X=0) versus interleaved data layout (X=1).
2. Generic (Y=0) versus subroutines for molecules of a specific type (Y=1).
3. Complete task separation (Z=0) versus partial task fusion (Z=1).

We implemented a solver based on a right-looking Cholesky factorization. We interleaved all the information representing groups of molecules and enforced suitable memory alignment. For all versions, we assisted the compiler with pragmas in order to vectorize specific loops. Moreover, we used the `restrict` qualifier to inform the compiler that different pointers do not alias. This allows for better code generation as there is no need to generate both scalar and vectorial versions,

Table 1. An alphabetical list of the molecules used in our experiments. The dimension of the matrix $\mathbf{A(r)} = \mathbf{Dg(r)Dg(r)}^T$ is equal to the number of bonds. The number of structural nonzeros on or below the main diagonal of $\mathbf{A(r)}$ is given in the column labeled "nnz". The flop count for one complete Newton step as implemented in newton000 is given for each molecule. Tetrahydrofuran is abbreviated as "THF".

molecule	structural information				flops						
	atoms	bonds	nnz	fill-in	$\sqrt{a}$	a/b	$a \cdot b$	$a+b$	$a-b$	$\|a\|$	total
acetone	10	9	24	0	9	42	252	93	255	9	660
acetonitrile	6	5	12	0	5	22	126	49	129	5	336
butanol	15	14	39	0	14	67	417	148	420	14	1080
chloroform	5	4	10	0	4	18	102	40	104	4	272
ethanol	9	8	19	0	8	37	223	82	226	8	584
methanol	6	5	12	0	5	22	126	49	129	5	336
THF	13	13	38	2	13	66	400	141	401	13	1034

and runtime checks for aliasing are avoided. We explored partial task fusion as described in Sect. 5. Experiments were carried out on a workstation with an Intel i7-4770 processor (Haswell microarchitecture, 3.4 GHz, 8 MB L3 cache) and 8 GB of RAM running Mageia 5 Linux (3.19.6-desktop-2.mga5 kernel).

We used the GCC compiler (version 4.9.2) with the following flags: -O3 -std=c11 -march=native -fno-math-errno which generate AVX2 instructions.

We selected 7 organic solvents for our numerical experiments: acetone, acetonitrile, butanol, chloroform, ethanol, methanol, and tetrahydrofuran, see Table 1. They are all produced and used on an industrial scale[1]. We simulated solvents composed of 1,600,000 molecules. For each solvent and variation of Newton's method we applied 10 iterations of Newton's method. For each execution we measured the wall-clock time. We made several repetitions of each experiment in order to ensure the reliability of the measurements.

6.2   Results

All results in this section are related to calculations which were performed using a single core and single precision floating point numbers. Since solvents typically consist of many copies of the same small molecule, parallelization across a multicore machine is straight forward and maximizing the single core performance is always a necessary first step.

We interleaved the molecules in groups of size 8, 16, 32, 64, and 128. With 8 codes, 7 molecules and 5 different values of the group size, there were 280 benchmarks to evaluate. In each case we did 20 repetitions. We measured the wall-clock time and computed the median run-time which is less sensitive to the

---

[1] Ethanol is frequently consumed by humans during festive occasions such as conference dinners.

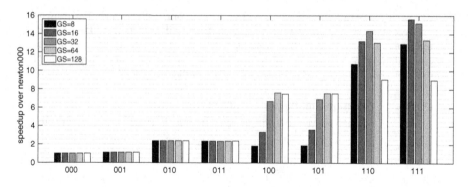

Fig. 5. This figure shows the speedup over the base code averaged over all molecules and computed for each implementation of Newton's method and group size (GS) separately.

effect of outliers. In the vast majority (265 of 280) cases the coefficient of variation was less than 5 % and it was less than 10 % in all cases. The speedup S_{XYZ} of newtonXYZ over the base code newton000 is computed as $S_{XYZ} = \frac{m(T_{000})}{m(T_{XYZ})}$, where $m(T_{XYZ})$ is the median runtime of newtonXYZ. Speedups S_{XYZ} corresponding to group size 16 are given in Fig. 2.

For each version of Newton's method and each group size we computed the average of the speedup for all molecules with respect to the base code newton000. These results are displayed in Fig. 5.

We also examined our codes using Intel's SDE (Software Development Environment). Deterministic counts for different types of instructions were determined for a benchmark involving 16,000 tetrahydrofuran (THF) atoms and 10 N iterations per molecule. Table 3 shows, for each version of Newton's method, the total number of executed instructions, the total number of floating-point operations (FLOPs), the fraction of FLOPs computed by scalar instructions, the fraction of FLOPs computed by vector instructions, and the FLOPs per instruction ratio.

The highest speedups are achieved through the application of all three optimization techniques. The code newton111 achieves speedups in the interval from 14.97 to 16.14, when the group size is 16, see Table 2. As the optimizations are applied the total number of instructions required to execute the THF benchmark is reduced by more than 97 %, from 1169.2 million instructions (MI) to a mere 33.6 MI, see Table 3.

Generic Subroutines (Y=0) Versus Specific Subroutines (Y=1). Specific subroutines achieve speedups between 2.1 and 2.4 with sequential data layouts (newton00Z versus newton01Z) and between 1.2 and 6.9 for the interleaved data layouts (newton10Z versus newton11Z) depending on the group size, see Fig. 5. Replacing generic codes newtonX0Z with specific codes newtonX1Z removes more than 80 % of the instructions, see Table 3. The deleted instructions

Table 2. Speedups S_{XYZ} for 8 different implementation of Newton's method `newtonXYZ` over the base code `newton000`. The codes are identified by their three digit binary extension `XYZ`. The molecules were interleaved in groups of size 16 for `newton1YZ`.

Molecule	Sequential layout				Interleaved layout			
	000	001	010	011	100	101	110	111
acetone	1.00	1.13	2.32	2.28	3.27	3.53	13.04	15.59
acetonitrile	1.00	1.11	2.28	2.25	3.31	3.63	13.58	15.80
butanol	1.00	1.14	2.29	2.44	3.41	3.66	13.27	16.04
chloroform	1.00	1.09	2.30	2.14	3.26	3.59	13.22	15.41
ethanol	1.00	1.13	2.42	2.35	3.46	3.71	13.56	16.14
methanol	1.00	1.10	2.22	2.19	3.32	3.65	13.03	15.26
THF	1.00	1.05	2.86	2.74	3.14	3.32	12.89	14.97

Table 3. Instruction and flop counts in millions for a benchmark consisting of 16,000 THF molecules and 10 N iterations per molecule. The group size was 16.

version	instr.	flops			flops/instr.
		total	scalar/total	vector/total	
000	1169.2	165.5	1	0	0.14
001	1130.4	165.5	1	0	0.15
010	226.2	134.7	1	0	0.60
011	226.1	134.7	1	0	0.60
100	282.9	165.5	0.152	0.848	0.58
101	267.7	165.5	0.139	0.861	0.62
110	52.2	135.4	0	1	2.59
111	33.6	135.4	0	1	4.03

include counter increments, comparisons, and jump instructions required for loops, as well as memory operations associated with indirect addressing, all of which are no longer necessary. The instruction count is reduced even further as the matrix expansion (Task 3) is avoided. The flop count is reduced as dummy operations involving zeros are avoided during the sparse factorizations.

Sequential Data Layout (X=0) Versus Interleaved Data Layout (X=1). For the sequential data layout versions `newton0YZ`, the compiler is not able to generate any vectorial code: all floating point operations are performed using scalar instructions. On the other hand, when interleaving the data layout `newton1YZ`, the compiler generates codes with high percentages of their FLOPs performed by vector instructions (around 85 % and 100 % for generic and specific codes, respectively). This reduces the number of executed instructions by more than 75 % in all cases, see Table 3. The performance of the vectorized codes peaks at a specific value of the group size, see Fig. 5. This happens due to a trade-off between vectorization profitability and the temporal reuse of vector registers and cache. For small group sizes, some loops are not vectorized because their iteration counts are too small. The generic code is less amenable to vectorization and require group sizes above 32 to show speedups close to the AVX SP vector length. On the other hand, when the group size is increased it is impossible to retain in vector registers all the values which could be reused, and the compiler has to generate spill code which increases L1 cache traffic. Eventually, the L1 data cache is exhausted and we experience a substantial drop in performance. A simple calculation can be offered in support of this second part of the argument. The THF molecule involves 13 atoms and 13 bonds. The amount of memory required to store the data necessary to formulate and solve the constraint equation can be computed as follows: 39 floating-point (FP) numbers for the spatial coordinates of the atoms, 13 FP numbers for the right hand side, and 40 FP numbers for the matrix, a total of 92 FP numbers or 368 bytes in single precision. If 128 molecules are interleaved, 47104 bytes are required and we exhaust the 32kB L1 data cache capacity of our i7-4770 CPU. When we interleave 64 THF molecules, we only require 23552 bytes, i.e. less than the L1 data cache capacity.

Complete Task Separation (Z=0) Versus Partial Task Fusion (Z=1). Task fusion has a significant effect on specific vectorized code (newton110) and small groups. For instance, when the group size is 16, it causes speedups in the interval from 1.16 to 1.21 depending on the molecule, see Table 2.

7   Conclusions

If you are solving a large number of identical sparse problems, then you should consider the simultaneous application of three distinct optimization techniques: the compiled code approach, partial task fusion, and interleaving the data describing different problems, as compiler technology has advanced to the point were a 15-fold increase in computational speed using AVX2 instructions may be possible.

We demonstrated speedups of this magnitude by solving the constraint equations for a solvent consisting of a large number of identical molecules of a specific type using Newton's method. We wrote 8 different implementations of Newton's method using a direct solver based on a right looking Cholesky factorization algorithm. We tested our codes on 7 different organic solvents which are produced on an industrial scale. The combined effect of all three optimization techniques is a single precision speedup between 14.97 and 16.14 for each of the different solvents.

Acknowledgments. The work is supported by eSSENCE, a collaborative e-Science programme funded by the Swedish Research Council within the framework of the strategic research areas designated by the Swedish Government. It is also supported in part by grants TIN2013-46957-C2-1-P and Consolider NoE TIN2014-52608-REDC (Spanish Gov.), gaZ: T48 research group (Aragón Gov. and European ESF), and HiPEAC-3 NoE (European FET FP7/ICT 287759). P.G. Risueño is funded by MPG. We would like to thank our reviewers as their comments made it possible to improve the clarity of our manuscript. We are grateful to our editor Roman Wyrzykowski who encouraged us to continue improving our code and allowed us to exceed the page limitation.

References

1. Adcock, S.A., McCammon, J.A.: Molecular dynamics: survey of methods for simulating the activity of proteins. Chem. Rev. **5**, 1589–1615 (2006)
2. Frenkel, D., Smit, B.: Understanding Molecular Simulations: From Algorithms to Applications, 2nd edn. Academic Press, San Diego (2002)
3. Moraitakis, G., Purkiss, A.G., Goodfellow, J.M.: Simulated dynamics and biological macromolecules. Rep. Prog. Phys. **66**, 383 (2003)
4. Liu, H., Sale, K.L., Holmes, B.M., Simmons, B.A., Singh, S.: Understanding the interactions of cellulose with ionic liquids: a molecular dynamics study. J. Phys. Chem. B **114**(12), 4293–4301 (2010)
5. Li, C., Tan, T., Zhang, H., Feng, W.: Analysis of the conformational stability and activity of candida antarctica Lipase B in organic solvents: insights from MD and QM simulations. J. Bio. Chem. **285**, 28434–28441 (2010)
6. Skoulidas, A.I., Sholl, D.S.: Self-diffusion and transport diffusion of light gases in metal-organic framework materials assessed using molecular dynamics simulations. J. Phys. Chem. B. **33**, 15760–15768 (2005)
7. García-Risueño, P., Echenique, P., Alonso, J.L.: Exact and efficient calculation of Lagrange multipliers in biological polymers with constrained bond lengths and bond angles: Proteins and nucleic acids as example cases. J. Comput. Chem. **32**, 3039–3046 (2011)
8. Ryckaert, J.P., Ciccotti, G., Berendsen, H.J.C.: Numerical integration of the Cartesian equations of motion of a system with constraints: molecular dynamics of n-alkanes. J. Comput. Phys. **23**, 327–341 (1977)
9. Andersen, H.C.: Rattle: a "velocity" version of the Shake algorithm for molecular dynamics calculations. J. Comput. Phys. **52**, 24–34 (1983)
10. Hess, B., Bekker, H., Berendsen, H.J.C., Fraaije, J.G.E.M.: LINCS: a linear constraint solver for molecular simulations. J. Comput. Chem. **18**, 1463–1472 (1997)
11. Barth, E., Kuczera, K., Leimkuhler, B., Skeel, R.: Algorithms for constrained molecular dynamics. J. Comput. Chem. **16**(10), 1192–1209 (1995)
12. Bailey, A.G., Lowe, C.P.: MILCH SHAKE: an efficient method for constraint dynamics applied to alkanes. J. Comput. Chem. **30**(15), 2485–2493 (2009)
13. Gustavson, F.G., Liniger, W., Willooughby, R.: Symbolic generation of an optimal Crout algorithm for sparse systems of linear equations. J. Assoc. Comput. Mach. **17**, 87–100 (1970)
14. Duff, I.S.: The impact of high-performance computing in the solution of linear systems: trends and problems. J. Comput. Appl. Math. **123**, 515–530 (2000)
15. Davis, T.A.: Direct Methods for Sparse Linear Systems. SIAM, Philadelphia (2006)

Massively Parallel Approach to Sensitivity Analysis on HPC Architectures by Using Scalarm Platform

Daniel Bachniak[1(✉)], Jakub Liput[2], Lukasz Rauch[1], Renata Słota[2,3], and Jacek Kitowski[2,3]

[1] Department of Applied Computer Science and Modelling, AGH University, Krakow, Poland
{bachniak,lrauch,rena,kito}@agh.edu.pl
[2] AGH University, ACC Cyfronet AGH, Krakow, Poland
j.liput@cyfronet.pl
[3] Department of Computer Science, AGH University, Krakow, Poland

Abstract. Sensitivity Analysis is widely used in numerical simulations applied in industry. The robustness of such applications is crucial, which means that they have to be fast and precise at the same. However, conventional approach to Sensitivity Analysis assumes realization of multiple execution of computationally intensive simulations to discover input/output dependencies. In this paper we present approach based on Scalarm platform, allowing to accelerate Sensitivity Analysis calculations by using modern e-infrastructures for distribution and parallelization purposes. The paper contains both description of the proposed solution and results obtained for a selected industrial case study.

Keywords: Sensitivity Analysis · e-Infrastructure · Scalarm · Industrial application

1 Introduction

The most of numerical models applied in virtual simulations of real problems are computationally or data intensive. In this form they are often used during evaluation of an objective function in multi-iterative optimization procedures, which results in unacceptable computing time. It is usually reduced by application of sensitivity analysis (SA) methods, allowing to determine the most influential input parameters of the model. Nevertheless, SA still requires multiple performance of investigated models with different input parameters (samples), specified by SA sampling method. Calculations of SA samples can be easily performed in distributed or parallel way. However, capabilities of currently available middleware are not sufficient to handle such functionality, since the most of the existing software is focused on management of computing jobs, but not on an additional functionality. Therefore, the main objective of this work is to design and implement the numerical procedures in form of the functional module of a

© Springer International Publishing Switzerland 2016
R. Wyrzykowski et al. (Eds.): PPAM 2015, Part I, LNCS 9573, pp. 172–181, 2016.
DOI: 10.1007/978-3-319-32149-3_17

selected middleware. In context of our research, support for a parameter sweep and its automation is desirable. The noticeable example of parameter sweep software is Nimrod [1], a tool supporting parameter studies over global computational grids. Although one of its module [2] supports experiments automation by optimization algorithms, it does not offer support for sensitivity analysis. Other popular example is OldMcData - The Data Farmer (OMD) [3], a software application designed to perform data farming [6] runs, supporting HTCondor [4] as a distributed computing backend. However, according to documentation of OMD v1.1, "the support for natural algorithms for search, optimization, and discovery (...) are planned" but, as of Oct. 2015, this functionality is still not implemented. On the other hand, a lot of SA methods have been already developed and implemented in distributed or parallel way. Many publications can be found, showing development of quite universal ([7,8]) and very narrow ([9]) SA approaches. In [5], Scalarm - the data farming platform is presented, implementing functionality similar to previous systems, but with an interactive approach. It offers support for data farming calculations, which, due to the parameterized input space and character of the performed calculations, include also typical SA methods. Taking into account the active development of the Scalarm platform, in contrast to other presented solutions, it appears to be the most suitable middleware for our needs. The proposed approach was validated by using numerical modelling created to simulate large crankshafts production process.

2   Sensitivity Analysis

SA investigates the model behavior for varying input parameters [10,11]. It determines how the variations of input data influence model outputs. The result of this investigation is a set of coefficients allowing to compare the effects of particular parameters. In this paper three global SA algorithms are taken into account and implemented as a part of HPC numerical library, i.e. Morris Design [12] and Factorial Design [13], which are less computationally demanding, and Sobol [14] which is more sophisticated procedure based on variance analysis.

2.1   Morris Design

Morris Design (MD) is a screening method which is based on the one-at-a-time approach. It requires small number of simulations, and its results are easily to interpret. In the algorithm, each model execution allows to measure local impact (relative output difference) of changing one parameter at a time. Estimation of global impact is a result of many local measurements. The elementary effect ξ_i is introduced as:

$$\xi_i(\mathbf{x}) = \frac{y(x_1, ..., x_{i-1}, x_i + \Delta_i, x_{i+1}, ..., x_k) - y(\mathbf{x})}{\Delta_i} \tag{1}$$

where: y is the model output, $\mathbf{x} \in \Omega \subset \mathbb{R}^k$ is the k-dimensional vector of model parameters x_i. The components $x_i, i = 1...k$ can take p discrete values in the set

$\{0, \frac{1}{p-1}, \frac{2}{p-1}, ..., 1\}$. The SA investigation domain is then a k-dimensional p-level grid. The value of Δ_i depends on p and can be the smallest element in the grid: $\Delta = \frac{1}{p-1}$ or its multiplication. Finally, two estimators of global impact based on local measurements have to be determined, i.e.: mean μ and standard deviation σ. Mean represents sensitivity of the model output with respect to i^{th} input parameter. High standard deviation indicates non-linearity of the parameter influence. The algorithm begins with values selection of each x_i components forming a starting point of investigation trajectory. Next, one component x_i of the starting point can be increased or decreased by Δ creating a new point. Any component i during creating a trajectory can be changed only once and each x_i produces only one elementary effect ξ_i. Presented procedure has to be repeated r times to obtain reliable results, and the cost of MD method can be calculated as $r(p+1)$ runs of the investigated model.

2.2   Factorial Design

Factorial Design (FD) allows investigation of multiple variables effect on a model response. It was proved [13] that FD can reduce the number of runs of the model by studying multiple factors simultaneously. This advantage (in comparison to one-at-a-time approach) is especially important when the investigated model is computationally intensive. The simplest factorial design investigates two factors (A, B), each at two levels (Fig. 1a). A_{low} represents the lower level of A and A_{high} represents the higher level, and similarly B_{low} and B_{high} represent two levels of factor B. The main effect of factor A is as a difference between the average response for high and low values of A. The result should be interpreted as a change in the response due to a change in the level of the factor.

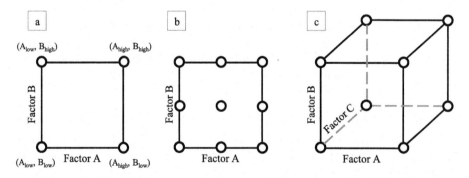

Fig. 1. Factorial Design, a) two factors, Two-level b) two factors, Three-level c) three factors, Two-level.

There are a number of different factorial designs (e.g. Three-level (Fig. 1b) or Mixed-level) which can return more information about the main effects in relatively few number of runs. However Two-level factorial design still can be considered as sufficient, especially if additional model runs are expensive or difficult to realize.

2.3   Sobol' Variance-Based Method

The Sobol' method is capable to estimate *Total Sensitivity Indices (TSI)*, which measures the main effects of factors and all interactions between input parameter. The investigated model response (y) is a function of k parameters and this relation can be written in following form:

$$y = f(x_1, x_2, x_3, ..., x_k). \tag{2}$$

In variance decomposition, the total model output variance can be represented as a sum of partial variances. According to Sobol, the number of the partial variances depends on the number of input factors and is equal to $2^k - 1$:

$$V(y) = \sum_i^k V_i + \sum_i^k \sum_{j>i}^k V_{ij} + ... + V_{12...k} \tag{3}$$

where $V(y)$ is the total unconditional variance, $\sum_i^k V_i$ represents the main effects of each parameters and $\sum_i^k \sum_{j>i}^k V_{ij}$ represents interactions of each two input parameters. The $V_{12...k}$ are the higher order influences (parameters interactions) on the model output. Equation (3) divided by the total variance of the model response $(V(y))$ gives the following equation:

$$\sum_i^k S_i + \sum_i^k \sum_{j>1}^k S_{ij} + ... + S_{12...k} = 1 \tag{4}$$

where $\sum_i^k S_i = \dfrac{\sum_i^k V_i}{V(y)}$ are the first order indices, $\sum_i^k \sum_{j>1}^k S_{ij} = \dfrac{\sum_i^k \sum_{j>i}^k V_{ij}}{V(y)}$ are the second order indices, and $S_{12...k}$ represents the higher order indices. The following equation was used to calculate first-order indices:

$$S_i = \frac{V[E(Y|x_i)]}{V(y)} \tag{5}$$

while the total-effect index for x_i as can be estimated as follows:

$$S_{T_i} = \frac{E(V(y|x_{\sim i}))}{V(y)} = 1 - \frac{V(E(y|x_{\sim i}))}{V(y)}. \tag{6}$$

To estimate first-order and total-effect indices usually a sampling numerical procedure based on Monte-Carlo approach is used. The method offers quite good evaluation of SA, however it is very time consuming. To reduce the number of model executions, the Sobol sequence was used. The advantage of Sobol quasi-random sequence is that it covers the input parameter space more uniformly (Fig. 2).

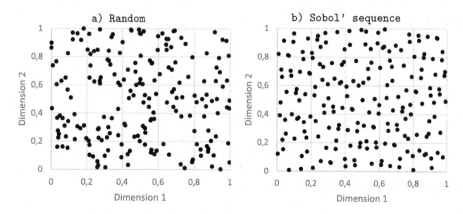

Fig. 2. 512 points of a two-dimensional a) standard random sequence and b) Sobol' quasi-random sequence.

3   Integration of Sensitivity Analysis with the Scalarm Platform

Scalarm [15,16] is a platform supporting parameter study approach, particularly the data farming process, providing a browser-based GUI and a RESTful API. The parameter study research in Scalarm consists in studied model simulation definition, followed by specification of parameter space in which the model will be evaluated. An execution of simulations with proper input data, results gathering, basic analysis and presentation are operated by Scalarm. The simulation associated with input parameter space which is executed and analyzed is referred to as an experiment. Studies can be conducted using a web browser only, giving the user flexibility in setting simulation parameter space and analyze result with general-purpose charts to reach a research goal. An example of such experiment is multi-agent behavioral study [17]. Many algorithms like SA methods are available to do advanced results analysis. Because simulations submission, experiment conduction and results download actions are supported by RESTful API, additional modules can be built to take advantage of advanced-analysis tools not supported by base Scalarm version, leaving tasks of simulations execution management to Scalarm.

3.1   Scalarm Achitecture Overview

Scalarm's architecture follows a master-worker concept. The master part of the system is responsible mainly for experiment management, i.a. in the field of simulation configurations preparation and results storage. In addition, the master part constitutes the platform interface, providing tools for experiment preparation and results analysis. The worker part realizes actual computation and consists of computational resources managed by Scalarm modules, which supervise simulations execution and cooperate with Scalarm's master part. The advantage of Scalarm is possibility to make use of heterogeneous computational resources, such as Grid, Cloud or private resources, in a unified manner. The additional module to extend Scalarm's analysis capabilities can be implemented as a RESTful client in any technology. Later in the text, it will be referred to as a Supervisor, because it can oversee and influence the experiment process.

3.2   Scalarm Sentisivity Analysis Experiment's Supervisor

The sensitivity analysis methods described in this paper was implemented in C# library, available as a DLL, able to run in the Mono [18] runtime for cross-platform usage. The Supervisor has been implemented in form of C# executable with use of methods from aforementioned library.

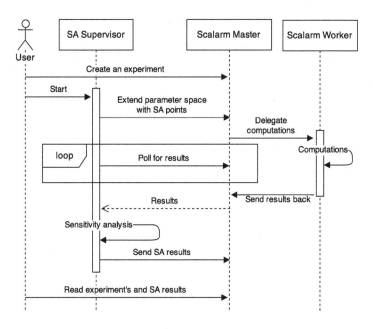

Fig. 3. Sequence diagram of SA Supervisor module cooperation with Scalarm

In Fig. 3 a sequence diagram of SA Supervisor module cooperation with Scalarm core has been presented. A user first creates a Scalarm experiment and then starts a SA Supervisor with the desired configuration. The SA library generates input space points, which are then added to Scalarm experiment's parameter space. Next, Scalarm delegates computations to computational resources and Supervisor waits for results. Results for scheduled simulation runs are fetched when they are available, and SA library performs calculations. Specific SA methods, selected before computing jobs submission, is now integrated within the Scalarm platform and realized as its internal functional module. The result of the analysis includes data, which is visualized in Scalarm's GUI or passed further

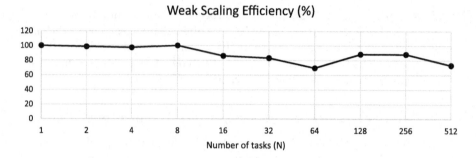

Fig. 4. Measurements of the weak scaling efficiency for Sensitivity Analysis conducted with use of Scalarm platform

to third-party software. The User can always browse archive experiments results gathered in Scalarm database by using Scalarm front-end.

4   Results

The solution proposed in this paper was validated in practice to obtain qualitative as well as quantitative results. An installation used for evaluation was made on PL-Grid resources. Referring to the diagram presented in Fig. 3, the SA Supervisor was launched on ACC Cyfronet User Interface machine, the Scalarm Master was running on PL-Grid Cloud virtual machine and the Scalarm Worker modules executed computations on Zeus cluster [19].

Ten Scalarm experiments were executed. A single experiment consisted in multiple model's evaluation with parameter space points generated by SA library. The simulation model always used 3 input parameters, and each experiment varied with number of SA's samples, which led to various number of parameter space points generation. The number of model's execution count is: $(n + 1) \times S$, where n is a parameters count and S is the number of samples, effectively giving four input space parameter points for one sample in the evaluated model.

In conducted experiments, the number of samples varied in $2^{0..9}$ range. Each single task (one task = four sequential executions of analyzed model) was executed on separate core. To measure a weak scaling efficiency, the number of used cores was increased proportionally to number of tasks. The largest experiment used 512 Zeus cluster cores and executed the model 2048 times with different parameters.

Measured experiment's execution time covered overall SA Supervisor operation time, including waiting in Zeus scheduling system queue. The weak scaling efficiency was expected to be close to 100 %, because the problem is embarrassingly parallel. Sequential part of experiment consisted in SA library computations, Scalarm communication and Zeus scheduler queue delays. Figure 4 presents weak scaling efficiency evaluated using the following formula:

$$E_N = \frac{t_1(d_1)}{t_N(d_N)} * 100 \% \tag{7}$$

where $t_1(d_1)$ is the time of referential measurements for $d_1 = 4$ model executions, $t_N(d_N)$ is the time measured for N cores and $d_N = N * d_1$. Referential single-sample experiment computed on single Zeus core took about 22 min. The execution time of many-samples computations oscillated in 22–33 min. range due to many cores used. Although there was an efficiency drop starting with 16 tasks, the fluctuations was caused by varied Zeus' queue-waiting time, sometimes increasing with many tasks submission simultaneously.

In case of qualitative aspect, the main attention was put on analysis of crankshafts manufacturing, where proper parameters of production process play crucial role in obtaining the optimal properties of final products. The most important part of the process is a multistage sequence of crankshaft cooling and heating. This part was parameterized to analyze influence of cooling times and rates on the angle of crankshaft deflection. As the input parameters, three following measures were taken into account i.e. flow stress (Sp20), Young modulus (E20) and temperature of the beginning of normalization (t_start). The first two parameters are related to mechanical properties of material, while the latter is process parameter defining temperature of annealing to obtain specific grain size in material mirostructure. The results obtained for two SA methods (Morris Design, Factorial Design) and all input parameters are presented in Fig. 5.

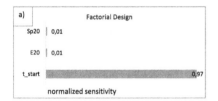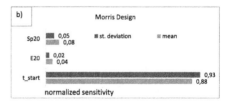

Fig. 5. Results of sensitivity analysis.

The results obtained from SA were convergent with technological assessment of the process, which proved the reliability of implemented methods.

5   Conclusions and Future Work

With a supervisor-module-as-a-client approach, we have been able to integrate a newly developed SA library with Scalarm without need to modify Scalarm's core services implementation. Moreover, the SA supervision script was implemented in pure .NET technology, not native for original Scalarm services, proving that the solution is very flexible.

The scalability of system consisting of SA library integrated with the Scalarm platform is good as expected – the drop of efficiency is caused mainly by Zeus queue delays, which is a multi-user cluster. Further improvements in efficiency could be achieved by using eg. queue reservations or non-shared resources.

The qualitative tests as well as convergence tests were performed on the results obtained from sensitivity analysis. Tests show that it was necessary to calculate at least 32 samples to obtain reliable information about impact of the model parameters to the results. In the case of considered problem, it was sufficient to evaluate mode at least 128 times, because further increase would not improve sensitivity analysis results. This shows the good convergence of the obtained results. The size of data should be chosen according to the complexity and the nature of the problem. Correct value can be found using analysis of convergence.

As some experiment supervision algorithms may find widespread application, extension with an ability to use some generic supervision modules out-of-the-box with public Scalarm web application is planned, providing a RESTful API allowing to start selected supervision modules on Scalarm server-side for a particular experiment via GUI.

Acknowledgments. This research is supported by the European Regional Development Fund program no. POIG.02.03.00-12-138/13 as part of the PLGrid NG. The creation of numerical simulations of cranckshaft cooling is supported by NCBiR project no. PBS1/B6/3/2012.

References

1. Buyya, R., Abramson, D., Giddy, J.: An economy driven resource management architecture for global computational power grids. In: International Conference on Parallel and Distributed Processing Techniques and Applications (PDPTA), Las Vegas, Nevada, USA, 26–29 June 2000
2. Abramson, D., Lewis, A., Peachy, T.: Nimrod/O: a tool for automatic design optimization. In: 4th International Conference on Algorithms & Architectures for Parallel Processing (ICA3PP 2000), Hong Kong, 11–13 December 2000
3. Upton, S.: Users Guide: OldMcData, the Data Farmer, Version 1.1. http://harvest.nps.edu/software.html. Accessed on 21 October 2015
4. Thain, D., Tannenbaum, T., Livny, M.: Distributed computing in practice: the Condor experience. Concur. Comput. Pract. Exper. **17**(2–4), 323–356 (2005)
5. Liput, J., Król, D., Słota, R., Kitowski J.: On scientific research using scalarm platform for modeling and simulation. In: International Conference Cybernetic Modelling of Biological Systems MCSB 2015. Bio-Algorithms and Med-Systems, 14–15 May 2015, Krakow, Poland, vol. 11, p. eA21 (2015)
6. Meyer, T., Horne, G.: NATO data farming report published in March 2014 launches new possibilities. In: Proceedings and Bulletin of the International Data Farming Community, Issue 15, Workshop 27, May 2014
7. Adams, B.M., Ebeida, M.S., Eldred, M.S., Jakeman, J.D., Swiler, L.P., Stephens, J.A., Vigil, D.M., Wildey, T.M., Bohnhoff, W.J., Dalbey, K.R., Eddy, J.P., Hu, K.T., Bauman, L.E., Hough, P.D.: Dakota: A Multilevel Parallel Object-Oriented Framework for Design Optimization, Parameter Estimation, Uncertainty Quantification, and Sensitivity Analysis: Version 6.2 Users Manual. https://dakota.sandia.gov/sites/default/files/docs/6.2/Users-6.2.0.pdf. Accessed on 8 May 2015

8. Hartwich, A., Stockmann, K., Terboven, C., Feuerriegel, S., Marquardt, W.: Parallel sensitivity analysis for efficient large-scale dynamic optimization. Optim. Eng. **12**(4), 489–508 (2011)
9. Ganesh, M., Hawkins, S.C.: A high performance computing and sensitivity analysis algorithm for stochastic many-particle wave scattering. SIAM J. Sci. Comput. **37**(3), A1475–A1503 (2015). doi:10.1137/140996069. Methods and Algorithms for Scientific Computing
10. Saltelli, A., Ratto, M., Andres, T., Campolongo, F., Cariboni, J., Gatelli, D., Saisana, M., Tarantola, S.: Global Sensitivity Analysis: The Primer. Wiley, New York (2008)
11. Szeliga, D., Kusiak, J., Rauch, L.: Sensitivity analysis as support for design of hot rolling technology of dual phase steel strips. Steel Res. Int., Special Issue, pp. 1275–1278 (2012)
12. Morris, M.D.: Factorial sampling plans for preliminary computational experiments. Technometrics **33**, 161–174 (1991)
13. Fisher, R.A.: The Design of Experiments, 9th edn. Macmillan, London (1971)
14. Sobol', I.M.: Sensitivity analysis for non linear mathematical models. Math. Model. Comput. Exp. **1**, 407–414 (1993)
15. Krol, D., Slota, R., Kitowski, J., Dutka, L., Liput, J.: Data farming on heterogeneous clouds. In: Kesselman, C., et al. (ed.) Proceedings of the IEEE 7th International Conference on Cloud Computing, Cloud 2014, 27 June–2 July 2014, Anchorage, Alaska. The Institute of Electrical and Electronics Engineers, pp. 873–880. doi:10.1109/CLOUD.2014.120
16. Krol, D., Kitowski, J.: Self-scalable services in service oriented software for cost-effective data farming. Future Gener. Comput. Syst. **54**, 1–15 (2016). http://dx.doi.org/10.1016/j.future.2015.07.003
17. Kvassay, M., Hluchy, L., Dlugolinsky, L., M., Schneider, B., Bracker, H., Tavcar, A., Gams, M., Krol, D., Wrzeszcz, M., Kitowski. J.: An integrated approach to mission analysis and mission rehearsal. In: Proceedings of the Winter Simulation Conference, p. 362. Winter Simulation Conference (2012)
18. http://www.mono-project.com. Accessed on 23 April 2015
19. http://www.cyfronet.krakow.pl/komputery/13345,artykul,zeus.html. Accessed on 11 May 2015

GPU Implementation of Krylov Solvers for Block-Tridiagonal Eigenvalue Problems

Alejandro Lamas Daviña and Jose E. Roman[✉]

D. Sistemes Informàtics i Computació, Universitat Politècnica de València,
Camí de Vera s/n, 46022 València, Spain
{alejandro.lamas,jroman}@dsic.upv.es

Abstract. In an eigenvalue problem defined by one or two matrices with block-tridiagonal structure, if only a few eigenpairs are required it is interesting to consider iterative methods based on Krylov subspaces, even if matrix blocks are dense. In this context, using the GPU for the associated dense linear algebra may provide high performance. We analyze this in an implementation done in the context of SLEPc, the Scalable Library for Eigenvalue Problem Computations. In the case of a generalized eigenproblem or when interior eigenvalues are computed with shift-and-invert, the main computational kernel is the solution of linear systems with a block-tridiagonal matrix. We explore possible implementations of this operation on the GPU, including a block cyclic reduction algorithm.

Keywords: GPU computing · Eigenvalue computation · Krylov methods · Block-tridiagonal linear solvers

1 Introduction

This paper is concerned with the computation of a few eigenpairs of an eigenvalue problem defined by one or two matrices with block-tridiagonal structure, using graphic processing units (GPU). We focus on Krylov methods, that will be competitive with respect to other methods when the percentage of wanted eigenvalues is small. We remark that our solvers are not restricted to symmetric matrices, but can work also in the non-symmetric case.

Given an $n \times n$ real matrix A, the standard eigenvalue problem is formulated as

$$Ax = \lambda x, \tag{1}$$

where λ is a scalar (eigenvalue) and $x \neq 0$ is an n-vector (eigenvector). There are n eigenpairs (x, λ) satisfying (1). If matrix A is symmetric, then all eigenvalues are real, otherwise eigenvalues are complex in general. In the generalized eigenvalue problem there are two intervening matrices,

---

This work was partially supported by the Spanish Ministry of Economy and Competitiveness under grant TIN2013-41049-P. Alejandro Lamas was supported by the Spanish Ministry of Education, Culture and Sport through grant FPU13-06655.

© Springer International Publishing Switzerland 2016
R. Wyrzykowski et al. (Eds.): PPAM 2015, Part I, LNCS 9573, pp. 182–191, 2016.
DOI: 10.1007/978-3-319-32149-3_18

$$Ax = \lambda Bx. \tag{2}$$

There are two major strategies for solving the above eigenproblems. One class of methods first reduce the matrices to a condensed form from which eigenvalues can be recovered more easily. These methods are generally more appropriate when all eigenvalues are required. In contrast, a second class of methods based on projecting the eigenproblem on a low-dimensional subspace are usually better suited for computing only a few eigenpairs. For large, sparse matrices, projection methods are the only viable strategy because they preserve sparsity, as opposed to transformation methods that produce fill-in during reduction to condensed form. In this work, we are targeting problems where matrices have a block-tridiagonal structure, whose blocks are not necessarily sparse, in which case transformation methods are in principle well suited, but we take the projection route since we are interested in just a few eigenvalues.

Consider first the standard eigenproblem (1). Transformation methods begin by reducing matrix A to either tridiagonal or upper Hessenberg form, in the symmetric or non-symmetric case, respectively. In terms of computational effort, this step is more expensive than the actual computation of eigenvalues, and hence many research efforts have been dedicated to improve its arithmetic intensity, specifically on GPUs [2,14]. Once the problem has been reduced to condensed form, various algorithms can be applied to compute the eigenvalues, such as the QR iteration, or specific iterations for symmetric tridiagonals, such as divide-and-conquer. Some of these algorithms are difficult to implement and have modest arithmetic intensity, and hence a hybrid CPU-GPU approach is often pursued [15]. In order to increase arithmetic intensity, some authors extend the algorithms to operate directly on a symmetric band matrix [3]. These methods compute all eigenvalues, which may be wasteful in some applications, and, optionally, all eigenvectors (requiring an additional computation).

One example of the projection methods is the Arnoldi algorithm: starting with v_1, $\|v_1\|_2 = 1$, the Arnoldi basis generation process can be expressed by the recurrence

$$v_{j+1}h_{j+1,j} = w_j = Av_j - \sum_{i=1}^{j} h_{i,j}v_i, \tag{3}$$

where $h_{i,j}$ are the scalar coefficients obtained in the Gram-Schmidt orthogonalization of Av_j with respect to v_i, $i = 1, 2, \ldots, j$, and $h_{j+1,j} = \|w_j\|_2$. The columns of V_j span the Krylov subspace $\mathcal{K}_j(A, v_1) = \mathrm{span}\{v_1, Av_1, A^2v_1, \ldots, A^{j-1}v_1\}$ and $Ax = \lambda x$ is projected into $H_j s = \theta s$, where H_j is an upper Hessenberg matrix with elements $h_{i,j}$ ($h_{i,j} = 0$ for $i \geq j + 2$). If the solutions of the projected eigenproblem are assumed to be (θ_i, s_i), $i = 1, 2, \ldots, j$, then the approximate eigenpairs $(\tilde{x}_i, \tilde{\lambda}_i)$ of the original problem are obtained as $\tilde{\lambda}_i = \theta_i$, $\tilde{x}_i = V_j s_i$.

Regarding the implementation of projection methods for eigenvalue problems on the GPU, particularly Krylov methods, the challenge is to reach high Gflops rate in the sparse matrix-vector product operation, see e.g. [10], since the rest of operations are quite simple as will be discussed in Sect. 3. In our case, we will not deal with sparse matrices, and the goal will be to implement highly efficient computational kernels for the block-tridiagonal matrix-vector product.

When addressing the generalized eigenproblem (2), a possible approach is to transform it to the standard form and apply the methods discussed above, for instance solve $B^{-1}Ax = \lambda x$ (provided B is non-singular). In the context of projection methods, where only a few eigenvalues are computed, the shift-and-invert transformation is commonly used,

$$(A - \sigma B)^{-1} Bx = \theta x, \tag{4}$$

where largest magnitude $\theta = (\lambda - \sigma)^{-1}$ correspond to eigenvalues λ closest to a given target value σ. Rather than computing matrix $(A - \sigma B)^{-1}B$ explicitly, Krylov methods normally operate implicitly by solving linear systems with $A - \sigma B$ when necessary. In our case, we need an efficient kernel to solve linear systems with a block-tridiagonal coefficient matrix on the GPU.

The rest of the paper is organized as follows. In Sect. 2 we describe the SLEPc library, in which we have developed our solvers, focusing on the support for GPU computing. Section 3 provides details of our implementation, paying special attention to the kernels for matrix-vector products and linear system solves with block-tridiagonals. Results of some computational experiments are shown in Sect. 4. Finally, we close with some concluding remarks.

2   SLEPc Solvers on GPU

SLEPc, the Scalable Library for Eigenvalue Problem Computations [5], is a software package for the solution of large-scale eigenvalue problems on parallel computers. It can be used to solve standard and generalized eigenproblems, (1) and (2), as well as other related problems. It can work with either real or complex arithmetic, in single or double precision. SLEPc provides a collection of eigensolvers, most of which are based on the subspace projection paradigm described in the previous section. In particular, it includes a robust and efficient parallel implementation of a restarted Krylov solver. It also supports the shift-and-invert transformation (4) with which interior eigenvalues can be computed making use of the linear solvers provided by PETSc[1].

In the development version, PETSc incorporates support for NVIDIA GPUs by means of Thrust and CUSP[2], performing vector operations and matrix-vector products through VECCUSP, a special vector class whose array is mirrored in the GPU, and a matrix class MATAIJCUSP, where data generated on the host is then copied to the device on demand. Later, support was extended for sparse matrix operations via CUSPARSE. The GPU model considered in PETSc uses MPI for communication between different processes, each of them having access to a single GPU [7]. The implementation includes mechanisms to guarantee coherence of the mirrored data-structures in the host and the device.

---

[1] http://www.mcs.anl.gov/petsc.

[2] Thrust is a C++ template library included in the CUDA software development toolkit that makes common CUDA operations concise and readable. CUSP is an open source library based on Thrust that targets sparse linear algebra.

In a previous work [11], preliminary support for GPU computing on SLEPc was analyzed in the context of an application arising from an integral equation. In this work, we extend the developments to general block-tridiagonal matrices including shift-and-invert.

3   Krylov Methods for Block-Tridiagonal Matrices

Krylov algorithms for eigenvalue computations are based on building an orthogonal basis of the Krylov subspace $\mathcal{K}_m(A, v_1)$ and then performing a Rayleigh-Ritz projection to extract approximate eigenpairs. Since convergence may be slow, it is necessary to restart the method, that is, discard part of the information contained in the subspace and extend the subspace again. We use the Krylov-Schur restart [13]. We will not describe the algorithm in detail here, just enumerate the main computational kernels:

1. Basis expansion. To obtain a new candidate vector for the Krylov subspace, a previous vector must be multiplied by A. In the generalized eigenproblem (2) the multiplication is by $B^{-1}A$ or, alternatively, by $(A - \sigma B)^{-1}B$ and hence linear system solves are required as discussed previously.
2. Orthogonalization and normalization of vectors. The jth vector of the Krylov basis must be orthogonalized against the previous $j - 1$ vectors. This can be done with the (iterated) modified or classical Gram-Schmidt procedure.
3. Solution of projected eigenproblem. A small eigenvalue problem of size m must be solved at each restart, for matrix $H = V^T A V$ (already available from previous steps).
4. Restart. The associated computation is VZ, where the columns of V span the Krylov subspace and Z is a small matrix of order $m \times r$, with $r < m$, formed by the Schur vectors of H (calculated on the previous step).

We assume that m is very small compared to the size of the matrices, n, and we can have for instance $m = 30$, $n = 100000$. The cost of item 3 in the above list is negligible compared to the rest, and hence it is not worth performing it on the GPU. Regarding items 2 and 4, they are currently done as vector operations (BLAS1) on the GPU. We plan to optimize these operations in the future so that they use BLAS level 2 (or even level 3 in some cases), but this is not very relevant in this paper because the dominant cost is by far the one associated with item 1. Next we discuss in detail the operations related to basis expansion.

3.1   Kernel: Matrix-Vector Product

Consider a block-tridiagonal matrix T of order n, with ℓ blocks of size k,

$$
T = \begin{bmatrix} B_1 & C_1 & & & \\ A_2 & B_2 & C_2 & & \\ & A_3 & B_3 & C_3 & \\ & & \ddots & \ddots & \ddots \\ & & & A_\ell & B_\ell \end{bmatrix}, \qquad \mathrm{rep}(T) = \begin{bmatrix} \square & B_1 & C_1 \\ A_2 & B_2 & C_2 \\ A_3 & B_3 & C_3 \\ \vdots & \vdots & \vdots \\ A_\ell & B_\ell & \square \end{bmatrix}, \qquad (5)
$$

where rep(T) stands for the memory representation of T (the $\square$ symbols indicate blocks with memory allocated but not being used).

The matrix-vector product $y = Tv$ can be computed by blocks as

$$y_i = [A_i \; B_i \; C_i] \begin{bmatrix} v_{i-1} \\ v_i \\ v_{i+1} \end{bmatrix}, \qquad i = 2, \ldots, \ell - 1, \tag{6}$$

with analog expressions for the first and last block row. Due to the arrangement of blocks in rep(T), it is possible to do the computation with a single call to BLAS _gemv per each block row. Similarly, a possible GPU implementation would call the corresponding CUBLAS [8] subroutine. We remark that when allocating memory for rep(T) on the GPU we use appropriate 2D padding for each block, to guarantee alignment of columns. Apart from the CUBLAS version, we have also implemented a customized kernel that performs the whole computation with a single kernel invocation.

3.2   Kernel: Linear System Solves

We now turn our attention to the solution of linear systems of equations

$$Tx = b. \tag{7}$$

This problem could be approached with LAPACK's general band factorization subroutines, but this is not available on the GPU. Hence, we focus on algorithms that operate specifically on the block-tridiagonal structure and are feasible to implement with CUDA. The (scalar) tridiagonal case was analyzed in [16], where the authors compare GPU implementations of several algorithms. We have extended two of the algorithms to the block case: Thomas and cyclic reduction.

Gaussian elimination on a tridiagonal matrix is sometimes referred to as the Thomas algorithm. We have implemented the block version of this method just for reference, because it is an inherently serial algorithm, with little opportunity of parallelism except for computations within one block. In the forward elimination phase, the algorithm computes $C_1 \leftarrow B_1^{-1}C_1$ and $b_1 \leftarrow B_1^{-1}b_1$, and

$$B_i \leftarrow B_i - A_i C_{i-1}, \tag{8}$$
$$C_i \leftarrow B_i^{-1}C_i, \tag{9}$$
$$b_i \leftarrow B_i^{-1}(b_i - A_i b_{i-1}), \tag{10}$$

for $i = 2, \ldots, \ell$; the backward substitution starts with $x_l \leftarrow b_l$ and runs for $i = \ell - 1, \ldots, 1$

$$x_i \leftarrow b_i - C_i x_{i+1}. \tag{11}$$

Steps (8)–(9) perform a block LU factorization, that needs to be computed only once. Subsequent right-hand sides only require steps (10)–(11). Note that the factorization is destructive, but we assume that the original matrix T is no longer needed. Factorization can be accomplished with a few calls to _gemm and LAPACK's _getrf/_getrs. With _getrf we compute the LU factorization with partial pivoting of the diagonal block B_i. We remark that since pivoting is limited to the diagonal block, this algorithm is numerically less robust than a full LU factorization, although in our tests the computed result was always accurate enough.

There are several alternative algorithms that try to increase the number of concurrent tasks and hence reduce the length of the critical path, although the cost in flops is increased. The cyclic reduction scheme [4] consists of $\log_2 \ell$ stages, where in every stage j all even blocks are eliminated in terms of the odd blocks, resulting in a system with a similar form but with halved number of unknowns. Matrix blocks of consecutive levels are related by

$$A_i^{(j+1)} = -A_{2i}^{(j)}(B_{2i-1}^{(j)})^{-1}A_{2i-1}^{(j)}, \tag{12}$$

$$B_i^{(j+1)} = B_{2i}^{(j)} - A_{2i}^{(j)}(B_{2i-1}^{(j)})^{-1}C_{2i-1}^{(j)} - C_{2i}^{(j)}(B_{2i+1}^{(j)})^{-1}A_{2i+1}^{(j)}, \tag{13}$$

$$C_i^{(j+1)} = -C_{2i}^{(j)}(B_{2i+1}^{(j)})^{-1}C_{2i+1}^{(j)}, \tag{14}$$

and analog recurrences for the right-hand side $b_i^{(j+1)}$ during forward elimination, and the solution vector $x_i^{(j+1)}$ during backward substitution. This algorithm also requires more memory, since $(B_{2i-1}^{(j)})^{-1}A_{2i-1}^{(j)}$ and $(B_{2i+1}^{(j)})^{-1}C_{2i+1}^{(j)}$ are necessary also during backward substitution. The other computed quantities can be stored in-place, so the storage requirements are at least 66 % higher than Thomas.

The GPU implementation requires a version of CUBLAS that provides the _getrf and _getrs operations, and more precisely *batched* versions of them that allow launching several factorizations/triangular solves simultaneously in order to increase potential parallelism. In this sense, the Thomas algorithm is very poor, since only one factorization or triangular solve can be done at a time.

An MPI parallel implementation of the cyclic reduction for block-tridiagonals was considered in [6,12]. A block cyclic reduction solver on GPU was used in [1] in the context of a CFD applications, with block sizes up to 32. We are interested in the case of much larger blocks, as was done in [9] in the context of a hybrid CPU-GPU solver based on MAGMA for the LU factorization.

4   Computational Results

The tests have been run on two computers:

Fermi 2 Intel Xeon E5649 processor (6 cores) at 2.53 GHz, 24 GB of main memory; 2 GPUs NVIDIA Tesla M2090, 512 cores and 6 GB GDDR per GPU. The operating system is RHEL 6.0, with GCC 4.6.1 and MKL 11.1.

Kepler 2 Intel Core i7 3820 processor (2 cores) at 3,60 GHz with 16 GB of main memory; 2 GPUs NVIDIA Tesla K20c, with 2496 cores and 5 GB GDDR per GPU. The operating system is CentOS 6.6, with GCC 4.4.7 and MKL 11.0.2.

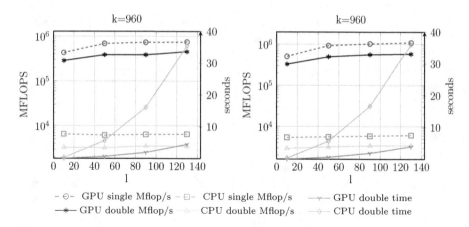

Fig. 1. Left y axis: Performance of the matrix-vector product operation for a fixed block size $k = 960$ and varying number of blocks ℓ for both CPU and GPU on the Fermi (left) and Kepler (right) machine, in single and double precision arithmetic. Right y axis: Eigensolve operation time for double precision arithmetic.

In both cases, the other software used is PETSc 3.6, SLEPc 3.6, CUDA 7.0 and CUSP 0.5.0.

Computational experiments have been conducted on random matrices, where we have varied the number of blocks ℓ and the block size k, up to the maximum storage space available on the GPU card. The matrices were generated in all cases on the CPU to use the exact same matrix on both runs (GPU/CPU).

We start discussing the matrix-vector product operation, implemented in the CPU with calls to BLAS and in the GPU with the ad-hoc CUDA kernel. We have computed the largest magnitude eigenvalue of random matrices, where the computation requires about 100 matrix-vector products. Figures 1 and 2 show the Mflop/s rate (left y axis) for our code running either on the CPU (with MKL and multi-thread enabled) or the GPU, and the total eigensolve operation time (right y axis). With a large block size (Fig. 1), we can see that performance does not depend too much on the number of blocks. In contrast, when we fix the number of blocks (Fig. 2) the performance is significantly lower for small block sizes. In any case, the benefit of using the GPU is evident since we are able to reach about 1 Tflop/s.

For assessing the performance of the shift-and-invert computation using the block oriented cyclic reduction algorithm, we have computed one eigenvalue closest to the origin ($\sigma = 0$) of random matrices of varying size. Results are shown in Figs. 3 and 4. In this case, the GPU version does not beat the CPU computation (using as many threads as computational cores in MKL operations). Nevertheless we can appreciate the sensitiveness of the GPU to the data size, as its performance improves when the number of blocks is increased for a large block size as well as when the block size is increased, while the CPU is only affected by the block size. The reported Mflop/s rate correspond to the LU factorization on double precision arithmetic, whereas the triangular solves only achieve

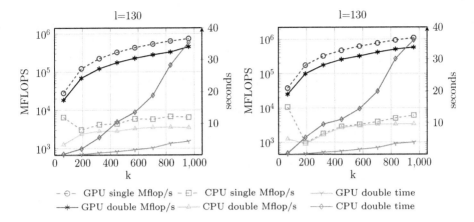

Fig. 2. Left y axis: Performance of the matrix-vector product operation for varying block size k and a fixed number of blocks $\ell = 130$ for both CPU and GPU on the Fermi (left) and Kepler (right) machine, in single and double precision arithmetic. Right y axis: Eigensolve operation time for double precision arithmetic.

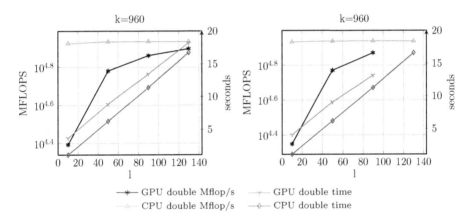

Fig. 3. Shift-and-invert case. Left y axis: Performance of factorization for a fixed block size k and varying number of blocks ℓ for both CPU and GPU on the Fermi (left) and Kepler (right) machine, in double precision arithmetic. Right y axis: Eigensolve operation time for double precision arithmetic.

a performance around 2.5-4 Gflop/s, both on CPU and GPU, and the single precision tests provided inaccurate results. The time shown corresponds to the total eigensolve operation, using the same number of restarts on both GPU and CPU runs. All in all, the performance is much worse than in the matrix-vector product case, as expected.

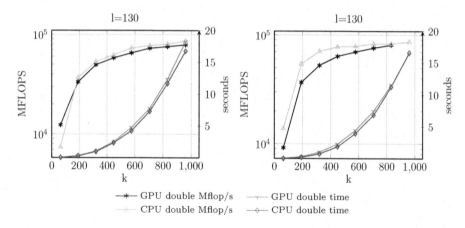

Fig. 4. Shift-and-invert case. Left y axis: Performance of factorization for varying block size k and a fixed number of blocks $\ell = 130$ for both CPU and GPU on the Fermi (left) and Kepler (right) machine, in double precision arithmetic. Right y axis: Eigensolve operation time for double precision arithmetic.

5   Conclusions

Computing a few eigenpairs of a block-tridiagonal matrix (or matrix pencil) is a computational problem that arises in different applications, e.g. in certain configurations of magnetohydrodynamic equilibrium solvers in the field of plasma physics [6]. We have explored GPU implementations in the context of the SLEPc library to compute either exterior or interior eigenvalues. In the former case, the main computational kernel is the matrix-vector product, that can be implemented with high Gflops rate. In the latter, it is necessary to perform a factorization of the matrix and then linear solves in each iteration of the eigensolver. The cyclic reduction algorithm on the GPU yields good performance for the factorization, but poorer in the case of triangular solves, as expected. Other algorithms for block-tridiagonal solves, such as those based on prefix sums, could be worth investigating to try to harness the full capability of the GPU.

In order to be able to address larger problems by exploiting several GPUs, either in the same node or in a compute cluster, we are currently developing MPI-based multi-GPU versions of the kernels, where each of the p MPI processes stores part of T and the local computations are carried out in the GPU. Hybrid CPU-GPU version could also be of interest. Also as a future work, we must optimize the orthogonalization of vectors on the GPU, since this cost becomes dominant once the time associated with the rest of operations has been reduced.

Acknowledgements. The authors are grateful for the computing resources provided by the Spanish Supercomputing Network (RES).

References

1. Baghapour, B., Esfahanian, V., Torabzadeh, M., Darian, H.M.: A discontinuous Galerkin method with block cyclic reduction solver for simulating compressible flows on GPUs. Int. J. Comput. Math. **92**(1), 110–131 (2014)
2. Bientinesi, P., Igual, F.D., Kressner, D., Petschow, M., Quintana-Ortí, E.S.: Condensed forms for the symmetric eigenvalue problem on multi-threaded architectures. Concur. Comput. Pract. Exp. **23**, 694–707 (2011)
3. Haidar, A., Ltaief, H., Dongarra, J.: Toward a high performance tile divide and conquer algorithm for the dense symmetric eigenvalue problem. SIAM J. Sci. Comput. **34**(6), C249–C274 (2012)
4. Heller, D.: Some aspects of the cyclic reduction algorithm for block tridiagonal linear systems. SIAM J. Numer. Anal. **13**(4), 484–496 (1976)
5. Hernandez, V., Roman, J.E., Vidal, V.: SLEPc: a scalable and flexible toolkit for the solution of eigenvalue problems. ACM Trans. Math. Softw. **31**(3), 351–362 (2005)
6. Hirshman, S.P., Perumalla, K.S., Lynch, V.E., Sanchez, R.: BCYCLIC: a parallel block tridiagonal matrix cyclic solver. J. Comput. Phys. **229**(18), 6392–6404 (2010)
7. Minden, V., Smith, B., Knepley, M.G.: Preliminary implementation of PETSc using GPUs. In: Yuen, D.A., Wang, L., Chi, X., Johnsson, L., Ge, W., Shi, Y. (eds.) GPU Solutions to Multi-scale Problems in Science and Engineering. Lecture Notes in Earth System Sciences, pp. 131–140. Springer, Heidelberg (2013)
8. NVIDIA: CUBLAS Library V7.0. Technical report, DU-06702-001_v7.0, NVIDIA Corporation (2015)
9. Park, A.J., Perumalla, K.S.: Efficient heterogeneous execution on large multicore and accelerator platforms: case study using a block tridiagonal solver. J. Parallel and Distrib. Comput. **73**(12), 1578–1591 (2013)
10. Reguly, I., Giles, M.: Efficient sparse matrix-vector multiplication on cache-based GPUs. In: Innovative Parallel Computing (InPar), pp. 1–12 (2012)
11. Roman, J.E., Vasconcelos, P.B.: Harnessing GPU power from high-level libraries: eigenvalues of integral operators with SLEPc. In: International Conference on Computational Science. Procedia Computer Science, vol. 18, pp. 2591–2594. Elsevier (2013)
12. Seal, S.K., Perumalla, K.S., Hirshman, S.P.: Revisiting parallel cyclic reduction and parallel prefix-based algorithms for block tridiagonal systems of equations. J. Parallel Distrib. Comput. **73**(2), 273–280 (2013)
13. Stewart, G.W.: A Krylov-Schur algorithm for large eigenproblems. SIAM J. Matrix Anal. Appl. **23**(3), 601–614 (2001)
14. Tomov, S., Nath, R., Dongarra, J.: Accelerating the reduction to upper Hessenberg, tridiagonal, and bidiagonal forms through hybrid GPU-based computing. Parallel Comput. **36**(12), 645–654 (2010)
15. Vomel, C., Tomov, S., Dongarra, J.: Divide and conquer on hybrid GPU-accelerated multicore systems. SIAM J. Sci. Comput. **34**(2), C70–C82 (2012)
16. Zhang, Y., Cohen, J., Owens, J.D.: Fast tridiagonal solvers on the GPU. In: Proceedings of the 15th ACM SIGPLAN Symposium on Principles and Practice of Parallel Programming. PPopp 2010, pp. 127–136 (2010)

Parallel Non-numerical Algorithms

Comparison of Large Graphs Using Distance Information

Wojciech Czech$^{(\boxtimes)}$, Wojciech Mielczarek, and Witold Dzwinel

Institute of Computer Science, AGH University of Science and Technology,
Kraków, Poland
{czech,dzwinel}@agh.edu.pl

Abstract. We present a new framework for analysis and visualization of large complex networks based on structural information retrieved from their distance k-graphs and B-matrices. The construction of B-matrices for graphs with more than 1 million edges requires massive BFS computations and is facilitated using Cassovary - an open-source in-memory graph processing engine. The approach described in this paper enables efficient generation of expressive, multi-dimensional descriptors useful in graph embedding and graph mining tasks. In experimental section, we present how the developed tools helped in the analysis of real-world graphs from Stanford Large Network Dataset Collection.

Keywords: Graph processing · Graph comparison · Graph visualization

1 Introduction

Large datasets being abundant in many fields of science and technology are frequently structured, that is they form graph patterns or networks which aggregate information about relations between objects and provide system-wise views of mechanisms generating data. Graph processing tasks involve searching state spaces, filtering, link prediction, generation of invariants, sub-structure matching, link recommendation, community detection and classification. As the size of graphs increases to millions of edges, their processing and analysis becomes challenging, both from computational and storage perspective. Additionally, specific structural properties of real-world graphs such as scale-free distribution of vertex degrees has a bad influence on efficiency of distributed computing. This is because the existence of densely connected vertices (hubs) makes graph partitioning highly problematic and has negative impact on communication between multiple machines. The computing node storing adjacency lists of hubs has to be contacted much more frequently than any other node. This causes serious delays, multiplicative with a number of edges crossing partitions.

In this work we discuss a specific subgroup of graph analysis tools, namely graph embedding that is a transformation of graph patterns to feature vectors also known as graph descriptors generation. Graph embedding is one of the most frequently used methods of graph comparison providing a bridge between

© Springer International Publishing Switzerland 2016
R. Wyrzykowski et al. (Eds.): PPAM 2015, Part I, LNCS 9573, pp. 195–206, 2016.
DOI: 10.1007/978-3-319-32149-3_19

statistical and structural pattern recognition [14]. After generating descriptors invariant under graph isomorphism, the power of well-established vector-based pattern recognition methods is brought to more complex structural world, which tackles with non-trivial combinatorial domain. Graph feature vectors enable computation of dissimilarity or similarity measures crucial in applications like image recognition, document retrieval and quantitative analysis of complex networks.

The rest of this work is organized as follows. In Sect. 2 we describe relation between graph embedding and complex network analysis. Next, in Sect. 3 the overview of tools for massive graph processing is presented. Section 4 introduces graph descriptors based on shortest-paths length and proposes new invariants well-suited for large graph analysis. In Sect. 5 we describe implementation details associated with application used in experiments on graph embedding. Later, in Sect. 6 we present selected results of analyzing complex networks from Stanford Large Network Dataset Collection. Section 7 concludes the paper offering final remarks and describing future work plans.

2   Complex Networks and Graph Comparison

Graph datasets appear frequently in various fields of contemporary science as data structures aggregating information about relations between numerous objects. Today, a variety of structured data is represented by different types of networks including cellular graphs describing regulatory mechanisms and metabolic reactions inside a cell, social networks encoding inter-personal links, spatial graphs representing proximity relations between elements embedded in metric space, web graphs gathering information about hyperlinks and many others. The non-trivial structure of real-world graphs studied extensively in recent decade gave rise to a new interdisciplinary research field of complex networks.

Starting from pioneering works by D.J. Watts [25] studying small-world phenomenon and A.L. Barabási [2] explaining scale-free degree distributions in real-world graphs, the analysis of complex networks revealed a set of interesting structural properties emerging from underlying dynamics [3]. The heavy-tailed distributions of vertex degrees have big influence on spreading processes on graphs and determine network resistance to random attacks at the same time inducing vulnerability to targeted attacks. The quantitative analysis of complex networks developed a set of graph invariants which allow to compare different networks and to make conclusions regarding their generating mechanism and future evolution.[1] Those descriptors reflect well-defined topological features of networks and their computational cost is rather practical - the challenge occurs at the scale of millions of edges.

In parallel to descriptors originating from the theory of complex networks, the structural pattern recognition community developed multiple alternative approaches to graph embedding [14]. The common characteristic of those

---

[1] Diameter, efficiency, characteristic path length, vertex betweenness, vertex closeness, vertex eccentricity, transitivity, clustering coefficient, assortativity [3].

descriptors is that they are typically multi-dimensional, encoding variety of topological features starting from local ones and moving to global. The goal is to capture most discriminative features of a graph and enable efficient classification and clusterization of graph patterns. The most recent studies in the field proposed descriptors based on different random walk models [13,23], spectral graph theory [26], prototype-based embedding [4,18] and substructure embedding [15]. Those descriptors were designed for graphs representing documents, molecules or images, which have considerably smaller size than a typical complex network with hundreds of thousands of edges. Therefore, they are less practical for analysis of large graphs due to considerable computational overhead. In the work [9] we introduced a new method of graph embedding which uses invariants of distance k-graphs and specifically B-matrices as a generic framework for graph descriptors generation. Here, we extend our study presenting distance-based invariants as a computationally feasible and robust alternative to spectral graph embedding. Based on B-matrices derived from shortest paths distributions we construct multi-dimensional representations of large complex networks and provide universal method of their visualization.

3   Massive Graph Processing

Generating vector representations of complex networks and specifically computing shortest-paths distance matrices for large graphs is a computationally expensive task which can be facilitated using different types of processing frameworks, both distributed and single-node. The tool which appears to be suitable for structured big data processing is Hadoop being state-of-art implementation of Map-Reduce parallel programming model. Unfortunately, the iterative nature of graph traversals such as BFS requiring multiple consecutive executions of heavy Map-Reduce jobs brings significant intrinsic computational cost to Hadoop-based implementations. Another negative factor stems from heavy-tailed vertex degree distributions typical for real-world graphs. It causes highly non-uniform workloads for reducers what badly affects overall performance.

To overcome those limitations, several frameworks optimized for iterative Map-Reduce processing were proposed. One of them is HaLoop [6], the modified version of Hadoop, designed to comply with specific requirements of iterative parallel algorithms. The framework reduces the number of unnecessary data shuffles, provides loop-aware task scheduling and caching of loop-invariant data. Twister [12] provides distributed environment, which brings several extensions improving efficiency of iterative Map-Reduce jobs, e.g., caching static data used by tasks, increasing granularity of tasks, introducing additional phase of computation (*combine*) or more intuitive API. PrIter framework [27] accelerates convergence of iterative algorithms by selective processing of certain data portions and giving higher priorities to related subtasks. This allows achieving considerable speedups comparing to Hadoop implementations.

Bulk-Synchronous Parallel (BSP) model is implemented by Google Pregel [22] (C++) or its open-source counterparts: Apache Giraph [1] (Java) and Stanford GPS [24] (Java). Here, the computation is divided into supersteps, which perform vertex-based local processing by evaluating functions defined on vertices and exchanging messages with neighbors. Synchronization barriers occur between supersteps imposing ordering required for on-time delivery of messages. Again, the hub vertices present in scale-free graphs cause communication peaks for certain workers in distributed BSP model. This problem cannot be overcome easily, as balanced graph partitioning is NP-complete task. The popular GraphLab framework [21] uses different Gather-Apply-Scatter (GAS) programming model, which enables execution without communication barriers (asynchronous mode). Moreover, it uses edge-based graph partitioning resulting in better-balanced communication for graphs with heavy-tailed degree distributions. The work [17] presents a comprehensive comparison of BSP and GAS frameworks performance.

Distributed processing as a way of dealing with large graphs brings fault tolerance and horizontal scalability but at the cost of implementation complexity and troublesome communication bursts caused by uneven data partitions. The different approach assumes in-memory processing of a whole graph on a single-machine. Due to increased availability of servers with considerable RAM size (256GB+) and existence of vSMP hypervisors enabling aggregation of memory from different machines, this design principle seems to be an interesting alternative to distributed graph processing. Assuming adjacency list as a graph representation, the unlabeled graph with 12 billions of edges can be processed on enterprise-grade server with 64GB RAM. SNAP (C++, Python) [20] is an example of robust graph library which accommodates in-memory single-server processing. An open-source Cassovary [16] is a Java/Scala library created and used by Twitter for computing of Who-To-Follow recommendations. We decided to extend Cassovary by adding module for graph embedding and visualization.

Our motivation for using Cassovary was twofold. Our first goal was to enable efficient generation of distance matrices and B-matrices for complex networks with millions of edges. This intermediate processing step allows further generation of multiple feature vectors, from scalars to robust multi-dimensional descriptors [9]. Our previous approach to B-matrix generation [7] was based on CUDA implementation of R-Kleene algorithm [10], which requires complete distance matrix to reside in memory. Despite significant speedups comparing to serial version, the usability of this method is limited by memory size of GPU used for processing, e.g., Tesla C2070 with 6GB memory allowed generation of distance matrix for graphs with 56281 vertices. The Cassovary helps to overcome this limitation. The second goal was to facilitate fast integration with existing *Graph Investigator* [8] software written in Java. This leads us towards production-proven, JVM-based platform with a good API.

4   Graph Descriptors from k-distance Graphs

The inter-vertex dissimilarity measures such as shortest path length or commute time provide comprehensive information about graph structure. In [7,9] we presented how to use shortest paths for constructing ordered set of distance k-graphs and generation of isomorphism invariants. Here, we briefly describe most important notions and introduce a new set of descriptors constructed by low-rank approximation of B-matrix.

Definition 1. *For an undirected graph $G = (V(G), E(G))$ we define vertex distance k-graph $G_k^{\mathcal{V}}$ as a graph with vertex set $V(G_k^{\mathcal{V}}) = V(G)$ and edge set $E(G_k^{\mathcal{V}})$ so that $\{u, v\} \in E(G_k^{\mathcal{V}})$ iff $d_G(u, v) = k$.*

$d_G(u, v)$ is dissimilarity measure between vertex u and v, in particular the length of the shortest path between u and v. It follows that $G_1^{\mathcal{V}} = G$ and for $k > diameter(G)$, $G_k^{\mathcal{V}}$ is an empty graph. For a given graph G the invariants of G-derived vertex k-distance graphs can be aggregated to form new descriptor of length $diameter(G)$. Moreover, the constant-bin histograms of selected vertex descriptor (e.g. degree) for $G_k^{\mathcal{V}}$ graphs form robust 2D graph representation called B-matrix.

Definition 2. *We define k-vertex-shell of vertex v as a subset of graph vertices at distance k from v (k-neighborhood of vertex v). The following equation defines vertex B-matrix of a graph. $B_{k,l}^{\mathcal{V}}$ = number of nodes that have l members in their k-vertex-shells.*

From this definition it follows that $B_{k,l}^{\mathcal{V}}$ is a number of vertices with degree l in a respective $G_k^{\mathcal{V}}$ graph. As presented in [9] the structure of vertex B-matrix reflects the structure of an underlying graph and provides robust method of graph visualization. Even more powerful B-matrices can be constructed using edge k-distance graphs.

Definition 3. *Let $G = (V(G), E(G))$ be an undirected, unweighted, simple graph. The distance from a vertex $w \in V(G)$ to an edge $e_{uv} = \{u, v\} \in E(G)$, denoted as $d_G^{\mathcal{E}}(w, e_{uv})$, is the mean of distances $d_G(w, u)$ and $d_G(w, v)$.*

For unweighted graphs $d_G^{\mathcal{E}}$ has integer or half-integer values.

Definition 4. *We define edge distance k-graph as a bipartite graph $G_k^{\mathcal{E}} = (U(G_k^{\mathcal{E}}), V(G_k^{\mathcal{E}}), E(G_k^{\mathcal{E}})) = (V(G), E(G), E(G_k^{\mathcal{E}}))$ such that for each $w \in V(G)$ and $e_{uv} \in E(G)$, $\{w, e_{uv}\} \in E(G_k^{\mathcal{E}})$ iff $d_G^{\mathcal{E}}(w, e_{uv}) = k$.*

The maximal value of k for which $G_k^{\mathcal{E}}$ is non-empty is $2 \times diameter(G)$. The descriptors of G built on edge distance k-graphs bring more discriminative information than ones obtained for vertex distance k-graphs.

Definition 5. *The k-edge-shell of vertex v is a subset of graph edges at distance k from v (k can have half-integer values). The following equation defines edge B-matrix of a graph. $B_{i,l}^{\mathcal{E}}$ = number of nodes that have l edges in their $\left(\frac{1}{2}i\right)$-edge-shells.*

In other words i-th row of $B^{\mathcal{E}}$ stores degree frequencies of vertices from $U(G_{0.5i}^{\mathcal{E}})$ set of bipartite $G_{0.5i}^{\mathcal{E}}$ graph. Being isomorphism invariant, the B-matrices provide a useful tool for visual comparison of networks and bring diversity of local and global graph features. As presented in [9] they can capture density, regularity, assortativity, disassortativity, small-worldliness, branching factor, transitivity, bipartity, closed walks count and many other structural properties. The space complexity for B-matrices is approximately $\mathcal{O}(n \times diameter(G))$, much less than $O(n^2)$ for distance matrices. This makes them feasible for embedding of complex networks (diameter of complex networks typically is proportional to $\log(n)$ or even $\log(\log(n))$).

The pre-selected rectangular fragments of B-matrices form long pattern vectors useful in machine learning tasks. For lower-dimensional representation the aggregated statistics of B-matrix rows (e.g. relative standard deviation) are applied to obtain feature vectors of size proportional to graph diameter.

5   Software for Constructing B-matrices

Motivated by the need of visualization and in-depth quantitative analysis of large graphs we have developed extension to Cassovary, which enables computing all-shortest-paths, save them as compressed distance matrix and construct a set of distance-based descriptors. The prepared module can be used for batch processing of graphs on a single node and producing data for *Graph Investigator* [8] application, which performs clusterization or classification of sets of graphs.

The computation is based on multi-threaded BFS algorithm utilizing FastUtil collections such as type-specific hash maps for optimized data access. The FastUtil was benchmarked as one of the fastest collections library for Java. The implemented module uses 2D hash maps for storing sparse B-matrices. The buffered parts of constructed distance matrices are gzipped to avoid accelerated disk space consumption. Apart from B-matrices the application can be configured to compute descriptors such as efficiency or vertex eccentricity (12 distance-based descritors available). It also provides efficient implementation of Betweenness Centrality for unweighted graphs based on methods described in [5]. This algorithm can be used for community detection on large graphs. The program accepts graph input in a form of adjacency matrix and gzipped list of edges and assumes that the graph is immutable after loading into memory. Fault tolerance is provided by replication of whole dataset on additional nodes.

The software contains modules responsible for inter-vertex distance computation, centrality computation and distance matrix compression. It can be accessed via Web page http://home.agh.edu.pl/czech/cassovary-plugin/.

6   Experiments

The performance of distance-based descriptor generation was tested on a set of complex networks from Stanford Large Network Dataset Collection [19]. We have selected 30 graphs of different size representing several types of structured data including social networks, networks with ground-truth communities, communication networks, citation networks, web graphs, co-purchasing networks, road networks, peer-to-peer networks and location based networks. All graphs were treated as undirected.

Our aim was to perform comparison of multi-level structural properties of networks from Stanford Dataset and provide their visualization in a form of B-matrices. The computed distance matrices and B-matrices enable in-depth analysis of the graphs and provide supplementary information to scalar descriptors generated using SNAP library and published on SNAP website [19]. As the networks differ in size (see Fig. 1) we perform additional processing of B-matrices by fixing l-shell bin size to extract same-size feature vectors and use them in unsupervised learning. To decrease horizontal resolution of B-matrix l-shell bin size can be set to number greater than 1. In this way coarse-grained B-matrices are obtained. In addition, logarithmic scaling of non-zero entries of $B^\star$ allowed us to reduce dynamic range effects, making heavy-tailed degree distributions of k-distance graphs more tractable and decreasing impact of graph size on the results of comparison.

Let $\star$ stand for $\mathcal{V}$ or $\mathcal{E}$ symbol, so that $B^\star$ denotes $B^\mathcal{V}$ or $B^\mathcal{E}$ matrix of a graph G. We perform graph embedding by packing rows or columns of B-matrices to form a long pattern vector.

$$
\begin{aligned}
D^\star_{long}(k_{min}, k_{max}, l_{min}, l_{max}) = [B^\star_{k,l}] \\
k_{min} \leq k \leq k_{max}, \; l_{min} \leq l \leq l_{max}
\end{aligned}
\tag{1}
$$

In order to extract most relevant data from B-matrices we decided to apply Singular Value Decomposition (SVD) and create truncated B-matrix $\tilde{B}^\star$. This post-processing step allows reducing noise in long pattern vectors obtained by row-column packing of B-matrices.

$$
B^\star = U \Sigma V^T = U \begin{bmatrix} D_{r \times r} & 0 \\ 0 & 0 \end{bmatrix} V^T = \sum_{i=1}^{r} \sigma_i u_i v_i^T,
\tag{2}
$$

$$
\tilde{B}^\star = \sum_{i=1}^{m} \sigma_i u_i v_i^T,
\tag{3}
$$

where $D = diag(\sigma_1, \sigma_2, \ldots, \sigma_r)$, $\sigma_1 \geq \sigma_2 \geq \ldots \geq \sigma_r$, $r = rank(B^\star)$, $m < r$ and u_i is i-th column vector from orthogonal matrix U (left singular vector),

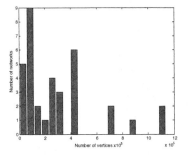
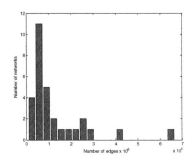

Fig. 1. Histogram of network sizes: number of vertices and number of edges.

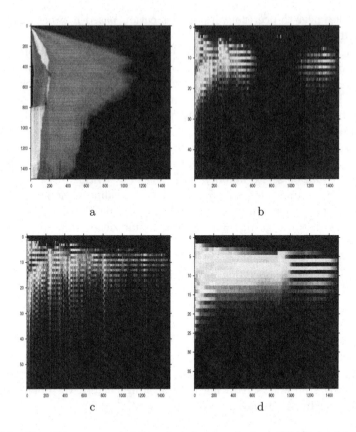

a   b

c   d

Fig. 2. Edge B-matrices generated for selected networks from SNAP database: a. road network of Pennsylvania (1088092 vertices, 1541898 edges), b. Web graph from University of Notre Dame (325729 vertices, 1090108 edges), c. Web graph from Berkeley and Stanford Universities (685230 vertices, 6649470 edges), d. Youtube online social network (1134890 vertices, 2987624 edges).

Fig. 3. Visualization of Stanford Web graph using *nr*-MDS [11].

v_i is i-th column vector from orthogonal matrix V (right singular vector). By truncating B-matrix with first m singluar values we extract only directions of the highest variance.

The edge B-matrices generated for selected networks are presented in Fig. 2. The patterns visible on coarse-grained B-matrices encode underlying structural properties and enable visual comparison of networks. The road network of Pennsylvania (see Fig. 2a) is a sparse graph with a big number of odd cycles of different size reflected by odd rows of edge B-matrix. The structure of road net B-matrices is much different than one exhibited by the rest of analyzed networks. The edge B-matrix of Web graph representing links among pages of Notre Dame University (Fig. 2b) follows community structure of this network being represented by disjoint areas and increased density islands. Assortative properties of Berkeley and Stanford Web graph are reflected by vertical lines on Fig. 2c. The clustered structure of this network is also visible in B-matrix being additionally confirmed by 3×10^5 graph nodes visualized employing nr-MDS method [11] (see Fig. 3). Youtube online social network Fig. 2d is a dense graph with high local clustering represented by a number of odd closed walks such as triangles or pentagons.

By selecting fixed parts of B-matrices and aggregating them into feature vectors we can build graph invariants encoding big part of information about graph structure. In the next experiment we embedded vectors obtained from vertex B-matrices into 2D space using Principal Component Analysis (PCA). In order to obtain same-size, comparable feature vectors we fixed l-shell bin size to 100 and used 100 such bins. The number of rows of vertex B-matrix was fixed to 20. This gave 2000-dimensional vector for each network in dataset. Before long vector extraction each B-matrix was scaled logarithmically and truncated based on first 10 singular values. The result of described 2D projection is presented in Fig. 4.

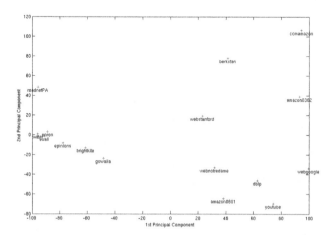

Fig. 4. The result of 2D embedding (PCA) of the long feature vectors representing subgroup of networks from SNAP dataset. Selected graphs were removed from the chart for better visibility.

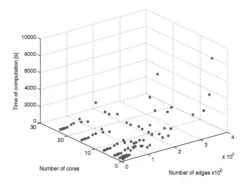

Fig. 5. Time of computations required for generation of edge and vertex B-matrices depending on number of edges and number of processor cores.

The comparison of SNAP networks reveals topological proximity between graphs sharing similar generating mechanism. Networks with ground-truth communities (*youtube*, *dblp*) are close to each other. The same is observed for communication networks (*enron*, *euall*, *twitter*). Web graphs (*webstanford*, *berkstan*, *webnotredame*) share similar 1st Principal Component (1PC) but are separated by 2nd Principal Component (2PC). It appears that 1PC distinguishes networks with explicit cluster structure from networks with less number of communities. In turn, 2PC appears to be related to local clustering coefficient (the value for *comamazon* network is 5 times bigger than the value for *youtube*).

We used resources of Academic Computer Centre Cyfronet AGH (http://www.cyfronet.krakow.pl/en, Zeus Cluster) to perform computations required for B-matrices generation. Figure 5 presents dependency of computation time on number of edges and number of used processor cores. The cluster nodes used in experiment have Intel Xeon E5645 2.40GHz CPU.

7   Conclusions

The paper describes the framework for large graph analysis based on information about all-pair shortest-paths. The main contribution is the software, which facilitates generation of distance matrices and B-matrices for real-world graphs with millions of edges. In addition, we proposed to use truncated B-matrices as a reduced-noise representation of graphs. Experiments show, that developed software can be efficiently used for large networks comparison based on feature vectors aggregating different structural properties of graphs. The B-matrices computed once for SNAP networks can be used for further analysis and visualization.

The main limitation of proposed approach stems from initial assumption of in-memory graph processing on a single computational node. For undirected networks with billions of edges this strategy seems acceptable, but the potential need to consider also edge labels or vertex labels increases memory requirements to a big extent and drastically worsens scalability factor.

Our future work will be focused on comparison of prepared software with SNAP library and its further integration with *Graph Investigator* application.

Acknowledgments. This research is supported by the National Centre Science Poland (NCN) DEC-2013/09/B/ST6/01549.

References

1. Avery, C.: Giraph: large-scale graph processing infrastructure on hadoop. In: Proceedings of the Hadoop Summit. Santa Clara (2011)
2. Barabási, A., Albert, R.: Emergence of scaling in random networks. Science **286**(5439), 509 (1999)
3. Boccaletti, S., Latora, V., Moreno, Y., Chavez, M., Hwang, D.: Complex networks: structure and dynamics. Phys. Rep. **424**(4–5), 175–308 (2006)
4. Borzeshi, E.Z., Piccardi, M., Riesen, K., Bunke, H.: Discriminative prototype selection methods for graph embedding. Pattern Recogn. **46**(6), 1648–1657 (2013)
5. Brandes, U., Pfeffer, J., Mergel, I.: Studying Social Networks: A Guide to Empirical Research. Campus Verlag, Frankfurt (2012)
6. Bu, Y., Howe, B., Balazinska, M., Ernst, M.D.: Haloop: efficient iterative data processing on large clusters. Proc. VLDB Endowment **3**(1–2), 285–296 (2010)
7. Czech, W.: Graph descriptors from B-matrix representation. In: Jiang, X., Ferrer, M., Torsello, A. (eds.) GbRPR 2011. LNCS, vol. 6658, pp. 12–21. Springer, Heidelberg (2011)
8. Czech, W., Goryczka, S., Arodz, T., Dzwinel, W., Dudek, A.: Exploring complex networks with graph investigator research application. Comput. Inform. **30**(2), 381–410 (2011)
9. Czech, W.: Invariants of distance k-graphs for graph embedding. Pattern Recogn. Lett. **33**(15), 1968–1979 (2012)
10. D'Alberto, P., Nicolau, A.: R-kleene: a high-performance divide-and-conquer algorithm for the all-pair shortest path for densely connected networks. Algorithmica **47**(2), 203–213 (2007)
11. Dzwinel, W., Wcisło, R.: Very fast interactive visualization of large sets of high-dimensional data. In: Proceedings of ICCS 2015, Reykjavik, 1–3 June 2015, Iceland, Procedia Computer Science (2015) (in print)
12. Ekanayake, J., Li, H., Zhang, B., Gunarathne, T., Bae, S.H., Qiu, J., Fox, G.: Twister: a runtime for iterative mapreduce. In: Proceedings of the 19th ACM International Symposium on High Performance Distributed Computing, pp. 810–818. ACM (2010)
13. Emms, D., Wilson, R.C., Hancock, E.R.: Graph matching using the interference of continuous-time quantum walks. Pattern Recogn. **42**(5), 985–1002 (2009)
14. Foggia, P., Percannella, G., Vento, M.: Graph matching and learning in pattern recognition in the last 10 years. Int. J. Pattern Recogn. Artif. Intell. **28**(01), 1450001 (2014)
15. Gibert, J., Valveny, E., Bunke, H.: Dimensionality reduction for graph of words embedding. In: Jiang, X., Ferrer, M., Torsello, A. (eds.) GbRPR 2011. LNCS, vol. 6658, pp. 22–31. Springer, Heidelberg (2011)

16. Gupta, P., Goel, A., Lin, J., Sharma, A., Wang, D., Zadeh, R.: Wtf: The who to follow service at twitter. In: Proceedings of the 22nd International Conference on World Wide Web, pp. 505–514. International World Wide Web Conferences Steering Committee (2013)

17. Han, M., Daudjee, K., Ammar, K., Ozsu, M.T., Wang, X., Jin, T.: An experimental comparison of pregel-like graph processing systems. Proc. VLDB Endowment **7**(12), 1047–1058 (2014)

18. Lee, W.-J., Duin, R.P.W.: A labelled graph based multiple classifier system. In: Benediktsson, J.A., Kittler, J., Roli, F. (eds.) MCS 2009. LNCS, vol. 5519, pp. 201–210. Springer, Heidelberg (2009)

19. Leskovec, J., Krevl, A.: SNAP Datasets: Stanford large network dataset collection. http://snap.stanford.edu/data

20. Leskovec, J., Sosič, R.: SNAP: A general purpose network analysis and graph mining library in C++. http://snap.stanford.edu/snap

21. Low, Y., Gonzalez, J.E., Kyrola, A., Bickson, D., Guestrin, C.E., Hellerstein, J.: Graphlab: a new framework for parallel machine learning (2014). arXiv:1408.2041

22. Malewicz, G., Austern, M.H., Bik, A.J., Dehnert, J.C., Horn, I., Leiser, N., Czajkowski, G.: Pregel: a system for large-scale graph processing. In: Proceedings of the 2010 ACM SIGMOD International Conference on Management of data, pp. 135–146. ACM (2010)

23. Qiu, H., Hancock, E.: Clustering and embedding using commute times. IEEE Trans. Pattern Anal. Mach. Intell. **29**(11), 1873–1890 (2007)

24. Salihoglu, S., Widom, J.: Gps: a graph processing system. In: Proceedings of the 25th International Conference on Scientific and Statistical Database Management, p. 22. ACM (2013)

25. Watts, D., Strogatz, S.: Collective dynamics of small-world networks. Nature **393**(6684), 440–442 (1998)

26. Xiao, B., Hancock, E., Wilson, R.: A generative model for graph matching and embedding. Comput. Vis. Image Underst. **113**(7), 777–789 (2009)

27. Zhang, Y., Gao, Q., Gao, L., Wang, C.: Priter: a distributed framework for prioritizing iterative computations. IEEE Trans. Parallel Distrib. Syst. **24**(9), 1884–1893 (2013)

Fast Incremental Community Detection on Dynamic Graphs

Anita Zakrzewska[(✉)] and David A. Bader

Computational Science and Engineering, Georgia Institute of Technology,
Atlanta, GA, USA
azakrzewska3@gatech.edu

Abstract. Community detection, or graph clustering, is the problem of finding dense groups in a graph. This is important for a variety of applications, from social network analysis to biological interactions. While most work in community detection has focused on static graphs, real data is usually dynamic, changing over time. We present a new algorithm for dynamic community detection that incrementally updates clusters when the graph changes. The method is based on a greedy, modularity maximizing static approach and stores the history of merges in order to backtrack. On synthetic graph tests with known ground truth clusters, it can detect a variety of structural community changes for both small and large batches of edge updates.

Keywords: Graphs · Community detection · Graph clustering · Dynamic graphs

1 Introduction

Graphs are used to represent a variety of relational data, such as internet traffic, social networks, financial transactions, and biological data. A graph $G = (V, E)$ is composed of a set of vertices V, which represent entities, and edges E, which represent relationships between entities. The problem of finding dense, highly connected groups of vertices in a graph is called community detection or graph clustering. In cases where communities are non-overlapping and cover the entire graph, the term community partitioning may also be used. Many real world graphs contain communities, such as groups of friends on social networks, lab colleagues in a co-publishing graph, or related proteins in biological networks. In this work, we present a new algorithm for incremental community detection on dynamic graphs.

1.1 Related Work

Most work in community detection has been done for static, unchanging graphs. Popular methods include hierarchical vertex agglomeration, edge agglomeration,

© Springer International Publishing Switzerland 2016
R. Wyrzykowski et al. (Eds.): PPAM 2015, Part I, LNCS 9573, pp. 207–217, 2016.
DOI: 10.1007/978-3-319-32149-3_20

clique percolation, spectral algorithms, and label propagation. Details of a variety of approaches can be found in the survey by Fortunato [11]. The quality of a community C is often measured using a fitness function. Modularity, shown in Eq. 1, is a popular measure that compares the number of intra-community edges to the expected number under a random null model [16]. Let k_{in}^C be the sum of edge weights for edges with both endpoint vertices in C and k_{out}^C be the sum of edge weights for edges with only one endpoint in C.

$$Q(C) = \frac{1}{|E|}(k_{in}^C - \frac{(k_{in}^C + k_{out}^C)^2}{4|E|})$$

(1)

CNM [6] is a hierarchical, agglomerative algorithm that greedily maximizes modularity. Each vertex begins as its own singleton community. Communities are then sequentially merged by contracting the edge resulting in the greatest increase in modularity. Riedy *et al.* present a parallel version of this algorithm in which a weighted maximal matching on edges is used [19].

Real graphs evolve over time. Many works study community detection on dynamic graphs by creating a sequence of data snapshots over time and clustering each snapshot. Chakrabarti *et al.* [5] introduce evolutionary clustering and GraphScope uses block modeling and information compression principles [21]. There are many other works which study the evolution of communities in changing graphs [9,14,22]. However, these are not incremental methods. A second category of algorithms incrementally updates previously computed clusters when the underlying graph changes. By reusing previous computations, incremental algorithms can run faster, which is critical for scaling to large datasets. This is especially useful for applications in which community information must be kept up to date for a quickly changing graph and low latency is desired. Moreover, incremental approaches can result in smoother community transitions.

Duan *et al.* [10] present an incremental algorithm for finding k-clique communities [18] in a changing graph and Ning *et al.* [17] provide a dynamic algorithm for spectral partitioning. Other algorithms search for emerging events in microblog streams [3,4]. Dinh *et al.* [7,8] present an incremental updating algorithm to modularity based CNM. First, standard CNM is run on the initial graph. For each graph update, every vertex that is an endpoint of an inserted edge is removed from its community and placed into a new singleton community. The standard, static CNM algorithm is then restarted and communities may be further merged to increase modularity.

The work most closely related to ours is by Görke *et al.* [13], in which the authors also present two dynamic algorithms that update a CNM based clustering. In their first algorithm, endpoint vertices of newly updated edges, along with a local neighborhood, are freed from their current cluster and CNM is restarted. The second algorithm stores the dendrogram of cluster merges formed by the initial CNM clustering. When the graph changes, the algorithm backtracks cluster merges using this dendrogram until specified conditions are met. If an intra-community edge is inserted, it backtracks till the two endpoint vertices are in separate communities. If an intra-community edge is deleted or an inter-community edge is inserted, it backtracks till both endpoints are their own singleton communities. No backtracking occurs on a inter-community edge deletion. After communities have been modified, CNM is restarted.

1.2   Contribution

In this work we present two incremental algorithms that update communities when the underlying graph changes. The first version, *BasicDyn* is similar to previous work by Görke *et al.* [13]. Our second algorithm, *FastDyn*, is a modification of *BasicDyn* that runs faster. We use various synthetic dynamic graphs to test the quality of our methods. Both *BasicDyn* and *FastDyn* are able to detect community changes.

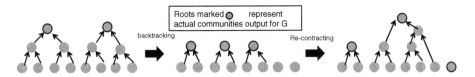

Fig. 1. The left image shows $Forest_G$ created after agglomeration. The center image shows the $Forest_G$ after backtracking undoes merges using *BasicDyn* or *FastDyn*. Right shows after re-starting agglomeration using *Agglomerate* or *FastAgglomerate*.

2   Algorithm

The basis of our algorithm is the parallel version of CNM presented by Riedy *et al.* [19]. We will denote this by *Agglomerate*. Each vertex in the graph G is initialized as its own singleton community, forming the community graph G_{comm}. Note that while G_{comm} is initially the same as G, each vertex of G_{comm} represents a group or cluster of vertices in G. The weighted degree of a vertex $c \in G_{comm}$ corresponds to k_{out}^c and the weight assigned to c (initially 0), corresponds to the intra-community edges k_{in}^c.

Next, the following three steps are repeated until the termination criterion is reached. (1) For each edge in G_{comm}, the change in modularity resulting from an edge contraction is computed. Given an edge (c_1, c_2) in G_{comm}, the change in modularity is given by the value $\Delta Q(c_1, c_2) = Q(c_1 \cup c_2) - (Q(c_1) + Q(c_2))$, which can be easily computed using Eq. 1. This score is assigned to the edge. (2) A weighted maximal matching is computed on the edges of G_{comm} using these scores. (3) Edges in the maximal matching are contracted in parallel. An edge contraction merges the two endpoint vertices into one vertex. Termination occurs when no edge contraction results in a great enough modularity increase.

All vertices that have been merged together by edge contractions correspond to a single community and so by repeatedly merging vertices in G_{comm}, we create ever larger communities in G. The history of these contractions can be stored as a forest of merges. Pairs of vertices in G_{comm} that were merged together by an edge contraction are children of the same parent. The root of each tree corresponds to the final communities found for G. We refer to this forest by $Forest_G$. Edge contraction scores can be computed in $\mathcal{O}(|E|)$ time. If the maximal matching is also computed in $\mathcal{O}(|E|)$, then the complexity of $Agglomerate$ is $\mathcal{O}(K * |E|)$, where K is the number of contraction phases. In the best case, the number of communities halves in each phase and the complexity is $\mathcal{O}(log(|V|) * |E|)$. In the worst case, the graph has a star formation and only one edge contracts in each phase, resulting in a complexity of $\mathcal{O}(|V| * |E|)$.

Our dynamic algorithm begins with the initial graph G and runs the $Agglomerate$ algorithm to create $Forest_G$. We then apply a stream of edge insertions and deletions to G. These updates can be applied in batches of any size. Small batch sizes are used for low latency applications, while aggregating larger batches can result in a relatively lower overall running time. For each batch of updates, our algorithm uses the history of merges in $Forest_G$ and undoes certain merges that previously had taken place. This breaks apart certain previous communities and un-contracts corresponding vertices in G_{comm}. Next, the $Agglomerate$ algorithm is restarted and new modularity increasing merges may be performed. Figure 1 shows this process. Next we describe two versions of the dynamic algorithm, each of which follows the basic steps described.

BasicDyn: When a new batch of edge insertions and deletions is processed, each vertex that is an endpoint of any edge in the batch is marked. Merges are then backtracked in $Forest_G$ until each marked vertex is in its own singleton community. After backtracking, $Agglomerate$ is restarted to re-merge communities.

FastDyn: The $FastDyn$ method is based on $BasicDyn$, but has two modifications that improve computational speed. The first is that backtracking of previous merges in $Forest_G$ occurs under more stringent conditions. Merges are only undone if the quality of the merge, as measured by the induced change in modularity, has significantly decreased compared to when the merge initially took place. Because merges may be performed in parallel using a weighted maximal matching on edges, not every vertex $c \in G_{comm}$ merges with its highest scoring neighbor $best(c)$. When a vertex $c \in G_{comm}$ merges at time t_i, we store both the score of that merge $\Delta Q_{t_i}(c, match(c))$ and the score of the highest possible merge c could have participated in $\Delta Q_{t_i}(c, best_{t_i}(c))$. After a batch of edge updates is applied at time t_{curr} and the graph G changes, the best and actual merge scores are rechecked. The score of a merge has significantly decreased if: $\Delta Q_{t_i}(c, best_{t_i}(c)) - \Delta Q_{t_i}(c, match(c)) + threshold < \Delta Q_{t_{curr}}(c, best_{t_{curr}}(c)) - \Delta Q_{t_{curr}}(c, match(c))$. This occurs if either the score of the merge taken has decreased or the score of the best possible merge has increased.

For each merge evaluated, $FastDyn$ checks the above condition and if it is met, $Forest_G$ is backtracked to undo the merge. Merges that are evaluated are those that occur between clusters that contain at least one member that is directly affected by the batch of edge updates (is an endpoint of an newly inserted

Data: G_{comm} and $Forest_{G,prev}$

```
1  while contractions occuring do
2  |   for v ∈ G_comm in parallel do
3  |   |   paired[v] = 0;
4  |   |   match[v] = v;
5  |   end
6  |   while matches possible do
7  |   |   for v ∈ G_comm s.t. paired[v] == 0 in parallel do
8  |   |   |   max = 0, match[v] = v;
9  |   |   |   for neighbors w of v s.t. paired[w] == 0 do
10 |   |   |   |   if ΔQ(v, w) > max then
11 |   |   |   |   |   max = ΔQ(v, w);
12 |   |   |   |   |   match[v] = w;
13 |   |   |   |   end
14 |   |   |   end
15 |   |   end
16 |   |   for v ∈ G_comm s.t. paired[v] == 0 in parallel do
17 |   |   |   if match[v] > 0 and (match[match[v]] == v or
        prevroot[v] == prevroot[match[v]]) then
18 |   |   |   |   paired[v] = 1;
19 |   |   |   end
20 |   |   end
21 |   end
22 |   for v ∈ G_comm in parallel do
23 |   |   Pointer jump match array to root;
24 |   |   dest[v]=root;
25 |   end
26 |   Contract in parallel all vertices in G_comm with same value of dest;
27 end
```

Algorithm 1. The *FastAgglomerate* algorithm used by *FastDyn*. *prevroot* labels each vertex by its root in $Forest_{G,prev}$, which corresponds to its previous community membership.

or deleted edge). In $Forest_G$, this corresponds to leaf nodes that represent such endpoint vertices as well as any upstream parent of such a leaf node.

The second modification of *FastDyn* occurs in the re-merging step. Instead of running *Agglomerate* after backtracking merges, *FastDyn* runs a modified method, which we will call *FastAgglomerate*. Because *Agglomerate* only allows two vertices to merge together in a contraction phase, star-like formations containing n vertices take n phases to merge, which results in a very long running time. We have found that backtracking merges creates such structures so that running *Agglomerate* after backtracking results in a very long tail of contractions, each of which only contacts a small number of edges. We address this problem with *FastAgglomerate*, which uses information about previous merges to speed up contractions. Instead of performing a maximal edge matching, *FastAgglomerate* allows more than two vertices of G_{comm} to contract together if in the previous time step these vertices eventually ended up contracted into the same community (if they correspond to nodes in $Forest_G$ that previously shared the same root). In the static case, merging several vertices together in one contraction phase could lead to deteriorating results. $FastAgglomerate$ is able to do this, however, because it uses information from the merges of the previous time step. Intuitively, merges that previously occurred are more likely to be acceptable later. Pseudocode is given in Algorithm 1. In the worst case, both $BasicDyn$ and $FastDyn$ will backtrack enough to undo many or most merges. In this case, the running time is the same as that of $Agglomerative$. For small updates, the dynamic algorithms run faster in practice. $FastDyn$ decreases the number of contraction steps needed by allowing several vertices to merge in a single step.

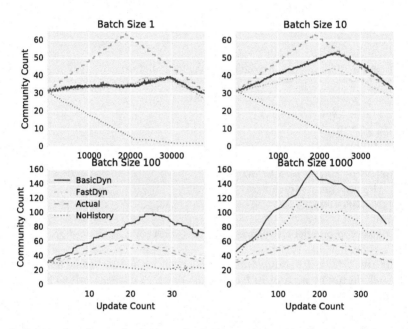

Fig. 2. Results for the synthetic test of splitting and merging clusters are shown. The actual and detected number of communities across time are plotted for batch sizes of 1, 10, 100, and 1000. An alternative approach is shown and labeled "NoHistory" (Color figure online).

3   Results

Dynamic community detection is only useful if the algorithm is able to detect structural changes. We evaluate our algorithm using two tests. First, we form a ring of 32 communities, each with 32 vertices. Each vertex in a community is on average connected to 16 other vertices in the cluster. Over time, edges are both inserted and removed so that each of the 32 clusters consecutively splits in two. Once each community has been split, edges are then updated so that

pairs of clusters merge together (different pairs than at the start), resulting in
32 clusters again. Figure 2 shows the number of communities detected over time
by *BasicDyn* with the solid blue line, the number detected by *FastDyn* with
the green dots and dashes, and the actual number with the dashed red line.
Because graphs are randomly generated, each curve is the average of 10 runs.
The x-axis represents the number of updates that have been made to the graph.
Batch sizes used are 1,10,100, and 1000. A batch size of n means that n edge
insertions and deletions are accumulated before the communities are incrementally updated. Since the curves rise and then fall, both algorithms are able to
detect first splits and then merges. As the batch size increases, the performance
of *FastDyn* improves, which is expected because a larger batch size allows more
communities to be broken apart during backtracking and the algorithm has more
options for how to re-contract communities. Therefore, the algorithm can output
results closer to those that would be found with a static recomputation beginning from scratch. We also compare the ability of *FastDyn* and *BasicDyn* to
detect community changes to the dynamic algorithm from [20], which also uses
greedy modularity maximization. After a batch of edges is inserted and removed,
this method removes each vertex with an updated edge from its current community into its own singelton community. Agglomeration is then restarted. The
results of this alternative approach are shown in Fig. 2 with the purple dotted
line labeled "NoHistory". Unlike the backtracking approaches, community splits
and merges are only detected for a large batch size of 1000.

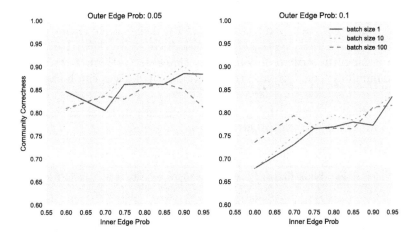

Fig. 3. Results for *FastDyn* on the second synthetic test using a shifting stochastic
block model are shown. The y-axis shows the quality of output and x-axis shows the
probability of an intra-community edge.

The second test is run on a graph formed using a stochastic block model [15].
We generate two separate graphs, blockA and blockB, each with two communities in which the probability of an intra-community edge and that of an inter-
community edge are both specified. BlockB uses the same parameters as blockA,

but with the vertex members of each community changed. Half of the vertices that were in community one in blockA move to the second community of blockB and half of those that were in community two of blockA move to community one of blockB. We begin by running *Agglomerate* on the entire blockA. Then, in a stream of updates, we remove edges from blockA and add edges from blockB. At the end of the stream, the graph will equal blockB. This test differs from the first test in an important way. In test one, the algorithm could always detect some number of distinct communities. Here, distinct communities only exist near the beginning and end of the stream, while in the middle the graph is a mixture of blockA and blockB. The updates to the graph in this test are randomly distributed across all vertices so that much of $Forest_G$ may be affected. In contrast, the updates in test one were applied to one cluster at a time. Lastly, test one specifically tests splits and merges, while test two uses a general rearrangement. Figure 3 shows how well *FastDyn* is able to distinguish the two communities of blockB at the end of the stream. Communities are detected for each batch, but we only evaluate at the end of the stream since that is when the graph returns to a known block structure. The x-axis varies the probability of an intra-community edge. Unsurprisingly, the algorithm performs better when the communities in blockA and blockB are denser. The y-axis shows the correctness of *FastDyn*, which is determined as follows. For each vertex v in the graph, we compare its actual community in blockB to its community output by *FastDyn*. The Jaccard index then measures the normalized overlap between the two sets, with 1 measuring perfect overlap, and 0 no overlap. The average Jaccard index across all vertices measures the algorithm correctness. Because blockA and blockB are randomly generated, each point plotted is an average of 50 separate runs on different graphs. The two synthetic tests described above show that *FastDyn* is able to distinguish structural community changes. Synthetic tests are useful because communities have ground truth and known changes can be applied.

Table 1. The time to compute the initial static community detection and average time to update with incremental algorithms is shown.

Graph	Initial	Batch size	FastDyn	BasicDyn
Facebook	3.5 s	10	0.1 s	0.2 s
		1000	0.7 s	4.9 s
DBLP	122 s	10	7.1 s	21.7 s
		1000	72 s	101 s
Slashdot	45 s	10	0.6 s	2.1 s
		1000	7.5 s	289 s

Table 1 shows the running time of the initial static community detection used by the dynamic algorithms and the average running time to process a batch of size 10 and 1000 edges for *FastDyn* and *BasicDyn*. The algorithms were tested on a Slashdot graph [12], a DBLP graph [1,24], and a Facebook wall post graph [2,23]. The *FastDyn* algorithm is faster than *BasicDyn* for all graphs and batch sizes. The speedup of using a dynamic algorithm is greater for small batch sizes because the incremental algorithms perform less work, while running time for static computation remains the same. Therefore, an incremental approach is most useful for monitoring applications where community results must be updated after a small number of data changes.

4   Conclusion

In this work, we present two incremental community detection algorithms for dynamic graphs. Both use hierarchical, modularity maximizing clustering as their base and update results by storing the history of previous merges in a forest. While *BasicDyn* is similar in nature to a previous approach, *FastDyn* has two improvements that significantly improve running time while maintaining clustering quality. *FastDyn* is faster than *BasicDyn* on all real graphs tested for both small and large update batch sizes. Incrementally updating allows for speedup over recalculating from scratch whenever the graph changes, especially for small batch sizes when low latency updates are needed. While the algorithm is based off a parallel static algorithm, and can be parallelized, the focus of this work was to study improvements in performance due to incremental updates. Parallel scalability is a topic for future work.

Acknowledgments. The work depicted in this paper was sponsored by Defense Advanced Research Projects Agency (DARPA) under agreement #HR0011-13-2-0001. The content, views and conclusions presented in this document do not necessarily reflect the position or the policy of DARPA or the U.S. Government, no official endorsement should be inferred.

References

1. Dblp co-authorship network dataset – KONECT, May 2015. http://konect.unikoblenz.de/networks/com-dblp
2. Facebook wall posts network dataset – KONECT, May 2015. http://konect.unikoblenz.de/networks/facebook-wosn-wall
3. Agarwal, M.K., Ramamritham, K., Bhide, M.: Real time discovery of dense clusters in highly dynamic graphs: identifying real world events in highly dynamic environments. Proc. VLDB Endowment **5**(10), 980–991 (2012)
4. Angel, A., Sarkas, N., Koudas, N., Srivastava, D.: Dense subgraph maintenance under streaming edge weight updates for real-time story identification. Proc. VLDB Endowment **5**(6), 574–585 (2012)
5. Chakrabarti, D., Kumar, R., Tomkins, A.: Evolutionary clustering. In: Proceedings of the 12th ACM SIGKDD International Conference on Knowledge Discovery and Data Mining, pp. 554–560. ACM (2006)
6. Clauset, A., Newman, M.E., Moore, C.: Finding community structure in very large networks. Physical Rev. E **70**(6), 066111 (2004)

7. Dinh, T.N., Shin, I., Thai, N.K., Thai, M.T., Znati, T.: A general approach for modules identification in evolving networks. In: Hirsch, M.J., Pardalos, P.M., Murphey, R. (eds.) Dynamics of Information Systems. Springer Optimization and Its Applications, vol. 40, pp. 83–100. Springer, Heidelberg (2010)

8. Dinh, T.N., Xuan, Y., Thai, M.T.: Towards social-aware routing in dynamic communication networks. In: 2009 IEEE 28th International Performance Computing and Communications Conference (IPCCC), pp. 161–168. IEEE (2009)

9. Duan, D., Li, Y., Jin, Y., Lu, Z.: Community mining on dynamic weighted directed graphs. In: Proceedings of the 1st ACM International Workshop on Complex Networks Meet Information & Knowledge Management, pp. 11–18. ACM (2009)

10. Duan, D., Li, Y., Li, R., Lu, Z.: Incremental k-clique clustering in dynamic social networks. Artif. Intell. Rev. **38**(2), 129–147 (2012)

11. Fortunato, S.: Community detection in graphs. Phys. Rep. **486**(3), 75–174 (2010)

12. Gómez, V., Kaltenbrunner, A., López, V.: Statistical analysis of the social network and discussion threads in slashdot. In: Proceedings of the 17th International Conference on World Wide Web, pp. 645–654. ACM (2008)

13. Görke, R., Maillard, P., Schumm, A., Staudt, C., Wagner, D.: Dynamic graph clustering combining modularity and smoothness. J. Exp. Algorithmics (JEA) **18**, 1–5 (2013)

14. Greene, D., Doyle, D., Cunningham, P.: Tracking the evolution of communities in dynamic social networks. In: 2010 International Conference on Advances in Social Networks Analysis and Mining (ASONAM), pp. 176–183. IEEE (2010)

15. Holland, P.W., Laskey, K.B., Leinhardt, S.: Stochastic blockmodels: first steps. Soc. Netw. **5**(2), 109–137 (1983)

16. Newman, M.E., Girvan, M.: Finding and evaluating community structure in networks. Phys. Rev. E **69**(2), 026113 (2004)

17. Ning, H., Xu, W., Chi, Y., Gong, Y., Huang, T.S.: Incremental spectral clustering by efficiently updating the eigen-system. Pattern Recogn. **43**(1), 113–127 (2010)

18. Palla, G., Derényi, I., Farkas, I., Vicsek, T.: Uncovering the overlapping community structure of complex networks in nature and society. Nature **435**(7043), 814–818 (2005)

19. Riedy, E.J., Meyerhenke, H., Ediger, D., Bader, D.A.: Parallel community detection for massive graphs. In: Wyrzykowski, R., Dongarra, J., Karczewski, K., Waśniewski, J. (eds.) PPAM 2011, Part I. LNCS, vol. 7203, pp. 286–296. Springer, Heidelberg (2012)

20. Riedy, J., Bader, D., et al.: Multithreaded community monitoring for massive streaming graph data. In: 2013 IEEE 27th International Parallel and Distributed Processing Symposium Workshops & PhD Forum (IPDPSW), pp. 1646–1655. IEEE (2013)

21. Sun, J., Faloutsos, C., Papadimitriou, S., Yu, P.S.: Graphscope: parameter-free mining of large time-evolving graphs. In: Proceedings of the 13th ACM SIGKDD International Conference on Knowledge Discovery and Data Mining, pp. 687–696. ACM (2007)

22. Tantipathananandh, C., Berger-Wolf, T., Kempe, D.: A framework for community identification in dynamic social networks. In: Proceedings of the 13th ACM SIGKDD International Conference on Knowledge Discovery and Data Mining, pp. 717–726. ACM (2007)

23. Viswanath, B., Mislove, A., Cha, M., Gummadi, K.P.: On the evolution of user interaction in facebook. In: Proceedings of the 2nd ACM Workshop on Online Social Networks, pp. 37–42. ACM (2009)
24. Yang, J., Leskovec, J.: Defining and evaluating network communities based on ground-truth. Nature **42**(1), 181–213 (2015)

A Diffusion Process for Graph Partitioning: Its Solutions and Their Improvement

Andreas Jocksch[(⊠)]

CSCS, Swiss National Supercomputing Centre, Via Trevano 131,
6900 Lugano, Switzerland
jocksch@cscs.ch

Abstract. We present a diffusion process on graphs for k-way partitioning. In this approach, various species propagate on the graph that cancel each other out, and every partition is represented by one species of the converged solution. The vertices and edges of the graph are reservoirs and resistances, respectively, and source terms are placed on the vertices. A distribution of these source terms on the graph is suggested and the resulting k-way partitioning of the diffusion process for basic graphs discussed. We present reference examples in which complex graphs are recursively bi-partitioned with a diffusion step and a subsequent Kernighan-Lin improvement step. For comparison the graphs are also partitioned with multilevel methods and a subsequent Kernighan-Lin improvement. For certain graphs the diffusion approach produces the best partitions.

Keywords: Graph partitioning · Diffusion

1 Introduction

The partitioning of graphs is the basis of many applications, such as integrated circuit design, image segmentation and load balancing in numerical simulations. Several heuristic approaches exist in order to solve this NP-hard problem, the most popular ones are the Kernighan-Lin (KL) heuristics [10], simulated annealing [11], spectral partitioning [6], partitioning based on geometry information [4] and diffusion. Several further developments to the Kernighan-Lin heuristics have been done. Fiduccia Mattheyses suggested a data handling for execution in linear time [5], an extension to k-way partitioning has been performed [17] and multiple levels of gain introduced [12]. One distinguishes between global methods as, e.g., spectral methods and methods applied locally as, e.g., Kernighan-Lin. Graph partitioning algorithms have been applied in a multilevel approach [8], which is the established method in order to obtain partitions in an efficient way. With a coarsening scheme multiple levels of coarser graphs are generated: the partitioning scheme is applied first to the coarsest level, then the solution is transferred to finer graphs and improved with refinement methods. Among others, a local refinement scheme based on the 'max-flow min-cut' algorithm has been successfully applied [18].

© Springer International Publishing Switzerland 2016
R. Wyrzykowski et al. (Eds.): PPAM 2015, Part I, LNCS 9573, pp. 218–227, 2016.
DOI: 10.1007/978-3-319-32149-3_21

Our focus is on diffusion-based graph partitioning at a single level. The results of the partitioning are improved with a randomized Kernighan-Lin algorithm. We see the main application of this contribution in load balancing for parallel computing, where the optimization target is to equalize the load of the processors and to minimize the communication costs. Thus, for the target the number of vertices at the partition boundaries had to be minimized, which represent the communication volume [15]. However, this goal is reached only indirectly, since most of the graph partitioning heuristics — as the one of Kernighan-Lin [10] — minimize the number of edge cuts.

2 Related Work

If we consider the graph of a network for load balancing, the exchange of load between neighboring processors can be described as a diffusion process [20], which has been shown to be analogous to the solution of the Poisson equation [3]. Diffusion on a random walk basis has been applied for determining coordinates of graphs, which were not only used for graph partitioning but also for dimensionality reduction and data set parametrization [13]. There are also deterministic approaches for the diffusion process for graph partitioning: In the 'bubble' algorithm [4] partitions grow in circles from an initial seed until they reach the border or other bubbles. In a loop the seed is replaced in the middle of the bubble and the process is repeated until convergence is reached. Motivated by the bubble algorithm, global diffusion processes were suggested, based on the solution of the diffusion equation. Pellegrini introduced the 'scotch' scheme [16] for bi-partitioning. Sources of two different species are placed at two dedicated vertices of the graph. The two different species propagate through the graph according to the diffusion equation. Every vertex acts as a small sink of the species and when the two different species meet they cancel out each other. The species of a particular vertex determines its partition. The scheme was applied for the refinement of parts of graphs. For efficiency reasons it was not computed until convergence but only a few iterations were performed.

Meyerhenke et al. [14] considered the case of many partitions, where the diffusion was performed in a loop of initial loads on center vertices or on whole partitions which were distributed by diffusion, and new centers or new partitions were determined [15]. Only a few iterations were executed for the single diffusion process. Thus, the distribution of the initial loads was a local operation on the graph with respect to the single partitions. Convergence for the whole problem was obtained by the loop over the single processes. For the solution of the diffusion process a multigrid scheme was applied.

We follow the concept of species canceling each other out [16], but for k species where all vertices act as source and we seek a fully converged solution.

3 Diffusion Algorithm

The diffusion equation is solved on the connected undirected graph $G = (V, E)$ with $|V| = n$ vertices and $|E| = m$ edges [3,16]. Vertex $v_i \in V$ and edge $e_{ij} =$

$(v_i, v_j) \in E$ have a weight of $c_i \in \mathbb{R}$ and $d_{ij} \in \mathbb{R}$, respectively. Furthermore the vertex has a load of $w_i \in \mathbb{R}$ and $\breve{w}_i \in \mathbb{N}$ is its species.

We seek a stationary solution, but for a convenient subsequent calculation we formulate the problem as an instationary process

$$\frac{\partial w_i}{\partial t} = \frac{\sum_k f_{ik} + s_i}{c_i}, \qquad\qquad w_i \geq 0, \quad e_{ik} \in E \qquad (1a)$$

$$\breve{w}_i = \begin{cases} \breve{w}_i & w_i > 0 \\ \breve{w}_{\arg\max_k(f_{ik})} & w_i = 0 \end{cases} \qquad\qquad (1b)$$

$$f_{ij} = \begin{cases} (w_i - w_j)\, d_{ij} & \breve{w}_i = \breve{w}_j \\ -(w_i + w_j)\, d_{ij} & \breve{w}_i \neq \breve{w}_j \end{cases} \qquad\qquad (1c)$$

where t is the time variable, f_{ij} the flux between vertices i and j, and s_i the source at the vertex i.

Every partition is represented by one species and the solution — the temporal derivative is supposed to vanish — specifies the final partitioning. Pellegrini's concept of different species canceling each other out [16] is extended here from two to many species. Thus only a scalar value has to be determined on the whole graph.

An alternative formulation for two species is without the species variable but with positive and negative load values — the sign indicates the species — which are fed with source terms in one partition and sink terms in the other, respectively. Thus, in the case of two species their development over time (Eq. 1) with constant source terms is a linear operation.

For the source term s_i different concepts can be chosen. The source can be localized to a single vertex [14]. The disadvantage of this procedure is that the centers tend to locate at vertices which are sparsely connected. We favor the concept that the total source S_p of partition p is distributed by a multiplicative scheme $\theta = 1$ or by an additive scheme $\theta = 0$; the latter represents equal distribution. In contrast to sources of the partitions at single vertices, changes to the strength are instantaneously distributed over the whole partition. The sources are determined according to

$$s_i = \frac{w_i c_i S_{\breve{w}_i}}{\sum_{k,\breve{w}_i = \breve{w}_k} w_k c_k}\, \theta + \frac{c_i S_{\breve{w}_i}}{\sum_{k,\breve{w}_i = \breve{w}_k} c_k}\,(1 - \theta)\,, \qquad\qquad (2a)$$

$$S_p = b_1 + b_2(N_{p,t} - N_p) \qquad\qquad (2b)$$

where $N_{p,t}$, N_p are the target and actual number of vertices of the partition p, respectively. The parameters b_1 and b_2 are determined heuristically where the damping is optimized. At the border between the species we interpolate linearly, according to the scheme

```
for k := 1 , n
    ips := 0
    for i := 1 , numedges[k]
        if w̆_i ≠ w̆_k and ( w_i/(w_i+w_k) - 1/2 ) < ips
            ips := ( w_i/(w_i+w_k) - 1/2 )
    end
    end
    N'_i := N'_i + ips · c_i
    N'_k := N'_k - ips · c_k
end
```

which considers for each vertex the edge with partition boundary closest to the vertex and ensures that the sum of the partition sizes always equals the weighted vertex number $N = \sum_k N_k = \sum_k c_k$. Thus

$$N_i = \sum_{k,\,\breve{w}_i=\breve{w}_k} c_k + \xi N'_i, \tag{3}$$

where for $\xi = 0$ no interpolation is done while for $\xi = 1$ linear interpolation is employed. It is a disadvantage of the linear interpolation that the solution does not trend to the exact non-interpolated target partition size. However, we see the benefit of a more exact solution due to the smoother function of the partition size.

Time integration can be performed by any time marching scheme. We apply the Euler forward integration scheme since it is descriptive. In order to ensure numerical stability, the timestep Δt for the Euler forward integration scheme is chosen to be $\Delta t = \min_i 1/2 \cdot (c_i / \sum_k d_{ik})$, based on a local estimate where the nonlinear term s_i/c_i is neglected. From the numerical stability of the time integration and consistency of Eq. 1 follows convergence according to the Lax-Richtmyer equivalence theorem. For the special case of two partitions and $s_i = const.$ the linear equation (Eq. 1) could be solved with any method, as an algebraic multigrid method [14] which accelerates the convergence. In the general case considered here a multilevel approach [8] can be used for convergence acceleration.

The algorithm can be used for improvement of an existing solution generated by other procedures, e.g. other diffusion algorithms [13,16], or standalone. However, in the latter case an initial solution is required and we use random assignment as the initial condition. Two different approaches are applied [15]. In the first, the number of vertices equal to the partition sizes is randomly selected, while in the second, an initial vertex is selected randomly for every partition. These initial vertices are assigned to their specific species, while the rest of the graph is assigned to no species, i.e., on these vertices s_i is chosen to be zero. In the initial phase the partitions grow until all unoccupied parts of the graph are assigned to a species. The solution proceeds further in time until convergence is reached. For the initial phase if no other species is reached this corresponds to a breadth first search (BFS) on the graph. The second approach leads very quickly to compact clusters of vertices.

4   Properties of the Solution and Convergence

If the graph presents an l-dimensional l times periodic homogeneous mesh the converged solution for a small partition and a large partition is a line segment, a circle and a sphere for dimension 1, 2 and 3, respectively, for both the multiplicative distribution of the source and the additive one. The minimum edge cut solution is an l-dimensional hypercube. Figure 1a shows the solution for partition sizes of ratio 1 : 15 for a two-dimensional mesh. The circle is stationary and its location depends on the random initial condition. The number of boundary points is smaller than the one of a square of the same partition size, which is the minimum edge-cut solution.

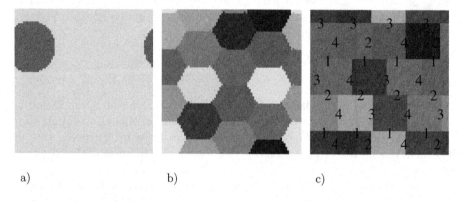

a)   b)   c)

Fig. 1. Diffusion. a) multiplicative scheme, 2 partitions of different size b) multiplicative scheme, 16 partitions of equal size c) recursive application of additive scheme, 16 partitions of equal size, steps of recursive partitioning 1..4 are indicated on their cuts

Figure 1b shows the result of the diffusion process for 16 species and equal partition sizes on a two-dimensional mesh where the multiplicative scheme is applied. As for the circle, the hexagon-like shapes (different edge lengths) of the species domains are the minimum surface-to-volume ratio solution which has been shown by disturbed diffusion [14].

For a recursive application of the diffusion process with multiplicative source the scheme does not find the minimum surface-to-volume ratio solution (Fig. 1c) but it finds it for an additive source. This solution is also the optimum edge cut solution for the global problem. However, it is not ideal in terms of communication costs since the partitions have more neighbors than necessary.

We determine the damping with the representative cases of two partitions, one smaller and one larger, for hypercubes of dimension 1 and 3. Figure 2 shows the development of the partition sizes over time for different damping constants b_1; b_2 is assumed to be 1. It strongly depends on the damping how fast the size converges to the target. Low damping shows initial oscillations, while high damping slows the approach to the solution. Zero damping causes temporal oscillations of the cube's solution (Fig. 2b). Although we compare graphs of approximately equal size it requires a much longer time until convergence for the line section (Fig. 2a) than for the cube (Fig. 2b). The number of iterations until convergence depends not only on the size of the graph but also on its structure. With increasing problem size the damping parameter for fastest convergence increases. Thus we make the conservative assumption that the actual graphs to be partitioned are line segments and choose a damping constant of 16 for complex graphs (Sect. 5).

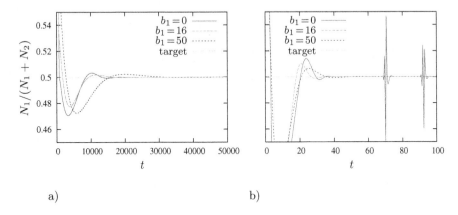

a)   b)

Fig. 2. Partition size in dependency of time for different damping constants b_1 with interpolation ($\xi = 1$) a) line segment, $n = 500$ b) cube, $n = 8^3$

Compact partitions are produced, although the solution might converge to a local optimum. The partitions should not be fragmented. If at initialization or at a certain integration time a fragmentation exists it should vanish in the solution process (with exceptions like symmetric problems).

For the 16-way partitioning example with equal partition sizes (Fig. 1b) the initial condition obtained from method two leads to a faster convergence; the fast compact clustering is advantageous in order to prevent fragmentation which needs to be resolved. In contrast, in the example with a partition size ratio of 1 : 15 (Fig. 1a), it is initial condition method one which leads to faster results, since the different species have approximately their final sizes from the outset, i.e., a temporary solution with equal partition sizes like method two would produce is avoided.

Our experience shows empirically that the solution converges within polynomial time.

5   Experiments

Although we seek primarily a minimum number of vertices at the partition boundary, the number of edge cuts is also a target, at least for comparison, since the performance of graph partitioning schemes is typically evaluated on the size of the partitions produced and the number of edge cuts. The diffusion process finds compact clusters but we want to improve its properties since it does

224   A. Jocksch

Table 1. Recursive partitions for different methods before and after improvement. Number triples indicate: max. partition size, max. number of edge cuts and max. number of boundary vertices, min. edge-cut solutions after improvement are in **bold**

graph	diffusion				multilevel KL			Metis			
	2 partitions										
4elt	7803	159	80	a	7803	158	80	7803	154	78	p
598a	55486	2520	679	m	55486	2465	673	55486	2504	689	p
auto	224413	11256	2940	a	224348	10512	2729	225169	10557	2724	k
fe_ocean	71769	590	432	m	71719	499	441	71819	536	403	k
fe_rotor	49991	2441	698	m	49809	2151	610	49809	2190	620	p
fe_sphere	8193	394	197	a	8193	424	212	8193	440	220	p
fe_tooth	39069	4029	1240	a	39068	4642	1438	39068	4297	1312	p
hammond	2360	110	58	a	2360	103	54	2376	97	49	k
	2 partitions with KL improvement										
4elt	**7803**	**145**	**73**	a	7803	154	78	7803	154	78	p
598a	**55486**	**2434**	**673**	m	55486	2465	673	55486	2504	689	p
auto	224348	10694	2807	a	**224348**	**10512**	**2729**	224348	10620	2764	k
fe_ocean	**71719**	**483**	**427**	m	71719	499	441	**71719**	**483**	**427**	k
fe_rotor	**49809**	**2141**	**609**	m	49809	2151	610	**49809**	**2141**	586	p
fe_sphere	**8193**	**394**	**197**	a	8193	408	204	8193	398	199	p
fe_tooth	**39068**	**3906**	**1210**	a	39068	4443	1392	39068	4146	1276	p
hammond	2360	99	51	a	**2360**	**90**	**46**	**2360**	**90**	**46**	k
	4 partitions with KL improvement										
4elt	**3902**	**201**	**102**	a	3902	213	107	3902	267	131	p
598a	27743	5840	1567	a	27743	6005	1631	**27743**	**5803**	**1583**	k
auto	112174	18968	4901	a	**112174**	**18497**	**4753**	112174	19368	5069	k
fe_ocean	35860	1407	1280	m	35860	1581	1416	**35860**	**1385**	**1132**	k
fe_rotor	**24905**	**5379**	**1458**	m	24905	5392	1470	24905	5380	1433	p
fe_sphere	**4097**	**422**	**209**	a	4097	428	213	4097	436	217	p
fe_tooth	**19534**	**3787**	**1158**	a	19534	5029	1544	19534	4200	1247	k
hammond	**1180**	**138**	**70**	a	1180	131	66	**1180**	**138**	71	k
	8 partitions with KL improvement										
4elt	**1951**	**169**	**85**	a	1951	206	102	1952	211	103	p
598a	**13872**	**5102**	**1382**	a	13872	5502	1461	13873	6872	1832	k
auto	56088	18766	4814	a	56088	21774	5618	**56088**	**18113**	**4679**	p
fe_ocean	**17930**	**1646**	**1449**	m	17930	1708	1502	17930	1700	1477	p
fe_rotor	12453	5200	1381	m	12453	5532	1459	**12453**	**5178**	**1360**	p
fe_sphere	**2049**	**356**	**175**	a	2050	376	185	2049	360	177	k
fe_tooth	**9767**	**4144**	**1178**	m	9767	4705	1427	9767	4492	1308	p
hammond	**590**	**130**	**62**	a	590	136	64	590	137	64	p

m multiplicative, *a* additive, *k* kmetis, *p* pmetis

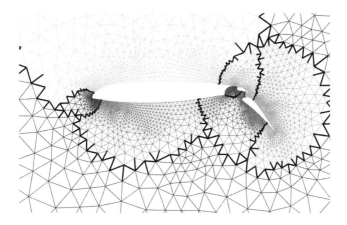

Fig. 3. Best min-cut solution of graph hammond (Table 1; mesh around a three element airfoil [6]; generated from data provided with Chaco [2]), split into 8 partitions, obtained with diffusion scheme and additive sources plus Kernighan-Lin improvement

not optimize for edge cuts as other schemes do. The edge cut related approach — we follow the Kernighan-Lin heuristics — tends to find local optima which are not well clustered. Thus the idea is to combine the diffusion process with the edge cut related approach. For initial partitioning we apply the diffusion method and afterwards the Kernighan-Lin scheme is used for improvement; for the latter we use the implementation of Chaco [7]. The diffusion, although it might find local minima itself, is less sensitive to local extrema and prevents the Kernighan-Lin scheme from iterating to its local minima. The smooth solution is transferred to a minimum edge cut solution by the improvement scheme. A comparison with other methods for initial partitioning, a Kernighan-Lin multigrid scheme (Chaco 2.2) and Metis 4.0.3 [8] is shown in Table 1. For both methods the Kernighan-Lin improvement scheme is applied. The graphs [1,2,7,19] have been partitioned recursively, including the improvement scheme into the recursion, with zero imbalance $\epsilon = 0$ after improvement. The time $t = 25000$ has been considered for the diffusion solution to be converged.

The number of edge cuts after the diffusion is not necessarily a measure of the quality of the diffusion solution but the solution after the improvement should be taken into account. Also the diffusion does not provide a perfect balance of partition sizes: this is obtained by the improvement scheme. The values shown in Table 1 are the best edge-cut results after improvement out of 10 realizations with different initial random seeds.

In general best edge-cut results are also good results for the number of vertices at the partition boundary. The best edge-cut solutions in Table 1 (bold) show also the minimum number of vertices at the partition boundary.

According to the particular graph and partition counts, each of the three different methods — diffusion, multilevel Kernighan-Lin or Metis — can produce the best partitions. The choice of the eight graphs (Table 1) introduces an element of random selection into our results; however, the fact that our approach compared to the multilevel schemes partially produces best min-cut solutions is interpreted as systematic. Thus we can consider that the diffusion method is indeed a useful method; however, the best known partitionings of the graphs [1,19] could not be improved.

Figure 3 shows the minimum edge-cut partitioning of the graph 'hammond' [6]. Seven out of the eight partitions are connected to the airfoil structure in the middle of the graph and contain part of the refined mesh.

Our implementation of the diffusion algorithm is a shared memory thread parallel version (OpenMP). The parallelization is straightforward, vertices and edges are simply distributed to the different threads, while the algorithm and its results remain the same as for the serial version. The speedup for an Intel® Haswell® processor from 1 thread to 12 threads is 3.6 for the bi-partitioning of the graph 'auto'.

6 Conclusions

A diffusion scheme for graph partitioning that is similar to that used in scotch [16], but different to that of Meyerhenke et al. [15], has been suggested, where the converged solution is used for the partitioning. Compared to the latter reference the approach has the advantage that the diffusion process is performed with only one real variable per vertex. We assume that the NP-hard graph partitioning problem is solved in polynomial time by the diffusion scheme.

It has been shown for the partitioning of a regular mesh that the diffusion scheme produces partitions on the graph with a trend to a minimum surface-to-volume ratio. For eight selected graphs the diffusion solution has been improved with the Kernighan-Lin heuristics. Samples with different initial random seed have been considered in order to exclude local minima found by the method. From the eight graphs each partitioned into 2, 4 and 8 partitions in 17 of 24 cases the best edge-cut solution was obtained by the diffusion method with improvement.

The method should be well suited for load balancing in parallel computing, where a high quality partitioning is required. It would be combined with other algorithms as part of a more complex partitioning suite such as other improvement schemes. Standalone, the partitioning costs are rather high compared to other methods, such as, e.g., Kernighan-Lin applied here, which would be no disadvantage if the effort for the partitioning is small compared to the overall costs of the computation.

The shared memory parallelization of the diffusion scheme is very straightforward.

Integrating for only a short time, rather than using the fully converged solution, is a possible option for efficiency improvement to be investigated. For a faster execution, and possibly also better partitioning, the method could be integrated in a multilevel approach [9].

Acknowledgements. The authors would like to thank T.W. Robinson and W. Sawyer for fruitful discussions.

References

1. http://www.corc.ieor.columbia.edu/meetings/ipcox/talks/kevin/export-ipco-talk/gparchive.html
2. https://cfwebprod.sandia.gov/cfdocs/CompResearch/templates/insert/software.cfm
3. Boillat, J.E.: Load balancing and Poisson equation in a graph. Concur. Pract. Exper. **2**(4), 289–313 (1990)
4. Diekmann, R., Preis, R., Schlimbach, F., Walshaw, C.: Shape-optimized mesh partitioning and load balancing for parallel adaptive FEM. Parallel Comput. **26**, 1555–1581 (2000)
5. Fiduccia, C.M., Mattheyses, R.M.: A linear-time heuristic for improving network partitions. In: 19th Design Automation Conference, pp. 175–181. IEEE (1982)
6. Hendrickson, B., Leland, R.: An improved spectral graph partitioning algorithm for mapping parallel computations. SIAM J. Sci. Comput. **16**(2), 452–469 (1995). SANDIA REPORT SAND92-1460 · UC-405 September 1992
7. Hendrickson, B., Leland, R.: A multilevel algorithm for partitioning graphs. In: Proceedings of Supercomputing 1995, vol. 1, pp. 626–657. ACM/IEEE (1995)
8. Karypis, G., Kumar, V.: A fast and high quality multilevel scheme for partitioning irregular graphs. SIAM J. Sci. Comput. **20**(1), 359–392 (1998)
9. Karypis, G., Kumar, V.: A parallel algorithm for multilevel graph partitioning and sparse matrix ordering. J. Parallel Distrib. Comput. **48**, 71–95 (1998)
10. Kernighan, B.W., Lin, S.: An efficient heuristic procedure for partitioning graphs. Bell Syst. Tech. J. **49**(2), 291–307 (1970)
11. Kirkpatrick, S., Gelatt Jr., C.D., Vecchi, M.P.: Optimization by simulated annealing. Science **220**(4598), 671–680 (1983)
12. Krishnamurthy, B.: An improved min-cut algorithm for partitioning VLSI networks. IEEE Trans. Comput. **c−33**(5), 438–446 (1984)
13. Lafon, S., Lee, A.B.: Diffusion maps and coarse-graining: a unified framework for dimensionality reduction, graph partitioning, and data set parametrization. IEEE Trans. Pattern Anal. **28**(9), 1393–1403 (2006)
14. Meyerhenke, H., Monien, B., Sauerwald, T.: A new diffusion-based multilevel algorithm for computing graph partitions of very high quality. In: Parallel and Distributed Processing, IPDPS, pp. 1–13. IEEE, Miami (2008)
15. Meyerhenke, H., Monien, B., Schamberger, S.: Graph partitioning and disturbed diffusion. Parallel Comput. **35**, 544–569 (2009)
16. Pellegrini, F.: A parallelisable multi-level banded diffusion scheme for computing balanced partitions with smooth boundaries. In: Kermarrec, A.-M., Bougé, L., Priol, T. (eds.) Euro-Par 2007. LNCS, vol. 4641, pp. 195–204. Springer, Heidelberg (2007)
17. Sanchis, L.A.: Multiple-way network partitioning. IEEE Trans. Comput. **38**(1), 62–81 (1989)
18. Sanders, P., Schulz, C.: Engineering multilevel graph partitioning algorithms. In: Demetrescu, C., Halldórsson, M.M. (eds.) ESA 2011. LNCS, vol. 6942, pp. 469–480. Springer, Heidelberg (2011)
19. Soper, A.J., Walshaw, C., Cross, M.: A combined evolutionary search and multilevel optimisation approach to graph-partitioning. J. Global Optim. **29**, 225–241 (2004)
20. Xu, C., Lau, F.C.M.: Load Balancing in Parallel Computers: Theory and Practice. Kluwer, Boston (1997)

A Parallel Algorithm for LZW Decompression, with GPU Implementation

Shunji Funasaka, Koji Nakano$^{(\boxtimes)}$, and Yasuaki Ito

Department of Information Engineering, Hiroshima University,
Kagamiyama 1-4-1, Higashihiroshima 739-8527, Japan
nakano@hiroshima-u.ac.jp

Abstract. The main contribution of this paper is to present a parallel algorithm for LZW decompression and to implement it in a CUDA-enabled GPU. Since sequential LZW decompression creates a dictionary table by reading codes in a compressed file one by one, its parallelization is not an easy task. We first present a parallel LZW decompression algorithm on the CREW-PRAM. We then go on to present an efficient implementation of this parallel algorithm on a GPU. The experimental results show that our parallel LZW decompression on GeForce GTX 980 runs up to 69.4 times faster than sequential LZW decompression on a single CPU. We also show a scenario that parallel LZW decompression on a GPU can be used for accelerating big data applications.

Keywords: Data compression · Big data · Parallel algorithm · GPU · CUDA

1 Introduction

A GPU (Graphics Processing Unit) is a specialized circuit designed to accelerate computation for building and manipulating images [4,7,14] Latest GPUs are designed for general purpose computing and can perform computation in applications traditionally handled by the CPU. Hence, GPUs have recently attracted the attention of many application developers [5,10]. NVIDIA provides a parallel computing architecture called *CUDA* (Compute Unified Device Architecture) [11], the computing engine for NVIDIA GPUs. CUDA gives developers access to the virtual instruction set and memory of the parallel computational elements in NVIDIA GPUs.

There is no doubt that data compression is one of the most important tasks in the area of computer engineering. In particular, almost all image data are stored in files as compressed data formats. In this paper, we focus on LZW compression, which is the most well known patented lossless compression method [15] used in Unix file compression utility "compress" and in GIF image format. Also, LZW compression option is included in TIFF file format standard [1], which is commonly used in the area of commercial digital printing. However, LZW compression and decompression are hard to parallelize, because they use dictionary

R. Wyrzykowski et al. (Eds.): PPAM 2015, Part I, LNCS 9573, pp. 228–237, 2016.
DOI: 10.1007/978-3-319-32149-3_22

tables created by reading input data one by one. Parallel algorithms for LZW compression and decompression have been presented [6]. However, processors perform compression and decompression in block-wise, it achieved a speed up factor of no more than 3. In [13], a CUDA implementation of LZW compression has been presented. But, it achieved only a speedup factor less than 2 over the CPU implementation using MATLAB. Also, several GPU implementations of dictionary based compression methods have been presented [9,12]. As far as we know, no parallel LZW decompression using GPUs has not been presented. In particular, decompression may be performed more frequently than compression; each image is compressed and written in a file once, but it is decompressed whenever the original image is used. Hence, we can say that LZW decompression is more important than the compression.

The main contribution of this paper is to present a parallel algorithm for LZW decompression and the GPU implementation. We first show that a parallel algorithm for LZW decompression on the CREW-PRAM [2], which is a traditional theoretical parallel computing model with a set of processors and a shared memory. We will show that LZW decomposition of a string of m codes can be done in $O(L_{max} + \log m)$ time using $\max(k, m)$ processors on the CREW-PRAM, where L_{max} is the maximum length of characters assigned to a code. We then go on to show an implementation of this parallel algorithm in CUDA architecture. The experimental results using GeForce GTX 980 GPU and Intel Xeon CPU X7460 processor show that our implementation on a GPU achieves a speedup factor up to 69.4 over a single CPU.

Suppose that we have a set of bulk data such as images or text stored in a storage of a host computer with a GPU. A user gives a query to the set of bulk data and all data must be processed to answer the query. To accelerate the computation for the query, data are transferred to the GPU through the host computer and they are processed by parallel computation on the GPU. For the purpose of saving storage space and data transfer time, data are stored in the storage as LZW compressed format. If this is the case, compressed data must be decompressed using the host computer or using the GPU before the query processing is performed. Note that, since the query is processed by the GPU, it is not necessary to transfer decompressed images from the GPU to the host computer. We will show that, since LZW decompression can be done very fast in the GPU by our parallel algorithm, it makes sense to store compressed data in a storage and to decompress them using the GPU.

2 LZW Compression and Decompression

The main purpose of this section is to review LZW compression/decompression algorithms. Please see Sect. 13 in [1] for the details.

The LZW (Lempel-Ziv & Welch) [16] compression algorithm converts an input string of characters into a string of codes using a string table that maps strings into codes. If the input is an image, characters may be 8-bit integers. It reads characters in an input string one by one and adds an entry in a string table (or a dictionary). At the same time, it writes an output string of codes by looking up the string table. Let $X = x_0x_1 \cdots x_{n-1}$ be an input string of characters and $Y = y_0y_1 \cdots y_{m-1}$ be an output string of codes. For simplicity of handling the boundary case, we assume that an input is a string of 4 characters a, b, c, and d. Let S be a string table, which determines a mapping of a string to a code, where codes are non-negative integers. Initially, $S(a) = 0$, $S(b) = 1$, $S(c) = 2$, and $S(d) = 3$. By procedure AddTable, new code is assigned to a string. For example, if AddTable(cb) is executed after initialization of S, we have $S(cb) = 4$. The LZW compression algorithm is described as follows:

[LZW compression algorithm]
```
1   for i ← 0 to n − 1 do
2       if(Ω · x_i is in S) Ω ← Ω · x_i;
3       else Output(S(Ω)); AddTable(Ω · x_i); Ω ← x_i;
4   Output(S(Ω));
```

In this algorithm, Ω is a variable to store a string. Also, "$\cdot$" denotes the concatenation of strings/characters.

Table 1 shows how the compression process for an input string $cbcbcbcda$. First, since $\Omega \cdot x_0 = c$ is in S, $\Omega \leftarrow c$ is performed. Next, since $\Omega \cdot x_1 = cb$ is not in S, Output($S(c)$) and AddTable(cb) are performed. More specifically, $S(c) = 2$ is output and we have $S(cb) = 4$. Also, $\Omega \leftarrow x_1 = b$ is performed. It should have no difficulty to confirm that 214630 is output by this algorithm.

Table 1. String table S, string stored in Ω, and output string Y for $X = cbcbcbcda$

i	0	1	2	3	4	5	6	7	8	-
x_i	c	b	c	b	c	b	c	d	a	
Ω	-	c	b	c	cb	c	cb	cbc	d	a
S	-	cb : 4	bc : 5	-	cbc : 6	-	-	cbcd : 7	da : 8	-
Y	-	2	1	-	4	-	-	6	3	0

Next, let us show LZW decompression algorithm. Let C be *the code table*, the inverse of string table S. For example if $S(cb) = 4$ then $C(4) = cb$. Initially, $C(0) = a$, $C(1) = b$, $C(2) = c$, and $C(3) = d$. Also, let $C_1(i)$ denote the first character of code i. For example $C_1(4) = c$ if $C(4) = cb$. Similarly to LZW compression, the LZW decompression algorithm reads a string Y of codes one by one and adds an entry of a code table. At the same time, it writes a string X of characters. The LZW decompression algorithm is described as follows:

[LZW decompression algorithm]
```
1   Output(C(y_0));
2   for i ← 1 to n − 1 do
3       if(y_i is in C) Output(C(y_i)); AddTable(C(y_{i−1}) · C_1(y_i));
4       else Output(C(y_{i−1}) · C_1(y_{i−1})); AddTable(C(y_{i−1}) · C_1(y_{i−1}));
```

Table 2 shows the decompression process for a code string 214630. First, $C(2) = c$ is output. Since $y_1 = 1$ is in C, $C(1) = b$ is output and AddTable(cb) is performed. Hence, $C(4) = cb$ holds. Next, since $y_2 = 4$ is in C, $C(4) = cb$ is output and AddTable(bc) is performed. Thus, $C(5) = bc$ holds. Since $y_3 = 6$ is not in C, $C(y_2) \cdot C_1(y_2) = cbc$ is output and AddTable(cbc) is performed. The reader should have no difficulty to confirm that $cbcbcbcda$ is output by this algorithm.

Table 2. Code table C and the output string for 214630

i	0	1	2	3	4	5
y_i	2	1	4	6	3	0
C	-	4 : cb	5 : bc	6 : cbc	7 : $cbcd$	8 : da
X	c	b	cb	cbc	d	a

3   Parallel LZW Decompression

This section shows our parallel algorithm for LZW decompression.

Again, let $X = x_0 x_1 \cdots x_{n-1}$ be a string of characters. We assume that characters are selected from an alphabet (or a set) with k characters $\alpha(0), \alpha(1)$, ..., $\alpha(k-1)$. We use $k = 4$ characters $\alpha(0) = a$, $\alpha(1) = b$, $\alpha(2) = c$, and $\alpha(3) = d$, when we show examples as before. Let $Y = y_0 y_1 \cdots y_{m-1}$ denote the compressed string of codes obtained by the LZW compression algorithm. In the LZW compression algorithm, each of the first $m-1$ codes $y_0, y_1, \ldots, y_{m-2}$ has a corresponding AddTable operation. Hence, the argument of code table C takes an integer from 0 to $k + m - 2$.

Before showing the parallel LZW compression algorithm, we define several notations. We define pointer table p using code table Y as follows:

$$p(i) = \begin{cases} \text{NULL} & \text{if } 0 \le i \le k - 1 \\ y_{i-k} & \text{if } k \le i \le k + m - 1 \end{cases} \tag{1}$$

We can traverse pointer table p until we reach NULL. Let $p^0(i) = i$ and $p^{j+1}(i) = p(p^j(i))$ for all $j \ge 0$ and i. In other words, $p^j(i)$ is the code where we reach from code i in j pointer traversing operations. Let $L(i)$ be an integer satisfying $p^{L(i)}(i) = \text{NULL}$ and $p^{L(i)-1}(i) \ne \text{NULL}$. Also, let $p^*(i) = p^{L(i)-1}(i)$. Intuitively, $p^*(i)$ corresponds to the dead end from code i along pointers. Further, let $C_l(i)$ $(0 \le i \le k + m - 2)$ be a character defined as follows:

$$C_l(i) = \begin{cases} \alpha(i) & \text{if } 0 \le i \le k - 1 \\ \alpha(p^*(i+1)) & \text{if } k \le k + m - 2 \end{cases} \tag{2}$$

It should have no difficulty to confirm that $C_l(i)$ is the last character of $C(i)$, and $L(i)$ is the length of $C(i)$. Using C_l and p, we can define the value of $C(i)$ as follows:

$$C(i) = C_l(p^{L(i)-1}(i)) \cdot C_l(p^{L(i)-2}(i)) \cdots C_l(p^0(i)). \tag{3}$$

Table 3 shows the values of p, p^*, L, C_l, and C for $Y = 214630$.

Table 3. The values of p, p^*, l, C_l, and C for $Y = 214630$

i	0	1	2	3	4	5	6	7	8	9
$p(i)$	NULL	NULL	NULL	NULL	2	1	4	6	3	0
$p^*(i)$	-	-	-	-	2	1	2	2	3	0
$L(i)$	1	1	1	1	2	2	3	4	2	-
$C_l(i)$	a	b	c	d	b	c	c	d	a	-
$C(i)$	a	b	c	d	cb	bc	cbc	$cbcd$	da	-

We are now in a position to show parallel LZW decompression on the CREW-PRAM. Parallel LZW decompression can be done in two steps as follows:

Step 1 Compute L, p^*, and C_l from code string Y.
Step 2 Compute X using p, C_l and L.

In Step 1, we use k processors to initialize the values of $p(i), C_l(i)$, and $L(i)$ for each i ($0 \leq i \leq k - 1$). Also, we use m processors and assign one processor to each i ($k \leq i \leq k + m - 1$), which is responsible for computing the values of $L(i), p^*(i)$, and $C_l(i)$. The details of Step 1 of parallel LZW decompression algorithm are spelled out as follows:

[Step 1 of the parallel LZW decompression algorithm]
1   for $i \leftarrow 0$ to $k - 1$ do in parallel // Initialization
2       $p(i) \leftarrow$ NULL; $L(i) = 1$; $C_l(i) \leftarrow \alpha(i)$;
3   for $i \leftarrow k$ to $k + m - 1$ do in parallel // Computation of L and p^*
4       $p(i) \leftarrow y_{i-k}$; $p^*(i) \leftarrow y_{i-k}$;
5   while($p(p^*(i)) \neq$ NULL)
6           $L(i) \leftarrow L(i) + 1$; $p^*(i) \leftarrow p(p^*(i))$;
7   for $i \leftarrow k$ to $k + m - 2$ do in parallel // Computation of C_l
8       $C_l(i) \leftarrow \alpha(p^*(i + 1))$;

Step 2 of the parallel LZW decompression algorithm uses m threads to compute $C(y_0) \cdot C(y_1) \cdots C(y_{m-1})$, which is equal to $X = x_0 x_1 \cdots x_{n-1}$. For this purpose, we compute the prefix-sums of $L(y_0), L(y_1), \ldots, L(y_{m-2})$ using $m - 1$ processors. In other words, $s(i) = L(y_0) + L(y_1) + \cdots + L(y_i)$ is computed for every i ($0 \leq i \leq m - 1$). For simplicity, let $s(-1) = 0$. After that, for each i ($0 \leq i \leq m - 1$), $L(y_i)$ characters $C_l(p^{L(y_i)-1}(y_i)) \cdot C_l(p^{L(y_i)-2}(y_i)) \cdots C_l(p^0(y_i))(= C(y_i))$ are copied from $x_{s(i-1)}$ to $x_{s(i)-1}$. Note that, the values of $p^0(y_i), p^1(y_i), \ldots, p^{L(i)-1}(y_i)$ can be obtained by traversing pointers from code i. Hence, it makes sense to perform the copy operation from $x_{s(i)-1}$ down to $x_{s(i-1)}$.

Table 4 shows he values of $L(y_i)$, $s(i)$, and $C(y_i)$ for $Y = 214630$. By concatenating them, we can confirm that $X = cbcbcbcda$ is obtained.

Table 4. The values of $L(y_i)$, $s(i)$, and $C(y_i)$ for $Y = 214630$

i	0	1	2	3	4	5
y_i	2	1	4	6	3	0
$L(y_i)$	1	1	2	3	1	1
$s(i)$	1	2	4	7	8	9
$C(y_i)$	c	b	cb	cbc	d	a

Let us evaluate the computing time. Let $L_{\max} = \max\{L(i) \mid 0 \le i \le k+m-1\}$. The for-loop in line 1 takes $O(1)$ time using k processors. Also, while-loop in line 5 is repeated at most $L(i) \le L_{\max}$ times for each i. Hence, for-loop in line 3 can be done in $O(L_{\max})$ time using m processors. It is well known that the prefix-sums of m numbers can be computed in $O(\log m)$ time using m processors [2]. Hence, every $s(i)$ is computed in $O(\log m)$ time using $m-1$ processors. After that, every $C(y_i)$ with $L(y_i)$ characters is copied from $x_{s(i)-1}$ down to $x_{s(i-1)}$ in $O(L_{\max})$ time using m processors. Therefore, we have

Theorem 1. *The LZW decomposition of a string of m codes can be done in $O(L_{\max} + \log m)$ time using $\max(k,m)$ processors on the CREW-PRAM, where k is the number of characters in an alphabet.*

4   GPU Implementation

The main purpose of this section is to describe a GPU implementation of our parallel LZW decompression algorithm. We focus on the decompression of TIFF image file compressed by LZW compression. We assume that a TIFF image file contains a gray scale image with 8-bit depth, that is, each pixel has intensity represented by an 8-bit unsigned integer. Since each of RGB or CMYK color planes can be handled as a gray scale image, it is obvious to modify gray scale TIFF image decompression for color image decompression.

As illustrated in Fig. 1, a TIFF file has *an image header* containing miscellaneous information such as ImageLength (the number of rows), ImageWidth (the number of columns), compression method, depth of pixels, etc. [1]. It also has *an image directory* containing pointers to the actual image data. For LZW compression, an original 8-bit gray-scale image is partitioned into *strips*, each of which has one or several consecutive rows. The number of rows per strip is stored in the image file header with tag RowsPerStrip. Each Strip is compressed independently, and stored as the image data. The image directory has pointers to the image data for all strips.

Next, we will show how each strip is compressed. Since every pixel has an 8-bit intensity level, we can think that an input string of an integer in the range $[0, 255]$. Hence, codes from 0 to 255 are assigned to these integers. Code 256 (ClearCode) is reserved to clear the code table. Also, code 257 (EndOfInformation) is used to specify the end of the data. Thus, AddTable operations assign

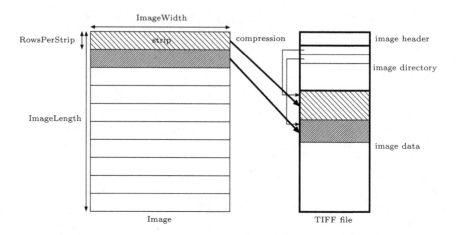

Fig. 1. An image and TIFF image file

codes to strings from code 258. While the entry of the code table is less than 512, codes are represented as 9-bit integer. After adding code table entry 511, we switch to 10-bit codes. Similarly, after adding code table entry 1023 and 2037, 11-bit codes and 12-bit codes are used, respectively. As soon as code table entry 4094 is added, ClearCode is output. After that, the code table is re-initialized and AddTable operations use codes from 258 again. The same procedure is repeated until all pixels in a strip are converted into codes. After the code for the last pixel in a strip is output, EndOfInformation is written out. We can think that a code string for a particular strip is separated by ClearCode. We call each of them a *code segment*. Except the last one, each code segment has $4094 - 258 + 1 = 3837$ codes. The last code segment for a strip may have codes less than that.

In our implementation, a CUDA block with 1024 threads is assigned a strip. A CUDA block decompresses each code segment in the assigned strip one by one. More specifically, a CUDA block copies a code segment of a strip stored in the global memory to a shared memory. After that, it performs Step 1 of the parallel LZW decompression. Tables for p, p^*, L, and C_l are created in the shared memory. We use 16-bit unsigned short integer for each element of the tables. Since each table has at most 4096 entries, the total size of tables is $4 \times 4096 \times 2 = 32$Kbytes. Each streaming multiprocessor has 96Kbytes and 2048 resident threads [11], two CUDA blocks are arranged at the same time and the occupancy can be 100 %. Since the table has 4095 entries, 1024 threads compute them in four iterations. In each iteration, 1024 entries of the tables are computed by 1024 threads. For example, in the first iteration, $L(i)$, $p^*(i)$, and $C_l(i)$ for every i ($0 \le i \le 1023$) are computed. After that, these values for every i ($0 \le i \le 1023$) are computed. Note that, in the second iteration, it is not necessary to execute the while-loop in line 5 until $p(p^*(i)) \neq$ NULL is satisfied. Once the value of $p^*(i)$ is less than 1024, the final resulting values of $L(i)$ and $p^*(i)$ are computed using those of $L(p^*(i))$ and $p^*(p^*(i))$. Thus, we can terminate the while-loop as soon as $p^*(i) < 1024$ is satisfied.

After the tables are obtained, the prefix-sums of s is computed in the shared memory for Step 2. Finally, the strings of characters of each code are written in the global memory. The prefix-sums can be computed by parallel algorithm for GPUs [3,8].

5 Experimental Results

We have used GeForce GTX 980 which has 16 streaming multiprocessors with 128 processor cores each to implement parallel LZW decompression algorithm. We also use Intel Xeon CPU X7460 (2.66GHz) to evaluate the running time of sequential LZW decompression.

We have used three gray scale images with 4096×3072 pixels (Fig. 2), which are converted from JIS X 9204-2004 standard color image data. They are stored in TIFF format with LZW compression option. We set RowsPerStrip = 16, and so each image has $\frac{3072}{16} = 192$ strips with $16 \times 4096 = 64$k pixels each. We invoked a CUDA kernel with 192 CUDA blocks, each of which decompresses a strip with 64k pixels. Table 5 shows the compression ratio, that is, "original image size: compressed image size." We can see that "Graph" has high compression ratio because it has large areas with constant intensity levels. On the other hand, the compression ratio of "Crafts" is small because of the small details. Table 5 also shows the running time of LZW decompression using a CPU and a GPU. In the table, T_1 and T are the time for constructing tables and the total computing time, respectively. To evaluate time T_1 of sequential LZW decompression, OUTPUT in lines 4 and 6 are removed. Also, to evaluate time T_1 of parallel LZW decompression on the GPU, the CUDA kernel call is terminated without computing the prefix-sums and writing resulting characters in the global memory. Hence, we can think that $T - T_1$ corresponds to the time for generating the original string using the tables. Clearly, sequential/parallel LZW decompression algorithms take more time to create tables for images with small compression ratio because they have many segments and need to create tables many times. Also, the time for creating tables dominates the computing time of sequential LZW decompression, while that for writing out characters dominates in parallel LZW decompression. This is because the overhead of the parallel prefix-sums computation is not small. From the table, we can see that LZW decompression for "Flowers" using GPU is 69.4 times faster than that using CPU.

"Crafts"   "Flowers"   "Graph"

Fig. 2. Three gray scale image with 4096×3072 pixels used for experiments

Table 5. Experimental results (milliseconds) for three images

Images	Compression ratio	Sequential(CPU)			Parallel(GPU)			Speedup ratio
		T_1	$T - T_1$	T	T_1	$T - T_1$	T	
"Crafts"	1.67 : 1	90.4	18.0	108	0.847	1.30	2.15	50.2 : 1
"Flowers"	2.34 : 1	70.9	16.6	87.5	0.541	0.719	1.26	69.4 : 1
"Graph"	38.0 : 1	27.2	19.3	46.5	0.202	1.59	1.79	26.0 : 1

We have evaluated the performance of three practical scenarios as follows:

Scenario 1: Non-compressed images are stored in the storage. They are transferred to the GPU thorough the host computer.

Scenario 2: LZW compressed images are stored in the storage. They are transferred to the host computer, and decompressed in it. After that, the resulting non-compressed images are transferred to the GPU.

Scenario 3: LZW compressed images are stored in the storage. They are transferred to the GPU through the host computer, and decompressed in the GPU.

The throughput between the storage and the host computer depends on their bandwidth. For simplicity, we assume that their bandwidth is the same as that between the host computer and the GPU. Note that since the bandwidth of the storage is not larger than that of the GPU in many cases, this assumption does not give advantage to Scenario 3. Table 6 shows the data transfer time for non-compressed and compressed files of the three images. Clearly, the time for non-compressed files is almost the same, because they have the same size. On the other hand, images with higher compression ratio take fewer time, because their sizes are smaller. It also evaluates the time for three scenarios. Clearly, Scenario 3, which uses parallel LZW decompression in the GPU, takes much fewer time than the others.

Table 6. Estimated running time (milliseconds) of three scenarios

Images	Compression ratio	Data Transfer		Scenario 1	Scenario 2	Scenario 3
		non-compressed	compressed			
"Crafts"	1.67 : 1	3.79	2.44	7.58	110	4.59
"Flowers"	2.34 : 1	3.83	1.73	7.66	89.2	2.99
"Graph"	38.0 : 1	3.80	0.167	7.60	46.7	1.96

6   Conclusion

In this paper, we have presented a parallel LZW decompression algorithm and implemented in the GPU. The experimental results show that, it achieves a speedup factor up to 69.4. Also, LZW decompression in the GPU can be used to accelerate the query processing for a lot of compressed images in the storage.

References

1. Adobe Developers Association: TIFF Revision 6.0. http://partners.adobe.com/public/developer/en/tiff/TIFF6.pdf
2. Gibbons, A., Rytter, W.: Efficient Parallel Algorithms. Cambridge University Press, Cambridge (1988)
3. Harris, M., Sengupta, S., Owens, J.D.: Parallel prefix sum (scan) with CUDA, Chap. 39. In: GPU Gems 3. Addison-Wesley (2007)
4. Hwu, W.W.: GPU Computing Gems Emerald Edition. Morgan Kaufmann, San Francisco (2011)
5. Kasagi, A., Nakano, K., Ito, Y.: Parallel algorithms for the summed area table on the asynchronous hierarchical memory machine, with GPU implementations. In: Proceedings of the International Conference on Parallel Processing (ICPP), pp. 251–250, September 2014
6. Klein, S.T., Wiseman, Y.: Parallel Lempel Ziv coding. Discrete Appl. Math. **146**, 180–191 (2005)
7. Man, D., Uda, K., Ito, Y., Nakano, K.: A GPU implementation of computing Euclidean distance map with efficient memory access. In: Proceedings of the International Conference on Networking and Computing, pp. 68–76, December 2011
8. Nakano, K.: Simple memory machine models for GPUs. In: Proceedings of the International Parallel and Distributed Processing Symposium Workshops, pp. 788–797, May 2012
9. Nicolaisen, A.L.V.: Algorithms for compression on GPUs. Ph.D. thesis, Technical University of Denmark, August 2015
10. Nishida, K., Ito, Y., Nakano, K.: Accelerating the dynamic programming for the matrix chain product on the GPU. In: Proceedings of the International Conference on Networking and Computing, pp. 320–326, December 2011
11. NVIDIA Corporation: NVIDIA CUDA C programming guide version 7.0, March 2015
12. Ozsoy, A., Swany, M.: CULZSS: LZSS lossless data compression on CUDA. In: Proceedings of the International Conference on Cluster Computing, pp. 403–411, September 2011
13. Shyni, K., Kumar, K.V.M.: Lossless LZW data compression algorithm on CUDA. IOSR J. Comput. Eng. **13**, 122–127 (2013)
14. Takeuchi, Y., Takafuji, D., Ito, Y., Nakano, K.: ASCII art generation using the local exhaustive search on the GPU. In: Proceedings of the International Symposium on Computing and Networking, pp. 194–200, December 2013
15. Welch, T.: High speed data compression and decompression apparatus and method. US patent 4558302, December 1985
16. Welch, T.A.: A technique for high-performance data compression. IEEE Comput. **17**(6), 8–19 (1984)

Parallel FDFM Approach for Computing GCDs Using the FPGA

Xin Zhou, Koji Nakano$^{(\boxtimes)}$, and Yasuaki Ito

Department of Information Engineering, Hiroshima University,
Kagamiyama 1-4-1, Higashi Hiroshima 739-8527, Japan
nakano@hiroshima-u.ac.jp

Abstract. The main contribution of this paper is to present an FPGA-targeted architecture called the hierarchical GCD cluster, that computes the GCDs of all pairs in a set of numbers. It is designed based on the FDFM (Few DSP slices and Few Memory blocks) approach and consists of 1408 processors equipped with one block RAM and one DSP slice each. Every processor works in parallel and computes the GCDs independently. We have measured the performance of our architecture to compute all pairs of two numbers in RSA moduli. Implementation results show that it runs $0.057\mu s$ per one GCD computation of two 1024-bit RSA moduli in a Xilinx Virtex-7 family FPGA XC7VX485T-2. It is 6.0 times faster than the best GPU implementation and 500 times faster than a sequential implementation on the Intel Xeon CPU.

Keywords: FDFM approach · Parallel algorithms · DSP slices · Block RAMs · RSA cryptosystem

1 Introduction

An FPGA (Field Programmable Gate Array) is an integrated circuit designed to be configured by a designer after manufacturing. It contains an array of programmable logic blocks, and the reconfigurable interconnects allow the blocks to be inter-wired in different configurations. Since any logic circuit can be embedded in an FPGA, it can be used for general-purpose parallel computing [2,11]. Recent FPGAs have embedded DSP slices and block RAMs. Xilinx Virtex-7 family FPGAs have DSP slices, each of which is equipped with a multiplier, adders/subtracters, logic operators, registers, etc. [16]. For example, the DSP slice has a two-input multiplier followed by multiplexers and a three input adder/subtracter/accumulator. It also has pipeline registers between operators to reduce the propagation time. A block RAM is an embedded dual-port memory supporting synchronized read and write operations, and can be configured as a 36k-bit or two 18k-bit dual port RAMs [18]. The main contribution of this paper is to present an architecture for computing the GCD (Greatest Common Divisor) of large two numbers using an FPGA. We employ *the FDFM (Few DSP slices and Few block Memories) approach* [1] to implement parallel GCD computation

© Springer International Publishing Switzerland 2016
R. Wyrzykowski et al. (Eds.): PPAM 2015, Part I, LNCS 9573, pp. 238–247, 2016.
DOI: 10.1007/978-3-319-32149-3_23

in the FPGA. The key idea of the FDFM approach is to use few DSP slices and few block RAMs for constituting a processor performing a specific computation. For example, hardware algorithms for RSA encryption/decryption have been implemented in the FPGA using the FDFM approach [7]. Their implementation using it is better than the conventional approach [12].

One of the applications for benchmarking GCD computation is breaking weak RSA keys. RSA [13] is one the most well-known public-key cryptosystems widely used for secure data transfer. RSA cryptosystem uses an encryption key open to the public and a secret decryption key. An encryption key is a pair (n, e) of modulus n and exponent e such that $n = pq$ for two distinct large prime numbers p and q, and e $(< (p - 1)(q - 1))$ and $(p - 1)(q - 1)$ are coprime. For example, for 1024-bit RSA cryptosystem, modulus n with 1024 bits is obtained by 512-bit prime numbers p and q. The decryption key for this encryption key is a pair (n, d) such that $de \equiv 1 \pmod{(p-1)(q-1)}$, that is, d is the multiplicative inverse of e $\pmod{(p - 1)(q - 1)}$. For a public encryption key (n, e), a message M $(0 \leq M \leq n - 1)$ is converted to the cipher message $C = M^e \bmod n$. Since $M \equiv M^{ed} \pmod n$ always holds for all messages M, the cipher message C can be converted to the original message M by computing $C^d \bmod n$. If the values of p and q are available, $d \equiv e^{-1} \pmod{(p-1)(q-1)}$ can be computed very easily by extended Euclidean algorithm [3]. However, to obtain p and q from an encryption key (n, e), we need to decompose n into p and q. Since the computation of factorization of large numbers is very costly, it is not possible to decompose n into p and q in practical computing time. RSA cryptosystem relies on the hardness of factorization of a large number.

Suppose that we have a set of many RSA encryption keys collected from the Web. If some of moduli in encryption keys are generated by inappropriate implementation of a random prime number generator, they may share or reuse the same prime number. We call RSA keys sharing a prime number *weak RSA keys*. Actually, several public keys collected from the Web includes weak RSA keys [10]. If two moduli share a prime number, they can be decomposed by computing the GCD (Greatest Common Divisor). More specifically, if two distinct moduli n_1 and n_2 share a moduli p then the GCD of n_1 and n_2 is equal to p. It is well known that the GCD can be computed very easily by Euclidean algorithms [8]. Once we have the GCD p, we can decompose n_1 into p and $\frac{n_1}{p}$ and n_2 into p and $\frac{n_2}{p}$. Hence, we may break weak RSA keys by computing the GCDs of all pairs of moduli in the Web. It has been shown that a complicated sequential algorithm can find a pair of weak RSA keys [10]. So, it makes no sense to perform straightforward pairwise computation of GCDs for RSA moduli for the purpose of breaking weak RSA keys. However, bulk computation of GCDs for RSA moduli is useful to measure the performance of the GCD computation. We designed and implemented 1408 GCD processors in a Xilinx Virtex-7 family FPGA XC7VX485T-2, that compute the GCDs in parallel. The implementation results show that, 1408 GCD processors can compute the GCD of one 1024-bit RSA moduli in expected $0.057\mu s$.

Several hardware implementations for computing the GCD on FPGAs have been presented [4,9]. However, they just implemented Binary Euclidean algorithm to compute the GCD using programmable logic blocks as it is. Hence, they can support the GCD computation for numbers with very few bits. On the other hand, several previously published papers have presented GPU implementations of Binary Euclidean algorithm in CUDA-enabled GPUs. Fujimoto [5] has implemented Binary Euclidean algorithm using CUDA and evaluated the performance on GeForce GTX285 GPU. The experimental results show that the GCDs for 131072 pairs of 1024-bit numbers can be computed in 1.431932 s. Hence, his implementation runs 10.9μs per one 1024-bit GCD computation. Scharfglass *et al.* [14] have presented a GPU implementation of Binary Euclidean algorithm. It performs the GCD computation of all 199990000 pairs of 20000 RSA moduli with 1024 bits in 2005.09 s using GeForce GTX 480 GPU. Thus, their implementation performs each 1024-bit GCD computation in 10.02μs. Later, White [15] has showed that the same computation can be performed in 63.0 s on Tesla K20Xm. It follows that it computes each 1024-bit GCD in 3.15μs. Quite recently, Fujita *et al.* have presented new Euclidean algorithm called Approximate Euclidean algorithm and implemented it in the GPU [6]. Approximate Euclidean algorithm performs perform each 1024-bit GCD computation in 0.346μs on GeForce GTX 780Ti and 28.6μs on Intel Xeon X7460 (2.66 GHz) CPU. Our implementation of the Hardware Binary Euclidean algorithm performs one 1024-bit GCD computation in 0.057μs which is 6.0 times faster than the GPU and 500 times faster than the CPU.

2 Euclidean Algorithms for Computing GCD

This section first reviews Fast Binary Euclidean algorithm for computing the GCD of two numbers X and Y. Please see [6] for the details. We then show Hardware Binary Euclidean algorithm by modifying Fast Binary Euclidean algorithm, which will be implemented it in an FPGA.

We assume that both input numbers X and Y are odd and $X \geq Y$ holds. Hence, the GCD of X and Y is always odd. If one of them is odd and the other is even, say X is odd and Y is even, then $\gcd(X, Y) = \gcd(X, \frac{Y}{2})$ holds. Thus, we can convert Y into odd numbers by removing consecutive 0 bits from the least significant bit of Y. Also, from $\gcd(X, Y) = 2 \cdot \gcd(\frac{X}{2}, \frac{Y}{2})$, we can obtain a factor of 2 in the GCD of X and Y if both X and Y are even. Thus, it should have no difficulty to modify GCD algorithms shown in this paper to handle even input numbers.

Let swap(X,Y) denote a function to exchange the values of X and Y. Further, let rshift(X) be a function returning the number obtained by removing consecutive 0 bits from the least significant bit of X. For example, if $X = 11010100$ in binary system, then rshift(X) = 110101. Using swap and rshift functions, we can write Fast Binary Euclidean algorithm as follows:

[Fast Binary Euclidean algorithm]
gcd(X,Y){
    do {
        $X \leftarrow$rshift($X - Y$);

```
        if(X < Y) swap(X, Y)
    } while (Y ≠ 0)
    return(X);
}
```

We will modify Fast Binary Euclidean algorithm to be implemented in the FPGA. We need to read all bits of X and Y to exchange them if we implement function swap as it is. Also, rshift function needs a large barrel shifter. Hence, we should avoid direct implementations of these functions in the FPGA. Instead of function rshift(X), we use rshift$_k(X)$, which removes at most k consecutive 0 bits from the least significant bit of X. In other words, if X has at most k consecutive 0 bits from the least significant bit, all of them are removed. If it has more than k 0 bits, then k 0 bits from the least significant bits are removed. For example, rshift$_2(11011000) = 110110$ and rshift$_2(11011010) = 1101101$ hold. If k is small, rshift$_k(X)$ can be implemented using a small barrel shifter. Using rshift$_k(X)$, we can design an Euclidean-based GCD algorithm for FPGAs as follows:

[Hardware Binary Euclidean algorithm]
```
gcd(X,Y){
    do {
        if(X is even) X ←rshift_k(X);
        else if (Y is even) Y ←rshift_k(Y);
        else if(X ≥ Y) X ←rshift_k(X − Y);
        else Y ←rshift_k(Y − X); // X < Y
    } while (X ≠ 0 and Y ≠ 0)
    if(X ≠ 0) return(X);
    else return(Y);
}
```

Note that rshift$_k$ function may return an even number. Hence, one of X or Y can be an even number. If this is the case, either X ←rshift$_k(X)$ or Y ←rshift$_k(Y)$ is executed until both of them are odd. Hence, both X and Y are odd, whenever rshift$_k(X − Y)$ is executed, Thus, the argument of rshift$_k$ is always even when it is executed.

If these Binary Euclidean algorithms are executed for two s-bit RSA moduli, the GCD is 1 or $\frac{s}{2}$-bit prime number. Hence, when either X or Y has less than $\frac{s}{2}$ bits during the execution of these Euclidean algorithms, the GCD must be 1. Thus, we can terminate the do-while loop when one of X or Y has less than $\frac{s}{2}$ bits. We call this *early-terminate* technique.

Table 1 shows the average number of iterations of the do-while loop 1024-bit RSA moduli for each values of k of rshift$_k$. Note that $k = \infty$ corresponds to Fast Binary Euclidean algorithm, which performs rshift function that removes all consecutive 0 bits. We can see that the early-terminate technique can reduce the number of iterations by half. Clearly, the number of iterations is smaller for large k. However, rshift$_k(X)$ needs a large barrel shifter for large k. We should select an appropriate value of k that balances the computing time and the used

Table 1. The number of iterations of the do-while loop for 1024-bit RSA moduli

k	Hardware Binary Euclidean								Fast Binary Euclidean
	1	2	3	4	5	6	7	8	
Non-terminate	1445.8	964.3	827.0	772.0	747.1	735.3	729.7	726.8	723.9
Early-terminate	721.1	481.1	412.7	385.2	372.9	367.0	364.2	362.8	361.4

hardware resources. For our FPGA implementation that we will show later, we have selected $k = 4$.

3   A GCD Processor for Computing the GCD

This section presents a *GCD processor*, which computes the GCD of two moduli by Hardware Binary Euclidean algorithm.

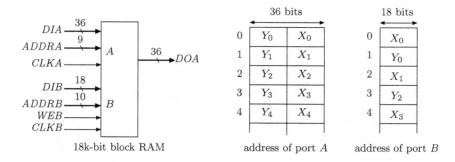

Fig. 1. A block RAM and the memory configuration

Our GCD processor uses one 18k-bit block RAM and one DSP slice in the FPGA. We use a block RAM as a simple dual-port memory [18] with ports A and B as illustrated in Fig. 1. Ports A and B are configured as read-only 36-bit mode and write-only 18-bit mode, respectively. In other words, the block RAM is a 512×36-bit memory for port A and 1024×18-bit for port B. Using these ports, we can read 36-bit data and write 18-bit data for different addresses at the same time. Two numbers X and Y of Hardware Binary Euclidean algorithm are stored as 18-bit words. If each of them has 1024 bits each, it is stored in $\lceil \frac{1024}{18} \rceil = 57$ words. Let $X_{56}X_{55} \cdots X_0$ denote 57 words representing X such that $X = \sum_{i=0}^{56} X_i \cdot 2^{18i}$ holds. Similarly, let $Y_{56}Y_{55} \cdots Y_0$ denote those representing Y. In the 18-bit mode, 18-bit data X_i and Y_i can be written in addresses $2i$ and $2i + 1$ ($0 \le i \le 56$), respectively, using port B, as illustrated in Fig. 1. They can be read from 36-bit port A such that 36-bit data Y_iX_i ($0 \le i \le 56$) is read from address i.

Figure 2 shows the architecture of a processor for computing the GCD by Hardware Binary Euclidean algorithm. The 36-bit output of the block RAM are connected to the DSP slice through multiplexers. The DSP slice has 18-bit two ports A and B. By two multiplexers, one of X and Y is given to port A, and one of X, Y, and zero (0) is given to port B. Since DSP computes the subtraction $A - B$, it can outputs one of $X - Y$, $Y - X$, $X - 0 (= X)$, and $Y - 0$ $(= Y)$. The subtraction is performed sequentially word-by-word. For example, suppose that we need to compute $Z = X - Y$. Let $Z_{56}Z_{55} \cdots Z_0$ denote 57 words representing Z and show how $X - Y$ are computed. First, X_0 and Y_0 read from the block RAM are transferred to ports A and B of the DSP slice, and $X_0 - Y_0$ is computed and stored in 19-bit register P. The MSB $P[18]$ corresponds to the borrow of the subtraction and the remaining 18 bits of P is equal to Z_0. After that, $X_1 - Y_1 - P[18]$ is computed and stored in P. This value is equal to Z_1. Repeating the same operation, the DSP slice outputs all words of Z one by one.

Note that, the resulting value of $Z = X - Y$ must be right-shifted appropriately, when it is stored in X or Y. We use a 4-bit shifter which can right-shift Z by 1 bit, 2 bits, 3 bits, or 4 bits. Function rshift$_4$ can be implemented as follows. We use a 17-bit register to store the output of P, which stores the value of Z. For example, let $Z_0 = z_{17}z_{16} \cdots z_0$ and $Z_1 = z_{35}z_{34} \cdots z_{18}$ be the first and the second outputs of P. In the next clock cycle just after Z_0 is stored in P, the register stores 17-bit $z_{17}z_{15} \cdots z_1$. In the same time P stores Z_1. Hence, we can have 21 consecutive bits $z_{21}z_{20} \cdots z_1$ of Z by picking four bits in $z_{21}z_{20}z_{19}z_{18}$ in Z_1. The 4-bit shifter selects consecutive 18 bits in 21 bits. For example, it selects $z_{18}z_{17} \cdots z_1$ for 1-bit shift and $z_{21}z_{20} \cdots z_4$ for 4-bit shift. It should be clear that the 4-bit shifter outputs rshift$_4(X - Y)$. This value is transferred to port B of the block RAM, and the value of X is updated.

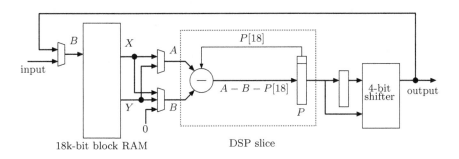

Fig. 2. The architecture of a GCD processor

Let us briefly confirm that the GCD processor can execute Hardware Binary Euclidean algorithm. It should be clear that, by configuring two multiplexers connecting the block RAM and the DSP, we can compute one of rshift$_4(X - Y)$, rshift$_4(Y - X)$, rshift$_4(X)$, and rshift$_4(Y)$. Also, the resulting value can be overwritten in X or Y through port B of the block RAM. The conditions "X is even" and "Y is even" can be determined by reading X_0 and Y_0 in the block RAM. Also, "$X \geq Y$" can be decided by reading X and Y from the MSB.

More specifically, we check if $X_{56} > Y_{56}$ holds, then "$X \geq Y$" is true. If $X_{56} <$ Y_{56}, it is false. If $X_{56} = Y_{56}$, then we need to read and compare X_{55} and Y_{55}. The same operation is repeated until "$X \geq Y$" can be determined. We should note that, with high probability, "$X \geq Y$" can be determined by reading several words of X and Y. Also, during the computation of Hardware Binary Euclidean algorithm, the number of bits of X and Y is decreased. Hence, we use registers to store the current numbers of bits of X and Y. By using them, we can easily determine if $X \neq 0$ and $Y \neq 0$. Further, for early-terminate, we can determine if the number of bits is less than $\frac{s}{2}$.

4   Hierachical GCD Cluster

This section presents a hierarchical parallel architecture that we call hierarchical GCD cluster. Since we can embed more than one thousand GCD processors in an FPGA, it makes sense to use multiple servers, each of which controls approximately one hundred GCD processors. The hierarchical GCD cluster consists of multiple GCD clusters, each of which involves multiple GCD processors as illustrated in Fig. 3. A single central server controls local servers, each of which maintains GCD processors in the same GCD cluster.

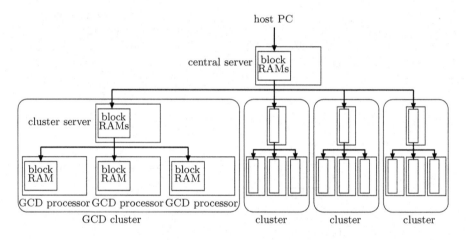

Fig. 3. The architecture of the Hierarchical GCD Cluster

We show how the hierarchical GCD cluster is used to break weak RSA keys. Suppose that we have a lot of moduli collected from the Web and they are stored in the main memory of the host PC. Our goal is to compute all pair of moduli using the hierarchical GCD cluster in an FGPA. For this purpose, we partition all moduli into groups with m moduli each. The host PC picks two groups and sends them to the central server in the FPGA. Let $N = \{n_0, n_1, \ldots n_{m-1}\}$ and $N' = \{n'_0, n'_1, \ldots n'_{m-1}\}$ denote two groups of m moduli each that the central server in the FPGA has been received. The hierarchical GCD cluster computes $\gcd(n_i, n'_j)$ for all pairs of i and j ($0 \leq i, j \leq m - 1$), and reports the GCDs larger than 1.

Next, we will show how the hierarchical GCD cluster computes the GCDs of N and N' using GCD clusters. Each group of m moduli is partitioned into b blocks of k moduli each, where $m = bk$. Let $N_i = \{n_{ik}, n_{ik+1}, \ldots, n_{(i+1)k-1}\}$ and $N'_i = \{n'_{ik}, n'_{ik+1}, \ldots, n'_{(i+1)k-1}\}$ $(0 \leq i \leq b-1)$ be two sets of k moduli in the i-th groups of sets N and N', respectively. Each cluster is assigned a task to compute the GCDs of all pairs X $(\in N_i)$ and Y $(\in N'_j)$ for a pair i and j $(0 \leq i,j \leq b-1)$. For this purpose, all moduli in N_i and in N'_j are copied from the block RAM in the central server to that in the local server of a GCD cluster. After the local server receives all moduli, the cluster starts computing the GCDs of all pairs X $(\in N_i)$ and Y $(\in N'_j)$. The local server then picks a pair X and Y and copies them to the block RAM of a GCD processor. Upon completion of the copy, the GCD processor starts computing the GCD of X and Y by the Hardware Binary Euclidean algorithm. This procedure is repeated for all GCD processors. If a GCD processor terminates the GCD computation, the local server sends a new pair to it. In this way, the GCDs of all pairs in N_i and N'_j are computed by a GCD cluster. When a GCD cluster completes the computation of all GCDs of a given pair of two groups, the central server picks a new pair i and j and sends all moduli in N_i and in N'_j to the local server. The same operation is repeated until the GCDs of all pairs N and N' are computed.

5   Implementation Results in the FPGA

We have implemented a GCD processor for 1024-bit, 2048-bit, 4096-bit, and 8192-bit moduli in VC707 evaluation board [19] equipped with the Xilinx Virtex-7 FPGA XC7VX485T-2. Note that one 18k-bit block RAMs can store up to two 9216-bit moduli. Table 2 shows the implementation results. Slice Registers and Slice LUTs (Look-Up-Tables) are hardware resources in CLB (Configurable Logic Block) [17], which are used to implement sequential logics. We can see that the used hardware resources are very few. In particular, we can confirm that only one 18k-bit block RAM and one DSP slice are used. Also, the clock frequency is about 350 MHz, which is sufficiently high.

Table 2. Implementation results of one GCD processor for 1024-bit, 2048-bit, 4096-bit, and 8192-bit moduli

Available	Slice Registers 607200	Slice LUTs 303600	DSP slices 2800	18kb block RAMs 2060	Clock Frequency (MHz)
1024-bit	155	179	1	1	345
2048-bit	160	185	1	1	355
4096-bit	165	192	1	1	343
8192-bit	170	230	1	1	350

For breaking 1024-bit moduli, a GCD cluster with a local cluster with eight 18k-bit block RAMs and 128 GCD processors is used. Since four 18k-bit block

RAMs can store $\lfloor \frac{4 \cdot 1024}{57} \rfloor = 71$ moduli with 1024 bits, each GCD cluster computes the GCDs of $71 \times 71 = 5041$ pairs of blocks stored in block RAMs. Hence, each GCD processor computes the GCDs for expected $\frac{5041}{128} = 39.4$ pairs of 1024-bit moduli. Also we arranged 64 block RAMs to the central server. Since a block of moduli is stored in four block RAMs, we can think that the central server has $8 \times 8 = 64$ pairs of blocks. Thus, each cluster computes the GCDs for moduli in expected $\frac{64}{11} = 5.8$ pairs of blocks. Table 3 shows the implementation results of clusters. Since a cluster server uses eight 18k-bit block RAMs, each GCD cluster with 128 GCD processors involves $128 + 8 = 136$ block RAMs. We have succeeded in implementing a hierarchical GCD luster with 11 GCD clusters and the central server, which uses $11 \cdot 128 = 1408$ DSP slices and $11 \cdot 136 + 64 = 1560$ block RAMs. Unfortunately, due to the overhead for the connection between the central server and GCD clusters, the clock frequency is decreased to 271 MHz.

Table 3. Implementation results of the GCD cluster and the hierarchical GCD cluster for 1024-bit moduli

Available	Slice Registers 607200	Slice LUTs 303600	DSP slices 2800	18kb block RAMs 2060	Clock Frequency (MHz)
GCD cluster	20548	26306	128	136	320
hierarchical GCD cluster	225176	288215	1408	1560	271

We have evaluated the number of clock cycles to compute all GCDs of $71 \times 71 = 5041$ pairs of 1024-bit moduli by one GCD cluster. For this purpose, we have used RSA moduli generated by OpenSSL Toolkit. By simulation, we have that it takes 859468 clock cycles to compute the GCDs of 5041 pairs. If a GCD cluster operates in 271 MHz as shown in Table 3, the expected computing time is $859468/271\text{MHz} = 3.17\,\text{ms}$. Also, it takes about $71 \times 2 \times 57 = 8094$ clock cycles to transfer a pair of two blocks of 71 moduli each and this overhead is negligible. Since the hierarchical GCD cluster can have up to 11 clusters, we can expect that the GCDs of $5041 \times 11 = 55451$ pairs in the same time. Thus, we can write that one GCD can be computed in expected $3.17\,\text{ms}/55451 = 0.057\mu s$.

6   Conclusions

We have presented an Euclidean-based hardware algorithm for computing the GCD. It was implemented in an FPGA using the FDFM approach such that each of 1408 GCD processors computes the GCD of 1024-bit moduli in RSA keys independently. From the implementation results, it can compute the GCD of two 1024-bit RSA moduli in expected $0.057\mu s$ on a Xilinx Virtex-7 family FPGA XC7VX485T-2.

References

1. Ago, Y., Ito, Y., Nakano, K.: An FPGA implementation for neural networks with the FDFM processor core approach. Int. J. Parallel Emergent Distrib. Syst. **28**(4), 308–320 (2013)
2. Bordim, J.L., Ito, Y., Nakano, K.: Accelerating the CKY parsing using FPGAs. IEICE Trans. Inf. Syst. **E86–D**(5), 803–810 (2003)
3. Cormen, T.H., Leiserson, C.E., Rivest, R.L., Stein, C.: Introduction to Algorithms. MIT Press, Cambridge (2001)
4. Devi, R., Singh, J., Singh, M.: VHDL implementation of GCD processor with built in self test feature. Int. J. Comput. Appl. **25**(2), 50–54 (2013)
5. Fujimoto, N.: High throughput multiple-precision GCD on the CUDA architecture. In: Proceedings of the International Symposium on Signal Processing and Information Technology, pp. 507–512, December 2009
6. Fujita, T., Nakano, K., Ito, Y.: Bulk GCD computation using a GPU to break weak RSA keys. In: Proceedings of the International Parallel and Distributed Processing Symposium Workshops, pp. 385–394, May 2015
7. Ito, Y., Nakano, K., Bo, S.: The parallel FDFM processor core approach for CRT-based RSA decryption. Int. J. Netw. Comput. **2**(1), 79–96 (2012)
8. Knuth, D.E.: The Art of Computer Programming, Volume 2: Seminumerical Algorithms. Addison-Wesley, Reading (1997)
9. Kohale, S.D., Jasutkar, R.W.: Power optimization of GCD processor using low power Spartan 6 FPGA family. Int. J. Conceptions Electron. Commun. Eng. **2**(1), 1–6 (2014)
10. Lenstra, A.K., Hughes, J.P., Augier, M., Bos, J.W., Kleinjung, T., Wachter, C.: Ron was wrong, Whit is right. Cryptology ePrint Archive, Report 2012/064 (2012). http://eprint.iacr.org/
11. Nakano, K., Yamagishi, Y.: Hardware n choose k counters with applications to the partial exhaustive search. IEICE Trans. Inf. Syst. **E88–D**(7), 1350–1359 (2005)
12. Nakano, K., Kawakami, K., Shigemoto, K.: RSA encryption and decryption using the redundant number system on the FPGA. In: Proceedings of the International Symposium on Parallel and Distributed Processing Workshops, pp. 1–8, May 2009
13. Rivest, R.L., Shamir, A., Adleman, L.: A method for obtaining digital signatures and public-key cryptosystems. Commun. ACM **21**, 120–126 (1978)
14. Scharfglass, K., Weng, D., White, J., Lupo, C.: Breaking weak 1024-bit RSA keys with CUDA. In: Proceedings of the Internatinal Conference on Parallel and Distributed Computing, Applications and Technologies, pp. 207–212, December 2012
15. White, J.R.: PARIS: A PArallel RSA-prime InSpection tool. Ph.D. thesis, California Polytechnic State University - San Luis Obispo, June 2013
16. Xilinx Inc.: 7 Series DSP48E1 Slice User Guide, November 2014
17. Xilinx Inc.: 7 Series FPGAs Configurable Logic Block User Guide, November 2014
18. Xilinx Inc.: 7 Series FPGAs Memory Resources User Guide, November 2014
19. Xilinx Inc.: VC707 Evaluation Board for the Virtex-7 FPGA User Guide (2014)

Parallel Induction of Nondeterministic Finite Automata

Tomasz Jastrzab[1]([✉]), Zbigniew J. Czech[1], and Wojciech Wieczorek[2]

[1] Institute of Informatics, Silesian University of Technology, Gliwice, Poland
{Tomasz.Jastrzab,Zbigniew.Czech}@polsl.pl
[2] Institute of Informatics, University of Silesia, Sosnowiec, Poland
wojciech.wieczorek@us.edu.pl

Abstract. The induction of a minimal nondeterministic finite automaton (NFA) consistent with a given set of examples and counterexamples, which is known to be computationally hard, is discussed. The paper is an extension to the novel approach of transforming the problem of NFA induction into the integer nonlinear programming (INLP) problem. An improved formulation of the problem is proposed along with the two parallel algorithms to solve it. The methods for the distribution of tasks among processors along with distributed termination detection are presented. The experimental results for selected benchmarks are also reported.

Keywords: Parallel algorithms · Nondeterministic finite automata · Integer nonlinear programming

1 Introduction

Grammatical inference is an intensively studied area with a wide variety of applications including machine learning, syntactic and structural pattern recognition, speech recognition and others. The core idea of grammatical inference is to provide, given some input data, some broader information on them. In the problem studied in this paper a finite sample $S = (S_+, S_-)$ consisting of examples and counterexamples is given. The task is to provide an automaton consistent with the given sample S, meaning that it is supposed to accept all examples (members of set S_+) and reject all counterexamples (members of set S_-). In particular we are interested in inducing a nondeterministic finite automaton (NFA) of minimum size in terms of a number of states.

The problem of inducing a minimal deterministic finite automaton has been proved NP-hard [1]. The decision version of the problem of converting a DFA to a minimal NFA is PSPACE-complete [2]. The induction of minimal regular expression is PSPACE-complete, as shown in [3]. In the view of the proof shown in [3] it is known that minimization of a regular expression for an arbitrary learning sample belongs to the class of NP-hard problems. So it is probable that induction of a minimal NFA is also of that complexity. The most popular

© Springer International Publishing Switzerland 2016
R. Wyrzykowski et al. (Eds.): PPAM 2015, Part I, LNCS 9573, pp. 248–257, 2016.
DOI: 10.1007/978-3-319-32149-3_24

algorithms for inducing finite automata (both deterministic and nondeterministic) for a given learning sample involve heuristics based on state merging. The commonly known state clustering algorithms are: evidence driven state merging (EDSM) [4] and regular positive and negative inference (RPNI) algorithms [5]. The key idea of these algorithms is to build first a prefix tree acceptor for a given input sample and then to remove redundant states by merging them together. The validity of states merging is decided based on whether the new automaton remains consistent with the sample. The algorithms for NFA induction have not gained that much attention and some of them are only theoretical [6].

A novel approach towards inducing NFAs is the idea of translating the problem of NFA induction into an instance of integer nonlinear programming (INLP) problem which we explore here. The motivation for our research is as follows. Namely, there are certain problems for which the induction of a minimal automaton and/or grammar is desired, particularly if the size (length) of input data is limited. Some examples of successful application of minimal grammars to the induction of ordinary generating functions are given in [7]. Also as stated in [8], such minimal representations help to analyze the structures of languages or mechanisms of grammars.

The main contributions of this paper are twofold. Firstly, based on our basic INLP problem formulation for the NFA induction [9], we introduce an improved version of this approach providing the significantly reduced constraint and variable sets, leading to a much smaller solution space. The actual size reduction factors are also shown. Secondly we propose two algorithms for solving the system of equations defined by the INLP. The first algorithm is theoretical in nature, while the second solves the problem under consideration in parallel. The parallel algorithm has been evaluated experimentally on some benchmarking test languages and has proved successful in solving all the test cases. The execution times of the parallel algorithm are measured and reported in the paper.

The paper is organized into five sections. Section 2 presents the basic terms and notions related to the theory of automata and languages. Section 3 discusses the INLP formulation of NFA induction problem and the algorithms to solve it. In Sect. 4 we report the experimental results for the benchmarking languages. The final section contains conclusions and further research perspectives.

2   Basic Terms and Notions

A word (string) w is a finite set of symbols over an alphabet Σ. The length of word w, denoted by $|w|$ is the number of symbols the word is built of, with the particular case of an empty word λ, in which no symbols occur. For a given alphabet Σ, the set Σ^* contains all possible strings over Σ and $\Sigma^+ = \Sigma^* - \{\lambda\}$. A language L, $L \subseteq \Sigma^*$, is a set of words over given alphabet Σ. Let $u, v, w \in \Sigma^*$ be words. Then word u is called a *prefix* of word w if $w = uv$ holds, where uv is a concatenation of words u, v. It is called a *proper* prefix if $v \neq \lambda$.

A nondeterministic finite automaton A is defined as a quintuple $A = \{Q, \Sigma, \delta, q_0, Q_F\}$, where Q denotes the finite set of states, Σ is the alphabet, δ is the transition function $\delta : Q \times \Sigma \rightarrow 2^Q$, $q_0 \in Q$ is the starting (initial) state and $Q_F \subseteq Q$ is the set of final (accepting) states. A word w belongs to the language L accepted by an automaton A if and only if w can be spelled out on at least one path starting in state q_0 and ending in any state $q \in Q_F$, where the path represents a series of transitions defined by δ.

The integer nonlinear programming problem (INLP) is a combinatorial optimization problem defined as follows. Given a set of nonlinear equations and inequalities (constraints) over input vector x find an assignment of integral (possibly binary) values to elements of x such that all constraints are satisfied and some objective function $f(x)$ is minimized (or maximized).

3   The Nondeterministic Finite Automata Induction as Integer Nonlinear Programming

3.1   The Overview

The problem of NFA induction can be converted into an instance of integer nonlinear programming problem, in which the number of states is minimized ($f(x) = |Q|$). However, in the paper we consider the INLP as a decision problem, in which we ask if, given an input sample S and an integer $k > 0$, there exists a k-state automaton consistent with S. By taking $k = 1, 2, \ldots$ we find the minimal NFA. We present here the definitions of the variable and constraint sets, to provide a basis for an improved INLP problem formulation. As will be shown the improvement relates to the sizes of constraint and variable sets, leading to the reduction of the solution space.

Let $A = \{Q, \Sigma, \delta, q_0, Q_F\}$, $S = (S_+, S_-)$ be defined as previously. Let $P(S)$ denote the set of prefixes of words $w \in S$. Then the three types of binary variables may be defined. First, the variables z_q denote whether a given state $q \in Q$ belongs to the set of final states $Q_F \subseteq Q$, i.e. z_q is 1 iff $q \in Q_F$. Secondly, the variables y_{apq} describe the transitions between states $p, q \in Q$, with label $a \in \Sigma$, i.e. y_{apq} is 1 iff $q \in \delta(p, a)$. Finally, the variables x_{wq} with $w \in P(S)$, $q \in Q$ specify whether the word w is spelled out on a path between starting state $q_0 \in Q$ and an arbitrary state q.

Given the definition of variables x_{wq}, y_{apq}, z_q, the problem of the NFA induction can be described with the following set of equations and inequalities [9]:

– if $\lambda \in S$ (empty word inclusion)

$$z_{q_0} = 1, \text{ if } \lambda \in S_+, z_{q_0} = 0, \text{ if } \lambda \in S_- \tag{1}$$

– for all $w \in S_+ - \{\lambda\}$ (acceptance of examples)

$$\sum_{q \in Q} x_{wq} z_q \geq 1 \tag{2}$$

– for all $w \in S_- - \{\lambda\}$ (rejection of counterexamples)

$$\sum_{q \in Q} x_{wq} z_q = 0 \tag{3}$$

– for all $w = a$, $w \in P(S)$, $a \in \Sigma$ (single-symbol prefixes inclusion)

$$x_{wq} - y_{aq_0q} = 0 \tag{4}$$

– for all $w = va$, $w \in P(S)$, $v \in \Sigma^+$, $a \in \Sigma$ (multi-symbol prefixes inclusion)

$$- x_{wq} + \sum_{p \in Q} x_{vp} y_{apq} \geq 0 \tag{5}$$

$$x_{wq} - x_{vp} y_{apq} \geq 0, p \in Q \tag{6}$$

It can be easily observed that the system of equations and inequalities (1) to (6) describes a nondeterministic finite automaton accepting all members of set S_+ and rejecting all members of set S_-.

Let $S = (S_+, S_-)$, x_{wq}, y_{apq}, z_q be defined as previously. Let Σ^B denote a logical (Boolean) sum. Then taking into account the binary nature of variables x_{wq}, y_{apq}, z_q and the rules of Boolean algebra, the equations and inequalities given by (2), (3), (5) and (6) can be restated as follows:

– for all $w \in S_+ - \{\lambda\}$ (acceptance of examples)

$$\Sigma^B_{q \in Q} x_{wq} z_q = 1 \tag{7}$$

– for all $w \in S_- - \{\lambda\}$ (rejection of counterexamples)

$$\Sigma^B_{q \in Q} x_{wq} z_q = 0 \tag{8}$$

– for all $w = va$, $w \in P(S)$, $v \in \Sigma^+$, $a \in \Sigma$ (multi-symbol prefixes inclusion)

$$- x_{wq} + \Sigma^B_{p \in Q} x_{vp} y_{apq} = 0 \tag{9}$$

Consequently, the complete reduced INLP comprises equations (1), (4), and (7) to (9). The correctness of the proposed formulation may be proved using the fact that for a binary variable b, the following hold: (i) $b + 0 = b$, (ii) $b + 1 = 1$ and (iii) $b + b = b$. Furthermore, note that the number of equations given in (7) to (9) is indeed reduced due to reconstruction of (5) and (6) into (9).

3.2   The Algorithms

The first algorithm relies on the observation that, assuming that the number of states of an induced NFA is set in advance, it is enough to determine the transition function δ and the set of final states Q_F. In terms of the INLP formulation it means that it is necessary to find the values of y_{apq} and z_q variables. Furthermore, the equations actually describing the NFA are only (7) and (8).

```
1: procedure INDUCTION( )
2:     buildEqs.(1), (4), (7) to (9)
3:     sort x_{wq} according to |w|
4:     for all x_{wq} do
5:         substitute x_{wq} with (4) or (9)
6:     end for
7:     while true do
8:         assign values to y_{apq}, z_q
9:         if Eqs. (7) and (8) are satisfied then
10:            return
11:        end if
12:    end while
13: end procedure
```

Fig. 1. The exhaustive search algorithm

The exhaustive search algorithm shown in Fig. 1 works in the following way. First, the initial set of equations is built according to Eqs. (1), (4), (7) to (9). Then the equations are rewritten so that only y_{apq} and z_q variables remain, which reduces the constraint set to Eqs. (1), (7) and (8) and the variable set to unknowns y_{apq} and z_q (lines 3–6). Such a representation can be obtained by using Eqs. (4) and (9) as definitions of variables x_{wq} and successively substituting them in the remaining constraints, consequently producing directly visible nonlinear expressions with respect to y_{apq}. The substitution procedure should select variables x_{wq} in ascending order of lengths of prefixes w, which minimizes the number of substitutions. In the final step of the algorithm (lines 7–12) binary values are assigned to y_{apq} and z_q variables and are checked against the set of modified equations (1), (7) and (8) until the solution is found. It is also worth noticing that the procedure in lines 7–12 can be easily parallelized since the sets of y_{apq} and z_q values are independent of each other.

Although we do not specify any particular algorithm for traversing the reduced solution space, let us note an important fact. We propose that z_q variables should have their values assigned prior to y_{apq}, since (7) and (8) will have the following general form $\Sigma^B(\Sigma^B(\prod y_{apq})z_q) = 1$ (Eq. 7) and $\Sigma^B(\Sigma^B(\prod y_{apq})z_q) = 0$ (Eq. 8), where $a \in \Sigma$ and $p, q \in Q$. Consequently, setting any z_q to zero makes the value of corresponding expression $\Sigma^B(\prod y_{apq})$ irrelevant in terms of the final result and conversely, if it is set to one then $\Sigma^B(\prod y_{apq})$ should have a specific value depending on the equation (either 1 in case of (7) or 0 in case of (8)).

The algorithm shown in Fig. 1 has several advantages. Firstly, it reduces significantly the cardinalities of constraint and variable sets, thus making the solution space much smaller than that in initial INLP formulation. Secondly, it allows for applying parallelism in the final phase, which can increase speedup of computation. What is more, in terms of algorithm implementation, modified Eqs. (7) and (8) can be easily expressed using bitwise operations. And finally, even for the expressions $\Sigma^B(\prod y_{apq})$ including many terms, their values may be known before setting all involved variables due to their binary nature. The crucial

```
 1: procedure INDUC_PARALLEL(level)
 2:     if level = COUNT(y) + COUNT(z) then
 3:         evaluate x_{wq}
 4:         if Eqs. (7) to (8) are satisfied then
 5:             found = true                          ▷ sent to others
 6:             return
 7:         end if
 8:     else
 9:         set next y_{apq}
10:         INDUC_PARALLEL(level + 1)
11:         if found then
12:             return
13:         end if
14:     end if
15: end procedure
```

Fig. 2. The parallel algorithm

drawback of the algorithm is that the substitution process rapidly increases the amount of memory required to store the modified equations and therefore makes the algorithm theoretical rather than applicable in practice.

The second algorithm (Fig. 2) is a modification of the exhaustive search algorithm from Fig. 1, allowing for a practical implementation by using parallel backtracking search. Backtracking is an exact method of solving combinatorial optimization problems. It traverses a solution space according to the rules of depth-first search. The algorithm uses the features introduced in Fig. 1, but does not perform the actual substitution step, except for the values resulting from (4). The majority of x_{wq} values are not replaced, however they are still treated as dependent variables and are only evaluated based on the values of y_{apq}, instead of being set directly. The recursive procedure INDUC_PARALLEL (Fig. 2) takes as an argument the current recursion level. Procedure COUNT, invoked in line 2, returns the number of occurrences of variables of given type. If all y_{apq} variables are set (line 2) the procedure evaluates x_{wq} in the order of increasing prefix length (line 3) and finally verifies current assignments against Eqs. (7) to (8) (lines 4–7), otherwise it sets the next variable y_{apq}. It is assumed that the z_q variables were set before entering the recursive procedure.

The distribution of computations among processors occurs mainly at the level of z_q combinations generation (before entering the recursive procedure). Each combination of z_q values, $q \in Q$, is assigned to a different group of processors to broaden exploration of a solution space. Such an approach is applied because every combination of z_q values is initially equally probable to produce a correct solution (except for the case of $z_q = 0$, for each $q \in Q$ which is never successful). The number of possible combinations, for a given number of states k, can be also restricted by Eq. (1). The combinations are generated independently by the processes and are assigned to them based on their ranks, thus forming groups of processes working on the same set of z_q variable assignments. The further

Table 1. Test languages

No.	Description
1	1^*
2	$(10)^*$
3	any string without an odd number of consecutive 0's after an odd number of consecutive 1's
4	any string without more than 2 consecutive 0's
5	any string of even length which, making pairs, has an odd number of (01)'s or (10)'s
6	any string such that the difference between the numbers of 1's and 0's is $3n$
7	$0^*1^*0^*1^*$

level of computations distribution takes place within each group in line 9. It basically assigns a different starting point of the search to each processor within the group, as proposed in [10].

Since only the first solution is searched for, upon finding it (line 6) the execution is terminated. Two distributed termination detection schemes were considered and implemented. The first one was based on the algorithm proposed in [11], where communication can occur between the master and worker processors, as well as between successive processors organized in a ring. The other termination algorithm adopts a star topology with the master in the middle. In the latter approach every processor has to send and receive a single message to/from the master process and may then terminate. However, the order in which these two operations occur is not set in advance. The probing for termination signal is performed periodically by each processor. However, as the experiments have shown the impact of this operation on the total execution time is not significant.

4   The Experiments

4.1   Experimental Setup

The experiments were conducted using a C++ implementation based on Message Passing Interface (MPI) library. To provide a means for comparison the Tomita benchmarking languages presented in Table 1 were selected [12].

4.2   Experimental Results

Let us first compare the sizes of constraint and variables sets (Table 2) obtained using the basic and reduced INLP formulations, after defining the equations and inequalities as bitwise operations. The columns of Table 2 represent respectively: k – number of states, $|c_i|$, $i \in \{B, R\}$ – cardinality of the constraint set for the basic (B) and reduced (R) INLP, $|v_i|$, $i \in \{B, R\}$ – cardinality of the variable set

Table 2. Constraint and variable set cardinalities

No.	k	$|c_B|$	$|c_R|$	$|v_B|$	$|v_R|$	$\Delta_{|c|}$	$\Delta_{|v|}$
1	1	59	35	24	3	59%	88%
2	2	214	75	74	10	65%	86%
3	4	1437	296	316	36	78%	89%
4	3	878	226	234	21	74%	91%
5	4	1214	249	272	36	78%	87%
6	3	622	162	168	21	74%	88%
7	4	1401	288	312	36	78%	88%

Table 3. Execution times for the parallel algorithms

No.	p_B	p_R	τ_B	τ_R
1	4	4	< 1	< 1
2	4	4	< 1	< 1
3	80	30	–	34
4	120	4	6	< 1
5	64	20	–	155
6	80	4	1	< 1
7	72	40	–	7287

for the basic (B) and reduced (R) INLP, $\Delta_{|c|}$ – reduction degree of a constraint set size, $\Delta_{|v|}$ – reduction degree of a variable set size. The variable set size is measured in terms of the number of variables that are directly set, before the solution is verified.

As can be observed based on the results presented in Table 2 the reduced INLP formulation provides significant improvement in the size of both the constraint and variable sets. The reduction rate for the constraint sets varies from 59 % up to 78 %, while for variable sets it is at the level of 86 % up to 91 % as compared to the initial formulation.

The results of running the parallel algorithm for the 7 benchmarking languages are reported in Table 3, where p_B, p_R represent the number of processors used for the basic and reduced INLP and τ_B, τ_R represent the algorithm execution times. The execution times are expressed in seconds and were measured using the MPI_WTIME function. They correspond to the total execution time of the parallel algorithm, including the initial step of building and partially rewriting the set of equations (for the reduced INLP). The timings for three out of seven languages for the basic approach (languages number 3, 5 and 7) are missing because the size of the solution space made the computations last unreasonably long, so the execution was aborted.

It can be seen from Table 3 that the algorithm performs very well for 6 out of 7 languages. The only exception from this is the 7th language, for which the execution time is significantly larger than in the six other cases. We presume that the reason for this may be that the combination of y_{apq} variables constituting the solution for this particular problem is not considered early in the process of combinations generation. This should be perceived as a suggestion for further research on the order in which variable values are assigned. It may be also worthwhile to consider using larger number of processors since the execution time for the 7th language was reduced almost by the factor of 2 with doubled number of processors (such results were obtained for the increase from 20 to 40 processors).

The reduced approach should be considered successful, since it allows for finding solutions for all Tomita languages, and with much smaller number of processors used as compared to the basic approach. Comparing the results with other algorithms, the approach also proves better than EDSM and RPNI[1].

5   Conclusions

The paper presents an improved formulation of the problem of nondeterministic finite automata induction by means of integer nonlinear programming. Two algorithms are proposed, one of them being theoretical in nature. The parallel algorithm, based on the theoretical considerations, has been implemented and evaluated experimentally on the Tomita benchmarking languages. The algorithm proves to be successful in solving the test cases and it also manages to outperform the well-known heuristic approaches.

We consider the proposed algorithms as an important contribution in the field of parallel algorithms applications, since they allow to solve the computationally hard NFA induction problem efficiently. Furthermore, the parallelism makes it possible to find the exact solutions for all benchmarking problems. The theoretical algorithm can be also treated as an important parallel "algorithmic skeleton" for solving similar problems.

In the future we plan to modify the order in which variables are set, as well as the ways of distributing the computations among processors. The preliminary considerations show that dependent variables may be evaluated in a more efficient way allowing for earlier solution space pruning. Moreover it would be worthwhile to test the algorithm against some other, possibly larger benchmarking sets.

Acknowledgment. Computations were carried out using the computer cluster Galera at the Academic Computer Center in Gdańsk (http://task.gda.pl/kdm) and using the computer cluster Ziemowit (http://ziemowit.hpc.polsl.pl) funded by the Silesian BIO-FARMA project No. POIG.02.01.00-00-166/08 in the Computational Biology and Bioinformatics Laboratory of the Biotechnology Centre in the Silesian University of Technology, Gliwice, Poland.

---

[1] The implementations of the RPNI and EDSM algorithms were taken from the open source project of grammatical inference tools (gitoolbox) available at https://code.google.com/p/gitoolbox. The toolbox is described in [13].

This research was supported by Grant No. DEC-2011/03/B/ST6/01588 from National Science Center of Poland and the Institute of Informatics research grant no. 525/RAU2/2014/9.

References

1. Gold, E.M.: Complexity of automaton identification from given data. Inf. Control **37**, 302–320 (1978)
2. Jiang, T., Ravikumar, B.: Minimal NFA problems are hard. SIAM J. Comput. **22**, 1117–1141 (1993)
3. Angluin, D.: An application of the theory of computational complexity to the study of inductive inference. Ph.D. thesis, University of California (1976)
4. Lang, K.J., Pearlmutter, B.A., Price, R.A.: Results of the abbadingo one DFA learning competition and a new evidence-driven state merging algorithm. In: Honavar, V.G., Slutzki, G. (eds.) ICGI 1998. LNCS (LNAI), vol. 1433, pp. 1–12. Springer, Heidelberg (1998)
5. Dupont, P.: Regular grammatical inference from positive and negative samples by genetic search: the GIG method. In: Carrasco, R.C., Oncina, J. (eds.) ICGI 1994. LNCS (LNAI), vol. 862, pp. 236–245. Springer, Heidelberg (1994)
6. García, P., Vázquez de Parga, M., Álvarez, G.I., Ruiz, J.: Learning regular languages using non-deterministic finite automata. In: Ibarra, O.H., Ravikumar, B. (eds.) Implementation and Applications of Automata, pp. 92–101. Springer, New York (2008)
7. Wieczorek, W., Nowakowski, A.: Grammatical inference in the discovery of generating functions. In: Gruca, A., Brachman, A., Czachórski, T., Kozielski, S. (eds.) Man–Machine Interactions 4. AISC, vol. 391, pp. 627–637. Springer International Publishing, Switzerland (2016)
8. Imada, K., Nakamura, K.: Learning context free grammars by using SAT solvers. In: International Conference on Machine Learning and Applications, ICMLA 2009, 13–15 December 2009, Miami Beach, FL, USA, pp. 267–272 (2009)
9. Wieczorek, W.: Induction of non-deterministic finite automata on supercomputers. In: Heinz, J., de la Higuera, C., Oates, T. (eds.) JMLR Workshop and Conference, Proceedings of the Eleventh International Conference on Grammatical Inference, ICGI 2012, 5–8 September 2012, University of Maryland, College Park, USA, vol. 21, pp. 237–242 (2012)
10. Quinn, M.J.: Parallel Programming in C with MPI and OpenMP. McGraw-Hill, New York (2004)
11. Dijkstra, E.W., Seijen, W.H., van Gasteren, A.J.M.: Derivation of a termination detection algorithm for distributed computations. Inf. Process. Lett. **16**, 217–219 (1983)
12. Tomita, M.: Dynamic construction of finite automata from examples using hillclimbing. In: Proceedings of the Fourth Annual Conference of the Cognitive Science Society, pp. 105-108 (1982)
13. Akram, H.I., de la Higuera, C., Xiao, H., Eckert, C.: Grammatical inference algorithms in MATLAB. In: Sempere, J.M., García, P. (eds.) ICGI 2010. LNCS, vol. 6339, pp. 262–266. Springer, Heidelberg (2010)

Parallel BSO Algorithm for Association Rules Mining Using Master/Worker Paradigm

Youcef Djenouri[1][(✉)], Ahcene Bendjoudi[2], Djamel Djenouri[2],
and Zineb Habbas[3]

[1] LRDSI, University of Saad Dahleb, Blida, Algeria
y.djenouri@gmail.com
[2] DTISI, CERIST Research Center, Algiers, Algeria
ahcene.bendjoudi@gmail.com, ddjenouri@acm.org
[3] LCOMS, University of Lorraine, Ile du Saulcy, 57045 Metz Cedex, France
zineb@univ-metz.fr

Abstract. The extraction of association rules from large transactional databases is considered in the paper using cluster architecture parallel computing. Motivated by both the successful sequential BSO-ARM algorithm, and the strong matching between this algorithm and the structure of the cluster architectures, we present in this paper a new parallel ARM algorithm that we call MW-BSO-ARM for master/worker version of BSO-ARM. The goal is to deal with large databases by minimizing the communication and synchronization costs, which represent the main challenges that faces any cluster architecture. The experimental results are very promising and show clear improvement that reaches 300 % for large instances. For examples, in big transactional database such as WebDocs, the proposed approach generates 10^7 satisfied rules in only 22 min, while a previous GPU-based approach cannot generate more than 10^3 satisfied rules into 10 h. The results also reveal that MW-BSO-ARM outperforms the PGARM cluster-based approach in terms of computation time.

Keywords: Master/Worker · Cluster architecture · BSO-ARM

1 Introduction

Association Rule Mining (ARM) problem can be formally be described as follows: Let I be the set of n items $\{I_1, I_2, ..., I_n\}$, D be the set of m transactions $\{d_1, d_2, ..., d_m\}$, then an association rule $(X \Rightarrow Y)$ is composed of two disjoint parts, the set of items, X, called antecedent part and the set of items, Y, called consequent part. ARM problem aims at finding all relevant association rules from the transactional database D. Two measures (support and confidence) have been used in most existing ARM algorithms for evaluating association rules. They are based on the frequency of the items that appeared in the given rule on the transactional database. The support of the rule is the number of transactions

© Springer International Publishing Switzerland 2016
R. Wyrzykowski et al. (Eds.): PPAM 2015, Part I, LNCS 9573, pp. 258–268, 2016.
DOI: 10.1007/978-3-319-32149-3_25

that contains its items over the number of all transactions and the confidence of the rule is the fraction between its support and the support of its antecedent part. The aim in effective association rules mining aims is to discover strong rules, with both confidence and support beyond user's specific threshold. This problem constitutes one of the more important problems in data mining, especially when dealing with large transactional databases. Indeed, the conventional sequential algorithms, such that APRIORI [1] and FPGROWTH [2] do not have the capacity to analyze such large databases, especially in terms of run-time performance.

To address this problem several researchers have exploited in various ways, the computing power of parallel machines.

- For years, the ARM community has mainly focused on exact parallel ARM algorithms. We can cite without being exhaustive Cuda-APRIORI [3] for GPU computing, PPCD and CCPD [4] for shared systems computing and CD, DD [5] for distributed systems computing. However, these algorithms suffer from many challenging problems. Buyya defined the cluster architecture in [6] as: "A cluster is a type of parallel or distributed processing system, which consists of a collection of interconnected and heterogenous computers cooperatively working together as a single computing resource". The same author in the same paper says that the cluster architecture is flexible by giving the possibility to adding or to removing nodes to a system. Many approaches have also been developed to solve ARM problem using cluster architecture. We can cite ([8], for the parallelization of Aprioiri, [9], for the parallelization of FPgrwoth).
- Concerning parallel metaheuristics for ARM, only few related papers are published this subject [7,14]. Moreover, the results obtained by application of these algorithms to big instances such as Webdocs benchmarks are not satisfactory.

Motivated by the principle of BSO-ARM algorithm recently proposed in [13], which is very appropriated for a client-server cluster architecture, we propose in this paper CS-BSO-ARM, parallel version of BSO-ARM algorithm based on cluster computing. To validate this new algorithm, several experiments have been carried out on the Ibnbadis cluster. Our experimental results are very promising. Indeed, we observed good efficiencies for medium instances and super efficiency reaching 300 % for large instances. Note that for big transactional database like WebDocs, our approach generates 10^7 satisfied rules into only 22 min, while previous well known approach cannot generate more than 10^3 satisfied rules into 10 hours. The results also reveal that CS-BSO-ARM outperforms the PGARM cluster-based approach in terms of computation time. The remainder of the paper is as follows: Sect. 2 presents the BSO algorithm proposed for the ARM problem, while Sect. 3 describes the proposed parallelization of the algorithm. The performance evaluation is described in Sect. 4, and finally, Sect. 5 draws the conclusions.

2   BSO-ARM Algorithm

BSO [10] simulates the foraging behavior of bees in nature. It is applied on many applications like information retrieval [11], SAT problem [10] and association rules mining [13]. In this section, we present the BSO-ARM algorithm (Bees Swarm Optimization for Association Rules Mining.

2.1   Presentation of BSO-ARM Algorithm

The BSO-ARM algorithm is first proposed in [12]. An extended version presented in [13], improves the quality of the obtained rules thanks to the proposition of three strategies of the search process called Modulo, Next or Syntactic. Formal and detailed definitions of these strategies are given in IJBIC, so they are deliberately omitted here. The aim of BSO-ARM is to find one part of association rules respecting minimum support and minimum confidence constraints in a reasonable time.

According to Algorithm 1, the detailed BSO-ARM is as follows:

In Line 5, the initial reference solution is generated randomly or using a heuristic. To avoid returning to previous solutions, the current Sref is inserted in the Tabou list, at each pass. In Line 10, from the current Sref, the search area of each bee is determined by using one of three possible strategies (Modulo, Next, Syntactic). Then, each bee explores its region thanks to neighborhood search operation (Lines $12 - 16$). After that, the communication is done between bees in order to select the best solution, which becomes the next Sref if it does not exist in Tabou list. Otherwise, the diversity criterion is performed by choosing randomly another solution from the search space and starting again on it. In order to keep the memory of the best solution in each pass, the best solution found is saved in the Best list. This process must be repeated until the maximum number of iterations is reached. At the end of the algorithm, we generate the rules from the Best List and display them to the data analyst in particular or to the user in general.

2.2   Experimental Limitation of BSO-ARM

In [13], BSO-ARM algorithm has been experimented on some standard data sets used in the evaluation of most state-of-the-art ARM algorithms. This experimental study revealed that the execution time of BSO-ARM grows exponentially when the number of transactions increases. When dealing with small and medium instances, BSO-ARM determines the final solution in reasonable time. However, for large instances such as webdocs data set that contains more than one million and half of transactions, BSO-ARM fails to give a result after 15 days of CPU-time. This large instance has been recently solved in [14], by a first parallel version of BSO-ARM on GPU architecture. However this approach takes 10 hours to generate 1000 satisfied rules, because only the evaluation of rules has benefited from GPU parallelism. In turn, The solutions are generated

Algorithm 1. BSO-ARM

1: Empirical parameters: K (Bees number), Flip, Distance, Max_Iter, a, b.
2: **Input**: Dataset transactions, MinSup, MinConf
3: **Output**: Set of association rules
4: $i \leftarrow 0$
5: Sref $\leftarrow$ the initial solution generated randomly or via an heuristic
6: Fmax (Sref, a,b, Minsup,Minconf)
7: $S* \leftarrow$ Sref
8: **while** $i < Max\_Iter$ **do**
9:     TabouList $\leftarrow$ TabouList + Sref
10:    Set_K_solutions $\leftarrow$ SearchArea(Sref)
11:    Assign one solution of set_k_solutions to each bee
12:    **for** each bee k **do**
13:        Fitness (k,a,b, Minsup,Minconf)
14:        neighborhood-search(k)
15:        Save the best solution found in dance_table
16:    **end for**
17:    Insert the best solution in Best
18:    Sref $\leftarrow$ the best solution in dance_table
19:    **if** Sref exists in TabouList **then**
20:        Sref $\leftarrow$ Criterion_Diversity(Sref)
21:    **end if**
22:    **if** Fmax(s) > Fmax(Sref) **then**
23:        $S* \leftarrow$ Sref
24:    **end if**
25:    $i \leftarrow i + 1$
26: **end while**
27: **for** each solution s in Best **do**
28:    Generate the rule from s
29: **end for**

sequentially. In this paper, we present a new parallel version of BSO-ARM, which will benefit from the power of the high perfomance cluster Ibnbadis[1].

3 MW-BSO-ARM: A Master/Worker BSO-ARM Algorithm

3.1 Motivation

In this section we present a parallel version of BSO ARM. There are several parallelization techniques that depend on different models of parallelism (task parallelism, data parallelism), different models of synchronization (message exchange, shared memory . . .). As mentioned in Sect. 1, we propose in this paper a master-worker parallel algorithm for two reasons. First, it is a simple and efficient model, mostly used in literature, to address large scale applications. Second, our main

---

[1] Ibnbadis is a cluster of CERIST research center, Algers, Algeria.

objective is precisely to extract an important number of significant rules from large databases.

3.2   Master Worker Architecture

A master-worker architecture can be simply described by Fig. 1

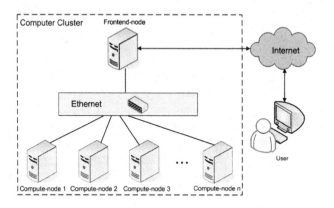

Fig. 1. Master-worker (obtained from [6])

According to this architecture, the user submits its job to the master via Internet. The master decomposes the problem to be solved into a collection of tasks, schedules and distributes the tasks among the compute-nodes (workers). When a worker completes its task, it sends the result to the master. The master merges all the results coming from the different workers in order to obtain the solution of the problem. The workers and master communicate via Ethernet, by using communication inter-processus model (MPI, PVM, etc. ...). Note that, this architecture is a mono-layer master-worker architecture (one master and multiple workers).

3.3   Presentation of CS-BSO-ARM

To parallelize the BSO-ARM algorithm in a client server model we have to define how the problem is decomposed by the server in order to generate the different tasks to be performed by the clients, the process performed by the server, that each client executes and how the client and server communicate. In our case, the clients and server communicate by using the MPI communication paradigm.

– **Generation of tasks by the server**: Thanks to the basic principle of BSO-ARM algorithm, its parallelization on a client-server architecture is quite direct. The server, determines the reference solution $(Sref)$ and generates K solutions using determination of search area strategy, according to the Modulo strategy presented in [13]. The tasks correspond then to the K regions to be explored by the K bees. The number of tasks in CS-BSO-ARM is static and corresponds to the number of bees. Note that CS-BSO-ARM is a task parallelism algorithm, unlike the parallel algorithm presented in [7] which is a data-parallelism algorithm. Indeed, in [7], the whole database is decomposed into several smaller databases and each smaller database is scanned by a given genetic algorithm.

– **The process performed by the server**: Once the server has generated the tasks, it distributes them among the different clients and waits the solution. When the server receives all solutions, it assigns to $Sref$ the best one. This process is repeated until a maximum number of iterations is reached.

– **The process performed by the client**: The client applies the local search process from its respective region using the neighborhood search computation strategy described in [13]. Each client returns the best solution it founds to the server. Note that this algorithm is highly synchronous, because all clients perform the same process, by exploring different regions. All the clients take the same CPU time to found the best local solution in each region.

As mentioned before, the Server is the initiator of the algorithm. Its main roles are to acquire computing resources from the different cores, to launch the different clients, to decompose the initial problem to be solved, and it is the process which decides on the termination of the CS-BSO-ARM. The main role of the clients is to request sub-problems and to solve them. They also evaluate solutions

3.4 Theoretical Properties of CS-BSO-ARM

Proposition 1. *The complexity of CS-BSO-ARM is defined by Eq. 1 as follows:*

$$O((\frac{IMAX \times K \times N}{FLIP}) + (IMAX \times N)) \tag{1}$$

If $IMAX$ is the number of maximum iterations of the algorithm, K the number of regions or bees in the colony, N is the number of items or the solution's size and $FLIP$ integer parameter using in the determination of search area strategy.

Remark 1. *The proof is deliberately omitted for the shortness of the paper.*

Comparing to the complexity cost of the sequential version already published in [13], which is equal to $O((\frac{IMAX \times K \times N}{FLIP}) + (IMAX \times N \times K))$, we remark that by increasing the solution's space, in other term, by increasing the number of regions K, the time complexity of the parallel version remains stable. However, when the number of K exceeds the number of processors, the time complexity is enhanced. For this reason, in Sect. 4, the number of bees is set to the number of processors.

4 Experimental Study

All implementations have been developed using C and all the experiments were run on the cluster Ibnbadis. The cluster Ibnbadis is composed of 32 nodes with 2 8-cores processors, offering a global computing power of up to 512 processors.

4.1   Benchmark Description

To validate CS-BSO-ARM algorithm, several experiments have been carried out using three types of well-known instances [13]. The first one is a collection of 5 medium instances, whose number of transactions varies between 6000 and 90000 and the number of items varies between 500 and 16000 items, while, the average size of transactions is between 2 and 50 items. The second one of instances is a collection of 2 large instances, whose the number of transactions varies between 100000 and 500000 transactions, the number of items varies between 1000 and 1600 items, while the average size of transactions is between 2 and 10 items. The third type is the big data *WebDocs* instance, the largest instance existing in the web, containing more than 1600000 transactions and more than 500000 items. Table 1 details the characteristics of these benchmarks.

Table 1. Data sets description

Instance type	Instance name	Transactions size	Item size	Average size
Meidum	Pumbs_star	40385	7116	50
	BMS-WebView-1	59602	497	2.5
	BMS-WebView-2	77512	3340	5
	Korasak	80769	7116	50
	Retail	88162	16469	10
Large	Connect	100000	999	10
	BMP POS	515597	1657	2.5
Big data context	WebDocs	1692082	526765	Not mentioned

4.2   Performance of CS-BSO-ARM Algorithm

Here, the performance of CS-BSO-ARM algorithm is analyzed according to the number of processors used with medium, large and big data instances described above. For the medium and large instances, the efficiency is computed according to the following formula: $Ef_i = \frac{t_{Seq}}{t_i \times i}$ Where t_{Sec}: is the runtime of sequential BSO-ARM published in [13], t_i is the runtime of CS-BSO-ARM algorithm using i processors. Ef_i is the efficiency of CS-BSO-ARM algorithm with i processors.

4.2.1   Medium Instances

Table 2 presents the efficiency of CS-BSO-ARM algorithm for medium instances. The column t_{Sec} is the runtime of sequential BSO-ARM proposed in [13]. The third column represents the number of satisfied rules, that means rules that verifiy the minimum support and the minimum confidence constraints Note that the number of satisfied rules for each instance is setting according to BSO-ARM algorithm (see [13]) by considering MinSup=10 % and MinConf=10 %. The aim is to attend this same number of satisfied rules by our parallel approach, in

Table 2. The Efficiency of the CS-BSO-ARM algorithm according to different number of processors using medium instances

Instance	t_{Seq}	Number of satisfied rules	Ef_8	Ef_{16}	Ef_{32}	Ef_{64}	Ef_{128}	Ef_{256}
Pumbs_star	3059,74	20000	0.42	0.44	0.48	0.53	0.58	0.59
BMSVIEW1	3550,27	24000	0.45	0.48	0.53	0.58	0.61	0.62
BMSVIEW2	4215,62	28000	0.47	0.49	0.54	0.60	0.63	0.64
Korasak	5013,68	32000	0.52	0.54	0.54	0.59	0.62	0.65
Retail	5890,69	32000	0.50	0.53	0.59	0.64	0.68	0.72

shorter time. According to this table, we remark that by increasing the number of processors from 8 to 256, the efficiency of CS-BSO-ARM is quasi often around 60 %, which is appreciable.

4.2.2 Large instances

Table 3 presents the efficiency of CS-BSO-ARM algorithm for large instances. According to this table, we remark that by increasing the number of processors from 8 to 256, the efficiency of CS-BSO-ARM is improved. Moreover, for the two instances, the super linear efficiency is met. Moreover, the efficiency reaches 300 % for the second large instances *BMP-POS*. In fact, not only the hardware resources help in the processing but the efficient software's design of this algorithm also help in exceeding linear efficiency.

Table 3. The Efficiency of the CS-BSO-ARM algorithm according to different number of processors using large instances

Instance	t_{Seq}	Number of satisfied rules	Ef_8	Ef_{16}	Ef_{32}	Ef_{64}	Ef_{128}	Ef_{256}
Connect	51840	64000	0.98	1.44	1.50	1.58	1.69	1.88
BMPPOS	259200	100000	2.28	2.89	2.92	2.98	3.03	3.27

4.2.3 Big Data instance

To test the stability of the proposed approach with big large instances, other experiments have been performed using the largest instance that exists on the web. The *WebDocs* instance containing more than 1500000 transactions and more than 500000 items. To the best of our knowledge, this huge instance has never been solved by any sequential approach. Our sequential BSO-ARM algorithm [13] gives no results after 15 days of processing. We recently solved this large instance by our GPU-based approach [14]. However, GPU-based approach takes 10 hours to generate just 1000 of satisfied rules. Table 4 shows the number of satisfied rules followed by the execution time (in *Sec*) for different number of iterations of the proposed approach. The number of nodes used for this experiment is set to 256 nodes. The results indicate that the number of satisfied rules increased simultaneously by increasing the number of iterations, or, the execution time increased lately. With 100000 iterations, 10 millions of rules can be generated in just 1309sec. These results confirms that the proposed approach outperforms our previously proposed GPU-based approach in terms of rule's quality and execution time. This is basically due to the advantages of the used cluster architecture.

Table 4. Scalability of the proposed approach using WebDocs instance

Number of iterations	Number of satisfied rules	Runtime (Sec)
10	1000	0.31
20	2000	0.44
50	5000	0.9
80	8000	1.16
100	10000	1.89
200	20000	3.03
500	50000	7.53
1000	100000	13.41
10000	1000000	131.2
100000	10000000	1309.22

4.3   Comparing CS-BSO-ARM Algorithm

The last experiments is to compare the proposed approach with PGARM [7]. Figure 2 presents the runtime of the two approaches (CS-BSO-ARM and PGARM

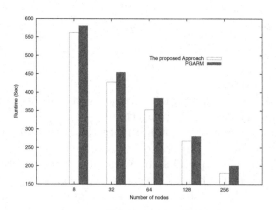

Fig. 2. CS-BSO-ARM Algorithm vs PGARM using WebDocs dataset (Sec)

in cluster architecture) using the Webdocs instance to generate 1 *Million* of association rules. By increasing the number of nodes from 8 to 256, our approach outperforms PGARM in terms of computation time. This results are obtained thanks to the efficient strategy in partitioning tasks which minimizes the communication cost between nodes.

5 Conclusion

In this paper, a new parallel ARM algorithm for high performance cluster computing is proposed. The whole process is divided among the bees, where each one explores independently the given region. Using such strategy, the communication and the synchronization operations are minimized compared to the existing parallel ARM algorithms. Indeed, the experimental study reveals that the suggested approach outperforms PGARM approach in terms of computation time. The results also reveal that by dealing the webdocs instance, our approach takes only 139 s instead of 10 h using GPU architecture. As perspective, we plane to integrate GPU host in the cluster hardware and to propose hybrid approach including GPU process.

References

1. Agrawal, R., Imielinski,T., Swami, A.N.: Mining association rules between sets of items in large databases. In: Proceedings of the ACM SIGMOD International Conference on Management of Data, pp. 207–216 (1993)
2. Han, J., Pei, J., Yin, Y.: Mining frequent patterns without candidate generation. ACM SIGMOD Rec. **29**(2), 1–12 (2000)
3. Silvestri, C., Orlando, S.: gpuDCI: exploiting GPUs in frequent itemset mining. In: 2012 20th Euromicro International Conference on Parallel, Distributed and Network-Based Processing (PDP). IEEE (2012)
4. Parthasarathy, S., Zaki, M.J., Ogihara, M., Li, W.: Parallel data mining for association rules on shared-memory systems. Knowl. Inf. Syst. **3**(1), 1–29 (2001)
5. Agrawal, R., Shafer, J.C.: Parallel mining of association rules. IEEE Trans. knowl. Data Eng. **8**(6), 962–969 (1996)
6. Buyya, R.: High Performance Cluster Computing. F'rentice, New Jersey (1999)
7. Taleb, A., Yahya, A., Taleb, N.: Parallel genetic algorithm model to extract association rules. In: Fifth International Conference on Advances in Databases, Knowledge, and Data Applications, DBKDA 2013, pp. 56–64, January 2013
8. Zhou, X., Huang, Y.: An improved parallel association rules algorithm based on MapReduce framework for big data. In: 2014 11th International Conference on Fuzzy Systems and Knowledge Discovery (FSKD), pp. 284–288. IEEE, August 2014
9. Li, H., Wang, Y., Zhang, D., Zhang, M., Chang, E.Y.: PFP: parallel FP-growth for query recommendation. In: Proceedings of the 2008 ACM Conference on Recommender Systems, pp. 107–114. ACM, October 2008
10. Sadeg, S., Yahi, S.: Cooperative bees swarm for solving the maximum weighted satisfiability. In: Proceeding of IWANN (2005)

11. Hadia, M.: Bees swarm optimization for web information retreival problem. In: 2012 IEEE/WIC/ACM International Conferences on Web Intelligence and Intelligent Agent Technology (WI-IAT), vol. 3. IEEE (2012)

12. Djenouri, Y., Drias, H., Habbas, Z., Mosteghanemi, H.: Bees swarm optimization for web association rule mining. In: 2012 IEEE/WIC/ACM International Conferences on Web Intelligence and Intelligent Agent Technology (WI-IAT), vol. 3, pp. 142–146. IEEE, December 2012

13. Djenouri, Y., Drias, H., Habbas, Z.: Bees swarm optimisation using multiple strategies for association rule mining. Int. J. Bio Inspired Comput. 6(4), 239–249 (2014)

14. Djenouri, Y., Bendjoudi, A., Mehdi, M., Nouali-Taboudjemat, N., Habbas, Z.: GPU-based bees swarm optimization for association rules mining. J. Supercomput. 71(4), 1318–1344 (2015)

Tools and Environments for
Parallel/Distributed/Cloud Computing

Distributed Computing Instrastructure as a Tool for e-Science

Jacek Kitowski[1,2]([⊠]), Kazimierz Wiatr[1,3], Lukasz Dutka[1], Maciej Twardy[1],
Tomasz Szepieniec[1], Mariusz Sterzel[1], Renata Słota[1,2], and Robert Pająk[1]

[1] AGH University, ACC Cyfronet AGH, Kraków, Poland
kito@agh.edu.pl
[2] Department of Computer Science, AGH University, Kraków, Poland
[3] Department of Electronics, AGH University, Kraków, Poland

Abstract. It is now several years since scientists in Poland can use the resources of the distributed computing infrastructure – PLGrid. It is a flexible, large-scale e-infrastructure, which offers a homogeneous, easy to use access to organizationally distributed, heterogeneous hardware and software resources. It is built in accordance with good organizational and engineering practices, taking advantage of international experience in this field. Since the scientists need assistance and close collaboration with service providers, the e-infrastructure is relied on users' requirements and needs coming from different scientific disciplines, being equipped with specific environments, solutions and services, suitable for various disciplines. All these tools help to lowering the barriers that hinder researchers to use the infrastructure.

Keywords: Distributed infrastructure · IT tools and services · Computing platforms · Clouds and grids

1 Introduction

The main goal of research is scientific discovery of unknown phenomena. Among typical three methodologies making new findings realistic: theoretical approaches by using sophisticated analytical methods, experimental investigations with (usually) big and expensive installations and computational studies, making wide use of information technology (IT), the last one has resulted in increasing popularity. Due to the complexity of most of the current problems this is a natural way to harness IT approach for both basic research, especially for extreme dimension/time scales and for versatile analysis of big data already existing or descended from experiments. Hence, computing infrastructures have led to ever-increasing contribution to e-Science research, while facing users with demanding technological obstacles, due to complicated IT stuff.

In order to prevent the users from the thorny technical problems and to offer them the most efficient and the most convenient way of making research on frontiers and challenges of current science – creation of a more flexible and easy to use ecosystem is required.

© Springer International Publishing Switzerland 2016
R. Wyrzykowski et al. (Eds.): PPAM 2015, Part I, LNCS 9573, pp. 271–280, 2016.
DOI: 10.1007/978-3-319-32149-3_26

In this paper we present assumptions and foundations of the distributed computing e-infrastructure as a tool for e-Science. The presented use case covers its implementation for Polish scientists.

2   Issues for e-Infrastructure Creation

Creation of an e-infrastructure needs synergistic effort in several dimensions:

1. **Meeting user demands in the field of grand challenges applications.** The activity toward a new e-infrastructure should be supported by a significant group of users with real scientific achievements and wide international collaboration as well as by well-defined requirements.
2. **Organizational**, which is probably the most important, though the most difficult in reality. Two perspectives are significant – horizontal and vertical – equally important and complementing each other.

   In the horizontal perspective, a federation of computer centres supporting the e-infrastructure with different kinds of resources and competences to cover interests of different groups of users is proposed. Some evident topics are to be addressed, like policy, collaboration rules, duties and privileges of each participant, for smooth and secure operation. Another feature to be attained is efficient use of federation resources by evaluation of computational projects from the community in order to grant them the most appropriate software and hardware environments.

   In the vertical perspective, organizational involvement of computer, computational and domain-specific experts into e-infrastructure operations is to be introduced for development of the most suitable hardware and software environments for the users, directly dedicated to their needs. Such a kind of collaboration provides the scientific community with necessary expertise, support from the structural, many level helpdesk and training for easy and efficient research using the e-infrastructure. A good example of such organization is Gauss Centre for Supercomputing [1].
3. **Technological**, which covers several issues including different computing hardware and software supported with scientific libraries, as well as a portfolio of middleware environments (e.g. gLite, UNICORE, QCG, generic cloud, like OpenNebula) and user-friendly platforms and portals. On that basis more sophisticated, tailored programming solutions can be developed.
4. **Energy awareness**, being a relative recent development. The problems faced are optimal scheduling strategies of computing jobs among federation resources to minimize energy consumption as a whole. As scale of resources and number of jobs increase, this problem becomes more critical than ever (e.g. [2]). Energy awareness is also a topic that influences selection of computing hardware.

3   Case Study of e-Infrastructure Conceptualization and Implementation

Due to large funding initiative in Poland and as a response to requirements of scientists, the Polish Grid Consortium was established in 2007, involving five of the largest Polish supercomputing centres: ACC Cyfronet AGH in Kraków (the coordinator), ICM in Warsaw, PCSS in Poznań, CI TASK in Gdańsk and WCSS in Wrocław. Members of the Consortium agreed to work as a federation to commence and jointly participate in the PLGrid Programme, to create a nationwide e-infrastructure and significantly extend the amount of computing resources provided to the scientific community.

Up-to-date fulfillment of the PLGrid Programme consists of several stages completed in subsequent projects.

- PL-Grid Project (2009–2012) aiming at providing the scientific community with basic IT platforms and computing services offered by the Consortium, initiating realization of the e-Science model in the various scientific fields. One of the measurements of success was ranking of all partners' resources by the TOP500 list (with total performance of 230 Tflops) in fall 2011, with Zeus cluster in Cyfronet located at 81st position.
- PLGrid Plus Project (2011–2015) focused on users, involving three kinds of contractors: computer, computational and domain-specific experts, which resulted in introducing 13 scientific domains with specialized software and hardware solutions, together with portals and environments. The total computational power offered by the Consortium was increased by additional 500 Tflops.
- PLGrid NG Project (2014–2015) targeting future development of the scientific domains by including into the project subsequent 14 scientific areas, due to rapid increase in demand for services for researchers in other fields. New domain-specific services cover a wide range of specialties – including provision of the specialized software, mechanisms of data storage, modern platforms integrating new type of tools and specialized databases – to speed up obtaining scientific results as well as streamline and automate the work of research groups.
- PLGrid Core Project (2014–2015) affirmed recognition of Cyfronet as a National Centre of Excellence, constituting the next step towards cloud computing and handling big data calculations. It aims not only at extension of hardware and software portfolio, but also dedicated accompanying facilities. One of them – a new backup Data Centre is on agenda. A new HPC asset has been installed, called Prometheus, with 1.7 Pflops, and put in operation in May 2015 for the community.

It is worth to mention the number of users close to 4000 currently, publishing regularly in highly ranked international journals, often with international collaborators, and many international projects ongoing with the help of the PLGrid infrastructure, funding by FP6, FP7, RFCS, EDA and other international agencies and collaborations. Two books on computing environments, portals, solutions and approaches developed during the Programme have been published by Springer Publisher [3,4].

4   PLGrid Platforms – Selected Examples

The computing infrastructure offered by the PLGrid infrastructure is not limited only to high performance computing clusters and large storage resources. A set of platforms, tools and services is provided, which hide the complexity of the underlying IT infrastructure and, at the same time, expose the actual functions that are important to perform the research. Within this section, capabilities of selected tools are described.

4.1 GridSpace – Web-Enabled Platform for Reproducible, Reviewable and Reusable Distributed Scientific Computing

GridSpace2 [5], as built on top of provided computing capabilities, enables scientists to easily create and run so-called *in silico* experiments that are featured by: (a) reproducibility – ability to effortlessly run the experiment at another time, by the other researcher or user, on the other computing capacity or using the alternative software, (b) reviewability – ability to effectively examine, verify, assess, test and scrutinize the experiment, (c) reusability – ability to smoothly apply the experiment to the other case, for the other purpose or in the other context.

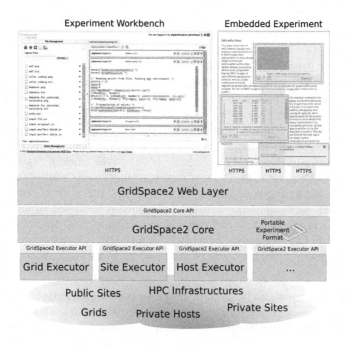

Fig. 1. GridSpace2 platform layers

GridSpace2 experiments are fully immersed in World Wide Web and structured as workflows composed of *code* and *data* items (see Fig. 1). Code items can be written in diverse programming languages and are interpreted by, so called, *interpreters*, which are implemented as *executables* and executed through *executors* on the underlying e-infrastructure. Executables installed on the e-infrastructure carry out computations while executors manage and orchestrate computation and data flow. Data items are simply file system elements that are processed, namely read and/or written, when executing code items. In the web layer, code and data items are embeddable as HTML *iframe* elements, which enables to create mash-up web pages that integrate content of various type and sources, including interactive GridSpace2 experiment items.

GridSpace2 is a generic and versatile platform that was applied in experiments from various scientific domains such as chemistry, material and urban engineering, physics and medicine. It was also adopted as a technology for executable scientific papers, namely, Collage Authoring Environment [6] that was integrated with the Elsevier ScienceDirect portal and empowered the first scientific journal issue featuring executable papers.

4.2   InSilicoLab – Science Gateway Framework

InSilicoLab [7] is a framework for building application portals, also called Science Gateways. The goal of the framework development is to create gateways that, on one hand, expose the power of large distributed computing infrastructures to scientists, and, on the other, allow the users to conduct *in silico* experiments in a way that resembles their usual work. The scientists using such an application portal can treat it as a workspace that organizes their data and allows for complex computations in a manner specific to their domain of science.

An InSilicoLab-based portal is designed as a workspace that gathers all that a researcher needs for his/her *in silico* experiments. This means: (a) capability of organizing data that is a subject or a product of an experiment, i.e., facilitating the process of preparation of input data for computations, possibility of describing and categorizing the input and output data with meaningful metadata as well as searching and browsing through all the data based on the metadata, (b) seamless execution of large-scale, long-lasting data- and computation-intensive experiments.

Every gateway based on the InSilicoLab framework is tailored to a specific domain of science, or even to a class of problems in that domain. The core of the framework provides mechanisms for managing the users' data – categorizing it, describing with metadata and tracking its origin – as well as for running computations on distributed computing infrastructures. Every InSilicoLab gateway instance is built based on the core components, but is provided with data models, analysis scenarios and an interface specific to the actual domain it is created for (see Fig. 2).

4.3   DataNet – Data and Metadata Management Service

DataNet [8] is a service built on top of the PLGrid high-performance computing infrastructure to enable lightweight metadata and data management. It allows creating data-models consisting of files and structured data to be deployed as specific repositories within seconds.

One of the main goals of DataNet is to make it usable from the largest set of languages and platforms possible. That is why the HTTP protocol was used as

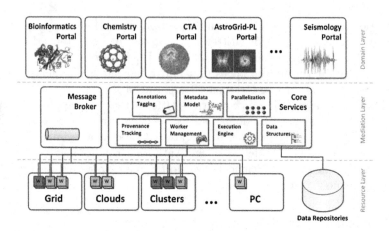

Fig. 2. Architecture of the InSilicoLab framework: domain layer, mediation layer with its core services, and resource layer. In the resource layer, Workers ('W') of different kinds (marked with colors) are shown.

a basis for transferring data between computing nodes and the service, together with the REST methodology applied to structure the messages sent to and from the repositories.

DataNet is fully integrated with the PLGrid's authentication and authorization system, so existing users can quickly gain access to the service with a fully automated registration process.

In order to ensure user data separation, each repository is deployed on a dedicated PaaS platform, which ensures scaling and database service provisioning for structured data. For high-throughput scenarios, it is possible to configure the system to expose several instances of a given repository to increase request processing rate.

Another aspect of using DataNet for data management is collaborative data acquisition, which is possible, because a given repository is identified by a unique URL. The URL can be shared among many computing infrastructures, software packages and different users or groups of users to acquire and process data within a single data model. For some collaboration efforts with large amounts of files this introduces structure and means to search the file space.

Figure 3 shows the layered architecture of the service.

4.4   Scalarm – a Platform for Data Farming

Executing a computer simulation many times, each with different input parameter values, is a common approach to studying complex phenomena in various science disciplines. Data farming is a methodology of conducting scientific research, considered as an extension of the task farming approach, combined with Design of Experiment (DoE) methods for parameter space reduction, and output data exploration techniques [9].

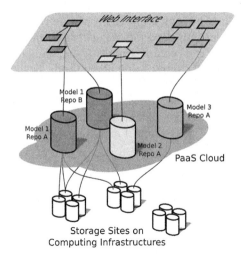

Fig. 3. DataNet architecture

A crucial requirement for efficient application of the data farming methodology is usage of dedicated tools supporting each phase of *in silico* experiments, following the methodology. Scalarm [10] is a complete platform, supporting the all above-mentioned data farming experiment phases, starting from experiment design, through simulation execution, to results analysis. All Scalarm functions are available to the user via GUI in a web browser (cf. Fig. 4).

To perform data farming experiment in Scalarm, a user prepares a simulation scenario, with input parameter types and an application specified. Through

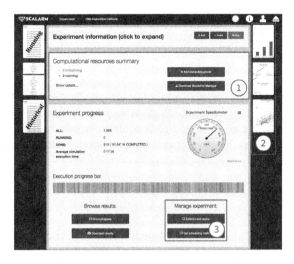

Fig. 4. Basic experiment progress view in Scalarm

the use of so-called adapters, any application can be run without modification, which allows to use Scalarm in the wide range of scientific disciplines, like metal processing technologies [11] or complex multi-agent simulations [12].

In addition to various scheduling systems for Grids, Scalarm supports several Cloud services [13] and user-defined servers. It also supports different results analysis methods with graphical presentation (see Fig. 4) as well as autonomous input space exploration methods, allowing to change parameter space without user intervention, to satisfy user-defined experiment goal.

4.5   Onedata – Uniform and Efficient Access to Your Data

Grid infrastructures consist of many types of heterogeneous distributed storage systems, managed locally [14]. Taking into account possible different requirements of a user, in terms of access to data [15], it is beneficial to provide a variety of storage systems, which poses challenges for unifying data access. Due to the independence of the centers in the grids, the management of storage systems (storage services) is decentralized.

The Onedata system [16] provides unified and efficient access to data stored in organizationally distributed environments, e.g. Grids and Clouds, and it is a complete response to the requirements of end-users, developers and administrators.

To offer the required functionalities, Onedata [17] merges and extends: (1) data hosting service, (2) high performance file system, (3) data management system and (4) middleware for developers.

While perceived as a high performance file system, Onedata provides access to data via a standard file system interface, offering coherent and uniform view on all data that can be distributed across the infrastructure of a geographically distributed organization.

Onedata is a data management system, which allows to manage various storage systems in a cost-effective manner without abandoning the uniform view on data and high performance. Its data management environment consists of: (a) monitoring systems, which gather information about storage utilization, (b) rules definition for automatic data management, (c) event-driven automatic data management based on the rules.

To provide high performance and scalability, Onedata is implemented in Erlang and in C language with noSQL database used. Information about metadata and the system state is stored in a fault-tolerant, high-performance, distributed noSQL database to avoid performance bottlenecks and guarantee data security.

5   Conclusions

The realization of the PLGrid Programme fits well with the need of development of an advanced IT infrastructure designed for the implementation of modern scientific research. The well-tailored PLGrid e-infrastructure fulfills researchers' needs for suitable computational resources and services. It also enables Polish scientific units collaboration with international research organizations, because vast range of services contribute to increase of cooperation between Polish scientists and international groups of specialists from twenty-seven different scientific domains of e-Science.

The essential fact is that anyone who is performing scientific research can be the user of the infrastructure. Access to the huge computational power, large storage resources and sophisticated services on a global level is free to Polish researchers and all those engaged in scientific activities associated with any university or research unit in Poland. To obtain an account in the PLGrid infrastructure, enabling access to its computing resources, one should only register in the PLGrid Portal [18].

Since 2010, the PLGrid infrastructure has been a part of the European Grid Infrastructure (EGI), which aims to integrate the national Grid infrastructures into a single, sustainable, production infrastructure. Further strong collaboration and exchange of ideas with EGI is foreseen.

Acknowledgements. This work was made possible thanks to the following projects: PLGrid Plus (POIG.02.03.00-00-096/10), PLGrid NG (POIG.02.03.00-12-138/13) and PLGrid Core (POIG.02.03.00-12-137/13), co-funded by the European Regional Development Fund as part of the Innovative Economy programme, including the special purpose grant from the Polish Ministry of Science and Higher Education.

References

1. Gauss Centre for Supercomputing (GCS) (2015). http://www.gauss-centre.eu/gauss-centre/EN/AboutGCS/aboutGCS_node.html
2. Gienger, M.: Towards energy aware scheduling between federated Data Centres. A presentation given at eChallenges International Conference Bristol (2014)
3. Bubak, M., Szepieniec, T., Wiatr, K. (eds.): Building a National Distributed e-Infrastructure - PL-Grid: Scientific and Technical Achievements. LNCS, vol. 7136. Springer, Heidelberg (2012)
4. Bubak, M., Kitowski, J., Wiatr, K. (eds.): eScience on Distributed Computing Infrastructure. LNCS, vol. 8500. Springer, Heidelberg (2014)
5. Ciepiela, E., Wilk, B., Harężlak, D., Kasztelnik, M., Pawlik, M., Bubak, M.: Towards provisioning of reproducible, reviewable and reusable in-silico experiments with the GridSpace2 platform. In: Bubak, M., Kitowski, J., Wiatr, K. (eds.) eScience on Distributed Computing Infrastructure. LNCS, vol. 8500, pp. 118–129. Springer, Heidelberg (2014)
6. Collage Authoring Environment (2015). https://collage.elsevier.com/
7. Kocot, J., Szepieniec, T., Wójcik, P., Trzeciak, M., Golik, M., Grabarczyk, T., Siejkowski, H., Sterzel, M.: A framework for domain-specific science gateways. In: Bubak, M., Kitowski, J., Wiatr, K. (eds.) eScience on Distributed Computing Infrastructure. LNCS, vol. 8500, pp. 130–146. Springer, Heidelberg (2014)
8. Harężlak, D., Kasztelnik, M., Pawlik, M., Wilk, B., Bubak, M.: A lightweight method of metadata and data management with DataNet. In: Bubak, M., Kitowski, J., Wiatr, K. (eds.) eScience on Distributed Computing Infrastructure. LNCS, vol. 8500, pp. 164–177. Springer, Heidelberg (2014)

9. Kryza, B., Król, D., Wrzeszcz, M., Dutka, Ł., Kitowski, J.: Interactive cloud data farming environment for military mission planning support. Comput. Sci. **13**(3), 89–99 (2012)
10. Król, D., Kryza, B., Wrzeszcz, M., Dutka, Ł., Kitowski, J.: Elastic infrastructure for interactive data farming experiment. In: Procedia Computer Science, Proceedings of ICCS 2012 International Conference on Computer Science, vol. 9 special issue, pp. 206–215. Omaha, Nebraska (2012)
11. Król, D., Słota, R., Rauch, Ł., Kitowski, J., Pietrzyk, M.: Harnessing Heterogeneous Computational Infrastructures for Studying Metallurgical Rolling Processes. In: Proceedings of eChallenges 2014 Conference, 29–30 October 2014. IIMC, Belfast (2014)
12. Laclavik, M., Dlugolinsky, S., Seleng, M., Kvassay, M., Schneider, B., Bracker, H., Wrzeszcz, M., Kitowski, J., Hluchy, L.: Agent-based simulation platform evaluation for human behavior modeling. In: Proceedings of ITMAS/AAMAS 2011, pp. 1–15. Taipei (2012)
13. Król, D., Słota, R., Kitowski, J., Dutka, Ł., Liput, J.: Data Farming on Heterogeneous Clouds. In: Proceedings of 7th IEEE International Conference on Cloud Computing (IEEE CLOUD 2014). IEEE (2014)
14. Słota, R., Dutka, Ł., Wrzeszcz, M., Kryza, B., Nikolow, D., Król, D., Kitowski, J.: Storage management systems for organizationally distributed environments PLGrid PLUS case study. In: Wyrzykowski, R., Dongarra, J., Karczewski, K., Waśniewski, J. (eds.) PPAM 2013, Part I. LNCS, vol. 8384, pp. 724–733. Springer, Heidelberg (2014)
15. Słota, R., Nikołow, D., Skałkowski, K., Kitowski, J.: Management of data access with quality of service in PL-Grid environment. Comput. Inform. **31**(2), 463–479 (2012)
16. Dutka, Ł., Słota, R., Wrzeszcz, M., Król, D., Kitowski, J.: Uniform and efficient access to data in organizationally distributed environments. In: Bubak, M., Kitowski, J., Wiatr, K. (eds.) eScience on Distributed Computing Infrastructure. LNCS, vol. 8500, pp. 178–194. Springer, Heidelberg (2014)
17. Wrzeszcz, M., Dutka, Ł., Słota, R., Kitowski, J.: VeilFS - a new face of storage as a service. In: Proceedings of eChallenges 2014 Conference, 29–30 October 2014. IIMC, Belfast (2014)
18. PLGrid Portal (2015). https://portal.plgrid.pl

A Lightweight Approach for Deployment of Scientific Workflows in Cloud Infrastructures

Bartosz Balis[1,2]([✉]), Kamil Figiela[1,2], Maciej Malawski[1,2], Maciej Pawlik[1,2], and Marian Bubak[1,2]

[1] Department of Computer Science, AGH University of Science and Technology, Krakow, Poland
{balis,kfigiela,malawski,bubak}@agh.edu.pl
[2] ACC Cyfronet AGH, AGH University of Science and Technology, Krakow, Poland
m.pawlik@cyfronet.pl

Abstract. We propose a lightweight solution for deployment of scientific workflows in diverse cloud platforms. In the proposed deployment model, an instance of a workflow runtime environment is created on demand in the cloud as part of the workflow application. Such an approach improves isolation and helps overcome major issues of alternative solutions, leading to an easier integration. The concept has been implemented in the HyperFlow workflow environment. We describe the approach in general and illustrate it with two case studies showing the integration of HyperFlow with the PLGrid infrastructure, and the PaaSage cloud platform. Lessons learned from these two experiences lead to the conclusion that the proposed solution minimizes the development effort required to implement the integration, accelerates the deployment process in a production system, and reduces maintenance issues. Performance evaluation proves that, for certain workflows, the proposed approach can lead to significant improvement of the workflow execution time.

Keywords: Scientific workflows · Cloud infrastructures · Application deployment

1 Introduction and Motivation

Scientific workflow systems commonly leverage cloud infrastructures for provisioning of computing resources [7,9,13]. Clouds provided by large commercial entities or computing centers are managed by diverse cloud middleware and cloud management platforms. Consequently, deployment of a scientific workflow in any such cloud infrastructure poses a significant challenge – challenge that is not merely a question of provisioning VM instances for a workflow application, but an integration with a complex ecosystem of users, middleware services, security requirements, etc.

We propose a novel approach to integration of scientific workflow systems with diverse cloud platforms in which the entire workflow runtime environment is deployed on-demand alongside workflow application components, so that the

© Springer International Publishing Switzerland 2016
R. Wyrzykowski et al. (Eds.): PPAM 2015, Part I, LNCS 9573, pp. 281–290, 2016.
DOI: 10.1007/978-3-319-32149-3_27

workflow engine and other runtime components are treated as part of the application itself. Benefits of this approach include: (1) *Improved isolation*: each workflow runs with its own instance of the workflow runtime system which eliminates many potential security holes. (2) *Easy integration*: the workflow system simply runs as part of the cloud application in a given cloud infrastructure. Consequently, adapting the workflow system to a new cloud platform does not require special integration effort, other than application deployment. (3) *Easier maintenance*: because of easy integration, new versions of the workflow system can be released and deployed very fast, e.g. by updating a Virtual Machine image. (4) *Improved performance*: workflow orchestration is more efficient thanks to leveraging fast instance-to-instance communication.

An important contribution of the proposed workflow deployment model is the integration with cloud platform services. Consequently, a cloud platform can be used not only for resource provisioning, but also for its more advanced capabilities, such as auto-scaling. The PaaSage platform case study (Sect. 4.2) demonstrates this in detail.

The solution we describe is based on our workflow engine HyperFlow.[1] In previous papers we described the HyperFlow workflow programming capabilities and its model of computation [4,5], as well as particular applications implemented with HyperFlow [2,3]. This paper focuses on the workflow deployment issues in the context of diverse cloud computing platforms.

This paper is organized as follows. Section 2 discusses related work. In Sect. 3 the proposed approach is described. Section 4 illustrates and validates our approach presenting two case studies of implementing support for scientific workflows in two different cloud infrastructures, and a performance evaluation experiment. Section 5 concludes the paper.

2   Related Work

Workflow engine is a crucial part of a workflow system orchestrating all application components in order to fulfill high-level business logic defined by the workflow graph. One can imagine three alternative deployment approaches of a workflow application in a cloud infrastructure.

First, the workflow engine can run outside the cloud as a 'client' coordinating the execution of application components deployed in the cloud. This approach, adopted by many workflow systems, for example in Pegasus [6], is effective but has some drawbacks. First, the set up of the workflow runtime environment is more involved as it needs to be done on the user machine. Second, the engine needs to communicate with all application components running in the cloud in order to monitor their execution progress and keep track of completed jobs. This may not always be possible for machines with private network addresses, so a front end communication proxy needs to be set up, complicating the application's architecture. For example, in Pegasus this is solved by using the Condor connection broker [17]. Morever, if the workflow engine is placed outside the

---

[1] https://github.com/dice-cyfronet/hyperflow.

cloud, while the workers are deployed in the cloud, their communication overhead is relatively high in comparison with instance-to-instance communication. In the case of fine-grained tasks this may cause a performance bottleneck. The Kepler system supports execution of workflows on Amazon EC2 instances [18]. In one of the supported usage modes, the Kepler engine can be deployed in the cloud using the provided Kepler AMI (Amazon machine image). However, a cloud workflow needs to utilize Kepler EC2 actors which explicitly manage EC2 instances. Consequently, the workflow is tightly coupled with the particular cloud infrastructure (EC2). HyperFlow, on the other hand, strictly separates the workflow business logic from its mapping onto computing resources.

In the second deployment approach the workflow engine is a permanently running service which can be part of a complex application platform hosted outside the cloud. Examples include the WS-PGRADE framework [1] and CLAVIRE [12]. Taverna workbench [14] also supports remote workflow execution through Taverna Server [19]. The advantage of such an approach is that the user does not need to set up the workflow system environment. The application platform typically provides tools which assist in workflow development and execution. It also can provide a unified cloud management layer which facilitates deployment of workflow components in diverse computing infrastructures. However, the disadvantage of such a platform is its complexity and maintenance issues. Adding support for a new computing infrastructure requires development of a new adapter implementing existing abstractions which is a major development effort. Frequent maintenance must be performed in order to adapt the software to inevitable changes of the APIs of the underlying systems. Furthermore, all existing deployments must be updated with such new releases. Consequently, deployment of such a platform as a scientific gateway on top of organization's computing infrastructure is a major business decision.

Finally, the third approach, adopted by HyperFlow, is to run the workflow engine as part of the workflow application, deployed on demand in the cloud alongside other application components. The following section explains this approach in detail.

3   A New Concept of Workflow Deployment in Clouds

3.1   Architecture

Figure 1 presents a conceptual workflow cloud deployment architecture based on the HyperFlow runtime environment. In our approach, the workflow together with its runtime environment are a *distributed application* that needs to be deployed in a concrete cloud, and configured so that the distributed components communicate with each other.

The deployed components of the workflow application, besides the actual domain-specific application components, comprise the workflow runtime environment: (1) **HyperFlow (HF) engine** enacting the workflow; (2) **Job queue** used for communication of job requests and results, implemented on the basis of the RabbitMQ message queue; (3) **HyperFlow Executors** running on application VMs.

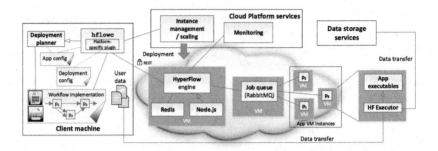

Fig. 1. General deployment architecture of a HyperFlow workflow in a cloud.

The HyperFlow engine enacts the workflow and once some workflow activities are ready to run, it creates appropriate job descriptions and sends them to the Job queue. The jobs are fetched by worker VMs hosting the application domain components. Thanks to such an architecture, we can easily scale up the application by starting multiple instances of a given worker. The HyperFlow Executors running on the VMs are responsible for invoking the application executables and passing the results back to the queue. Typically these results will not contain the actual payloads, only information about job's completion and references to job's outputs uploaded to the storage service by the HF Executor.

In addition to workflow enactment, the HyperFlow engine also acts as a HTTP server exposing a REST API to the external world. This API is used by the `hflowc` client to monitor the status of the workflow application. It can also be used to manage the workflow and send it new data for processing. The workflow model adopted by HyperFlow is based on Process Networks [10], so each workflow activity is a continuous 'process' (but not in the sense of an OS process) which consumes input data, invokes a procedure and emits results to its output channels.

3.2 Workflow Development, Configuration and Deployment

In order to create an executable workflow, the user needs to do the following steps[2]: (1) *Prepare the application components.* The user needs to prepare VM images that contain software packages invoked from the workflow. (2) *Implement the workflow.* The developer needs to specify the workflow structure using a simple HyperFlow JSON-based format (easily generated in the case of large-scale workflows). The HyperFlow workflow model goes beyond a simple DAG of activities allowing the developer to express complex workflow patterns using a simple syntax [5]. (3) *Configure the workflow.* In this step the user needs to provide the following information: which VM images should be started, their configuration (e.g. available memory), and the mapping of workflow activities onto application programs installed on the VM images. Alternatively, workflow

---

[2] We are working on tools to help automate these steps; currently they need to be done manually.

activities can be implemented as JavaScript functions that will be executed by the HyperFlow engine in the context of the *Node.js* runtime. Consequently, the user gains advanced programming capabilities offered by the mainstream *Node.js* ecosystem.

The user creates a workflow instance in the cloud by invoking command `hflowc setup`, where `hflowc` (HyperFlow Client) is an auxiliary tool that aids in deployment of the workflow application and communicates with this application at runtime. The way the deployment of the workflow application is done highly depends on the particular cloud platform, so the `hflowc` client loads an appropriate plugin which implements the platform-specific deployment procedure.

3.3   Run-Time Integration

Besides deployment-time integration, there are platform-specific issues that need to be solved which occur at runtime of the workflow application. The first one is the integration with the monitoring service of a cloud platform. This is achieved by using a module for the HyperFlow engine which implements the platform-specific integration. This module is loaded as a plugin at runtime, based on the platform-specific configuration.

The second issue is related to scheduling and autoscaling of the workflow application. Currently we are working on a *Deployment planner* tool which can be invoked by the HyperFlow client in order to prepare a deployment plan and possibly also autoscaling rules. The way this plan and these rules are enacted is, again, highly dependent on a particular cloud platform.

Finally, the user can configure storage services for staging input and output data as the workflow runs. This configuration is passed to the HyperFlow Executors that take care of data transfers. Currently we support a shared file system, GridFTP, Amazon S3, and PLG-Data, a data movement service specific to PLGrid.

4   Concept Validation

In this section, we study two cases of integration of the HyperFlow workflow runtime environment with cloud platforms. The first case is the PLGrid computing infrastructure in which part of the computing resources are managed by a cloud middleware. The second case is the PaaSage cloud management platform. We not only describe how the integration was implemented, but also discuss the alternative solutions that we considered, their tradeoffs, and finally how our chosen workflow deployment model contributed to overcoming major problems and an easier integration.

4.1   Case Study: PLGrid

PLGrid, a Polish National computing infrastructure for e-Science, provides resources and high-level services for scientists in diverse disciplines, including astrophysics, high energy physics, nanotechnologies, life and health sciences, and computational chemistry [11]. The computing resources of PLGrid, originally mainly managed by the grid middleware, are increasingly made available in the IaaS cloud provisioning model.

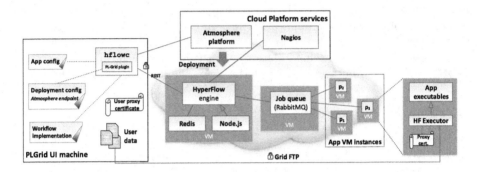

Fig. 2. Support for HyperFlow scientific workflows in the PLGrid infrastructure.

In PLGrid, HyperFlow has been integrated as a service enabling the users to program and run scientific applications, an example being a multi-frontal solver for finite-element meshes [2]. From the user's point of view, such a service must be accessible either from a graphical interface in the PLGrid portal, or as a command line tool on a PLGrid *User Interface* machine (on which the users have shell accounts with PLGrid credentials and PLGrid client software). Moreover, the access to the service can be limited based on the membership in a particular user group. From the administrator perspective, the availability of the service needs to be monitored using the standard PLGrid monitoring services including visualization on the monitoring dashboard. Finally, the access to the service must be blocked if the user exceeds the resource usage quota managed by the PLGrid user grants system [16].

Figure 2 shows the architecture which implements the support for scientific workflows in PLGrid. PLGrid utilizes the OpenStack cloud middleware managed by the Atmosphere cloud platform [15]. Atmosphere provides a REST API allowing clients to manage instances of Virtual Machines based on previously prepared images. Each instance can have its unique configuration which can be injected via the REST API. PLGrid uses X.509 certificates for user authentication and delegation of rights. User proxy (temporary) certificates are automatically transferred to VM instances where they are used by a HF Executor plugin to support secure data movement over Grid FTP.

The proposed solution not only fulfills all user and administrator requirements, but also simplifies the integration and helps avoid multiple major problems. For example, the original concept for the integration was to host the HyperFlow engine as a core service of the PLGrid middleware. In such a case the engine would have to be hosted and maintained as a permanent server. Morever, it would have to implement multitenancy in order to support multiple users and control access to the service. By contrast, in the adopted solution workflow engine instances run with individual user credentials, so HyperFlow does not need to support multitenancy and control resource usage quotas – these functions are handled by the cloud provisioning middleware. This leads to a better isolation of the workflow runtime environment and elimination of many potential security holes. Consequently, new versions of the workflow service can be deployed and made available to the users much faster because they do not need to undergo a thorough security audit.

4.2 Case Study: PaaSage

The second use case shows the integration of the HyperFlow environment with the PaaSage cloud middleware [8]. PaaSage is an advanced cloud management platform supporting design, development and runtime optimization of cloud applications, following the principles of model-driven software engineering.

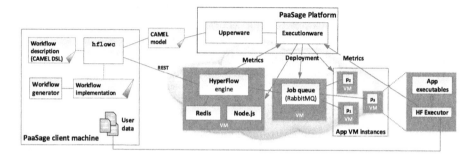

Fig. 3. Integration of HyperFlow with the PaaSage cloud platform.

The architecture of the integration of HyperFlow with PaaSage is shown in Fig. 3. PaaSage utilizes the Cloud Modeling and Execution Language (CAMEL), a domain-specific language (DSL) describing various aspects of the application, including its deployment model and scalability rules. The first step towards workflow deployment is the translation of the HyperFlow workflow description (Workflow graph) into the CAMEL model. This model is then consumed by the PaaSage Upperware component which generates a deployment plan, passed to PaaSage Executionware which performs the actual deployment.

The Executionware is also responsible for autoscaling of the application by dynamically adjusting the number of VM instances. This adjustment is governed by application-specific scaling rules triggered by events derived from monitoring of the application and infrastructure. In the case of workflows, we are currently working on an approach where scaling rules relate the number of VM instances with the workflow *execution stage*. Such an approach is particularly effective in a relatively common case where the workflow can be clearly divided into multiple distinct phases, each performing different tasks and having specific resource requirements. The scaling rule can thus be as follows: *as soon as the workflow reaches stage 2, start 5 instances of virtual machine vm24.* The workflow is divided into stages based on the analysis of its graph and previous executions. At runtime, the completion of a stage is detected based on the monitoring data reported by the HyperFlow engine to the PaaSage Executionware.

The HyperFlow workflow deployment model contributed to simplification of the concept for the integration with the PaaSage platform. Our goal was to extend the PaaSage platform with support for scientific workflows, so that scientific applications can benefit from the advanced capabilities of the platform, such as multi-cloud deployment and autoscaling. We were considering three possible solutions: (i) HyperFlow engine as a core service of the PaaSage platform; (ii) HyperFlow engine as a client of the platform coordinating the workflow from outside the cloud; (iii) HyperFlow engine as part of the application deployed and managed by the PaaSage platform. The first solution would be effective but also would require substantial modifications of the PaaSage middleware. The second solution turned out to be problematic: for example, the HyperFlow engine running outside the cloud would not be able to report relevant monitoring data to the PaaSage middleware. Consequently, the third option was the most natural and involved the least development effort. Furthermore, this solution enables the reconfiguration of the HyperFlow runtime components themselves. For example, the same Job queue can be shared by multiple HF engines of the same user, or it can be deployed separately for each one, depending on the load.

4.3 Performance Evaluation

We have performed experiments to measure the performance gain of our approach. To this end, we ran the same workflow in two deployment variants: (a) HyperFlow runtime deployed inside the cloud; (b) HyperFlow runtime deployed remotely, on an Amazon EC2 instance (Ireland region). Both variants used exactly the same configurations of worker nodes (one with 4 VMs, another with 1 VM, each with 6 vCPUs). In the experiment we used a Montage workflow with 1432 tasks.

The results of the evaluation are shown in Table 1. The total workflow execution time in the remote deployment variant was significantly higher, reaching up to 150 % of the execution time in the local deployment variant. Let us note that this difference was only due to the remote communication of many small orchestration messages between the Job queue and HF Executors deployed on application VMs (compare Fig. 1).

Table 1. Results of performance evaluation. Workflow with 1432 tasks. Local variant: HyperFlow runtime deployed inside the cloud (avg ping rtt $0.25ms$; Remote variant: HyperFlow runtime deployed remotely on Amazon EC2 (Ireland, avg ping rtt $52ms$).

Variant	Execution time (4x6vCPU)	Execution time (1x6vCPU)
Local wf engine	$5\,min48\,s \pm 3\,s$	$25\,min7\,s$
Remote wf engine	$8\,min43\,s \pm 2\,s$ (150 %)	$26\,min2\,s$ (104 %)

5   Conclusion

We proposed a novel approach to deployment and integration of scientific workflow applications in cloud infrastructures, where an instance of the workflow runtime environment is deployed in the cloud as part of the workflow application. We have described two cases of very different cloud platforms: PLGrid leveraging an IaaS cloud provisioning model, and PaaSage which provides high-level services for deployment of cloud applications. The experiments validating the concept have proven that the proposed deployment model fits well diverse cloud platforms, fulfilling user requirements and contributing to simplified integration which avoids major integration and maintentance challenges faced by alternative approaches. Performance evaluation experiments prove that the proposed approach can lead to significant improvement of workflow execution times, especially for workflows with many relatively fine-grained tasks.

HyperFlow has been integrated in the PLGrid infrastructure and is available as a production workflow system to Polish scientists, granting access to federated cloud infrastructures resources of several Polish computing centers. Adaptation of several scientific applications as HyperFlow workflows in PLGrid is already a work in progress. Future work involves the finalization of the HyperFlow integration with PaaSage, experiments with real large-scale applications in a production environment, and integration with workflow scheduling and execution optimization services.

Acknowledgments. This work is partially supported by the PLGrid Core project (POIG.02.03.00-00-096/10); and by the EU-FP7 project PaaSage; AGH grant no. 11.11.230.124 is also acknowledged.

References

1. Balasko, A., Farkas, Z., Kacsuk, P.: Building science gateways by utilizing the generic WS-PGRADE/gUSE workflow system. Comput. Sci. J. **14**(2), 307–325 (2013)
2. Balis, B., Figiela, K., Malawski, M., Jopek, K.: Leveraging workflows and clouds for a multi-frontal solver for finite-element meshes. Procedia Comput. Sci. **51**, 944–953 (2015)
3. Balis, B., Kasztelnik, M., Malawski, M., Nowakowski, P., Wilk, B., Pawlik, M., Bubak, M.: Execution management and efficient resource provisioning for flood decision support. Procedia Comput. Sci. **51**, 2377–2386 (2015)
4. Balis, B.: Hyperflow: a model of computation, programming approach and enactment engine for complex distributed workflows. Future Gener. Comput. Syst. **55**, 147–162 (2015)
5. Balis, B.: Increasing scientific workflow programming productivity with Hyper-Flow. In: Proceedings of 9th Workshop on Workflows in Support of Large-Scale Science. WORKS 2014, pp. 59–69. IEEE Press (2014)
6. Deelman, E., Vahi, K., Juve, G., Rynge, M., Callaghan, S., Maechling, P.J., Mayani, R., Chen, W., da Silva, R.F., Livny, M., et al.: Pegasus, a workflow management system for science automation. Future Gener. Comput. Syst. **46**, 17–35 (2014)

7. Hoffa, C., Mehta, G., Freeman, T., Deelman, E., Keahey, K., Berriman, B., Good, J.: On the use of cloud computing for scientific workflows. In: eScience, 2008. eScience 2008, pp. 640–645. IEEE (2008)
8. Jeffery, K., Horn, G., Schubert, L.: A vision for better cloud applications. In: Proceedings of the 2013 International Workshop on Multi-Cloud Applications and Federated Clouds, pp. 7–12. ACM (2013)
9. Juve, G., Deelman, E.: Scientific workflows and clouds. Crossroads 16(3), 14–18 (2010)
10. Kahn, G.: The semantics of a simple language for parallel programming. In: Information Processing, pp. 471–475. North-Holland(1974)
11. Kitowski, J., Turała, M., Wiatr, K., Dutka, Ł.: PL-Grid: foundations and perspectives of national computing infrastructure. In: Bubak, M., Szepieniec, T., Wiatr, K. (eds.) PL-Grid 2011. LNCS, vol. 7136, pp. 1–14. Springer, Heidelberg (2012)
12. Knyazkov, K.V., Kovalchuk, S.V., Tchurov, T.N., Maryin, S.V., Boukhanovsky, A.V.: CLAVIRE: e-Science infrastructure for data-driven computing. J. Comput. Sci. 3(6), 504–510 (2012)
13. Malawski, M., Juve, G., Deelman, E., Nabrzyski, J.: Cost-and deadline-constrained provisioning for scientific workflow ensembles in IaaS clouds. In: Proceedings of International Conference on High Performance Computing, Networking, Storage and Analysis, p. 22. IEEE Computer Society Press (2012)
14. Missier, P., Soiland-Reyes, S., Owen, S., Tan, W., Nenadic, A., Dunlop, I., Williams, A., Oinn, T., Goble, C.: Taverna, reloaded. In: Gertz, M., Ludäscher, B. (eds.) SSDBM 2010. LNCS, vol. 6187, pp. 471–481. Springer, Heidelberg (2010)
15. Nowakowski, P., Bartyński, T., Gubała, T., Harężlak, D., Kasztelnik, M., Meizner, J., Bubak, M.: Managing cloud resources for medical applications. In: Proceedings of the Cracow Grid Workshop, vol. 2012 (2012)
16. Radecki, M., Szymocha, T., Piontek, T., Bosak, B., Mamoński, M., Wolniewicz, P., Benedyczak, K., Kluszczyński, R.: Reservations for compute resources in federated e-infrastructure. In: Bubak, M., Kitowski, J., Wiatr, K. (eds.) eScience on Distributed Computing Infrastructure. LNCS, vol. 8500, pp. 80–93. Springer, Heidelberg (2014)
17. Vöckler, J.S., Juve, G., Deelman, E., Rynge, M., Berriman, B.: Experiences using cloud computing for a scientific workflow application. In: Proceedings of 2nd International Workshop on Scientific Cloud Computing, pp. 15–24. ACM (2011)
18. Wang, J., Altintas, I.: Early cloud experiences with the Kepler scientific workflow system. Procedia Comput. Sci. 9, 1630–1634 (2012)
19. Wolstencroft, K., et al.: The Taverna workflow suite: designing and executing workflows of Web Services on the desktop, web or in the cloud. Nucleic Acids Research (2013)

Distributed Execution of Dynamically Defined Tasks on Microsoft Azure

Piotr Wiewiura[1], Maciej Malawski[1]($^{(\boxtimes)}$), and Monika Piwowar[2]

[1] Department of Computer Science, AGH University of Science and Technology,
Krakow, Poland
malawski@agh.edu.pl
[2] Department of Bioinformatics and Telemedicine, Jagiellonian University,
Krakow, Poland

Abstract. Microsoft Azure is a relatively new public cloud service which has potentially much to offer to the scientific community, but has not yet been widely used for research applications. This paper evaluates suitability of Microsoft Azure as a platform for execution of computational applications in three scenarios: we evaluate dynamic horizontal scaling, distributed execution of a CPU-intensive application – POV-Ray ray-tracer, and distributed execution of a bioinformatics application – Exon-Visualizer. To this end, we created a Distributed Task Library (DTL), due to lack of free, simple solution for distributed execution of dynamically defined tasks in .NET. In conclusion, we show that while dynamic horizontal scaling is quite slow, Microsoft Azure is a worthy platform for computational applications, offering, in conjunction with DTL, an easy way to speed up CPU-intensive, embarrassingly parallel problems.

Keywords: Cloud computing · Azure · Distributed processing · Bioinformatics

1 Introduction

Cloud computing has gained a lot of attention over the last few years and has been a subject to research and evaluation by scientific community [1,4]. Microsoft Azure (formerly known as Windows Azure), is a relatively young public cloud platform. Azure has potentially much to offer to the scientific community, but has not yet been widely used for research applications, and since the platform is rapidly evolving, thus new studies like this one are needed.

The main goal of this paper is to evaluate suitability of Microsoft Azure as a computational applications execution platform. We decided to check Azure's computational capabilities as such study may be helpful to scientists and other people who seek high-performing cloud platform. Our earlier paper [4] showcases component-based approach to cloud applications. In the scope of this research

M. Malawski—This research was partially supported by EU FP7-ICT project PaaSage (317715), Polish grant 3033/7PR/2014/2 and AGH grant 11.11.230.124.

© Springer International Publishing Switzerland 2016
R. Wyrzykowski et al. (Eds.): PPAM 2015, Part I, LNCS 9573, pp. 291–301, 2016.
DOI: 10.1007/978-3-319-32149-3_28

we wanted to set up an architecturally comparable software environment and start from performing similar tests and then, extend them. We selected three use cases to evaluate: **Dynamic horizontal scaling** – Azure is not free and user pays for every minute of running node. This creates a need for dynamic changing of number of running instances according to the current and expected workload. The goal was to test how quickly Azure responds for scale out requests. **CPU-intensive application** – **POV-Ray** – Many scientific methods rely on pure CPU power, so it is natural to check how Azure copes with such problems. Moreover, we wanted to check how difficult it is to take an existing, proprietary application and wrap it in such a way that it can be executed in a distributed manner on Azure. **Real-world bioinformatics application** – **ExonVisualizer** – The goal was to take a scientific application with available source code, developed in .NET and make it work on Azure. For this purpose we selected the ExonVisualizer application [13]. In this way we could check how much impact on existing code has the addition of cloud support. Additionally, ExonVisualizer is not as CPU-bound as POV-Ray, so it provides a different perspective on distributed execution.

To perform our evaluation, after the analysis of existing tools, we decided to create a lightweight library which would allow to execute arbitrary tasks with arbitrary data on Azure. The **Distributed Task Library** (DTL) is a lightweight, simple, portable .NET library allowing for distributed execution of tasks defined at runtime.

The paper is organized as follows. After presentation of related work in Sects. 2 and 3 describes the design and architecture of the DTL and the most interesting challenges encountered and solved. Subsequently, Sects. 4, 5 and 6 describe the performance evaluation of horizontal scaling, POV-Ray benchmark application and ExonVisualizer application from bioinformatics domain. Finally, the conclusions of the paper are presented along with suggested future work.

2   Related Work

One of the first attempts to evaluate Azure [2] focuses on performance in the areas including: Blob Storage, Table Storage, Queue Storage, Dynamic horizontal scalability, and SQL Azure Database. They measured e.g. that it took about 600 s, on average, to instantiate first Small Instance virtual machine and 80 s for subsequent machines. Blob Storage bandwidth for a single client was above 13 MB/s download and 6 MB/s upload speed. In the second study [6], performed 2 years later they observed a 200 s improvement in startup times.

A comprehensive study of scientific projects on Azure [1] observes that after initial problems, Azure has been successfully used in solving such problems as watershed modeling, metagenomics, analyzing fMRI scans, GIS processing and many others. The network performance, however, is a bottleneck, especially for MPI applications. Moreover, VM deployment takes much longer than many scientific programmers expect. Also, uploading large amounts of data is problematic, but can be addressed by Azure Import/Export service [9]. On the other hand, an interesting feature is ability of sharing VM images using VM Depot service [8].

Aneka [5] is a commercial PaaS solution and a framework which facilitates development of distributed applications in the cloud. Aneka provides a complete SDK for .NET Framework applications and also comes with rich set of tools. It can be used to develop and run applications on so-called Aneka Clouds which can span multiple physical or virtual infrastructures including public or private clouds. Aneka is a potentially powerful platform which could be used to not only execute but also manage all other aspects of scientific applications in a cloud. However, the fact that this is a commercial software forced us to seek other solutions.

Regarding solutions offered by Microsoft, the Azure HPC Scheduler, proposed to support MPI, SOA and Parametric Sweep Applications, has been recently replaced by a preview of Azure Batch [7]. It aims to offer a complete batch processing capability as a service, so it may be of interest for scientific applications in the future.

3   Distributed Task Library – Design and Implementation

To evaluate Microsoft Azure and to be able to port our existing ExonVisualizer application, we decided to create the Distributed Task Library (DTL). It is a lightweight, simple, portable .NET library allowing for distributed execution of tasks defined at runtime. It supports also basic management operations like horizontal scaling. The list of DTL requirements is as follows:

- **Distributed execution** – DTL should be able to execute a given task on multiple machines in parallel, distribute data required for this task and collect results.
- **Install-and-forget worker service** – Machines used for execution (workers) should have only a small, once-installed worker service responsible for task execution.
- **Dynamic deployment and execution** – Workers should be generic, while task definition and all the data and binary dependencies should be distributed by DTL.
- **Simplicity** – DTL must be easy to use. Executing a task in a distributed manner with default behavior should take no more than a few lines of code.
- **Dynamic horizontal scaling** – ability to add, remove, stop and start workers at runtime.
- **Portability and independence from execution platform** – DTL should work with Azure but it must be extensible to allow switching to another platform.
- **Written in C#** – to facilitate the integration of existing applications.
- **Support for Linux (Mono)** – since many scientific applications are Linux-only.
- **Worker service auto-update** – Updating workers should not require any manual actions, as update process on a large number of machines would be very tedious.

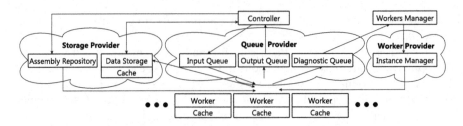

Fig. 1. Architecture of the distributed task library

The main concept was to develop an execution platform which would accept a job defined by user and distribute its execution over multiple nodes using a master-worker paradigm, supporting a parametric sweep model of execution. The resulting library should be comparably powerful and provide as simple API as Microsoft's Task Parallel Library [3], but designed to work in a distributed environment, hence the name – Distributed Task Library.

3.1 DTL Components and Implementation

Figure 1 shows the high-level architecture overview diagram. The Workers, Queues and Storage providers can use external services, so it is possible to e.g. use Amazon SQS, Microsoft Azure Storage and a mix of workers from different clouds.

Controller is a main component of DTL. It accepts job requests, prepares them for execution and sends to other components. It also collects the results and makes them available to the user. It uses Queues to send tasks and receive results from Workers. It is also responsible for transferring data from and to Storage and Assembly Repository.

Worker executes queued tasks. Uses Queues, Storage and Assembly Repository to obtain task definitions, input data and upload results. Runs as a daemon and can have multiple instances. Workers do not need connectivity with Controller, only to Queue and Storage Providers and they can use local Common Data Cache for reusing data between tasks. The first idle worker fetches the first available task from the queue, which results in a dynamic load-balancing of work among the workers.

Queue Provider gives access to queue infrastructure. Since publicly hosted queue systems have relatively low message size limit (64 KB for Azure Queues), while task input data may be often much larger, DTL queue messages contain only reference to the data kept in Data Storage. The Queue provider uses the Input and Output Queue for scheduled tasks and results, respectively, and a Diagnostic Queue used by Workers Manager to receive information about number of workers and their status.

Storage Provider stores and provides API to access binary data (BLOBs) that contain a payload of tasks and their results. **Data Storage** stores arbitrary collections of bytes, and assigns ID to each BLOB to and returns it to uploader. This ID is used by Common Data Cache mechanism to avoid reuploading the same data. **Assembly Repository** stores all the .NET assemblies used by tasks so they could be reused in future, which is useful for many .NET applications that use common, popular assemblies.

Workers Manager manages workers from all execution sites: deals with requesting change of the state of workers (create, start, stop, delete) based on current system state, execution site properties (e.g. VM performance and cost) and strategies set by user. It is an optional component, not needed if users manage the workers by themselves.

Worker Provider is a platform which is able to provide workers in form of physical or virtual machines running DTL worker service/daemon. It includes the Instance Manager, responsible for interfacing with specific Worker Provider and exposing standardized API for basic instance operations (create, start, stop, delete).

One of our goals within DTL development was to make it a lightweight library with a small set of external dependencies. Fortunately, .NET Framework is quite mature and provides many solutions out of the box. Microsoft .NET 4.5 and C# 5.0 introduced async/await keywords and new asynchronous programming patterns coming with them that simplified the implementation of many cloud operations like e.g. managing virtual machines or transferring data.

3.2   Challenges and Their Solutions

Dynamic Task Definition and Execution. One of our main goals was to support dynamic task definition and execution. It means that workers should not need any prior knowledge about tasks being executed. A natural and convenient solution for .NET developers is to allow them to take their part of code, wrap it into a single Execute method and let the DTL take care of the task distribution. This is possible thanks to the concept of .NET assemblies which are essentially .NET executables or dll libraries. DTL accepts tasks in a form of a class which implements ICommand generic interface. Upon execution, DTL detects in which assembly the task is implemented and what are its dependencies. Then it gathers all these assemblies and uploads them to our Assembly Repository, if they are not already present there. The task message which is sent to the workers contains binary serialized command instance, list of required assemblies and references to common and input data stored in Data Storage. Workers can then ask Assembly Repository for the listed assemblies, download them and keep in cache for later reuse. To address the problem of loading multiple versions of the same assembly at the same time, we decided to execute each task is in separate application domain where it loads its required assemblies. After execution, the application domain is unloaded. We consider the dynamic task definition and execution as one of the major accomplishments of the implementation part of this paper and a feature which is hard to come by in such form.

Data Format. To make data handling flexible and simple, we decided that the only requirement for user is to provide a stream of bytes to represent both input data and common data. This led to creation of IDataSource generic interface and a base class for all data sources. Moreover, we created a few

DataSource implementations which may come handy in many scenarios, such as BytesDataSource, ObjectDataSource, FileDataSource, MetadataDataSource and StorageDataSource – for accessing bytes, object streams, files, metadata and data existing in the cloud storage, respectively.

Linux (Mono) Support and Using Microsoft Azure REST API. As Linux support was one of our goals, this ruled out making use of the official Azure SDK for .NET which does not work for Mono. Azure REST API became our API of choice for accessing compute, queue and storage services. DTL is designed to allow replacing it in the future, e.g. by using wrappers based on native SDK. The latest Mono edition (3.6.0) has no problems with running DTL, even though it uses such features as application domains combined with remoting, marshalling and binary serialization. As of now, DTL worker for Linux comes only in a form of a command-line executable but there should be no problem in turning it into a full-fledged daemon.

Worker Auto-update. Pursuing the simplicity and minimum maintenance goal we chose to implement auto-update of DTL worker. In the distributed environment, where DTL workers are deployed on dozens of machines it would be inconvenient to manually update worker on each of them. Each worker checks for update information on startup under specified URL. If update is found, it is downloaded, extracted, the updater replaces worker files, requiring the restart of the worker. This process is highly flexible thanks to the updater which is always downloaded and thus update process can be modified with a new release if needed.

4   Evaluation of Horizontal Scaling

Ability to scale the number of running nodes (also called scaling out or scaling horizontally) robustly is a very important feature of every cloud service. The goal of the test was to measure the relation of number of requested machines to time of their instantiation and deletion. We were creating instances from a captured image: this method is slower than starting stopped instances, but does not incur storage costs when machines are not used. The test procedure involved requesting instantiation of n machines and then deleting them all at once. A custom VM image was prepared, based on Windows Server 2012 VM with DTL Worker Service installed and set up to start at boot. The image was kept in the same affinity group as the instantiated VMs (West Europe) and had the size of 127 GB, as Azure provides no way to request smaller OS disk. Additionally, a single Cloud Service was created upfront and each time we used Small instance type.

Figure 2a shows instantiation times when the given number of VMs were requested. First machine was up and running in about 235 s and each subsequent 45 s after a previous one. These times are reasonable considering the image size. Surprisingly, the plot is nearly linear, while it might be expected that the all

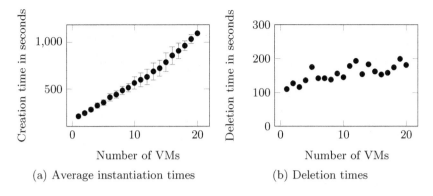

Fig. 2. Creating 1 to 20 instance VMs from 127 GB Windows Server 2012 image

machines will start within similar time frame. However, requesting instantiation of i machines is not atomic from perspective of Azure REST API. There is a separate request issued to add each of the VMs, since Azure does not allow for more than one operation on the given deployment at the same time. The new request is rejected until the currently requested machine goes into "Starting" state. The overall result is that a set of 20 machines is instantiated in 1096 s instead of about 235 s if simultaneous operations were allowed. The second plot Fig. 2b shows how long it takes to completely remove n VMs, which includes deleting their disks from storage. It takes about 156 s to remove all machines, on average (with standard deviation around 25 s).

The presented data comes from a single set of results, since several times we encountered problems that one or more machines got stuck in "Running (Provisioning)" state. It was also impossible to connect to the problematic VM with Remote Desktop. The main observation from this test is that although the provisioning times of Azure are improving, the reliability of the service may still be an issue.

5 CPU-intensive Benchmark Application - POV-Ray

As an example a of CPU-intensive application we chose rendering a scene in POV-Ray open-source ray tracing program, often used for benchmarking [10]. This choice was made also to test how easy it is to employ DTL to distribute workload of existing, popular software without modifying its source. For parallel execution the image area can be split into arbitrary number of rectangular segments. The tests were performed using POV-Ray 3.7 and Small Instance Azure VMs with Windows Server 2012 R2 and preinstalled DTL Worker Service. The tests had four stages – using 1, 2, 4 and 16 VMs that were already instantiated and started before the test. Each stage consisted of five cases – image was divided into 1, 2, 4, 8 and 16 equally-sized segments. Rendering each of these segments was a separate DTL task. Entire image resolution of the benchmark scene was 128 x 128 pixels. Each case was repeated three times.

Figure 3 shows speedup of the application. Overall time is a time span between the moment when user issues rendering request and the moment when the entire image is displayed on the screen, so it includes all the overheads of distributed processing in DTL. In the ideal case the speedup should be linear, but the execution overheads are especially pronounced when number of workers is high. The optimal number of segments is equal to the number of workers. In Fig. 3 it is clearly visible at which point adding more workers does not reduce rendering times but can even increase them. One reason behind this is a fairly constant time of POV-Ray startup and shutdown (i.e. time when process is running but nothing is being rendered).

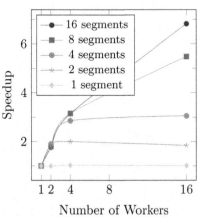

Fig. 3. Speedup of rendering for 1, 2, 4 and 16 Small Instance workers combined

Actual rendering phase takes 318 s (97 % of the whole run). The most of the remaining time of the test (5.19 s) is spent on uploading common data, i.e. POV-Ray package of 3.29 MB. The same data is downloaded to VMs in about 0.33 s (10 MB/s download speed). Sending and receiving queue messages is fast due to their small size. Input message contains more data – not only links to stored blobs but also embedded serialized parameter objects and list of required assemblies. Installation of POV-Ray, which takes 0.22 s on average, is also negligible.

Our tests show that it is possible to use DTL to run 3rd party application in a distributed manner (provided it has built-in support for splitting its workload) with no significant effort. Azure copes quite well with CPU-intensive tasks and we did not encounter any reliability issues with connection to blob and queue services.

6   Real-World Bioinformatics Application - ExonVisualizer

To evaluate the DTL on a real application, we chose the Exon Visualizer, as an example of a Web app requiring some background processing on server side. ExonVisualizer [13] is a tool for visualization of protein regions (2D and 3D structures) encoded by particular exons and uses 3rd party BLAST 2 Sequences (bl2seq) application. It is written in C# using ASP .NET and is intended to be used through web browser where user can provide input data and analyze output which is generated on a web server.

Identification of 3D regions of the proteins encoded by the exons allows to evaluate how the structure of the gene (with its specific exons) translates into an important feature as regards protein areas in their structures. From the viewpoint of further biological studies, it is important whether exon fragments can be considered as carriers of specific information about the protein eg. about hydrophobicity [11,12]. This can be the key to understanding the process of alternative splicing. Another aim of the analysis [13] may be to explore exons for accumulation of mutations. This may suggest that the part of the genetic information (exons) are more (or less) susceptible to accumulate mutations. Such an application that allows the projection of genetic information on the protein level is an important step for analysis outlined above, especially through the analysis on large-scale (on a pool of proteins and genes from different organisms).

Input to ExonVisualizer consists of GenBank Flat File with nucleotide data and PDB file with protein structure data. There are two operation modes: **Single task** - user provides a single pair of GenBank and PDB files, and **Batch task** - user provides a zip archive with multiple file pairs. Finally, user must specify three alignment properties: PDB domain, Gap cost and Substitution matrix. Application output consists of identified and visualised exons found in the provided protein structure, available in HTML, text, PyMol and VMD formats.

Since the goal was to add distributed processing to ExonVisualizer and test it afterwards, we decided to augment batch mode only, as it takes much more time when the input archive contains a large number of files. We decoupled the core part of application which governed reading data, processing and aggregating results, from the web UI and ASP .NET-related code. The most compute-intensive part is parsing the input files and BLAST execution. DTL command accepts the original user input data, i.e. two files and alignment parameters and returns object containing result data. This means that we had to extract the input zip file, find all matching file pairs, pack them again and send each pair as a single input item. Therefore, each pair could be processed in parallel on separate nodes.

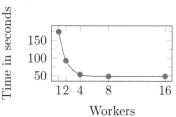

Fig. 4. ExonVisualizer execution time for 1, 2, 4, 8 and 16 Small Instance workers

To test how well will ExonVisualizer and DTL work with various numbers of workers, we prepared a console application in C# serving as a test driver. We used 1, 2, 4, 8 and 16 Azure Small Instance VMs, based on Windows Server 2012 R2 image with preinstalled DTL worker. Test data was 64 pairs of GenBank and PDB files, zipped into an 6.7 MB archive. Each piece of input data consisted of zipped, matching GenBank and PDB file pairs. Their sizes ranged from 25 KB to around 1 MB. Alignment parameters were left at default: GapOpenCost = 11, GapExtendCost = 1, SubstitutionMatrix = Blosum62. Each scenario was executed three times.

Figure 4 shows that if there is only one worker available, the overall execution takes about 174 s. Adding another worker approximately halves that time. The speedup limit at 8 workers can be clearly seen: above that number the overheads result in a slowdown. Figure 5 shows detailed data in scenario with one machine, and 64-pair set of input data. Total execution time takes over 150 s. Uploading and downloading common data and assemblies take only 5 s and they

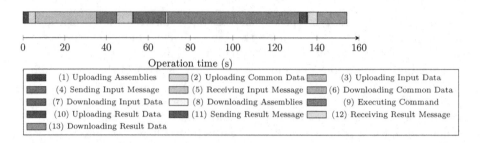

Fig. 5. Decomposition of overall ExonVisualizer execution time into phases.

do not dominate, because they are done only once for the whole process. Much time is spent on uploading (30 s), downloading the input data (15 s), executing the DTL command (62 s) and downloading results (15 s).

Implementation of DTL into ExonVisualizer proved to be feasible and resulted in a fully-functional solution. Presented graphs show that distributing the workload shortens the overall execution time considerably. However, we realize that employing DTL and Azure for ExonVisualizer yields gains from user's perspective only for large data sizes. Figure 5 shows that command execution took around 60 s, while at the peak of our speedup graph (8 workers) the total execution time is about 50 s. It is a negligible difference and the main problem is that the actual processing is fast, while transmitting files is slow. There are, however, other ways to potentially speed up ExonVisualizer, by e.g. using multiple cores, or by sending the entire batch archive as one DTL task. Nevertheless, scenarios with request streams from multiple users can benefit from the current solution, since the number of workers can be (manually) adjusted to meet the demand.

7   Conclusions and Future Work

In this paper we evaluated suitability of Microsoft Azure for computational applications, and to achieve this goal, we created a .NET **Distributed Task Library**. Thanks to it, we could quite easily adapt tested applications to Azure. DTL can be useful not only in this research, but also to serve as a tool for other users. We came to the following conclusions regarding our three scenarios. **Dynamic horizontal scaling** of Azure, particularly its low speed (nearly 4 min to create a new VM), may be problematic. Scientific applications, however, often work in batch mode and their demands may be better predicted beforehand. Testing the **CPU-intensive application – POV-Ray** proved that such applications scale very well with the number of workers. It also showed that using DTL, it is not difficult to take existing, proprietary software and execute it on multiple VMs. Finally, we managed to modify **real-world bioinformatics application – ExonVisualizer**, and test it on Azure. Although we did not observe a high speedup, our evaluation gives us important knowledge about the characteristics of the application to fully take advantage of Azure capabilities. All in all, Microsoft Azure is definitely a worthy platform for computational applications, especially for embarrassingly parallel and high-throughput workloads. For future work, it would be interesting to test Linux VMs on Azure. Autoscale (automatic horizontal scaling) and Batch services have been added to Azure recently, so it would be worth to evaluate their performance using a workload of scientific applications. There are also a few areas which should be improved before the DTL is ready for general use, including ability to detect failed tasks, better sandboxing and supporting multiple workers on a single machine.

References

1. Gannon, D., Fay, D., Green, D., Takeda, K., Yi, W.: Science in the cloud: Lessons from three years of research projects on Microsoft Azure. In: Proceedings of the 5th ACM Workshop on Scientific Cloud Computing, pp. 1–8. ScienceCloud 2014. ACM (2014)
2. Hill, Z., Li, J., Mao, M., Ruiz-Alvarez, A., Humphrey, M.: Early observations on the performance of Windows Azure. In: HPDC 2010 Proceedings of the 19th ACM International Symposium on High Performance Distributed Computing, pp. 367–376. ACM, June 2010
3. Leijen, D., Schulte, W., Burckhardt, S.: The design of a task parallel library. In: Proceedings of the 24th ACM SIGPLAN Conference on Object Oriented Programming Systems Languages and Applications, pp. 227–242. Microsoft Research. ACM, October 2009
4. Malawski, M., Meizner, J., Bubak, M., Gepner, P.: Component approach to computational applications on clouds. In: Proceedings of the International Conference on Computational Science, ICCS 201, vol. 4, pp. 432–441. Elsevier, May 2011
5. Manjrasoft Pty Ltd: Aneka: Enabling .NET-based enterprise grid and cloud computing. http://www.manjrasoft.com/products.html (Accessed 07 August 2014)
6. Mao, M., Humphrey, M.: A performance study on the VM startup time in the cloud. In: CLOUD 2012 Proceedings of the 2012 IEEE Fifth International Conference on Cloud Computing, pp. 423–430. IEEE Computer Society Washington, DC, USA 2012, June 2012
7. Corporation, M.: Windows Azure HPC. http://azure.microsoft.com/en-us/solutions/big-compute/ (Accessed 15 May 2015)
8. Microsoft Open Technologies: VM depot - find, deploy and share images for Windows Azure. http://vmdepot.msopentech.com (Accessed 07 August 2014)
9. Myers, T., et al.: Microsoft Azure import/export service. http://azure.microsoft.com/en-us/documentation/articles/storage-import-export-service/
10. Persistence of Vision Raytracer Pty. Ltd.: Benchmarking with POV-Ray. http://www.povray.org/download/benchmark.php (Accessed 26 July 2014)
11. Piwowar, M., Banach, M., Konieczny, L., Roterman, I.: Structural role of exoncoded fragment of polypeptide chains in selected enzymes. J. Theor. Biol. **337C**, 15–23 (2013). http://www.ncbi.nlm.nih.gov/pubmed/23896319
12. Piwowar, M., Banach, M., Konieczny, L., Roterman, I.: Structural role of exons in hemoglobin. Bio-Algorithms Med-Syst. **9**(2), 81–90 (2013)
13. Piwowar, M., Krzysztof, P., Piotr, P.: ExonVisualiser - application for visualization exon units in 2D and 3D protein structures. Bioinformation **8**(25), 1280–1282 (2012)

Scalable Distributed Two-Layer Block Based Datastore

Adam Krechowicz$^{(\boxtimes)}$, Stanisław Deniziak, Mariusz Bedla, Arkadiusz Chrobot, and Grzegorz Łukawski

Department of Computer Science, Kielce University of Technology, Kielce, Poland
{a.krechowicz,s.deniziak,m.bedla,a.chrobot,g.lukawski}@tu.kielce.pl

Abstract. Modern distributed systems require fast and scalable datastores for efficient processing of huge amounts of information. Traditional DBMSs are insufficient for such a purpose. Hence, new datastore models were developed. In the paper we present the design and implementation of a scalable distributed two-layer block based datastore. We also discuss experimental results of comparing the performance of our datastore with MongoDB, a widely used NoSQL datastore.

Keywords: Scalable Distributed Data Structures · Distributed datastore · MongoDB · NoSQL datastore

1 Introduction

Rapidly increasing amount of information processed and stored on the Internet requires effective and reliable data storage systems. Huge data sets are associated with business applications, social networking and multimedia services. Even short periods of data outages can be expensive and may result in loss of productivity, as well as financial consequences, while permanent data loss can be catastrophic. Therefore, reliable and efficient datastores are important components of most IT infrastructures.

Much of the data is still stored in the most versatile format, the flat file. The growing collections of multimedia files, text documents, graphics, spreadsheets etc. are created and stored in datacenters. Storing of such a large number of files requires distributed storage systems that provide scalability and fast access. NoSQL datastores appeared to be the best solutions for this purpose.

In this paper we propose an efficient distributed datastore for storing a huge number of files. The architecture of the datastore is based on Scalable Distributed Two-Layer Data Structures (SD2DS) [13] supporting key-value access. The SD2DS enable storing data in a distributed environment and they are very efficient for storing medium- and big-size files. We show that for files larger than 1MB our datastore is more efficient than MongoDB, one of the most popular distributed datastores.

The rest of the paper is organized as follows. Next section reviews the related work. Section 3 presents the motivation for the paper. In Sect. 4 the architecture

© Springer International Publishing Switzerland 2016
R. Wyrzykowski et al. (Eds.): PPAM 2015, Part I, LNCS 9573, pp. 302–311, 2016.
DOI: 10.1007/978-3-319-32149-3_29

of our datastore is described. Section 5 contains experimental results. The paper ends with conclusions.

2   Related Work

Existing NoSQL solutions implement a variety of data models like: columns, key-value, documents and graphs. Thus, only some of them can be used to store flat files. MongoDB [12] is one of the most known document-based datastores. Documents are stored in BSON format which is binary version of JSON. MongoDB stores the data on reliable media like disk drives by utilizing journals but it also uses cache mechanisms to speed up the access to the most recently used data. MongoDB offers atomic access for a single document. One of the most interesting fact about MongoDB is that it can distribute data on many servers within a cluster by using sharding technique. A shard can be a standard MongoDB instance which is responsible for managing a portion of the whole data. The Config Server is a special MongoDB instance which is responsible for managing the whole set of shards. It allows for splitting the overloaded shard into two separate shards and removing the faulty shards from cluster. The MongoDB client does not have any information about the current state of shards in the cluster. It needs to communicate through the special element called Mongos which uses the information from the Config Servers and routes the requests from clients to appropriate MongoDB instance. In a situation where many clients need to use single Mongos instance it may cause serious bottlenecks. The main purpose of the MongoDB is to store documents in BSON format. However, it also allows for efficiently storing data of any format using special API called GridFS. The data managed by GridFS are stored in two separate collections. One, which is usually called fs.files, stores the metadata of each element. The other, usually called fs.chunks, stores the actual data in small parts, typically of the size equal to 255 KiB.

Memcached [3,11] is an open source in–memory caching system that implements Distributed Hash Table (DHT), to store data items. It is used by such companies as YouTube or Twitter to reduce database load in their services. Memcached uses the key–value concept for data addressing. A unique identifier is associated with each stored data item. For performance reasons the maximal size of a single data item should not exceed 1 MiB. The caching system may be used in a centralized or distributed environment. In the latter case the Memcached servers are located on different nodes of a cluster and they do not communicate with each other. Only the client software knows exact location of servers and it uses the consistent hashing algorithm to assign data items to them. Such a hashing method provides a good load balance and some degree of fault tolerance. Internally, each Memcached server divides its memory into different zones called classes [1]. Each class stores data items of a particular size. If server does not have enough space for a new data item, it uses the LRU algorithm to discard one of the previously stored.

Memcachedb is a distributed high available key–value datastore [4]. It is compatible with Memcached but unlike the latter it stores data items in a persistent manner. To provide data coherence Memcachedb supports transactions. The fault tolerance and availability is achieved by data replication.

Hazelcast [7] is a library of Java collections and other utilities for distributed computing environments. It has many applications, like in-memory data grid, caching, web session clustering and NoSQL key-value datastore. Hazelcast is implemented in Java but provides support for other programming languages like C/C++ and also for .NET framework. It is used in many branches of industry, like financial services, telecommunications and software development. Hazelcast is organized as peer-to-peer. Data are stored in partitions distributed over cluster of servers. It scales to new servers added to a cluster increasing capacity and throughput. It provides high performance by storing data in a RAM. Hashing function is used for addressing the data. Backup servers guarantee that no single point of failure exists. Additionally, data may be stored to a persistent datastore and Hazelcast may be used in a transactional context. Hazelcast provides many data structures supporting distributed programming. There are distributed implementations of collections (a map, a queue, a set, a list), a mechanism for publishing messages, selected concurrent classes (for example Lock), an id generation support and a replicated map. New collections can be also created using Service Programming Interface (SPI). Data stored in Hazelcast may be processed in many ways. User defined tasks can be executed by distributed implementation of executor service. An entry processor may be used to run code processing the entries. The collections may be queried according to Criteria API and Distributed SQL Query. It can be also processed using the Hazelcast MapReduce framework.

Cassandra [2,6] is one of the most popular column datastores. It is an open source Java application that is used in the following industries: financial services, entertainment, social networking, education and others. Data items are identified by a primary key and stored as rows in tables. The key makes it possible to localize a node storing the data. A partition key is a first component of a primary key. Data is stored in denormalized form and Java collections may be used to express relation between the data. Every node may contain many virtual nodes. A replication factor defines a number of replicas stored on different nodes in the same or different data centers. There is no single point of failure. Cassandra do not support full ACID transaction like RBDMS. Instead it provides either atomic, isolated, and durable transactions with eventual/tunable consistency or lightweight transactions ("compare and set"). Cassandra can be accessed using Cassandra Query Language (CQL), similar to SQL but without joins and subqueries, except batch analysis. When a client sends a request to a node in a cluster, the node becomes a coordinator for the request. It takes part in the communication between the client and nodes storing the requested data. Nodes periodically exchange information using a gossip protocol.

3   Motivation

The main goal of our research is to develop an efficient distributed datastore for data organized in flat files. We assume that the primary features of our datastore should be the scalability and fast access to data. Recently we have designed and developed an SD2DS architecture [13], as a low level mechanism for management of distributed data. Since SD2DS is highly scalable, key–value oriented and stores data in memory, it seems to be especially useful and attractive for such a purpose.

There is no single point in SD2DS that could become a bottleneck for client–server communication, even if many clients make requests simultaneously. The two layer structure allows for better memory utilization when data records of a large size are stored. Short data access time is achieved by keeping the headers buckets in RAMs and applying the distributed linear hashing algorithm for addressing them. The SD2DS–based data store could be easily modified in order to add persistence or throughput scalability.

We believe that the unique features of the SD2DS make it possible to obtain SD2DS-based datastore with a much better performance than the existing ones. To verify this expectation we evaluate the performance of our datastore and compare the results with the results obtained for other commonly used storage systems like MongoDB. This allows us not only to assess the advantages of SD2DS-based datastore but also to identify shortcomings that we should address in our future works.

4   The SD2DS Data Store

Scalable Distributed Two–Layer Data Structures (SD2DS) were developed to efficiently store large amounts of data in a distributed RAM of a multicomputer. Although the basic idea of SD2DS was inspired by Scalable Distributed Data Structures (SDDS) [9,10], it overcomes some of the ancestor's disadvantages (like slow splitting operations) by using double layered structure. In fact, the SDDS LH* is very useful and efficient for management of the first layer of SD2DS, as shown in the paper. Moreover, the basic SD2DS architecture may be expanded for improving load balancing, fault tolerance and so on [8,13].

4.1   Basic SD2DS Architecture

The data stored in SD2DS consist of *components* of constant or variable size. Each component is uniquely identified by an unique *key*, thus SD2DS is suitable for building a *key–value* type datastore. Components stored in SD2DS are split into two parts:

- *Header*, consisting of the key and so called *locator* (memory pointer, URL/IP address, etc.), pointing where the second part of the component (body) is stored. Generally, the header stores the metadata of a component,
- *Body*, consisting of the key and the component data.

Headers are stored in the first layer of SD2DS, called *the file*, while bodies are stored in the second layer, called *the storage*. Both layers are managed independently and many different algorithms may be applied for the first and the second layer (e.g. hashing, B–trees, LH* [10] and RP* [9]). Moreover, headers and bodies may be supplemented with additional information such as checksums, multiple locators, counters and so on.

A single instance of SD2DS contains at least three types of elements:

- *Buckets*, containing both, the headers and bodies. Depending on a particular implementation, there may be separated buckets for storing headers and bodies, or a bucket may store both layers simultaneously.
- *Clients*, processing the data stored in buckets. A client may be supplemented with additional mechanism for addressing of the first layer of SD2DS, such as SDDS LH* file images. This implies that there is no need for any element that is required for addressing so there is no risk of bottlenecks or a single point of failures.
- *Split Coordinator*, responsible for managing state of both layers, crucial for ensuring scalability of the SD2DS. Depending on the implementation, there may be multiple Coordinators or none, if layers manage their state and scalability on their own.

4.2   Scalability of the SD2DS

An empty SD2DS consists of one bucket. As more and more components are inserted into SD2DS, the structure expands by attaching new buckets. If SDDS scheme is applied for the first layer, it expands by performing splits [10], thus there is a need to move data between buckets. Fortunately, the process if fast due to the small size of component headers.

In case of the second layer, once inserted, component bodies stay in the same bucket during the whole lifetime of SD2DS. This is a big advantage, as there is no need to transfer large amounts of data during evolution of the structure. Because both, the first and the second layer, may use the same set of servers, a whole component may be stored on a single server, what is the most optimistic case. Figure 1 presents a sample SD2DS file, starting from single, empty bucket with address (a). All inserted components are stored in the single bucket and locators in their headers are set to (a) (Fig. 1b). As more and more components are inserted, more buckets are attached to the file (Fig. 1d-e). Headers are stored with respect to the LH* rules, while bodies are inserted into buckets having available space.

Every bucket stores so called Redirection Bucket Address (RBA), which is used to store component bodies. Initially, the RBA is set to NULL, thus all bodies are inserted locally. If a bucket overloads, its RBA is adjusted to point a bucket with available space. The Split Coordinator is responsible for keeping a pool of buckets with available space and for adjusting RBA of every bucket.

Two–layer design implies an indirect addressing, thus before reaching the data, a client must contact the first layer for obtaining the corresponding locator. However, our experiments showed that the need of indirect addressing doesn't affect much the total efficiency of SD2DS, especially for relatively large portions of data (starting from about 1MiB). Moreover, as each component body stays in the same bucket as it was firstly inserted, locators may be stored and re–used by clients (depending on the particular implementation of SD2DS).

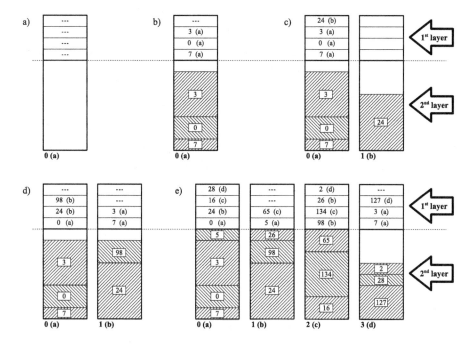

Fig. 1. Sample SD2DS file evolution

4.3   Implementation

Our previous experiments showed that SDDS LH* is especially useful for management of the first layer of SD2DS. Moreover, both layers may be supplemented with additional elements like checksums (for fault tolerance), reference counters (for load balancing) and so on. Thus, the general idea of SD2DS can be tailored to specific needs. The datastore presented in the paper uses basic SD2DS architecture, where LH* is used for management of the first layer and a simple expansion algorithm (buckets are filled with component bodies one by another) for the second layer.

We utilize SD2DS conception to develop efficient data store mostly for storing large multimedia files like photos, videos etc. We choose C++ language for developing the core structure of our system, mostly because it allows efficient memory management which is crucial in our project.

The original concept of SD2DS assumes that the data set is located entirely in the main memory of the cluster nodes. In the real world environment it is often unacceptable because of the risk of data loss, that may occur even after slight failure like memory bit flips. To preserve the data from unexpected loss, we decided to store data on hard disks of the cluster nodes. To preserve the high efficiency we prepared a simple caching mechanism for storing the subset of the data also in the main memory. We used basic Last Recently Used algorithm to perform this task. In our future work we plan to develop more advanced algorithms which will be more suitable for our needs.

5   Experimental Results

We choose MongoDB for comparison with our datastore for many reasons. First of all, both systems are implemented in the similar way (both are written in C++). They allow similar data distribution models by utilizing hashing or partition ranges. But most importantly, both stores the metadata separately from the actual data, in different containers that can even be located on separate locations. Moreover, the architectures of MongoDB and SD2DS are very similar. Both datastores use special element for managing the expansion and shrinking of the structure (Split Coordinator in case of SD2DS and Config Server in case of MongoDB). According to DB-ENGINES site [March 2015] MongoDB is the most popular document store. Moreover, it has the greatest popularity from all NoSQL stores and is in the fourth position in general ranking. We used sharded GridFS API for evaluating MongoDB.

For the purpose of the experiments we developed a simple benchmark utility for carrying out the performance analysis. The existing benchmarking utilities, like Yahoo! Cloud Serving Benchmark (YCSB) [5], are focused mostly on small pieces of data. Java was used to prepare this tool mainly because it is still very popular language for enterprise and web applications. Moreover, both MongoDB and SD2DS offer good drivers for Java. This utility allows us to measure performance in multi-client environments. In the future we plan to extend this tool in such a way, that it will allow us to execute more advanced tests, using wider variety of NoSQL solutions.

We conduct our real world experiments on a cluster, consisting of 16 machines, on which the data were stored. We also used additional machine for running SD2DS Split Coordinator and MongoDB Config Servers. We used 10 machines for running clients. For testing MongoDB we additionally used up to 5 machines to run mongos instances which are crucial for running clients. A single

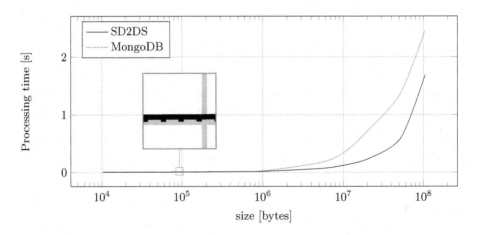

Fig. 2. Time comparison of adding components using single client

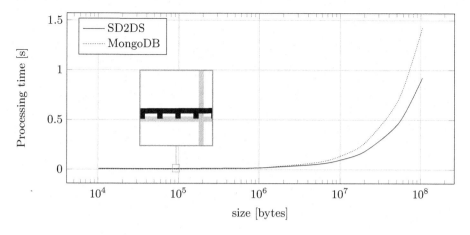

Fig. 3. Time comparison of getting components using single client

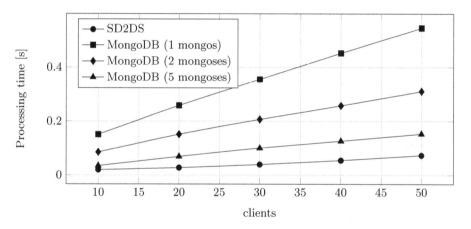

Fig. 4. Time comparison of getting components of fixed (1MiB) size

node in the cluster consisted of two 8 cores processors with 8GiB of RAM and was running on Centos 6. The servers were connected using 1GiB Ethernet.

We started our experiments with simple comparison of transmission times, for different sizes of data sent between clients and datastores. We used just one client that operated on datastores. The results are presented in Figs. 2 and 3 for adding and getting data, respectively. This simple tests shows us that despite very good caching mechanisms in MongoDB SD2DS allows better transmission times and as a result serves clients faster when using very big components. Smaller components (<1MiB) are still faster served by MongoDB. It is caused by the specific architecture of SD2DS in which the two layers are completely disjoint. Hence, the client is forced to communicate with servers twice in one query and it affects the effectiveness for small components. The next tests we performed in multi-client environments. We used up to 50 clients operating on

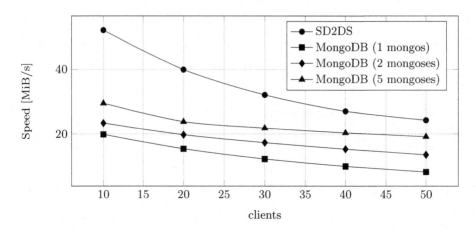

Fig. 5. Speed comparison of getting components of different sizes (0.5MiB - 1.5MiB)

components of fixed size of 1MiB. Figure 4 presents average processing times. The results for MongoDB are strictly correlated with the number of mongos instances. Using many clients connected to one instance of mongos seriously affects the effectiveness of the storage. The multimedia data, like photos, usually vary in sizes, thus the final tests were run with variable sizes of components. We have chosen sizes of components in the range from 0.5MiB to 1.5MiB, that corresponds to the sizes of our stored photos. The results are presented in Fig. 5. Like in the previous experiments we achieved better performance than MongoDB, even with 5 mongos instances. Because of the variable sizes of the components we needed to measure speed instead of the processing time.

6   Conclusions

In this paper we have presented SD2DS–based scalable distributed block-based datastore. The results of experimental evaluation indicate that the datastore outperforms MongoDB for files of size bigger then 1 MiB. Moreover, results of other experiments showed that the SD2DS-based datastore is more scalable with the respect to the number of clients.

The tests allowed us to discover that the performance of the SD2DS–based datastore needs improvement for smaller files. This problem will be addressed in our future works. We also intend to add new functions to the datastore in order to process information in more advanced ways than just adding and getting data.

Acknowledgement. The research used equipment funded by the European Union in the Innovative Economy Programme, MOLAB - Kielce University of Technology.

References

1. Carra, D., Michiardi, P.: Memory partitioning in Memcached: an experimental performance analysis. In: IEEE ICC, pp. 1154–1159 (2014)
2. Cassandra: The Apache Cassandra Project. http://cassandra.apache.org (Accessed 14 April 2015)
3. Chidambaram, V., Ramamurthi, D.: Performance analysis of memcached. http://citeseerx.ist.psu.edu/viewdoc/download?doi=10.1.1.409.411&rep=rep1&type=pdf (Accessed 13 April 2015)
4. Chu, S.: Memcachedb: The Complete Guide. http://memcachedb.org/memcachedb-guide-1.0.pdf (Accessed 13 April 2015)
5. Cooper, B.F., Silberstein, A., Tam, E., Ramakrishnan, R., Sears, R.: Benchmarking cloud serving systems with YCSB. In: Proceedings of the 1st ACM Symposium on Cloud Computing, pp. 143–154 (2010)
6. DataStax: DataStax Documentation Apache Cassandra™2.1. http://docs.datastax.com/en/cassandra/2.1/ (Accessed 14 April 2015)
7. Hazelcast: Hazelcast The Leading In–Memory Data Grid. http://hazelcast.com (Accessed 14 April 2015)
8. Krechowicz, A., Deniziak, S., Łukawski, G., Bedla, M.: Preserving dataconsistency in scalable distributed two layer data structures. In: Kozielski, S., Mrozek, D., Kasprowski, P., Małysiak-Mrozek, B., Kostrzewa, D. (eds.) BeyondDatabases, Architectures and Structures (BDAS). Communications in Computer and Information Science, vol. 521, pp. 126–135. Springer, Heidelberg (2015)
9. Litwin, W., Neimat, M.A., Schneider, D.: RP*: a family of order preserving scalable distributed data structures. In: Proceedings ofthe Twentieth International Conference on Very Large Databases (VLDB), pp. 342–353 (1994)
10. Litwin, W., Neimat, M.A., Schneider, D.A.: LH* – a scalable, distributed data structure. ACM Trans. Database Syst. (TODS) **21**(4), 480–525 (1996)
11. Memcached: Memcached – A Distributed Memory Object Caching System. http://memcached.org (Accessed 13 April 2015)
12. MongoDB: The MongoDB 3.0 Manual. http://docs.mongodb.org/manual/ (Accessed 14 April 2015)
13. Sapiecha, K., Łukawski, G.: Scalable Distributed Two-layer DataStructures (SD2DS). Int. J. Distrib. Syst. Technol. (IJDST) **4**(2), 15–30 (2013)

Metadata Organization and Management for Globalization of Data Access with Onedata

Michał Wrzeszcz[1,2], Krzysztof Trzepla[1], Rafał Słota[1], Konrad Zemek[1],
Tomasz Lichoń[1], Łukasz Opioła[1], Darin Nikolow[1,2], Łukasz Dutka[1],
Renata Słota[1,2(✉)], and Jacek Kitowski[1,2]

[1] ACC Cyfronet AGH, AGH University, Kraków, Poland
{wrzeszcz,darin,rena,kito}@agh.edu.pl, dutka@cyfronet.pl
[2] Department of Computer Science, AGH University, Kraków, Poland

Abstract. The Big Data revolution means that large amounts of data have not only to be stored, but also to be processed to unlock the potential of access to information and knowledge for scientific research. As a result, scientific communities require simple and convenient global access to data which is effective, secure and shareable. In this article we analyze how researchers use their data working in large scientific projects and show how their requirements may be satisfied with our solution called Onedata. Major technical mechanisms of metadata management and organization are described.

Keywords: Data management system · Mass storage systems · Grid · Cloud

1 Introduction

People exhibit a profound need to networking which drives the development of social networks like Facebook, Myspace, YouTube, ResearchGate and other similar environments used in everyday life. Similar trend is observed in modern e-science, where the exchange of data, information and knowledge is an essential factor accelerating scientific discoveries. The importance of data and especially its processing for scientific discoveries have been recognized and a paradigm of science called the fourth paradigm [1] has been identified. The 4th paradigm relates to scientific research based on processing and analysis of large amounts of data. This large amounts of data not only have to be stored, but also have to be processed to unlock the potential of access to such amount of information. This problem is often called the Big Data approach [2].

There are various data storage systems in today world which scientists can choose from. Those storage systems provide different advantages which often results in distributing data between many storage systems or services. Examples of storage system include personal storage (HDD), enterprise NAS/SAN, cluster storage, HSM storage, Cloud storage. Moreover, the storage systems might be owned by different organizations, which can be geographically distributed [3].

© Springer International Publishing Switzerland 2016
R. Wyrzykowski et al. (Eds.): PPAM 2015, Part I, LNCS 9573, pp. 312–321, 2016.
DOI: 10.1007/978-3-319-32149-3_30

The data itself can be structured or non-structured and kept in databases or filesystems respectively. A researcher often needs to access more data pieces of possibly different types at once for doing his work which is inconvenient and results in creation of barriers between the user or his applications and the data.

Global data storage is a paradigm where distributed pieces of user's structured and non-structured data are virtually integrated into a common data store allowing for unified, easy and efficient access to organizationally and geographically distributed data from multiple locations. Building such solution is challenging. Onedata[1] is a solution for global data storage being developed by the authors facing those challenges. In this paper we focus on the metadata organization and management aspects of Onedata allowing to meet the global storage requirements typically announced by scientific communities.

The rest of the paper is organized as follows. State of the art is presented in the following section. Section 3 describes researchers requirements for global data storage. The key aspects of metadata organization and management for global data access in Onedata are presented in Sect. 4 while relevant implementation details are depicted in Sect. 5. Section 6 includes experimental result discussion and the last section concludes the paper.

2   State of the Art

Storing data in a Cloud based services is relatively new alternative applying the pay-as-you-go model and providing easy anytime/anyplace data access. This alternative is commonly referred as StaaS (Storage-as-a-Service). Examples of popular StaaSs include Amazon S3, Dropbox, OneDrive, Google Drive [4]. This is convenient for typical business or personal use but in the case of data intensive HPC scientific computing there are some drawbacks/disadvantages [5]. The main problem is accessing data with performance sufficient for running HPC applications. The problem is partially addressed by Amazon and Google clouds by providing high performance storage and clusters of virtual machines for demanding clients but it does not solve the problem of efficient accessing to their storage from outside of their clouds and vice versa. The mentioned services do not provide features allowing for integration with external data storage which forces the user to copy and keep the data on the cloud storage or use additional access methods and services.

In order to improve users' comfort of data access, data management system that hides storage systems complexity can be used. In this context systems like Parrot [6], iRODS [7], Syndicate Drive[2] and Storj[3] should be listed. QoS oriented data management can be used to improve user's experience in accessing data [8,9]. Still, efficiency is a price to pay for simplification in these systems. While they implement the concept of location transparency at some level, the mechanisms that provide this transparency generate significant overhead and

---

[1] http://onedata.org/.

[2] http://syndicatedrive.com/.

[3] http://storj.io/.

can be considered as bottlenecks. Moreover, some of the systems do not support fragment access and require complete data download before usage which is inefficient and thus unacceptable in computing infrastructures intended for HPC.

The fact that none of the presented systems fully implements the global data storage paradigm has motivated us to describe the mechanisms which allow to meet requirements of scientific communities for a global storage solution with our novel approach.

3   Global Data Access Requirements - A Researcher View

Many scientific applications (e.g. astronomy, climate research, high energy physics) process data from different storage systems or services. Usually, all the needed files are copied to local scratch storage prior to running a job. After the job is done the output files are transferred back. Depending on the job's data access pattern some files may not be fully exploited resulting in inefficient usage of storage and network. The transfers of input and output data are explicitly defined by the user. Sometimes scientists are doing computations on different infrastructures belonging to different organization which additionally complicates job operation. Other problems faced by researchers concern data sharing, difficult access caused by security issues, heterogeneity of storage.

A global storage system should help researchers to overcome those problems. On the other hand, the management of such a system should be easy to make it usable. Thus, we define the following requirements for a global data storage for science: (a) transparent access with standard and popular methods like BUI (Browser User Interface), Unix/POSIX filesystems, CDMI (Cloud Data Management Interface); (b) easy access (single-sign-on) to data organizationally distributed among various storage resource providers; (c) easy management allowing for fine-grain monitoring of storage resources; and (d) high performance.

4   Key Aspects of Data Access Globalization in Onedata

The mentioned requirements are addressed in our solution named Onedata. The development of Onedata has started as part of the PL-Grid Plus project, which needed a federated storage solution for their users [3]. Further development concerned globalization aspects [10]. Onedata is deployed in the PL-Grid infrastructure and is going to be used in EU projects EGI Engage and Indigo Data Cloud. We discuss the key aspects of Onedata globalization in the following sections.

4.1   Global Data Organization

In order to hide the complexity of data distribution in Onedata we introduce the concepts of *spaces* and *providers*. Spaces are logical containers holding data stored by associated *users*. Providers are organizations providing storage resources to Onedata. Additionally, to simplify administration and access control, the users may be associated in *groups*. Figure 1 gives an overview on the relations between those entities. Such design allows for easier metadata management since each provider take care only about the metadata of supported spaces.

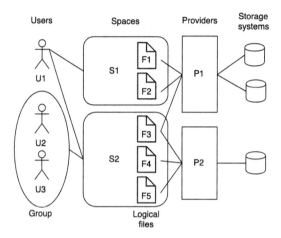

Fig. 1. Onedata entity relations

User. Each Onedata user owns an account containing basic personal information. Such accounts are essentially cached credentials provided by external OpenID providers, extended with information of space and group adherence. The user can be a member of multiple groups and can be associated with multiple spaces.

Group. Groups are introduced to simplify administrative tasks. Associating a group with a space is equivalent to associating each user from the group with this space. Similarly, granting permission to group is equivalent to granting permission to every group member. In Fig. 1, users *U2* and *U3* share a group. This group is a associated with space *S2*.

Space. The space can be supported by many providers and shared by many users. It hides the complexity of storage interfaces and unifies access to data. Files collected in a space may be stored on multiple heterogeneous storage systems, but remain transparent to the users in terms of data location. In Fig. 1, space *S1* is supported by provider *P1* while *S2* is supported by *P1* and *P2*. A closer look at the file *F3* indicates that files may be stored on multiple providers. Such a feature gives possibility to configure data replication and leads to better performance on read operations.

Provider. Providers contribute to Onedata with storage resources. In the given example, provider *P1* operates on two underlying storage systems containing data of files *F1*, *F2* and *F3*. Due to fact that *P1* supports both spaces, it can serve files *F4* and *F5* too. In such a case, first read operation requires provider *P1* to download necessary data from provider *P2*, so consecutive requests would read the local copy. On the other hand, as *P2* does not support *S1*, each request concerning files *F1* and *F2* has to be redirected to *P1*.

The most important feature of the above mentioned concept is the support of spaces by providers. It means that a provider supports particular scientific activity rather than users/groups. The user can do research transparently using various storage systems from different providers.

4.2   Globally Distributed Metadata

The metadata are distributed between providers which may not trust each other. In order to make the cooperation of those providers possible additional component trusted by providers supporting a given space is needed. We call it Global Registry (see Sect. 5.1). Global Registry allows cooperation of independent providers to offer users globally unified space for their geographically distributed data (see Fig. 2).

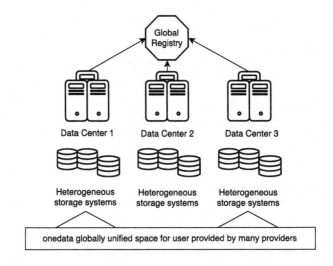

Fig. 2. Onedata concept of data access globalization using Global Registry

Each provider keeps locally the metadata of supported spaces. The metadata of a space are synchronized with the other providers supporting the same space. The list of providers supporting a space is provided by Global Registry (see Sect. 5.1).

A client of Onedata provides native filesystem access to a space via FUSE[4]. The client connects to a provider to read and modify metadata according to user's requests. By issuing a DNS request, the client receives from the Global Registry the most appropriate provider to connect to.

---

[4] Filesystem in User Space, http://fuse.sourceforge.net.

Caching is a technique allowing for fast access to data used in the past and is present in each level of data flow within Onedata. For example, a provider caches data results coming from querying Global Registry about supported spaces while the client caches the metadata of files being accessed. Caching of distributed data introduces cache coherency problems which are minimized by using feedback channels to flush or update caches.

5 Implementation of Data Access Globalization with Onedata

A distinct feature of Onedata is providing high quality data access solution for organizationally distributed storage environments comparable with the quality of solutions managed by single organization. Details about organizing and management of metadata for data access globalization are prezented below.

5.1 Global Registry

Global Registry acts as an entry point to Onedata providing user authentication functionality based on the OpenID model. In order to support cooperation between providers Global Registry stores only essential information about users, spaces and providers. After logging the communication goes between the users and providers directly mostly omitting Global Registry. The design allows for scalable grow of the Global Registry cluster. The mentioned caching offloads Global Registry.

Global Registry provides a REST interface that providers can use to manipulate metadata, either their own (authenticating with Global-Registry-issued provider certificate) or user's ones (authorizing with OpenID Connect protocol). Global Registry also plays role as a certificate authority. Anyone can install Onedata software and become a provider, including large computer centers and private machines with storage space. Because of that, the providers do not trust anyone except Global Registry.

The requests between modules of Onedata triggered for a file request are shown in Fig. 3. The requested file $F4$ (see Fig. 1) belongs to space supported by providers $P1$ and $P2$. The file data blocks reside in $P2$ and the FUSE Client is attached to $P1$. The FUSE Client exchanges user credentials with Global Registry (req. 2) and sends file request to $P1$ (req. 3). $P1$ verifies that FUSE Client is authorized (req. 4) and sends request to $P2$ (req. 5). $P2$ verifies that $P1$ is allowed to access the data (req. 6). Every actor in the system is now sure of others' identities and authorizations, and can proceed with the data transfer. The responses of Global Registry are cached temporally to decrease the delay of subsequent communication attempts.

5.2 File Metadata Description

File metadata are stored in provider's local NoSQL database (DB). A single file metadata is kept in a single document. The document is synchronized between

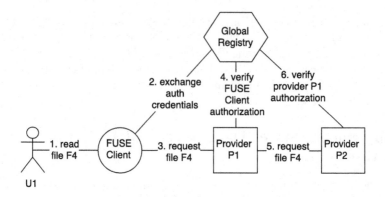

Fig. 3. Requests in Onedata system presented in Fig. 1.

providers supporting the space which is hosting the file. Since the file may have fragments distributed and replicated between providers it is important to know which providers holds the given file fragment. This information is kept in the metadata. Although the current version of Onedata does not explicitly support replica creation, implicit data replication may occur when more providers are reading the same file fragments. We do not remove those replicas immediately for performance reasons. The replicas get removed after some time of inactivity depending on provider's configuration. When a fragment is updated by a provider it informs the other providers to invalidate their replicas and metadata is updated pointing the actual fragment holder. Rarely, it may occurs that two providers independently modify a file fragment. In this case the generated data conflict is automatically resolved by Onedata (see Sect.5.4).

5.3   Local Data Usage and Meta-Data Management

Onedata supports direct access to storage which means that the client can access the storage omitting the provider. In this case the same storage, e.g., Lustre filesystem, is mounted on the providers machines and client machines. This approach is used because data are transferred between the client/FUSE buffers and the storage with minimal overhead and the client communicates with providers for exchanging metadata only.

Each client maintains its own metadata cache which is updated when the client writes or modifies files in the given space. In this case the client notifies the provider about the changes. Next, the provider modifies its metadata and propagates the modifications to the appropriate providers. In order to minimize the network load caused by the metadata changes propagation the changes are aggregated and bulk transferred. By configuring aggregation parameters a balance between metadata refreshing latency and server and network load can be settled.

When the requested file fragment does not reside in the storage of the provider which is currently serving the client (usually this is the local provider) then the missed fragment is transferred from the remote provider holding the fragment to the local provider storage and then accessed from there.

5.4 Meta-Data Synchronization

Onedata implements an efficient method of synchronizing metadata of a given space supported by many providers. As mentioned before metadata are kept by the provider in its own DB. A special component deployed on each provider called *DBSync* implements the synchronization of metadata documents between providers databases. Some implementation details are present below.

Reading changes from DB. DB management system (DBMS) notifies *DBSync* about every change in the database. *DBSync* aggregates changes over a time interval to decrease communications. This interval is relatively short to keep the probability of conflicts low. It should be mentioned that conflicts are unavoidable without use of locks.

Propagating changes. Changes are efficiently propagated to the relevant providers using the following method. First the list of providers (recipients) supporting the given space is obtained from Global Registry. Next the sending provider (sender) divides recipients into two groups. From each group, the sender selects two providers that will be responsible for propagating those changes to their group. This way, whenever the provider needs to propagate changes to other providers that support the space, the provider sends only 2 messages (and additional 2 for keeping redundancy).

Resolving conflicts. *DBSync* provides simple conflict management based on native conflict resolver of DBMS. This way each provider has the same metadata.

6   Results

Onedata design of global metadata distribution allows provision of unified and coherent view on data, and on the other hand allows for achieving high performance data access by introducing low overhead. A file is represented as a set of blocks where each block can be stored in multiple heterogeneous storage systems, which may belong to different providers. Blocks can be replicated when the same data is accessed via different providers for performance reason as shown in Fig. 4 presenting a screenshot of Onedata's file manager. Updates of metadata, including metadata that contain information about the blocks, are handled using events. To set balance between efficiency, time needed to achieve coherency of system and system load, three parameters may be used: events aggregation strategy, size of block, and blocks' replication type. More aggressive events aggregation, larger blocks and wide replication decreases load of the system in terms of computing power but results in larger time needed to achieve system's consistency after last change and higher storage systems utilization.

Performance tests using sequential writes for single user/process have been conducted to evaluate the overhead introduced by Onedata. The tests were done

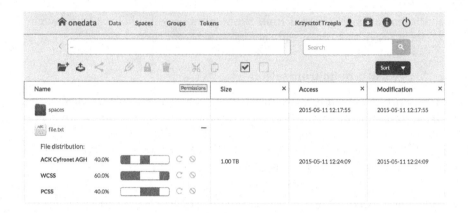

Fig. 4. Block distribution visualization with WebGUI

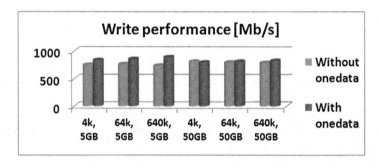

Fig. 5. Onedata performance overhead for sequential writes.

in the PL-Grid production environment using GPFS as underlying storage system for Onedata. We used the dd Unix command to write files using different block and file sizes directly to GPFS and to the same GPFS through Onedata. The test results are presented in Fig. 5. There are cases in which the write performance of GPFS with Onedata is slightly higher than the performance of the bare GPFS. It is due to the additional buffering layer used by the FUSE module. However, it should be noted that the GPFS was used as it is without any performance tuneup.

7 Conclusions

A system providing efficient global data access needs appropriate data organization and management. The access to metadata has to be fast therefore the metadata should be as close to client as possible. On the other hand, the computational overhead for metadata management and replication should be acceptable. A key for design of a global data access system is understanding that it is impossible to make the system that has minimal overhead, storage systems utilization, computing power and network bandwidth demand simultaneously. A good system should be able to find compromises between various requirements to provide high quality of service, at the same time being able to work in environment where storage, computing and network resources are limited. We believe that Onedata sets this balance properly, offering users an efficient global data storage solution. Our future work will focus on adopting the replication-based data access optimization methods developed in our previous studies [11,12].

Acknowledgments. This research is supported partly by the European Regional Development Fund program no. POIG.02.03.00-12-137/13 as part of the PLGrid Core and AGH-UST grant no. 11.11.230.124.

References

1. Hey, A.J.G., Tansley, S., Tolle, K.M.: The Fourth Paradigm: Data-intensive Scientific Discovery. Microsoft Research Redmond, Redmond (2009)
2. Mills, S., Lucas, S., Irakliotis, L., Data, D.B., et al.: A Practical Guide to Transforming the Business of Government. Technical report (2012). http://www.ibm.com/software/data/demystifying-big-data/
3. Dutka, L., Słota, R., Wrzeszcz, M., Król, D., Kitowski, J.: Uniform and efficient access to data in organizationally distributed environments. In: Bubak, M., et al. (eds.) eScience on Distributed Computing Infrastructure. LNCS, vol. 8500, pp. 178–194. Springer, Heidelberg (2014)
4. Oliveira, T., Mendes, R., Bessani, A.: Sharing files using cloud storage services. In: Lopes, L., et al. (eds.) Euro-Par 2014, Part II. LNCS, vol. 8806, pp. 13–25. Springer, Heidelberg (2014)
5. Palankar, M.R., Iamnitchi, A., Ripeanu, M., Garfinkel, S.: Amazon S3 for science grids: a viable solution?. In: Proceedings of DADC, pp. 55–64. ACM (2008)
6. Thain, D., Livny, M.: Parrot: transparent user-level middleware for data-intensive computing. Scalable Comput. Pract. Experience **6**(3), 9–18 (2005)
7. Hunich, D., Muller-Pfefferkorn, R.: Managing large datasets with iRODS - a performance analysis. In: Proceedings of IMCSIT, pp. 647–654. IEEE (2010)
8. Słota, R., Nikolow, D., Skalkowski, K., Kitowski, J.: Management of data access with quality of service in PL-Grid environment. Comput. Inform. **31**(2), 463–479 (2012)
9. Nikolow, D.: Semantic-based storage QoS management methodology - case study for distributed environments. Comput. Inform. **31**(6), 1345–1366 (2012)
10. Dutka, L., Wrzeszcz, M., Lichon, T., Słota, R., Zemek, K., Trzepla, K., Opioła, Ł., Słota, R., Kitowski, J.: Onedata - a step forward towards globalization of data access for computing infrastructures. Procedia Comput. Sci. **51**, 2843–2847 (2015)
11. Słota, R., Skitał, Ł., Nikolow, D., Kitowski, J.: Algorithms for automatic data replication in grid environment. In: Wyrzykowski, R., et al. (eds.) PPAM 2005. LNCS, vol. 3911, pp. 707–714. Springer, Heidelberg (2006)
12. Słota, R., Nikolow, D., Skitał, Ł., Kitowski, J.: Implementation of replication methods in the grid environment. In: Sloot, P.M.A., et al. (eds.) EGC 2005. LNCS, vol. 3470, pp. 474–484. Springer, Heidelberg (2005)

Hypergraph Based Abstraction for File-Less Data Management

Bartosz Kryza[(⊠)] and Jacek Kitowski

Department of Computer Science, Faculty of Computer Science,
Electronics and Telecommunications, AGH University of Science
and Technology, Krakow, Poland
bkryza@agh.edu.pl

Abstract. Data management is currently one of the predominant issues in both large scale as well as consumer computing systems. While most data is still stored in regular files and managed by various filesystems, current trends show that users no longer treat their data as files but rather objects, which is particularly evident on mobile devices and Cloud based applications, powered by distributed, highly scalable databases.

In this paper we present our attempt to consider the possibility of abandoning the concept of files and filesystems in favor of more elastic and elementary data management and storage approach, by proposing a new aproach to distributed data modelling which does not require filesystems. In our opinion, filesystems are a legacy paradigm not adhering to modern data management use cases.

Keywords: File systems · Data management · Hypergraphs

1 Introduction

Data in computer systems has been organized into files managed by filesystems since their inception. This made sense as long as data was mainly stored in block based storage systems such as tape or hard drives. Currently however, SSD based memories as well as emerging NVRAM technologies do not require data to be stored in sequentially accessed blocks. More importantly however, files introduce several issues and bottlenecks which limit the way applications and operating systems could be developed.

First of all, data in files is unnaturally clustered, causing high duplication [9,14] and unnecessary transfer overhead (e.g. consider a document containing various images and charts, when we only need a single formula from selected section).

Furthermore, filesystems themselves organize files in large tree-based namespaces. In case of small numbers of files they can be comprehensible, however beyond certain number of directories paths become useless from the point of view of actually finding information, as it is not possible to remember all possible branches where certain file is stored. Thus modern operating systems

© Springer International Publishing Switzerland 2016
R. Wyrzykowski et al. (Eds.): PPAM 2015, Part I, LNCS 9573, pp. 322–331, 2016.
DOI: 10.1007/978-3-319-32149-3_31

provide their users with real-time file scanning tools, such as Spotlight, which enable to find files using keyword based search and without traversing manually directory structure.

Another important issue is the fact that files create a certain barrier for the operating systems, which makes it impossible or at least difficult to address specific data item inside the files, as the operating system would have to understand all possible file formats. This makes it difficult to achieve in a generic way such functionality as for instance accessing and referencing metadata stored inside the multimedia files.

In our previous work [17,18], we have proposed a novel approach to data management and representation, called Filess, where the building blocks managed by the operating system are not files but objects, interrelated in the form of hypergraph, a generalization of graphs where edges can connect more than 2 nodes.

Our vision assumes the following objectives. First of all there no files, neither in the application nor operating system layers. All data items are represented in the form of objects interrelated using relations forming a hypergraph structure. Data and metadata exist at the same level, for instance there is no difference between the 'Image' object and the object describing its author or authorization policy. Filess is not a metadata formalism such as Dublin Core or even Semantic Web, which impose high restriction on how the data can be modeled. Data replication is managed by Filess middleware it is not necessary for users to copy and store the information for either security or efficiency reasons. As a consequence data redundancy is minimized. Objects can be split into subgraphs representing their content in more detail and multiple objects can be aggregated into other objects by connecting them using edges. In this position paper we focus on presenting the hypergraph data model for storing the data in the Filess based systems. The paper discusses motivation and theoretical model, as implementation and performance details are not yet the focus of this research.

2   Related Work

Most research in the area of making the existing directory based file systems more flexible can be classified into the area of semantic file systems [10], i.e. file systems where files have attached meaning. This paper sketches a vision of file systems where files can be annotated in some way, and the basic file system operation such as copy or delete don't take directory paths as arguments but the *semantic* description of the files. The problem with these solutions is that still all the information is either fragmented or clustered into files, and the semantics deal only with meta data attached to these files in the form of some attributes. Nevertheless, these solutions are very important for our work as these approaches address important issues, mainly of how information can be found in file based systems. One of the formal attempts at file system implementation based on set theoretical basis is a file system using Formal Concept Analysis [7], which employs the FCA formal model of classification, neighborhood estimation and Boolean querying. A similar approach, although still bounded by the constraints of regular files, is the Logical File System project [22]. The basic role of this file system is to allow searching for files using first-order logic formulas instead of conventional directory paths. Unfortunately the use of first-order logic inference can seriously impair the scalability of the system in highly distributed settings. Until now, one major industrial attempt at abstracting the file concept from the operating system was the WinFS (Windows File System), which is a research effort from Microsoft [11]. Its basic assumption is to store all information about data in the system, including what would usually be referred to as file in a relational database. With respect to high scalability, an interesting approach is represented by the Google BigTable system [4], which allows to store up to Peta bytes of information about URLs indexed by Google. However the information is stored in tables, columns and rows and is accessed on a key-value basis in order to be optimized for storing information about URLs and the implementation itself is based on the Google File System, thus still all the information is inherently chunked into files. With respect to more flexible user interfaces, which could naturally evolve from the proposed solution of file-less environments, several research projects have already addressed that issue, although they were still limited by the file-centered nature of current information systems. NEPOMUK [12] is a project whose goal is provision of a semantic desktop based on Semantic Web technologies to knowledge workers, by extending most popular applications with ability to process semantic annotations of data. Currently several new formalisms and technologies begin to emerge, such as technologies related with Semantic Web [3], including RDF and OWL [13] and various knowledge base solutions which allow to annotate web 'files', that is web pages [16,19,20]. Although created for the purpose of annotating existing information, these technologies by themselves could possibly be used to provide a basis for our vision. Additionally several formal models of information categorization and abstraction have emerged, which can be useful in this research. One such example is Rough Set theory [23], which provides means of automatic reduction of attribute space required for generating an equivalent classification of objects, and thus could be used for some form of optimization of indexing of information within the system.

3 Data Model

The underlying model proposed in this research for universal representation of data is hypergraph [2,8]. Hypergraphs are a generalization of regular graphs with the property that each edge can connect any number of vertices, i.e. edges are simply subsets of the vertex set of the hypergraph (undirected graph) or pairs of subsets of vertices (directed hypergraph). This enables more natural modelling of complex relations like n-m joins between objects in the data model. All data in this model is stored in data objects which are vertices of the hypergraph, while relations and properties are linked using hyperedges.

Each node's value can be any of the selected data types (Fig. 1):

- *Number* - this is a union data type which allows to store any numeric data type while providing users with a simple API, which handles actual data type identification on the library level,
- *String* - this data object provides means for storing any text in UTF8,
- *List* - most graph data modeling frameworks do not provide lists or arrays, which can be very inefficient when modeling using graph nodes. This data object provide simple means for compositing a set of data objects into ordered structure,
- *Binary* - this data object provides means for storing large binary data such as videos, where the actual data is hashed and stored in a separate distributed key-value store,
- *Composite* - composite data objects are objects which do not need to store any actual value in their node, but provide links for other data objects. Any object containing a value can also be a Composite object, in which case the value represents a flattened representation of the objects structure. This situation can occur during decomposition of an object into a graph,
- *Stream* - buffer objects provide abstraction over I/O functionality of the operating system, these objects cannot be transferred between nodes, and are volatile, i.e. their state and value cannot be synchronized and no version information for these objects is maintained, only read or write operations are allowed. These objects enable complete removal of file and filesystem concepts from the applications code.

In order to support maximum flexibility, no high level typing mechanism is enforced by the data model, custom typing can be achieved using application specific properties attached to the graph nodes.

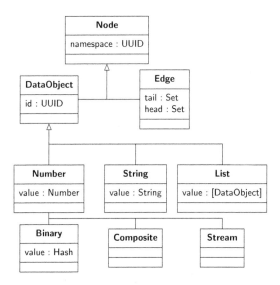

Fig. 1. Data model structure and properties

4   Example Hypergraph Representation of Simple Data Model

In order to present how hypergraphs can be used to represent various data models, let's consider a simple model, with 3 representations: relational, JSON and XML. The model contains a very simple relationship between the people and their hobbies. In relational model (see Fig. 2) [5], this requires an intermediate relation which enables to create n-m mapping between these entities. The main concepts of this model are relation names and column names, which is a subset of a Cartesian product of a sequence of sets. A given relation has a name and provides a list of names for the subsets (columns) in the Cartesian product of the sets which are the domain for the relation (table).

Person

id	name
1	John
2	Mary
3	Sam

Hobbies

personId	hobbyId
1	1
2	1
2	2
3	2
3	3

Hobby

id	name
1	sailing
2	photography
3	movies

Fig. 2. Relational representation of the simple model

JSON is a text format used to represent key-value pairs, where keys are always strings, and values can be any of the following types: Number, String, Boolean, Array, Object and `null`. The JSON representation of the same model is presented in Fig. 3. It is more flexible than relational as it allows for nesting multiple subobjects within the properties of the object at the price of duplication of information.

XML (eXtensible Markup Language) is a W3C recommendation [6] which is a tree based model for representing structure data on the Internet. In contrast to JSON, it provides means for specifying unique namespaces for all elements, ordering of the nodes as well as assigning attributes to nodes (unordered). Similarly, the XML representation of our simple model is presented in Fig. 4.

Finally the hypergraph representation of the model is presented below. According to the definition a hypergraph is a pair of sets representing the vertices and edges. The vertex set contains all entities from the model. The *id* properties have been skipped as they are not important in this model.

$$H = (V, E)$$

$$V = \{"John", "Mary", "Sam",$$
$$"sailing", "photography", "movies", "Person", "Hobby"\}$$

```
[{
  id: 1,
  type: "Person",
  name: "John",
  hobby: [{id: 1, type: "Hobby", name: "sailing"}]
},
{
  id: 2,
  type: "Person",
  name: "Mary",
  hobby: [{id: 1, type: "Hobby", name: "sailing"},
          {id: 2, type: "Hobby", name: "photography"}]
},
{
  id: 3,
  type: "Person",
  name: "Sam",
  hobby: [{id: 2, type: "Hobby", name: "photography"},
          {id: 3, type: "Hobby", name: "movies"}]
}]
```

Fig. 3. JSON representation of the model

```
<Person id="1">
  <name>John</name>
  <hobbies>
    <Hobby id="1">
      <name>sailing</name>
    </Hobby>
  </hobbies>
</Person>
<Person id="2">
  <name>Mary</name>
  <hobbies>
    <Hobby id="1"> <name>sailing</name> </Hobby>
    <Hobby id="2"> <name>photography</name> </Hobby>
  </hobbies>
</Person>
<Person id="3">
  <name>Sam</name>
  <hobbies>
    <Hobby id="2"> <name>photography</name> </Hobby>
    <Hobby id="3"> <name>movies</name> </Hobby>
  </hobbies>
</Person>
```

Fig. 4. XML representation of the model

The edge set is divided into subset based on the label of the edge.

$$E = E_{type} \cup E_{hobby} \cup E_{people}$$

$$E_{type} = \{e_1 = (\{"John","Mary","Sam"\}, \{"Person"\}),$$
$$e_2 = (\{"sailing","photography","movies"\}, \{"Hobby"\})\}$$

$$E_{hobby} = \{e_3 = (\{"John","Mary"\}, \{"sailing"\}),$$
$$e_4 = (\{"Mary","Sam"\}, \{"photography"\}),$$
$$e_5 = (\{"Sam"\}, \{"photography","movies"\})\}$$

Finally we have to connect the objects composing the model into a set:

$$E_{people} = \{e_6 = (\{"John","Mary","Sam"\}, \varnothing)\}$$

From this representation, we can see that directed hyperedges represent relations between nodes in such a way that it is not necessary to repeat the nodes in different relations which is the case in the JSON and XML representations. Furthermore, it does not require multiple edges to connect nodes which are the same, as the hyperedge can directly connect all source and target nodes which are in a particular relation.

In general the mapping from these formats to hypergraph model can be achieved as follows. For relational model, each row is simply mapped to a single composite data object with edges representing the columns and their particular values as target data objects. Each relation (i.e. table) can be represented as a set of data objects representing rows. More interesting is the case of foreign key dependecies. In case of relational model it is impossible to directly create $n{:}m$ relations. The mapping of XML data into directed hypergraph can be achieved as follows. All simple tags (containing only values) are converted to simple data objects. All complex tags, which contain children tags are converted to composite data objects. All tag attributes are added to respective data objects using edges. In case of JSON Boolean values can be modelled using Number data object type, Array's by creating lists and `null` values can be achieved using hyperedges with empty head sets. Object values can be directly represented using Composite data objects. One issue is that of namespaces, as the edges created from the JSON key's must be attached to some namespace in order to disambiguate them from other edges.

5   Filess Prototype

Filess prototype is developed currently using available technologies enabling evaluation of the idea on the proof-of-concept level. The implementation is created in Java language. We have evaluated several solutions including [15,21]. HypergraphDB provides a native directed hypergraph databases implemented in Java

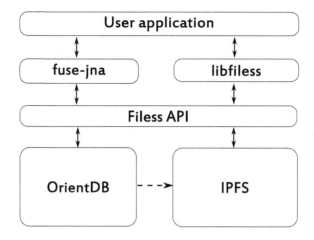

Fig. 5. Architecture of the Filess prototype

with backend based on BerkeleyDB. Finally we chose OrientDB, which is a multi-document database enabling modeling using document, key-value as well as graph paradigms simultaneously. In order to support legacy applications, an intermediate FUSE filesystem plugin was implemented which allows applications to access the information in the form of files which are composed on demand from the underlying graph when applications try to gain access to the data object. The implementation is based on *fuse-jna* Java Fuse provider, which allowed us to use direct OrientDB Java bindings.

Filess exposes an abstract API enabling basic operations on the data objects such as searching, creating and opening, which is available to applications through *libfiless* library. As mentioned above, each user sees the global data space from their own perspective, which is identical on all devices from which they access the system. Current Filess prototype is implemented as an intermediate layer between user applications and distributed graph database backend (see Fig. 5).

Binary, read-only data objects are stored in a separate distributed database called IPFS (Interplanetary File System) [1], which provides efficient hashing and distribution of large binary files between multiple nodes by diving them into blocks and maintaining a tree structure based on the blocks hash values.

6   Conclusions

In this paper we have presented a novel approach to data management and representation, where we propose to abandon the concept of files and filesystems altogether from the future IT infrastructures. Files and filesystems introduce several issues such as unnatural clustering of data, are a barrier for operating systems and generic services to operate directly on data and require users to navigate large hierarchical namespaces.

Presented approach has the potential to enable much more natural access to information, while minimizing the redundancy and data transfer on a global scale, allowing at the same time for highly fine grained access control, not based on files, but on actual data elements, which will enable creation of much more sophisticated and natural computing infrastructures able to handle information processing tasks on a global scale.

Future work will focus on further evaluation of the prototype, design and implementation of security mechanism enabling multiple users to securely share and operate on the global data graph and design of an operating system without filesystem.

Acknowledgment. This research has been funded by Polish National Science Centre grant *File-less architecture of large scale distributed information systems* number: DEC-2012/05/N/ST6/03463.

References

1. Benet, J.: IPFS - content addressed, versioned, P2P file system. CoRR abs/1407.3561 (2014). http://arxiv.org/abs/1407.3561
2. Berge, C.: Hypergraphs Combinatorics of Finite Sets, vol. 45. North-Holland, North-Holland Mathematical Library (1989)
3. Berners-Lee, T., Hendler, J., Lassila, O.: The semantic web. Sci. Am. **284**(5), 34–43 (2001)
4. Chang, F., Dean, J., Ghemawat, S., Hsieh, W.C., Wallach, D.A., Burrows, M., Chandra, T., Fikes, A., Gruber, R.E.: Bigtable: a distributed storage system for structured data. In: Proceedings of the 7th USENIX Symposium on Operating Systems Design and Implementation, OSDI 2006, vol. 7, p. 15. USENIX Association, Berkeley, CA, USA (2006)
5. Codd, E.F.: A relational model of data for large shared data banks. Commun. ACM **13**(6), 377–387 (1970)
6. Cowan, J., Tobin, R.: Xml information set 2nd edn. Technical report, February 2004. http://www.w3.org/TR/2004/REC-xml-infoset-20040204
7. Ferré, S., Ridoux, O.: A file system based on concept analysis. In: Palamidessi, C., Moniz Pereira, L., Lloyd, J.W., Dahl, V., Furbach, U., Kerber, M., Lau, K.-K., Sagiv, Y., Stuckey, P.J. (eds.) CL 2000. LNCS (LNAI), vol. 1861, pp. 1033–1047. Springer, Heidelberg (2000)
8. Gallo, G., Longo, G., Pallottino, S., Nguyen, S.: Directed hypergraphs and applications. Discrete Appl. Math. **42**(2–3), 177–201 (1993)
9. Gantz, J., Reinsel, D.: The digital Universe in 2020: big data, bigger digital shadows, and biggest growth in the far east. International Data Corporation, December 2010
10. Gifford, D.K., Jouvelot, P., Sheldon, M.A., O'Toole Jr., J.W.: Semantic file systems. SIGOPS Oper. Syst. Rev. **25**(5), 16–25 (1991)
11. Grimes, R.: Code name WinFS: revolutionary file storage system lets users search and manage files based on content. MSDN Magazine 19(1) (2004). http://msdn. microsoft.com/msdnmag/issues/04/01/WinFS/

12. Groza, T., Handschuh, S., Moeller, K., Grimnes, G., Sauermann, L., Minack, E., Mesnage, C., Jazayeri, M., Reif, G., Gudjonsdottir, R.: The NEPOMUK project - on the way to the social semantic desktop. In: Pellegrini, T., Schaffert, S. (eds.) Proceedings of I-Semantics 2007, pp. 201–211. JUCS (2007)

13. Horrocks, I., Patel-Schneider, P.F., van Harmelen, F.: From SHIQ and RDF to OWL: the making of a web ontology language. Web Semantics: Science, Services and Agents on the World Wide Web **1**(1), 7–26 (2003). http://www.sciencedirect.com/science/article/pii/S1570826803000027

14. IDC iView,: The Digital Universe Decade - Are You Ready? International Data Corporation, Framingham, MA, USA (2010). http://www.emc.com/digital_universe

15. Iordanov, B.: HyperGraphDB: a generalized graph database. In: Shen, H.T., Pei, J., Özsu, M.T., Zou, L., Lu, J., Ling, T.-W., Yu, G., Zhuang, Y., Shao, J. (eds.) WAIM 2010. LNCS, vol. 6185, pp. 25–36. Springer, Heidelberg (2010)

16. Kryza, B., Pieczykolan, J., Kitowski, J.: Grid organizational memory: a versatile solution for ontology management in the grid. In: e-Science 2006, Second IEEE International Conference on e-Science and Grid Computing, 2006, p. 16 (2006)

17. Kryza, B., Kitowski, J.: Filess - file-less architecture for future information systems. In: 2014 IEEE Fourth International Conference on Big Data and Cloud Computing, BDCloud 2014, Sydney, Australia, pp. 281–282. 3–5 December 2014

18. Kryza, B., Kitowski, J.: Comparison of information representation formalisms for scalable file agnostic information infrastructures. Comput. Inf. **34**(2), 473–494 (2015)

19. Kryza, B., Slota, R., Majewska, M., Pieczykolan, J., Kitowski, J.: Grid organizational memory-provision of a high-level grid abstraction layer supported by ontology alignment. Future Gener. Comput. Syst. **23**(3), 348–358 (2007)

20. Mylka, A., Mylka, A., Kryza, B., Kitowski, J.: Integration of heterogeneous data sources in an ontological knowledge base. Comput. Inf. **31**(1), 189–223 (2012)

21. Orien Technologies: OrientDB project website. http://www.orientechnologies.com

22. Padioleau, Y., Ridoux, O.: A logic file system. In: Proceedings of the General Track: 2003 USENIX Annual Technical Conference, San Antonio, Texas, USA. pp. 99–112. 9–14 June 2003

23. Pawlak, Z.: Rough set approach to knowledge-based decision support. Eur. J. Oper. Res. **99**, 48–57 (1995)

Using Akka Actors for Managing Iterations in Multiscale Applications

Katarzyna Rycerz[1]([✉]) and Marian Bubak[1,2]

[1] Institute of Computer Science, AGH, Al. Mickiewicza 30, 30-059 Kraków, Poland
{kzajac,bubak}@agh.edu.pl
[2] Academic Computer Centre – CYFRONET, Nawojki 11, 30-950 Kraków, Poland

Abstract. In this work we investigate possibilities given by the modern technologies as Akka actor platform and Docker toolkit for managing execution of multiscale applications. We present fine-grained (i.e. iteration) level approach to this problem which is rather difficult to address in standard HPC environments with coarse-grained queuing systems. We propose the solution that uses Docker containers as wrappers for computational modules that are components of multiscale applications. This approach allows for a fine-grained assignment of resources to such modules. We also investigate possibilities of Akka actor framework for building distributed steering systems for such applications with the REST protocol offered by the Spray toolkit. Results of an experiment validation shows that thanks to Docker features, sheduling application on a fine graned iteration level is possible. They also show that using Akka dispatcher as a scheduler for computational tasks (iterations) is a promising approach; moreover the Akka and Spray toolkits do not introduce a significant overhead.

Keywords: Multiscale applications · Actors · Akka · Iteration graphs

1 Introduction

Multiscale applications are used in various fields of science such as biomedicine [25], material science [27] or astrophysics [20]. Different coupling templates of multiscale applications were identified and described in [4] which was a motivation for elaboration building and execution tools [6,24]. In this paper, we focus on support for execution of iterations of multiscale applications. We present a fine-grained (iteration level) approach to the problem which is difficult to address due to the coarse grained nature of HPC queuing and reservation systems.

Figures 1 and 2 show general structures of macro-micro interactions: in Fig. 1, which presents a concurrent execution in a loop, after each iteration the Macro A1 module sends data to the Micro B1 counterpart and waits for the results. The values from iterations i, which comprise internal states of the Macro A1 and Micro B1 modules, are then used in their iterations $i + 1$.

In Fig. 2, which presents execution with dynamic creation of modules, in each iteration Macro A2 model creates several Micro B2 modules, sends data to them

© Springer International Publishing Switzerland 2016
R. Wyrzykowski et al. (Eds.): PPAM 2015, Part I, LNCS 9573, pp. 332–341, 2016.
DOI: 10.1007/978-3-319-32149-3_32

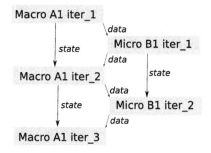

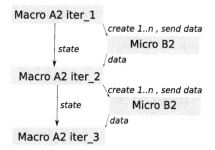

Fig. 1. Typical structure of interactions between micro and macro scale modules in the multiscale simulations - a cuncurrent execution

Fig. 2. Typical structure of interactions between micro and macro scale modules in the multiscale simulations - a dynamics creation of modules

and waits for the results. In that case, Micro B2 module is each time created with different initial values and finishes by sending final values to Macro A2 module. In both cases, if the macro module needs to use different resources than micro module, which is often the case, there is an issue of the idle time each module has to wait wasting resources. In this work, we present results that are steps towards the solution that allows to more efficiently use resources during these idle times [21]. In particular, we focus on the possibilities of new technologies: Akka actors [2] and Docker [28] for that purpose.

Akka is a modern toolkit that implements actor model firstly introduced in [1]. It is a mathematical model of concurrent computation with entities called actors as the universal primitives of concurrent computation. Actors communicate with asynchronous messages: after receiving a message they can perform actions which include: making local decisions, creating more actors, sending more messages, and/or determining how to respond to the next message received. The actor model is used not only for a theoretical understanding of computation, but also as a basis for practical implementations of concurrent systems like Akka. Akka is a modern technological solution that uses actor model to support concurrent, distributed, and resilient message-driven applications on the Java Virual Machine.

Docker framework [28] allows a user to wrap up an arbitrary piece of software in an separate container that contains all required dependencies. Then, the container can run in different execution environment. From the perspecive of multiscale application developer, Docker is very useful for legacy modules of multiscale applications with many dependences such as numerical libraries etc. Additionally, Docker containers can be easly steered from outside, they can be monitored and reassigned resources during runtime. Therefore, as we show in this paper, it is possible to schedule multiscale applications on the iteration level.

This paper is organized as follows. In Sect. 3 we summarize Akka features that can be used for efficient management of multiscale applications on the iteration level. In Sect. 4 we present the overall concept of this management system, Sect. 5 presents the experiment and its results of applying Akka lightweight actors and dispatcher solution for scheduling calculations of artifical application performing LU factorization in iterations. We conclude in Sect. 6.

2   Related Work

Many tools were developed to support programming multiscale applications: there are environments as Multiscale Application Designer [22] that cooperates with GridSpace [6] to support composing and running multiscale applications as in-silico experiment. Another tool is MUSCLE [11] that allows to build multiscale simulations from modules executed concurrently and communicating directly, there is also well known MPI standard or frameworks built above it like MCT [15] or AMUSE [20]. There are also solutions that build modules using specialized scientific simulation software (e.g. LAMMPS molecular dynamics simulator [14], Gaussian electronic structure programm [9] or MATLAB [17] numerical computing environment) and combine them through files. Additionally, the distributed multiscale applications often use parallel programming solution like MPI. There are also available other parallel libraries such as Charm++ [13], Chapel [5] or data processing libraries like Sector/Sphere [10]. Thanks to Docker usage our solution is independent on specyfic application dependencies and software usage.

Up to our best knowledge, all these propositions focus on multiscale applications that can be described as a set of connected single-scale modules, i.e. modules that implement single-scale process models [12]. Therefore, the existing support is on the coarse-grained level with single scale modules as grains. However, the analysis of such applications [4] showed that most of the single scale modules that comprise multiscale applications can be described as set of *send-calculate-receive* iterations executed in a loop.

We propose to support execution of multiscale application on iterations granurality. In this paper, we focus on scheduling of such iterations, which is practically not possible in available coarse-grained HPC resource manager systems like PBS [18] or LSF [16]. Such systems are deployed and ready to use in a number of by e-infrastructures as PLGrid [19]. The main reason is that running each iteration as a separate job is unpractical and would produce significant overhead as the average waiting time in a resource manager queue on production systems are longer that idle time between iterations in typical multiscale applications.

3   Analysis of Akka Features

As a technology we have chosen Akka framework [2], because of its ability to dynamic creation of many leightweight entities called actors. In this section we summarize our analisys of Akka possibilities as toolkit and runtime designed for building concurrent, distributed, message-driven applications for managing multiscale applications on the iteration level. The basic Akka features and their possible support for execution management are shown in Table 1 and include:

Table 1. Akka features and their possible support for multiscale application execution management.

Akka features	Possible execution support for multiscale applications
Dynamic creation of many lightweight actors	Possibility of modeling a single iteration as one actor
Actor behaviour and state mechanisms	Possibility of modeling a behaviour of single scale modules as a state machines
Asynchronous communication between actors	Effective steering application execution by messages
Scalability by akka routing concept	Effective communication between one to many connection of modules (e.g. one macro module with many micro modules)
Managing actor execution by dispatcher and mapping actors to thread pools	Managing calculation iterations using dispatcher concepts
Support for distributed actors with REST protocol (Spray toolkit [26])	Support for bulding distributed multiscale applications
Actor persystence and supervision	Support for fault tolerance mechanims in distributed applications

implementation of actor model in general [1] that supports: dynamics creation of lightweight actors, programming of actor behaviours (i.e. state) changes and asynchronous communication in concurrent environment. Additionally, Akka supports scalability and fault-tolerance achieved by persistance and supervision mechanisms. Moreover, Akka can be used with Spray toolkit [26] designed for building REST/HTTP-based integration layers on top of actors. We particulary focus on investigating possibility of modelling single iterations as separate actors and management of such wrapped computing tasks by a custom dispatcher used as a scheduling (queuing) mechaninisms. We also investigate possibilities given by Spray toolkit that can be used for bulding REST-based distributed multi-scale applications management system. The support for standard networking communication like REST/HTTP is especially important if we want to manage execution of distributed application, where each single scale module needs different resources often located in different e-infratructures with their own access policies.

4   Management of Multiscale Applications

Basing on generic structures presented in Figs. 1 and 2 we propose a management system for multiscale applications that schedules execution on the iteration level and uses both Akka and Docker features (Fig. 3).

The goal of *Resource Manager* is to group modules of similar demands on the same resources. This allows to efficiently manage idle times between their

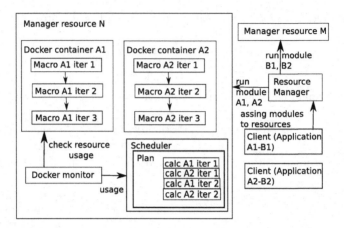

Fig. 3. Architecture of the system grouping and scheduling calculations according to resource requirements.

iterations: the modules do not necessary have to belong to the same application, so the module of one application can be scheduled to execution while the module from the other application waits for the results from the other components (on other resources) of its own application. Docker containers are used to separate environments of each module. The system monitors containers resources usage. Basing on the *Monitor* information, the *Scheduler* changes shares of the resources between containers to realize the plan: one module is executed in the idle time of the others. The *Managers*, *Monitor* and *Scheduler* and implemented as Akka actors the communication of *Managers* and *Clients* with the Spray toolkit.

5   Results of Experiments

In this section we present results of two different experiments that test various management aspects of applications with communication structure as in the Figs. 1 and 2. As a basic module for our test application we used implementation of LU factorization algorithm with parametrized matrix size which was useful to steer CPU time used by the simulation. Next we built artificial application from such core computing modules. LU factorization was implemented with lapack library routine wrapped by Ruby linalg library run using Scala.sys.process package.

5.1   Minimizing Idle Times by Runtime Management of Docker

In the first experiment, we check the ability of using Docker for iteration level scheduling. We have built two separate applications, each having two basic modules A and B communicating using distributed Ruby simulating behaviour of macro and micro module of typical multiscale application with communication scheme as described in Sect. 1. In each iteration, module B asked module A to perform calculation (LU factorization of 3000 x 3000 matrix) and send back the short summary of result. After receiving the answer, module B performed its own calculation (also LU factorization). The application consisted of 10 such iterations. Therefore, we build two such applications with modules A1, B1 and A2, B2. Each module was separately wrapped in Docker container. We have performed our experiments on four cores, each Intel Core i7-2600 CPU 3.40 GHz. Thanks to Docker features we were able to assign each module to a specyfic core and monitor cpu usage.

We have compared three different cases:

1. Separate run of single application as a reference run,
2. Run of two applications at the same time, modules A1 and A2 sharing the same core, modules B1 and B2 on separate cores without additional scheduling,
3. Run of two applications at the same time, modules A1 and A2 sharing the same core with scheduling algorithm, modules B1 and B2 on separate cores.

The idea behing our scheduling algorithm was to assign more CPU shares to that A module that uses more CPU. In this way we force the other A module to wait until the first module finishes its iteration and to "jump" into the idle time of the first module. The scheduling algorithm we use was as follows:

1. check cpu usage of module A1 and A2
2. if cpu usage of A1 is greater then cpu usage A2 then assign all CPU shares to the Docker container containing A1
3. if cpu usage of A2 is greater then cpu usage A1 then assign all CPU shares to the Docker container containing A2

In the case without scheduling where both A1 and A2 modules equally shared CPU, their iterations took twice as much time as running separately. Then, after finishing their iteration both of them waited for their corresponding modules B1 and B2 wasting the time. The results are shown in the Table 2. For the avarage time of a single iteration for all modules the $\sigma = 0.5$, for the total execution time for all modules the $\sigma = 10$. The experiment was run 10 times. The results show that scheduling algorithm reduced execution time of both A1 and A2 execution modules sharing the same resource (CPU) as expected. It also

Table 2. Comparison of running two test applications with and without scheduling algorithm.

	Avr iteration time, s				Total avr time, s			
	A1	A2	B1	B2	A1	A2	B1	B2
Reference run	16.1	16.1	17.3	17.3	316	316	325	325
Without scheduling	30.6	29.9	17.9	18.1	474	481	485	484
With scheduling	18.3	16.9	17.6	17.1	392	414	360	343

shows that thanks to Docker features sheduling application on a fine graned iteration level is possible.

5.2   Testing Akka Dispatcher

The architecture of the second experiment is shown in the Fig.4. For the experiment, we have used LU algorithm with matrix size set to 1600. Such computing module was wrapped as an actor that was created as a result of receiving *calculate* POST message. The actor creation was steered by a client communicating with the wrapping service. After each request for an independent calculation, the manager actor inside a wrapping service created a single calculation actor only for this CPU intensive calculation. The quening mechanism was build on Akka customized dipatcher with *thread-pool-executor* type of executor with settings of one thread per core. Since each single calculation request was realized by the separate actor, their execution was scheduled by the dispatcher in the way assuring one cpu-intensive calculation per core, one after another. In this way one computational intensive task did not interfere the other which was reflected by its execution time which was fixed (around 1.5 s) and independent of the number of calculation requests. The communication between client and wrapping service was realized by REST protocol. The experiment was performed with Akka 2.3.11 for Scala 2.11 on Intel Core i7-2600 CPU @3.40 GHz. This scenario should be useful for running independent computational tasks that belong to different applications, but are grouped on the same resource in order to use idle times of the others. The results are shown in the Fig. 5 where we have compared execution of the varied (1–10000) number of calculations on the same CPU resource by running them concurrently (naive method) and one after another with Akka-based queueing mechanisms. As can be seen, the execution time of a single tasks is fixed for Akka-based solution and grows lineary for naive solution. The figure shows average results from 10 runs; for each result, σ does not exceeds 8 percent.

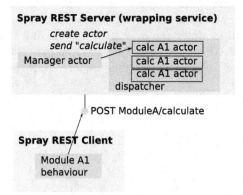

Fig. 4. The architecture of the experiment of using Akka actors and dispatcher for quening calculations.

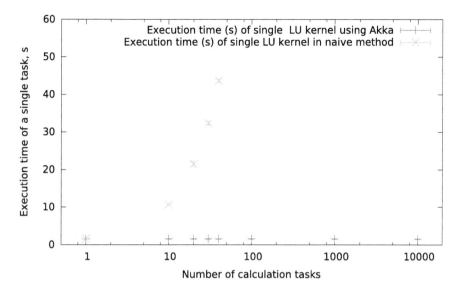

Fig. 5. Comparision of execution time of a single calculation using the Akka dispatcher and naive method.

The presented results show that scheduling computations by Akka dispatcher is a promising approach and can be used for iteractions of multiscale application modules.

6   Conclusions and Future Work

The presented experiments confirmed that using Docker and Akka technologies, we should be able to build the system that improves execution of multiscale applications consisting the modules that execute set of *send-calculate-receive* iterations in a loop. The usage of Docker allows not only to wrap legacy application module with its depencencies, but also to reassign resources it uses during runtime, which, as was demonstrated in the experiment, allows to efficiently schedule its execution. The results are also promising for building the Akka-based framework managing execution of the applications consisting of distributed modules communicating in iterations. The results showed that scheduling calculations using Akka dispatcher is possible with acceptable overhead. This would not be possible with PBS queue systems on much coarse-grained level where each module is threated as a single grain [3]. In our opinion, thanks to their expressiveness the new languages and technologies like Akka are going towards the same role as Problem Solving Environments in the past [8].

In the future, we plan to extend our experiments according to the full structure of multiscale applications. Additionally, as a resources, apart from HPC available by PLGrid Core services [7], we also plan to use cloud resources and mechanism of lightweight virtualization containers. The main challenges are to build a solution for a set of legacy multiscale applications [23].

Acknowledgements. The research presented in this paper has been partially supported by the European Union within the European Regional Development Fund program no. POIG.02.03.00-12-137/13 as part of the PLGrid Core project. It was also partly supported by the AGH grant no 11.11.230.124

References

1. Agha, G.: Actors: a model of concurrent computation in distributed systems, Series in artificial intelligence. MIT Press **11**(12), 12 (1986)
2. Allen, J.: Effective akka. O'Reilly Media (2013). http://www.worldcat.org/isbn/1449360076
3. Borgdorff, J., Bona-Casas, C., Mamonski, M., Kurowski, K., Piontek, T., Bosak, B., Rycerz, K., Ciepiela, E., Gubała, T., Harezlak, D., Bubak, M., Lorenz, E., Hoekstra, A.G.: A distributed multiscale computation of a tightly coupled model using the multiscale modeling language. In: Procedia Computer Science, vol. 9, pp. 596–605 (2012). http://www.sciencedirect.com/science/article/pii/S1877050912001858
4. Borgdorff, J., Falcone, J.L., Lorenz, E., Bona-Casas, C., Chopard, B., Hoekstra, A.G.: Foundations of distributed multiscale computing: formalization, specification, and analysis. J. Parall. Distrib. Comput. **73**(4), 465–483 (2013)
5. Chamberlain, B.L., Callahan, D., Zima, H.P.: Parallel programmability and the chapel language. Int. J. High Perform. Comput. Appl. **21**(3), 291–312 (2007)
6. Ciepiela, E., Harezlak, D., Kocot, J., Bartyński, T., Kasztelnik, M., Nowakowski, P., Gubała, T., Malawski, M., Bubak, M.: Exploratory programming in the virtual laboratory. In: Proceedings of IMCSIT 2010, Wisla, Poland (2010)
7. Core, P.: PLGrid Core Home Site (2014). http://www.plgrid.pl/projekty/core
8. Gallopoulos, E., Houstis, E., Rice, J.: Computer as thinker/doer: problem-solving environments for computational science. IEEE Comput. Sci. Eng. **1**(2), 11–23 (1994)
9. Gaussian. http://www.gaussian.com/
10. Gu, Y., Grossman, R.L.: R.: Sector and sphere: the design and implementation of a high performance data cloud. philosophical transactions a special issue associated with the uk e-science all hands meeting (2008)
11. Hegewald, J., Krafczyk, M., Tölke, J., Hoekstra, A.G., Chopard, B.: An agent-based coupling platform for complex automata. In: Bubak, M., Albada, G.D., Dongarra, J., Sloot, P.M.A. (eds.) ICCS 2008, Part II. LNCS, vol. 5102, pp. 227–233. Springer, Heidelberg (2008)
12. Hoekstra, A., Kroc, J., Sloot, P. (eds.): Simulating Complex Systems by Cellular Automata. Understanding Complex Systems. Springer, Heidelberg (2010). http://springer.com/9783642122026
13. Kale, L.V., Krishnan, S.: Charm++: a portable concurrent object oriented system based on c++. In: OOPSLA 1993, Proceedings of the Eighth Annual Conference on Object-oriented Programming Systems, Languages, and Applications, NY, USA, pp. 91–108. ACM, New York (1993). http://doi.acm.org/10.1145/165854.165874
14. LAMMPS: large-scale atomic/molecular massively parallel simulator (2011). http://lammps.sandia.gov/
15. Larson, J., Jacob, R., Ong, E.: The model coupling toolkit: a new Fortran90 toolkit for building multiphysics parallel coupled models. Int. J. High Perform. Comput. Appl. **19**(3), 277–292 (2005)
16. LSF: Platform Computing. Platform Load Sharing Facility (LSF). http://www.platform.com/

17. Matlab. http://www.mathworks.com/products/matlab
18. PBS: Torque web page. http://www.adaptivecomputing.com/products/open-source/torque/
19. PLGrid: plus home site (2014). http://www.plgrid.pl/en
20. Zwart, S.P., Mcmillan, S., Harfst, S., et al.: A multiphysics and multiscale software environment for modeling astrophysical systems. New Astron. **14**(4), 369–378 (2009)
21. Rycerz, K., Bubak, M.: Managing iterations graphs in multiscale applications using Akka. In: KU KDM 2015 Eighth ACC Cyfronet AGH Users Conference, 11–13 March 2015
22. Rycerz, K., Bubak, M., Ciepiela, E., Harelak, D., Gubaa, T., Meizner, J., Pawlik, M., Wilk, B.: Composing, execution and sharing of multiscale applications. Future Gener. Comput. Syst. **53**, 77–87 (2015). http://www.sciencedirect.com/science/article/pii/S0167739X15002034
23. Rycerz, K., et al.: Enabling multiscale fusion simulations on distributed computing resources. In: Kitowski, J., Wiatr, K., Bubak, M. (eds.) eScience on Distributed Computing Infrastructure. LNCS, vol. 8500, pp. 195–210. Springer, Heidelberg (2014). http://dx.doi.org/10.1007/978-3-319-10894-0_14
24. Rycerz, K., Ciepiela, E., Dyk, G., Groen, D., Gubala, T., Harezlak, D., Pawlik, M., Suter, J., Zasada, S., Coveney, P., Bubak, M.: Support for multiscale simulations with molecular dynamics. In: 2013 International Conference on Computational Science, Procedia Computer Science, vol. 18, pp. 1116–1125 (2013). http://www.sciencedirect.com/science/article/pii/S1877050913004201
25. Sloot, P.M., Hoekstra, A.G.: Multi-scale modelling in computational biomedicine. Briefings Bioinf. **11**(1), 142–152 (2010). http://bib.oxfordjournals.org/content/11/1/142.abstract
26. Spray.io: Web page. http://spray.io
27. Suter, J.L., Anderson, R.L., Christopher Greenwell, H., et al.: Recent advances in large-scale atomistic and coarse-grained molecular dynamics simulation of clay minerals. J. Mater. Chem. **19**, 2482–2493 (2009). http://dx.doi.org/10.1039/B820445D
28. Turnbull, J.: The docker book. Number v1 3 (2014)

Application of Parallel Computing

Synthetic Signature Program
for Performance Scalability

Javier Panadero$^{(\boxtimes)}$, Alvaro Wong, Dolores Rexachs, and Emilio Luque

Computer Architecture and Operating System Department,
Universidad Autonoma of Barcelona, Barcelona, Spain
{javier.panadero,alvaro.wong}@caos.uab.es,
{dolores.rexachs,emilio.luque}@uab.es

Abstract. Due to the complexity of message-passing applications, prediction of the scalability is becoming an increasingly complex goal. To
make an efficient use of the system, it is important to predict the application scalability in a target system. Based on prediction models, such
as PAS2P (Parallel Application Signature for Performance Prediction),
we propose to create a Synthetic Signature (SS) program that allows us
to predict the application performance using a limited set of resources
and in a bounded analysis time. The SS uses the Scalable Logical Traces
(SLT) as input, containing the relevant behavior of the communications
and compute of the application. We model this information given by the
process's small-scaled PAS2P signatures to generate a Scaled Trace for
N number of processes. Basically, the SS will be executed per iterations
in order to obtain the performance prediction. The prediction error was
3.59 % for all applications tested using 4 nodes of the system.

Keywords: Scalability prediction · MPI application · HPC systems

1   Introduction

Due to the complex interaction between parallel applications and HPC systems,
many applications may suffer performance inefficiencies when they scale. In order
to achieve an efficient use of the system, it is critical to predict the performance
application before executing it with a large number of resources.

Benchmarks are used to measure the application performance. They provide
the users with information about the performance over system resources and
never give information about what would happen if the user scales the application using more processes without executing.

Parallel applications consist of a set of phases (segments of code) that are
repeated throughout the application [1]. The phases were written in the application code using specific communication and computing patterns, which follow determined behavioral rules. To identify these phases in a transparent and

J. Panadero—This research has been supported by the MINECO (MICINN) Spain
under contract TIN2011-24384.

Fig. 1. Generating the PAS2P Signature using PAS2P tool.

automatic way, we use the PAS2P tool [2]. As shown in Fig. 1, for a parallel application, PAS2P generates its Signature (instrumented executable) containing the relevant phases (segments of parallel code delimited by communication events) and their repetition rates (weights). Then it predicts the application performance on target machines.

It is important to note that the Signature will predict only for the number of processes that will be constructed. If we want to know the application performance of a larger number of processes without using all the system resources, we need to make an analysis and a model of the application based on the number of processes. This is why we create the Parallel Program Prediction Scalability (P3S) model [3]. P3S models the information given by the process's small scaled signatures, which is to say, it models the communication pattern (phase communications), the computation pattern (instructions of the phase) and the weight of each phase creating a Scalable Logical Trace (SLT) for N number of processes.

To create the SLT, we checked that when the application increases its number of processes, the number of the application phases remains constant, but its patterns change their behavior following behavioral rules, being functionally constant. The first step is to model the phase structure it will have for N number of processes. As shown in Fig. 2, P3S extracts the phase structure taking into account the communication pattern, the computation pattern and the weight. In a second step, we model the computing time behavior curve for each phase, which will be explained in the following sections.

In this global context, we propose a method that allows us to predict the performance of MPI parallel applications with more processes than the number of resources used to predict the application performance. The method is focused on generating the Synthetic Signature (SS), that will use the SLT as an input. Once processed, this allows us to predict the application performance using the number of processes that the SLT was created for.

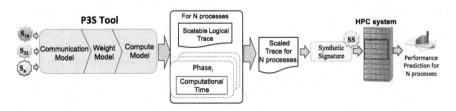

Fig. 2. P3S generates the SLT and the computational time for N processes, used by the SS to predict the application performance in a small set of system resources.

We focus on explaining the SS program using a reduced number of system resources, to predict the application performance for a greater number of processes. For all the tested applications, the prediction error was less than 3.59 %.

This paper is organized as follows: Sect. 2 presents the related work, Sect. 3 presents the P3S tool, Sect. 4 presents the SS Program, Sect. 5 presents the experimental validation and finally Sect. 6, the conclusions and future work.

2   Related Works

There are other works related to the automatic generation of codes to predict the application performance. Xu et al. [4] propose a method to generate an application skeleton with which to predict the performance application, while Xu et al. [5,6] propose generating a communication benchmark to predict the performance. Both methods are based on the application trace for the number of processes to predict, to generate their proposals. Our proposal differs from these works, because we generate the SS, using small scale executions and we project the application behavior for a larger number of processes. Similar approaches [7,8] are based on reducing the application to a minimal form that preserves key properties of the original one. The difference with our work is that we execute the SS, using a reduced number of cores to predict the application performance.

There are also works based on the use of trace-driven simulators [9–11] to predict the application performance. These types of simulators present as limitation that they have not implemented the hardware characteristics of all HPC systems. Our work differs from these approaches, because we create and execute the SS in the system in order to predict the performance.

3   P3S Tool

The main goal of the tool is to model the parallel application to analyze the strong scalability behavior on a given system, by analyzing the information given

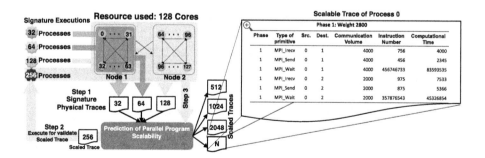

Fig. 3. Scaled Traces generated by P3S tool.

by the execution of a limited set of application signatures, on a small scale, as is shown in Fig. 3.

The tool consists of two steps, a first step of application modeling, where the communication pattern and the weight of each phase of the signature are modeled obtaining the SLT, and a second stage in which we modeled the computation pattern, where the computational time between two communication events is predicted generating the Scaled Trace for N processes.

We checked that the number of differents phases remains constant when we increase the number of processes, which means that the computing work to be carried out is the same, distributed among different number processes, because we are working with strong scalability and it is the same code segment. The communications (number of messages and destinations), the communication volume and the number of instructions can change, modifying only the structure of the phases. To relate the phases of the signature for a different number of processes, P3S uses functional similarity. In the first stage, the P3S tool models general behavior equations of communication, computation and weight of each phase, to project their behaviors for the SLT that we want to generate.

Once the general equations have been modeled, P3S carries out the second stage, where the computational time of each phase is predicted and it is included in the SLT. To predict the computation time, P3S uses a regression-based model by phase, which uses the computational time of each phase of the initial set of PAS2P Signatures as input data. As is shown in Fig. 3, the output of the tool is the Scaled Trace for N number of processes. This trace is composed of the communication events, the number of instructions, the computational time and the weight of each phase.

4   Synthetic Signature Program

In order to achieve our goal, we have focused on how to execute the Scaled Traces generated by the P3S tool in a limited number of system resources. We designed the SS program, which allows us to execute the information of the Scaled Trace by loading ranges of processes. As shown in Fig. 4, the SS program loads range

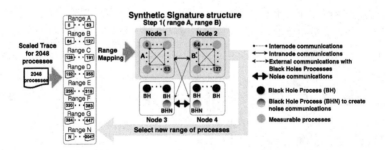

Fig. 4. Synthetic Signature structure.

A that corresponds with the Scaled Traces of process 0 to 63 and range B that corresponds with the processes 64 to 127 (first iteration).

The goal of the SS program is execute the communication events at an appropriate frequency to measure the communication time of the phases to predict the application execution time. To do this, the SS generates the same type of communications (MPI_Datatype), the source and destination of the message and the communication volume which would be generated by the application in its execution. On the other hand, to emulate the computational time given by the Scaled Trace, the SS calls to nanosleep(), function of the operative system, per each event executed.

Due to the limited resources, we define a system configuration to execute the SS program. As shown in Fig. 4, the SS will use Nodes 1 and 2 in the system which we want to predict to measure the communications and Nodes 3 and 4 to solve the communications that depend on the ranges of Scaled Traces. Once the communications are measured, the SS changes the ranges of processes. In order to explain how the SS works, we created the following definitions:

- *Range of Scaled Traces processes (RangesST):* Ranges of processes that will measure their communication events and will be loaded in Nodes 1 and 2.
- *Measurable Process (MeasurableP):* A process that loads a Scaled Trace of a process in order to measure their communications. Some MeasurablePs will communicate with the Black Hole processes when their communications are out of the RangesST processes.
- *Black Hole process (BH):* A process that sends or receives communications from MeasurablePs, the SS creates a number of BH processes depending on the number of MeasurableP that communicate out of the RangesST processes.
- *Black Hole Noise processes (BHN):* Pair of processes that generate the same number of messages in transit on the interconnection network per phase.

The following definitions corresponds to the MeasurableP:

- *Internode communications:* Communications between two processes of the same system node (Node 1 or Node 2). These messages will be measured using timers of the operative system. An example is shown in Fig. 5: Process 0 sends a message to process 63.
- *Intranode communications:* Communications between two processes using the interconnection network from Node 1 to Node 2 or vice versa. An example is shown in Fig. 5: Process 0 sends a message to process 64.

In Fig. 5 we show the next type of messages that correspond to the BH processes, emphasizing that the following communications will not be measured:

- *Control Message:* There are two types of Control Messages. The first type, seen in point A in Fig. 5, when a message is received by BH process from MeasurableP, it means that the MeasurableP is requesting a message with some characteristics. This happens because a MeasurableP needs to receive a message from a process which is not within the established RangesST processes. For this reason, the BH process generates a message with specific buffer size,

MPI datatype, and MPI_TAG for the MeasurableP that requests it, as shown
in point B of Fig. 5. The second type of Control Message occurs when any BH
process sends a message to the BHN of the same Node in order to enable the
noise communications when a phase starts.

- *External Communications (EComms):* The communication is executed
  between the MeasurableP and the BH process, in order to solve the communication dependencies of the phase.
- *Noise Communications:* Messages between two BHN processes. The SS generates the same number of communications in transit, as the application does.

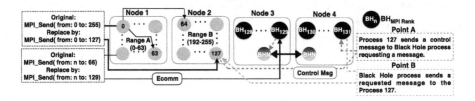

Fig. 5. Communications with the BH processes of the Synthetic Signature.

4.1   Setup the RangesST and Black Holes Processes

Once we have defined the structure of the SS, we have to execute the SS to predict
the application performance. We need to apply the same mapping policies that
the application will use to execute, to do this, these mapping policies have to be
provided to the RangesST processes to execute in each iteration and the SS has
to make n as many iterations as the number of RangesST processes there are.

For every iteration of SS, we need to load different RangesST processes in
the MeasurablePs. As an example, for a total of 256 processes, if we are going
to execute the SS in a system with 64 core nodes, the SS will use two nodes of
64 cores for the MeasurablePs and two more nodes for the BHs processes. For
each iteration, the SS will make 4 RangesST processes, whose ranges values are
(startProcess–EndProcess): 0–63, 64–127, 128–191 and 192–256.

To update the RangesST processes after measures, the MeasurablePs will
load different Scaled Traces of processes and the SS will modify the message
destination which will depend on the loaded RangesST processes. For example,
if we launch the SS using 2 nodes of 64 cores, the MPI world contains 128
processes (from 0 to 127 process). Now, if we select the RangesST processes
from 0 to 63 it will be executed in Node 1, and for the other RangesST processes
from 192 to 256 it will be executed it in the Node 2. In order to communicate
the processes from both RangesST, the messages' destinations will be labeled
with processes belonging to the MPI world, using the MPI_Comm_Rank of the
SS, as shown in Fig. 5.

Once we have selected the range of processes to execute in current iteration,
the Scaled Traces of each process selected have to be analyzed in order to know

the total number of the EComms (out of the RangesST processes) and the total number of BH processes.

To select the number of BH processes to be created, we use the Eq. 1, where we create one BH process for each MeasurableP that has an EComm. The destination of these messages will be changed to the rank of the BH process, as shown in Fig. 5. In the worst case, we would have to create the same number of MeasurableP for each BH process.

$$Number\ of\ BH\ Processes = \sum_{i=1}^{m}(MeasurableP\ with\ EComms_i) \qquad (1)$$

4.2   Execution of the SS Program

In this section, we describe how the SS is executed with the selected resources in the previous section to obtain the execution time of each phase in a bounded time. After this, we used the PAS2P Eq. 2 to predict the application performance for the number of processes for which the Scale Trace was generated.

We illustrate the procedure we followed in order to execute the SS with an example in Fig. 6. As we can see, we used a system with 64 core nodes and a Scaled Trace generated for 512 processes, thus, we have 8 RangesST processes. We have executed the SS with 192 processes using 4 nodes of 64 cores (limited resources), 128 MeasurableP and 64 BH processes (32 BH processes in Node 3 and Node 4). As we have said before, the number of the BH processes has been selected in function of the Ecomms of the RangesST processes.

It is noteworthy that before we start the measures, the SS executes one first iteration (step 0) to warm-up all the machine components. In this example, we focus on phase 1 of process 0. As we can see in Fig. 6, for the first iteration (step 1), we have selected the RangesSTs processes 0–63 for Node 1, and 64–127 for the Node 2. We have selected the same mapping processes, in function of the application mapping used, to execute the application with all the resources (512 processes) and seeking to maintain the physical distances of the interconnection network. To measure the communication events, we start when the communication event occurs in one MeasurableP, and we stop when the same communication event ends.

The step 1, as seen in the table contained in Fig. 6, all events start with the flag "Executed" as false. The same figure shows that the events with the ID 3, 4 and 5 exchange information with the process 255 (Ecomms). Since process 255 is not involved in the RangesST processes of current iteration, we changed the destination of these messages to the BH process, which in this case (the BH process 128), is allocated in Node 3. The objective is to execute the Ecomms but not measure them, in order to create the same number of messages in transit. If we need more messages, the BHNs create the missing messages to reflect the same number of messages on the application.

The communication events with the ID 0, 1 and 2 exchange messages with the process 64 (intracommunications), therefore, we execute these communications to measure their execution time and it is saved in the event structure, and we

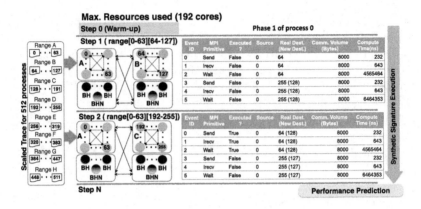

Fig. 6. Execution of the Synthetic Signature

change the flag "Executed" as true, seen in the Executed column of the table contained in Fig. 6 (see step 2). Additionally, whenever an event is executed, we emulate the computational time of the application (see Compute Time column in table contained in Fig. 6) using the nanosleep() function of the OS.

When all the communications events have been executed, the SS checks if all the communication events have been measured. Since there are 3 events which have not been measured in the iteration, we need to carry out a new iteration. As we can see in Fig. 6, for this iteration (step 2), we selected a new RangesST processes (192 to 255) for the Node 2. We have to identify the external communications and assign them a BH process. In this second iteration, since the first 3 communications have already been measured in the previous iteration (step 1), they are executed but not measured. In this iteration, these communications are Ecomms, thus, they exchange information with the BH process.

Once all the communications events have been measured. To obtain its predicted execution time, we follow the procedure until all the processes have been measured. Then, we select the physical process of the SS with a higher execution time in order to apply the PAS2P Eq. 2, so as to obtain the Predicted Execution Time, where PET is the Predicted application Execution Time, m is the number of phases, $PhaseETi$ is the Phase i Execution Time and Wi is the weight of the phase i.

$$PET = \sum_{i=1}^{m} (PhaseET_i)(W_i) \tag{2}$$

5   Experimental Validation

In this section we describe the experimental framework for the performance prediction we carried out, using limited system resources in order to validate that the SS allows us to predict the application execution time. Then, in the report, we explain the results of experiments which compare our predictions with the PAS2P Signature and the performance of real programs.

Table 1. Prediction for Sweep3D using the SS and the PAS2P Signature.

Phase ID	Weight (W)	Synthetic Signature PhaseET (Sec.)	(PhaseET)* W(Sec.)	PETE (%)	PAS2P Signature PhaseET (Sec.)	(PhaseET)* W(Sec.)	PETE (%)	AET (Sec.)
0	100	10.9437	1094.37		10.0477	1004.77		
1	100	0.1209	12.09		0.1158	11.58		
2	99	14.9401	1479.13		14.8701	1472.13		
3	100	10.9942	1099.42		10.1899	1018.99		
4	100	10.799	1079.90		10.729	1072.90		
		PET:	4764.91	4.16	PET:	4580.37	0.13	4574.32
PhaseET: Phase Execution Time					PET: PredictedExecution Time			
PETE: Predicted ExecutionTime Error					AET: Application Execution Time			

To carry out the experimental validation we used Cluster Dell, whose architecture is composed by 8 node with 64 cores AMD Opteron 6262 with 48 GB RAM interconnected by ConnectX IB Mellanoxcard. We are presenting the results obtained for the CG, BT and SP from the NPB using Class E with 1000 iterations, Sweep3D using an input 250 with 100 iterations, and N-body using 4,000,000 atoms. In order to evaluate the prediction quality, we repeat each experiment 30 times to obtain the arithmetic average.

As we said before, one of our goals is using limited system resources. In Table 2, the "System Cores" columns show a comparison between the number of cores used by the SS and the number of cores used by the PAS2P Signature to predict performance. It is important to note that the SS uses less than half of the system resources (512 cores). For all the SS executions, we used 128 MesurableP and 91 BH processes at most, this means, 4 Nodes from Cluster Dell.

To obtain the Prediction Execution Time (PET), we execute the SS and PAS2P Signature on Cluster Dell to measure the execution time of each phase and multiply it by its weight. Then, we execute the whole application to compare the PETs with the Application Execution Time (AET) to obtain the Prediction Execution Time Error (PETE). As shown in Table 1, to predict the execution time of Sweep3D, we execute the SS in order to obtain the PhaseET. After obtaining these times, the SS multiplies the PhaseET by the weight of each phase to obtain the PET. We carried out the same procedure with the PAS2P Signature, with the difference that the PAS2P Signature uses the resources that the application will need to predict.

Table 2 shows the predictions obtained using, on one hand, the SS and on the other hand, the PAS2P Signature. When comparing columns 2 (SYET) and 5 (SET), which represent the execution time obtained when using the SS and the PAS2P Signature to obtain the PET, the results are similar because the Signatures have the same number of phases to be measured. When we compare these times with the last column (AET), it can be seen how they have been significantly reduced when compared with the AET, therefore, we are validating

Table 2. Prediction using the SS and the PAS2P Signature.

Program	Synthetic Signature				PAS2P Signature				
	SYET (Sec.)	PET (Sec.)	PETE (%)	System Cores	SET (Sec.)	PET (Sec.)	PETE (%)	System Cores	AET (Sec.)
SP	133.98	6927.00	0.98	219	134.98	6901.51	1.34	484	6995.66
BT	221.84	11642.86	2.46	219	229.24	11992.94	0.46	484	11937.58
N-Body	184.92	5090.77	1.77	130	192.23	5009.26	3.34	512	5182.65
Sweep3D	308.02	4764.91	4.16	160	307.12	4580.37	0.13	512	4574.32
CG	578.34	11108.52	8.59	160	514.47	10648.84	4.10	512	10229.20
SYET: Synthetic Execution Time									

that the signature represents a small fraction of the AET. Columns 3 and 7 show the Predicted Execution Time (PET) for each Signature. Finally, columns 4 and 8 (PETE) present the Prediction Execution Time Error regarding the AET.

Overall, the results show that we can predict the application performance using a limited set of resources through the SS. For these results, the SS represents 3.94 % of the AET. We also see that the prediction quality of the SS has an average accuracy of over 3.59 %.

6   Conclusions and Future Work

We propose a method to construct a Synthetic Signature which executes the Scaled Trace obtained by the P3S tool. The SS allows us to analyze and predict strong scalability of parallel applications on a given system, which strives to use a reduced set of resources. The method is focused on generating the SS of the application, containing the application phases and their weights. The SS is executed by ranks of processes, in an iterative way, using reduced system resources in order to predict the application performance.

For future work, due to the high number of time-constraint applications, we consider that it would be very interesting and useful to extend the general methodology in order to be able to predict weak scalability. A limitation of the SS is that it only executes point to point communications. We are currently analyzing how to include the collective communications in the SS.

References

1. Wong, A., Rexachs, D., Luque, E.: Parallel application signature for performance analysis and prediction. IEEE Trans. Parallel Distrib. Syst. (TPDS) 26(7), 2009–2019 (2015)
2. Panadero, J., Wong, A., Rexachs, D., Luque, E.: A tool for selecting the right target machine for parallel scientific applications. ICCS 18, 1824–1833 (2013)
3. Panadero, J., Wong, A., Rexachs, D., Luque, E.: Scalability of parallel applications: an approach to predict the computational behavior. In: Proceedings of the International Conference on Parallel and Distributed Processing Techniques and Applications, PDPTA 2015 (2015)

4. Wu, X., Deshpande, V., Mueller, F.: Scalabenchgen: auto-generation of communication benchmarks traces. In: 2012 IEEE 26th International Parallel Distributed Processing Symposium (IPDPS), pp. 1250–1260 (2012)
5. Xu, Q., Subhlok, J., Zheng, R., Voss, S.: Logicalization of communication traces from parallel execution. In: 2009 IEEE International Symposium on Workload Characterization, IISWC 2009, pp. 34–43 (2009)
6. Xu, Q., Subhlok, J.: Construction and elevation of coordinated performance skeleton. In: International Conference on High Performance Computing, pp. 73–86 (2008)
7. Van Ertvelde, L., Eeckhout, L.: Dispersing proprietary applications as benchmarks through code mutation. SIGOPS Oper. Syst. Rev. **42**(2), 201–210 (2008)
8. Zhai, J., Sheng, T., He, J., Chen, W., Zheng, W.: Fact: fast communication trace collection for parallel applications through program slicing. In: Proceedings of the Conference on High Performance Computing Networking, Storage and Analysis, pp. 1–12 (2009)
9. Prakash, S., Bagrodia, R.L.: MPI-SIM: using parallel simulation to evaluate MPI programs. In: Proceedings of the 30th Conference on Winter Simulation, pp. 467–474 (1998)
10. Tikir, M.M., Laurenzano, M.A., Carrington, L., Snavely, A.: PSINS: an open source event tracer and execution simulator for MPI applications. In: Sips, H., Epema, D., Lin, H.-X. (eds.) Euro-Par 2009. LNCS, vol. 5704, pp. 135–148. Springer, Heidelberg (2009)
11. Ridruejo Perez, F.J., Miguel-Alonso, J.: INSEE: an interconnection network simulation and evaluation environment. In: Cunha, J.C., Medeiros, P.D. (eds.) Euro-Par 2005. LNCS, vol. 3648, pp. 1014–1023. Springer, Heidelberg (2005)

FEniCS-HPC: Automated Predictive High-Performance Finite Element Computing with Applications in Aerodynamics

Johan Hoffman[1,2], Johan Jansson[1,2]([✉]), and Niclas Jansson[1]

[1] Computational Technology Laboratory, School of Computer Science
and Communication, KTH, Stockholm, Sweden
{jhoffman,jjan,njansson}@kth.se
[2] BCAM - Basque Center for Applied Mathematics, Bilbao, Spain

Abstract. Developing multiphysics finite element methods (FEM) and scalable HPC implementations can be very challenging in terms of software complexity and performance, even more so with the addition of goal-oriented adaptive mesh refinement. To manage the complexity we in this work present *general* adaptive stabilized methods with *automated* implementation in the FEniCS-HPC *automated* open source software framework. This allows taking the weak form of a partial differential equation (PDE) as input in near-mathematical notation and automatically generating the low-level implementation source code and auxiliary equations and quantities necessary for the adaptivity. We demonstrate new optimal strong scaling results for the whole adaptive framework applied to turbulent flow on massively parallel architectures down to 25000 vertices per core with ca. 5000 cores with the MPI-based PETSc backend and for assembly down to 500 vertices per core with ca. 20000 cores with the PGAS-based JANPACK backend. As a demonstration of the power of the combination of the scalability together with the adaptive methodology allowing prediction of gross quantities in turbulent flow we present an application in aerodynamics of a full DLR-F11 aircraft in connection with the HiLift-PW2 benchmarking workshop with good match to experiments.

Keywords: FEM · Adaptive · Turbulence

1   Introduction

As computational methods are applied to simulate even more advanced problems of coupled physical processes and supercomputing hardware is developed towards massively parallel heterogeneous systems, it is a major challenge to manage the complexity and performance of methods, algorithms and software implementations. Adaptive methods based on quantitative error control pose additional challenges. For simulation based on partial differential equation (PDE) models, the finite element method (FEM) offers a general approach to numerical discretisation, which opens for automation of algorithms and software implementation.

© Springer International Publishing Switzerland 2016
R. Wyrzykowski et al. (Eds.): PPAM 2015, Part I, LNCS 9573, pp. 356–365, 2016.
DOI: 10.1007/978-3-319-32149-3_34

In this paper we present the FEniCS-HPC open source software framework with the goal to combine the generality of FEM with performance, by optimisation of generic algorithms [2, 4, 13]. We demonstrate the performance of FEniCS-HPC in an application to subsonic aerodynamics.

We give an overview of the methodology and the FEniCS-HPC framework, key aspects of the framework include:

1. **Automated discretization** where the weak form of a PDE in mathematical notation is translated into a system of algebraic equations using code generation.
2. **Automated error control**, ensures that the discretization error e = u − U in a given quantity is smaller than a given tolerance by adaptive mesh refinement based on duality-based a posteriori error estimates. An a posteri error estimate and error indicators are automatically generated from the weak form of the PDE, by directly using the error representation.
3. **Automated modeling**, which includes a residual based implicit turbulence model, where the turbulent dissipation comes only from the numerical stabilization, as well as treating the fluid and solid in fluid-structure interaction (FSI) as one continuum with a phase indicator function tracked by a moving mesh and implicitly modeling contact.

We demonstrate new optimal strong scaling results for the whole adaptive framework applied to turbulent flow on massively parallel architectures down to 25000 vertices per core with ca. 5000 cores with the MPI-based PETSc backend and for assembly down to 500 vertices per core with ca. 20000 cores with the PGAS-based JANPACK backend. We also present an application in aerodynamics of a full DLR-F11 aircraft in connection with the HiLift-PW2 benchmarking workshop with good match to experiments.

1.1 The FEniCS Project and State of the Art

The software described here is part of the FEniCS project [2], with the goal to automate the scientific software process by relying on general implementations and code generation, for robustness and to enable high speed of software development.

Deal.II [1] is a software framework with a similar goal, implementing general PDE based on FEM in C++ where users write the "numerical integration loop" for weak forms for computing the linear systems. The framework runs on supercomputers with optimal strong scaling. Deal.II is based on quadrilater (2D) and hexahedral (3D) meshes, whereas FEniCS is based on simplicial meshes (triangles in 2D and tetrahedra in 3D).

Another FEM software framework with a similar goal is FreeFEM++ [3], which has a high-level syntax close to mathematical notation, and has demonstrated optimal strong scaling up to ca. 100 cores.

2   The FEniCS-HPC Framework

FEniCS-HPC is a problem-solving environment (PSE) for automated solution of PDE by the FEM with a high-level interface for the basic concepts of FEM: weak forms, meshes, refinement, sparse linear algebra, and with HPC concepts such as partitioning, load balancing abstracted away.

The framework is based on components with clearly defined responsibilities. A compact description of the main components follows, with their dependencies shown in the dependency diagram in Fig. 1:

FIAT: Automated generation of finite element spaces V and basis functions $\phi \in V$ on the reference cell and numerical integration with FInite element Automated Tabulator (FIAT) [12,13]

$$e = (K, V, \mathcal{L})$$

where K is a cell in a mesh $\mathcal{T}$, V is a finite-dimensional function space, $\mathcal{L}$ is a set of degrees of freedom.

FFC+UFL: Automated evaluation of weak forms in mathematical notation on one cell based on code generation with Unified Form Language (UFL) and FEniCS Form Compiler (FFC) [11,13], using the basis functions $\phi \in V$ from FIAT. For example, in the case of the Laplacian operator

$$A_{ij}^K = a_K(\phi_i, \phi_j) = \int_K \nabla\phi_i \cdot \nabla\phi_j dx = \int_K lhs(r(\phi_i, \phi_j)dx)$$

where A^K is the element stiffness matrix and $r(\cdot, \cdot)$ is the weak residual.

DOLFIN-HPC: Automated high performance assembly of weak forms and interface to linear algebra of discrete systems and mesh refinement on a distributed mesh $\mathcal{T}_\Omega$ [10].

$$A = 0$$
$$\text{for all cells } K \in \mathcal{T}_\Omega$$
$$A \mathrel{+}= A^K$$

$$Ax = b$$

Unicorn: Automated Unified Continuum modeling with Unicorn choosing a specific weak residual form for incompressible balance equations of mass and momentum with example visualizations of aircraft simulation below left and turbulent FSI in vocal folds below right [4].

$$r_{UC}((v, q), (u, p)) = (v, \rho(\partial_t u + (u \cdot \nabla)u) + \nabla \cdot \sigma - g) + (q, \nabla \cdot u) + LS((v, q), (u, p))$$

where LS is a least-squares stabilizing term described in [7].

A user of FEniCS-HPC writes the weak forms in the UFL language, compiles it with FFC, and includes it in a high-level "solver" written in C++ in DOLFIN-HPC to read in a mesh, assemble the forms, solve linear systems, refine the mesh, etc. The Unicorn solver for adaptive computation of turbulent flow and FSI is developed as part of FEniCS-HPC.

FEniCS-HPC

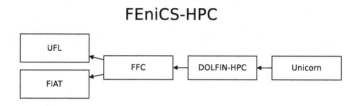

Fig. 1. FEniCS-HPC component dependency diagram.

2.1 Solving PDE Problems in FEniCS-HPC

Poisson's Equation. To solve Poisson's equation in weak form $\int_{\Omega}(\nabla u, \nabla v) - (f, u) = 0 \quad \forall v \in V$ in the framework, we first define the weak form in a UFL "form file", closely mapping mathematical notation (see Fig. 2). The form file is then compiled to low-level C++ source code for assembling the local element matrix and vector with FFC. Finally we use DOLFIN-HPC to write a high-level "solver" in C++, composing the different abstractions, where a mesh is defined, the global matrix and vector are assembled by interfacing to the generated source code, the linear system is solved by an abstract parallel linear algebra interface

```
Q = FiniteElement("CG", "tetrahedron", 1)

v = TestFunction(Q)  # test basis function
u = TrialFunction(Q) # trial basis function
f = Coefficient(Q) # function

# Bilinear and linear forms
a = dot(grad(v), grad(u))*dx
L = v*f*dx
```

```
// Define mesh, BCs and coefficients
PoissonBoundary boundary;
PoissonBoundaryValue u0(mesh);
SourceFunction f(mesh);
DirichletBC bc(u0, mesh, boundary);

// Define PDE
PoissonBilinearForm a;
PoissonLinearForm L(f);
LinearPDE pde(a, L, mesh, bc);

// Solve PDE
Function u;
pde.solve(u);

// Save solution to file
File file(''poisson.pvd'');
file << u;
```

Fig. 2. Poisson solver in FEniCS-HPC with the weak form in the UFL language (left) and the solver in C++ using DOLFIN-HPC (right).

(using PETSc as back-end by default), and then the solution function is saved to disk. The source code for an example solver is presented in Fig. 2.

The Incompressible Navier-Stokes Equations. We formulate the General Galerkin (G2) method for incompressible Navier-Stokes equations (1) in UFL by a direct input of the weak residual. We can automatically derive the Jacobian in a quasi-Newton fixed-point formulation and also automatically linearize and generate the adjoint problem needed for adaptive error control. These examples are presented in Fig. 3.

```
V = VectorElement("CG", "tetrahedron", 1)
Q = FiniteElement("CG", "tetrahedron", 1)

v = TestFunction(V); q = TestFunction(Q)
u_ = TrialFunction(V); p_ = TrialFunction(Q)
u = Coefficient(V); p = Coefficient(Q)
u0 = Coefficient(V); um = 0.5*(u + u0)

# Momentum and continuity weak residuals
r_m = (inner(u - u0, v)/k + \
  ((nu*inner(grad(um), grad(v)) + \
  inner(grad(p) + grad(um)*um, v))))*dx + LS_u*dx
r_c = inner(div(u), q))*dx + LS_p*dx

# Newton's method Ju_i+1 = Ju_i - F(u_i)
a = derivative(r_m, u, u_)
L = action(a, u) - r_m
```

```
# Adjoint problem (stationary part) for r_m
a_adjoint = adjoint(derivative(r_m - inner(u, v)/k*dx, u))
L_adjoint_c = derivative(action(r_c, p), u, v)
L_adjoint = inner(psi_m, v)*dx - L_adjoint_c
```

Fig. 3. Example of weak forms in UFL notation for the cG(1)cG(1) method for incompressible Navier-Stokes equations (left) together with the adjoint problem (right).

3   Parallelization Strategy and Performance

The parallelization is based on a fully distributed mesh approach, where everything from preprocessing, assembly of linear systems, postprocessing and refinement is performed in parallel, without representing the entire problem or any pre-/postprocessing step on a single core.

Initial data distribution is defined by the graph partitioning of the corresponding dual graph of the mesh. Each core is assigned a set of whole elements and the vertex overlap between cores is represented as ghosted entities.

3.1   Parallel Assembly

The assembling of the global matrix is performed in a straightforward fashion. Each core computes the local matrix of the local elements and add them to the global matrix. Since we assign whole elements to each core, we can minimize data dependency during assembly. Furthermore, we renumber all the degrees of freedom such that a minimal amount of communication is required when modifying entries in the sparse matrix.

3.2   Solution of Discrete System

The FEM discretization generates a non-linear algebraic equation system to be solved for each time step. In Unicorn we solve this by iterating between the velocity and pressure equations by a Picard or quasi-Newton iteration [6].

Each iteration in turn generates a linear system to be solved. We use simple Krylov solvers and preconditioners which scale well to many cores, typically BiCGSTAB with a block-Jacobi preconditioner, where each block is solved with ILU(0).

3.3   Mesh Refinement

Local mesh refinement is based around a parallelization of the well known recursive longest edge bisection method [15]. The parallelization splits up the refinement into two phases. First a local serial refinement phase bisects all elements marked for refinement on each core (concurrently) leaving several hanging nodes on the shared interface between cores. The second phase propagates these hanging nodes onto adjacent cores.

The algorithm iterates between local refinement and global propagation until all cores are free of hanging nodes. For an efficient implementation, one has to detect when all cores are idling at the same time. Our implementation uses a fully distributed termination detection scheme, which includes termination detection in the global propagation step by using recursive doubling or hypercube exchange type communication patterns [10]. Also, the termination detection algorithm does not have a central point of control, hence no bottlenecks, less message contention, and no problems with load imbalance.

Dynamic Load Balancing. In order to sustain good load balance across several adaptive iterations, dynamic load balancing is needed. DOLFIN-HPC is equipped with a scratch and remap type load balancer, based on the widely used PLUM scheme [14], where the new partitions are assigned in an optimal way by solving the maximally weighted bipartite graph problem. We have improved the scheme such that it scales linearly to thousands of cores [8,10].

Furthermore, we have extended the load balancer with an a priori workload estimation. With a dry run of the refinement algorithm, we add weights to a dual graph of the mesh, corresponding to the workload after refinement. Finally, we repartition the unrefined mesh according to the weighted dual graph and redistribute the new partitions before the refinement.

4   Strong Scalability

To be able to take advantage of available supercomputers today the entire solver in FEniCS-HPC needs to demonstrate good strong scaling to at least several thousands of cores. For planned "exascale" systems with many million cores, strong scalability has to be attained for at least hundreds of thousands of cores.

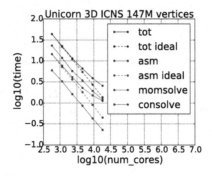

Fig. 4. Strong scalability test for the full G2 method for incompressible turbulent Navier-Stokes equations (assemble linear systems and solve momentum and continuity) in 3D on a Cray XC40.

In this section we analyze scaling results using the PETSc parallel linear algebra backend based on pure MPI and the JANPACK backend based on PGAS.

In Fig. 4 we present strong scalability results with the PETSc pure MPI backend for the full G2 method for turbulent incompressible Navier-Stokes equations (1) (assemble linear systems and solve the momentum and continuity equations) in 3D on a mesh with 147M vertices on the Hornet Cray XC40 computer. We observe near-optimal scaling to ca. 4.6 kcores for all the main algorithms (assembly and linear solves). Going from 4.6 kcores to 9.2 kcores we start to see a degradation in the scaling with a speedup of ca. 0.7, and from 9.2 kcores to 18.4 kcores the speedup is 0.5. It's clear that it's mainly the assembly that shows degraded scaling.

In Fig. 5 we present results for assembling four different equations using the JANPACK backend, where FEniCS-HPC is running in a hybrid MPI+PGAS

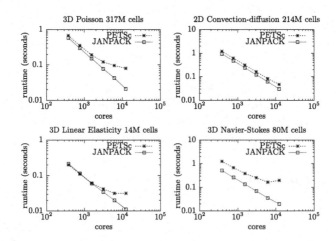

Fig. 5. Sparse matrix assembly timings for four different equations on a Cray XC40.

mode. We observe that for large number of cores, the low latency one-sided communication of PGAS languages in combination with our new sparse matrix format [9] greatly improves the scalability.

5   Unicorn Simulation of a Full Aircraft

In the Unicorn component we implement the full G2 method and fix the weak residual to the cG(1)cG(1) stabilized space-time method for incompressible Navier-Stokes equations (or a general stress for FSI).

In a cG(1)cG(1) method [7] we seek an approximate space-time solution $\hat{U} = (U, P)$ which is continuous piecewise linear in space and time (equivalent to the implicit Crank-Nicolson method). With I a time interval with subintervals $I_n = (t_{n-1}, t_n)$, W^n a standard spatial finite element space of continuous piecewise linear functions, and W_0^n the functions in W^n which are zero on the boundary Γ, the cG(1)cG(1) method for constant density incompressible flow with homogeneous Dirichlet boundary conditions for the velocity takes the form: for $n = 1, ..., N$, find $(U^n, P^n) \equiv (U(t_n), P(t_n))$ with $U^n \in V_0^n \equiv [W_0^n]^3$ and $P^n \in W^n$, such that

$$
\begin{aligned}
r((U, P), (v, q)) &= ((U^n - U^{n-1})k_n^{-1} + (\bar{U}^n \cdot \nabla)\bar{U}^n, v) + (2\nu\epsilon(\bar{U}^n), \epsilon(v)) \\
&\quad - (P, \nabla \cdot v) + (\nabla \cdot \bar{U}^n, q) + LS = 0, \ \forall \hat{v} = (v, q) \in V_0^n \times W^n
\end{aligned} \tag{1}
$$

where $\bar{U}^n = 1/2(U^n + U^{n-1})$ is piecewise constant in time over I_n and LS a least-squares stabilizing term described in [7].

We formulate a new general adjoint-based method for adaptive error control based on the following error representation and adjoint weak bilinear and linear forms with the error $\hat{e} = \hat{u} - \hat{U}$, adjoint solution $\hat{\phi}$, output quantity ψ and the hat signifying the full velocity-pressure vector $\hat{U} = (U, P)$, with $r_G = r - LS$:

$$
(\hat{e}, \psi) = \overline{r'}(\hat{e}, \hat{\phi}) = r_G(\hat{U}; \hat{\phi}) \quad a_{adjoint}(v, \hat{\phi}) = \overline{r'}(v, \hat{\phi}) \quad L_{adjoint}(v) = (v, \psi) \tag{2}
$$

We have used our adaptive finite element methodology for turbulent flow and FEniCS-HPC software to solve the incompressible Navier-Stokes equations of the flow past a full high-lift aircraft model (DLR-F11) with complex geometry at realistic Reynolds number for take-off and landing. This work is an extension of our contributed simulation results to the 2nd AIAA CFD High-Lift Prediction Workshop (HiLiftPW-2), in San Diego, California, in 2013 [5].

In the following results we focus on the angle of attack $\alpha = 18.5°$. To quantify mesh-convergence we plot the coefficients and their relative error compared to the experimental values (serving as the reference) versus the number of vertices in the meshes, and plot meshes and volume renderings of quantities related to the adaptivity in Fig. 6.

We see that our adaptive computational results come very close to the experimental results on the finest mesh, with a relative error under 1 % for cl and cd. For other angles we observe similar results presented in [5].

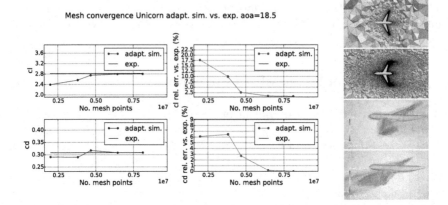

Fig. 6. Plots for the aircraft simulation at $\alpha = 18.5°$. Lift coefficient, cl, and drag coefficient, cd, vs. angle of attack, α, for the different meshes from the iterative adaptive method (left). Slice aligned with the angle of attack showing the tetrahedra of the starting mesh versus the finest adaptive mesh (top right). Volume rendering of the velocity residual and adjoint velocity magnitude (bottom right).

6   Summary

We have given an overview of the general FEniCS-HPC software framework for automated solution of PDE, taking the weak form as input in near-mathematical notation, with automated discretization and a new simple method for adaptive error control, suitable for parallel implementation. On the Hornet Cray XC40 supercomputer we demonstrate new optimal strong scaling results for the whole adaptive framework applied to turbulent flow on massively parallel architectures down to 25000 vertices per core with ca. 5000 cores with the MPI-based PETSc backend and for assembly down to 500 vertices per core with ca. 20000 cores with the PGAS-based JANPACK backend.

Using the Unicorn component in FEniCS-HPC we have simulated the aerodynamics of a full DLR-F11 aircraft in connection with the HiLift-PW2 benchmarking workshop. We find that the simulation results compare very well with experimental data; moreover, we show mesh-convergence by the adaptive method, while using a low number of spatial degrees of freedom.

Acknowledgments. This research has been supported by EU-FET grant EUNISON 308874, the European Research Council, the Swedish Foundation for Strategic Research, the Swedish Research Council, the Basque Excellence Research Center (BERC 2014-2017) program by the Basque Government, the Spanish Ministry of Economy and Competitiveness MINECO: BCAM Severo Ochoa accreditation SEV-2013-0323 and the Project of the Spanish MINECO: MTM2013-40824.

We acknowledge PRACE for awarding us access to the supercomputer resources Hermit, Hornet and SuperMUC based in Germany at The High Performance Computing Center Stuttgart (HLRS) and Leibniz Supercomputing Center (LRZ), from the Swedish National Infrastructure for Computing (SNIC) at PDC – Center for High-Performance Computing and on resources provided by the "Red Española de Supercomputación" and the "Barcelona Supercomputing Center - Centro Nacional de Supercomputación".

We would also like to acknowledge the FEniCS and FEniCS-HPC developers globally.

References

1. Bangerth, W., Hartmann, R., Kanschat, G.: deal.II — a general-purpose object-oriented finite element library. ACM Trans. Math. Softw. **33**(4), 1–27 (2007)
2. FEniCS. FEniCS project (2003). http://www.fenicsproject.org
3. Hecht, F.: New development in freefem++. J. Numer. Math. **20**, 251–266 (2012)
4. Hoffman, J., Jansson, J., Vilela de Abreu, R., Degirmenci, N.C., Jansson, N., Müller, K., Nazarov, M., Spühler, J.H.: Unicorn: parallel adaptive finite element simulation of turbulent flow and fluid-structure interaction for deforming domains and complex geometry. Comput. Fluids **80**, 310–319 (2013)
5. Hoffman, J., Jansson, J., Jansson, N., Vilela De Abreu, R.: Towards a parameter-free method for high reynolds number turbulent flow simulation based on adaptive finite element approximation. Comput. Meth. Appl. Mech. Eng. **288**, 60–74 (2015)
6. Hoffman, J., Jansson, J., Stöckli, M.: Unified continuum modeling of fluid-structure interaction. Math. Mod. Meth. Appl. S. **21**, 491 (2011)
7. Hoffman, J., Johnson, C.: Computational Turbulent Incompressible Flow. Applied Mathematics: Body and Soul, vol. 4. Springer, Heidelberg (2007)
8. Jansson, N.: High Performance Adaptive Finite Element Methods: With Applications in Aerodynamics. Ph.D. thesis, KTH Royal Institute of Technology (2013)
9. Jansson, N.: Optimizing sparse matrix assembly in finite element solvers with one-sided communication. In: Daydé, M., Marques, O., Nakajima, K. (eds.) VECPAR 2012. LNCS, vol. 7851, pp. 128–139. Springer, Heidelberg (2013)
10. Jansson, N., Hoffman, J., Jansson, J.: Framework for massively parallel adaptive finite element computational fluid dynamics on tetrahedral meshes. SIAM J. Sci. Comput. **34**(1), C24–C41 (2012)
11. Kirby, R.C., Logg, A.: A compiler for variational forms. ACM Trans. Math. Softw. **32**(3), 417–444 (2006)
12. Kirby, R.C.: Algorithm 839: fiat, a new paradigm for computing finite element basis functions. ACM Trans. Math. Softw. (TOMS), 502–516 (2004)
13. Logg, A., Mardal, K.-A., Wells, G.N., et al. (eds.): Automated Solution of Differential Equations by the Finite Element Method. Lecture Notes in Computational Science and Engineering, vol. 84. Springer, Heidelberg (2012)
14. Oliker, L.: PLUM parallel load balancing for unstructured adaptive meshes. Technical report RIACS-TR-98-01, RIACS, NASA Ames Research Center (1998)
15. Rivara, M.C.: New longest-edge algorithms for the refinement and/or improvement of unstructured triangulations. Int. J. Numer. Meth. Eng. **40**, 3313–3324 (1997)

Accelerating NWChem Coupled Cluster Through Dataflow-Based Execution

Heike Jagode[1]([✉]), Anthony Danalis[1], George Bosilca[1],
and Jack Dongarra[1,2,3]

[1] University of Tennessee, Knoxville, USA
jagode@icl.utk.edu
[2] Oak Ridge National Laboratory, Oak Ridge, USA
[3] University of Manchester, Manchester, UK

Abstract. Numerical techniques used for describing many-body systems, such as the Coupled Cluster methods (CC) of the quantum chemistry package NWCHEM, are of extreme interest to the computational chemistry community in fields such as catalytic reactions, solar energy, and bio-mass conversion. In spite of their importance, many of these computationally intensive algorithms have traditionally been thought of in a fairly linear fashion, or are parallelised in coarse chunks.

In this paper, we present our effort of converting the NWCHEM's CC code into a dataflow-based form that is capable of utilizing the task scheduling system PARSEC (Parallel Runtime Scheduling and Execution Controller) – a software package designed to enable high performance computing at scale. We discuss the modularity of our approach and explain how the PARSEC-enabled dataflow version of the subroutines seamlessly integrate into the NWCHEM codebase. Furthermore, we argue how the CC algorithms can be easily decomposed into finer grained tasks (compared to the original version of NWCHEM); and how data distribution and load balancing are decoupled and can be tuned independently. We demonstrate performance acceleration by more than a factor of two in the execution of the entire CC component of NWCHEM, concluding that the utilization of dataflow-based execution for CC methods enables more efficient and scalable computation.

Keywords: PaRSEC · Tasks · Dataflow · DAG · PTG · NWChem · CCSD

1 Introduction

Simulating non-trivial physical systems in the field of Computational Chemistry imposes such high demands on the performance of software and hardware, that it comprises one of the driving forces of high performance computing. In particular, many-body methods, such as Coupled Cluster [1] (CC) of the quantum chemistry package NWCHEM [15], come with a significant computational cost, which stresses the importance of the scalability of nwchem in the context of real science.

ⓒ Springer International Publishing Switzerland 2016
R. Wyrzykowski et al. (Eds.): PPAM 2015, Part I, LNCS 9573, pp. 366–376, 2016.
DOI: 10.1007/978-3-319-32149-3_35

On the software side, the complexity of these software packages – with diverse code hierarchies, and millions of lines of code in a variety of programming languages – represents a central obstacle for long-term sustainability in the rapidly changing landscape of high-performance computing. On the hardware side, despite the need for high performance, harnessing large fractions of the processing power of modern large scale computing platforms has become increasingly difficult over the past couple of decades. This is due both to the increasing scale and the increasing complexity and heterogeneity of modern (and projected future) platforms. This paper is centered around code modernization, focusing on adapting the existing NWChem CC methods to a dataflow-based approach by utilizing the task scheduling system PaRSEC. We argue that dataflow-driven task-based programming models, in contrast to the control flow model of coarse grain parallelism, are a more sustainable way to achieve computation at scale.

The Parallel Runtime Scheduling and Execution Control (PaRSEC) [2] framework is a task-based dataflow-driven runtime that enables task execution based on holistic conditions, leading to a better computational resources occupancy. PaRSEC enables task-based applications to execute on distributed memory heterogeneous machines, and provides sophisticated communication and task scheduling engines that hide the hardware complexity from the application developer. The main difference between PaRSEC and other task-based engines lies in the way tasks, and their data dependencies, are represented. PaRSEC employs a unique, symbolic description of algorithms allowing for innovative ways of discovering and processing the graph of tasks. Namely, PaRSEC uses an extension of the symbolic Parameterized Task Graph (PTG) [3,4] to represent the tasks and their data dependencies to other tasks. The PTG is a problem-size-independent representation that allows for immediate inspection of a task's neighborhood, regardless of the location of the task in the Directed Acyclic Graph (DAG). This contrasts all other task scheduling systems, which discover the tasks and their dependencies at run-time (through the execution of skeleton programs) and therefore cannot process a future task that has not yet been discovered, or face large overheads due to storing and traversing the DAG that represents the whole execution of the parallel application.

In this paper, we describe the transformations of the NWChem CC code to a dataflow version that is executed over PaRSEC. Specifically, we discuss our effort of breaking down the computation of the CC methods into fine-grained tasks with explicitly defined data dependencies, so that the serialization imposed by the traditional linear algorithms can be eliminated, allowing the overall computation to scale to much larger computational resources.

Despite having in-house expertise in PaRSEC, and working closely and deliberately with computational chemists, this code conversion proved to be laborious. Still, the outcome of our effort of exploiting finer granularity and parallelism with runtime/dataflow scheduling is twofold. First, it successfully demonstrates the feasibility of converting TCE generated code into a form that can execute in a dataflow-based task scheduling environment. Second, it demonstrated that utilizing dataflow-based execution for CC methods enables more efficient and scalable computations. We present a thorough performance evaluation and demonstrate that the modified CC component of NWCHEM outperforms the original by more than a factor of two.

2   Implementation of Coupled Cluster Theory

The Coupled Cluster theory is considered by many to be the gold standard for accurate quantum-mechanical description of ground and excited states of chemical systems. Its accuracy, however, comes at a significant computational cost. An important role in designing the optimum memory vs. cost strategies in Coupled Cluster implementations is played by the automatic code generator, the Tensor Contraction Engine (TCE) [6]. In the first subsection, we highlight the basics necessary to understand the original parallel implementation of CC through TCE. We then describe our design decisions of the dataflow version of the CC code.

2.1   Coupled Cluster Theory Through TCE

Especially important in the hierarchy of the CC formalism is the iterative CC model with Single and Double excitations (CCSD) [13], which is the base for many accurate perturbative CC formalisms. Our starting point for the investigation in this paper is the CCSD version that takes advantage of the alternative task scheduling, and the details of these implementations have been described in [7].

In NWCHEM, the CCSD code (among other kernels) is generated through the TCE into multiple sub-kernels that are divided into so-called "T1" and "T2" subroutines for equations that determine the T1 and T2 amplitude matrices. These amplitude matrices embody the number of excitations in the wave function, where T1 represents all single excitations and T2 represents all double excitations. The underlying equations of these theories are all expressed as contractions of many-dimensional arrays or tensors (generalized matrix multiplications). There are typically many thousands of such terms in any one problem, but their regularity makes it relatively straightforward to translate them into FORTRAN code – parallelized with the use of *Global Arrays* (GA) [11] – through the TCE.

Structure of the CCSD Approach. For the iterative CCSD code, there exist 19 T1 and 41 T2 subroutines, and all of them highlight very similar code structure and patterns. Figure 1 shows the pseudocode FORTRAN code for one of the generated T1 and T2 subroutines, highlighting that most work is in deep loop nests. These loop nests consist of three types of code:

- Local memory management (i.e., MA_PUSH_GET(), MA_POP_STACK()),
- Calls to functions (i.e., GET_HASH_BLOCK(), ADD_HASH_BLOCK()) that transfer data over the network via the GA layer,

```
my_next_task = SharedCounter()
DO h7b = 1,noab
  DO p3b = noab+1,noab+nvab
    IF (int_mb(k_spin+h7b).eq.int_mb(...)) THEN
      call MA_PUSH_GET(f(p3b,h7b),..., k_c)

      DO p5b = noab+1,noab+nvab
        DO h6b = 1,noab
          call GET_HASH_BLOCK(dbl_mb(k_b),...,f(p3b,p5b,h7b,h6b))
          call TCE_SORT_4( dbl_mb(k_b),...,f(p3b,p5b,h7b,h6b))
          ...
          call DGEMM( ..., f(p3b,p5b,h7b,h6b))
        END DO
      END DO

      call ADD_HASH_BLOCK(dbl_mb(k_c), ...)
      my_next_task = SharedCounter()
    END IF
  END DO
END DO
```

Fig. 1. Pseudocode of one CCSD subroutine as generated by the TCE.

– Calls to the subroutines that perform the actual computation on the data
  GEMM() and SORT() (which performs an $O(n)$ remapping of the data, rather
  than an $O(n * log(n))$ sorting).

The control flow of the loops is parameterized, but static. That is, the induction variable of a loop with a header such as "DO p3b = noab+1,noab+nvab" (i.e., p3b) may take different values between different executions of the code, but during a single execution of CCSD the values of the parameters noab and nvab will not vary; therefore every time this loop executes it will perform the same number of steps, and the induction variable p3b will take the same set of values. This enables us to restructure the body of the inner loop into tasks that can be executed by PARSEC. That is, tasks with an execution space that is parameterized (by noab, nvab, etc.), but constant during execution.

Parallelization of CCSD. Parallelism of the TCE generated CC code follows a coarse task-stealing model. The work inside each T1 and T2 subroutine is grouped into chains of multiple matrix-multiply kernels (GEMM). The GEMM operations within each chain are executed serially, but different chains are executed in a parallel fashion. However, the work is divided into levels. More precisely, the 19 T1 subroutines are divided into 3 different levels and the execution of the 41 T2 subroutines is divided into 4 different levels. The task-stealing model applies only within each level, and there is an explicit synchronization step between the levels. Therefore the number of chains that are available for parallel execution at any time is a subset of the total number of chains.

Load balancing within each of the seven levels of subroutines is achieved through shared variables (exemplified in Fig. 1 through SharedCounter()) that are atomically updated (read-modify-write) using GA operations. The use of shared variables, that are atomically updated is bound to become inefficient at large scale, becoming a bottleneck and causing major overhead.

Also, the notion of *task* in the current CC implementation of NWCHEM and the notion of *task* in PARSEC are not identical. As discussed before, in NWCHEM, a *task* is a whole chain of GEMMs, executed serially, one after the other. In our PARSEC implementation of CC, each individual GEMM kernel is a task on its own, and the choice between executing them as a chain, or as a reduction tree, is almost as simple as flipping a switch. *In summary, the most significant impact of porting CC over PARSEC is the ability to eliminate redundant synchronizations between the levels and to break down the algorithms into finer grained tasks with explicitly defined dependencies.*

2.2   Coupled Cluster Theory over PaRSEC

PARSEC provides a front-end compiler for converting canonical serial codes into the PTG representation. However, due to computability limits, this tool is limited to polyhedral codes, i.e., loops, branches, and array indexes that only depend on affine functions of the loop induction variables, constant variables, and numeric literals. The CC code generated by TCE is neither organized in pure tasks – i.e., functions with no side-effects to any memory other than arguments passed to the function itself – nor is the control flow affine. For example, branches such as "IF(int_mb(k_spin+h7b-1)...)" (see Fig. 1) are very common. Such branches make the code not only non-affine, but statically undecidable since their outcome depends on program data, and thus it cannot be resolved at compile time.

While the behavior of the CC code depends on program data, this data is constant during a given execution of the code. Therefore, the code can be expressed as a parameterized DAG, by using lookups into the program data, either directly or indirectly. In our implementation we access the program data indirectly by builting meta-data structures in a preliminary step. The details of this "first step" are described later in this section.

In the work described in this paper, we implemented a dataflow form for all functions of the CCSD computation that are associated with calculating parts of the T2 amplitudes, particularly the ones that perform a GEMM operation (the most time consuming parts). More precisely, we converted a total of 29^1 of the 41 T2 subroutines – which we refer to under the unified moniker of "GA:T2" for the original version, and "PaRSEC:T2" for the dataflow version of the subroutines.

Design Decisions. The original code of our chosen subroutines consists of deep loop nests that contain the memory access routines as well as the main computation, namely SORT and GEMM. In addition to the loops, the code contains several IF statements, such as the one mentioned above. When CC executes, the

---

[1] All subroutines with prefix "icsd_t2_" and suffices: 2_2_2_2(), 2_2_3(), 2_4_2(), 2_5_2(), 2_6(), lt2_3x(), 4_2_2(), 4_3(), 4_4(), 5_2(), 5_3(), 6_2_2(), 6_3(), 7_2(), 7_3(), vt1ic_1_2(), 8(), 2_2_2(), 2_4(), 2_5(), 4_2(), 5(), 6_2(), vt1ic_1, 7(), 2_2(), 4(), 6(), 2().

code goes through the entire execution space of the loop nests, and only executes the actual computation kernels (SORT and GEMM) if the multiple IF branches evaluate to true. To create the PARSEC-enabled version of the subroutines (PaRSEC:T2), we decomposed the code into two steps:

The first step traverses the execution space and evaluates all IF statements, without executing the actual computation kernels (SORT and GEMM). This step uncovers sparsity information by examining the program data (i.e., int_mb(k_spin+h7b-1)) that is involved in the IF branches, and stores the results in custom meta-data vectors that we defined.

The custom meta-data vectors merely hold information regarding the actual loop iterations that will execute the computational kernels at run-time, i.e., iterations where all the IF statements evaluate to true. This step significantly reduces the execution space of the loop nests by eliminating all entries that would not have executed. In addition, this step probes the GA library to discover where the program data resides in memory and stores these addresses into the meta-data structures as well.

The second step is the execution of the PTG representation of the subroutines. Since the control flow depends on the program data, the PTG examines our custom meta-data vectors populated by the first step; this allows the execution space of the modified subroutines over PARSEC to match the original execution space of GA:T2. Also, using the meta-data structures, PARSEC accesses the program data directly from memory, without using GA.

Parallelization and Optimization. One of the main reasons we are porting CC over PARSEC is the ability of the latter to express tasks and their dependencies at a finer granularity, as well as the decoupling of work tasks and communication operations that enables us to experiment with more advanced communication patterns than serial chains. Since matrix addition is an associative and commutative operation, the order in which the GEMMs are performed does not bear great significance as long as the results are atomically added. This enables us to perform all GEMM operations in parallel and sum the results using a binary reduction tree. Clearly, in this implementation there are significantly fewer sequential steps than in the original chain [10]. In addition, the sequential steps are matrix additions, not GEMM operations, so they are significantly faster, especially for larger matrices. Reductions only apply to GEMM operations that execute on the same node, thus avoiding additional communication.

The original version of the code performs an atomic accumulate-write operation (via calls to ADD_HASH_BLOCK()) at the end of each chain. Since our dataflow version of the code computes the GEMMs for each chain in parallel, we eliminate the **global** atomic GA functionality and perform direct memory access instead, using **local** atomic locks within each node to prevent race conditions.

```
SUBROUTINE ccsd_energy_loc()
  start = ga_wtime()

c Initialize PaRSEC
  call parsec_init()

  DO iter=1,maxiter

c    Calculate t1 amplitudes of CCSD
     call icsd_t1()

c    Calculate t2 amplitudes of CCSD
     call icsd_t2()

     call tce_residual_t1()
     call tce_residual_t2()

  ENDDO

c Finalize PaRSEC
  call parsec_finalize()

  end = ga_wtime()-start
END
```

```
SUBROUTINE icsd_t2()
c Unchanged icsd_x subroutines
  call icsd_t2_1()
  call icsd_t2_2_1()
  call icsd_t2_2_2_1()
  ...

c Bridge code:
c Metadata of changed subroutines
  call populate_metadata()

c NWChem-PaRSEC Handshake:
c Execute tasks of changed subroutines
  call parsec_start_execution()

c Free metadata of changed subroutines
  call free_metadata()
END
```

Fig. 2. High level view of PARSEC code in NWCHEM.

3   Performance Evaluation

In this section we present the performance of the entire CCSD code using the dataflow version "PaRSEC:T2" of the 29 CC subroutines and contrast it with the performance of the original code "GA:T2". Figure 2 depicts a high level view of the integration of the PARSEC-enabled code in NWCHEM's CCSD component. The code that we timed (see **start** and **end** timers in Fig. 2) includes all 19 T1 and 41 T2 subroutines as well as additional execution steps that set up the iterative CCSD computation. The only difference between the original NWCHEM runs and our modified version is the replacement of the 29 original T2 subroutines "GA:T2" with their dataflow version "PaRSEC:T2" and the prerequisites discussed in Sect. 2.2; these prerequisites include: meta-data vector population, initialization, and finalization of PARSEC. Also, in our experiments we allow for all iterations of the iterative CCSD code to reach completion.

3.1   Methodology

As input, we used the beta-carotene molecule ($C_{40}H_{56}$) in the 6-31G basis set, composed of 472 basis set functions. In our tests, we kept all core electrons frozen, and correlated 296 electrons. Figure 3a shows the relative workload of different subroutines (omitting those that fell under 0.1 %). To calculate this load we sum the number of floating point operations of each GEMM that a subroutine performs (given the sizes of the input matrices). Additionally, Fig. 3b shows the distribution of chain lengths for the five subroutines with the highest workload in the case of beta-carotene. The different colors in this figure are for readability only. As can be seen from these statistics, the subroutines that we targeted for our dataflow conversion effort comprise approx. 91 % of the execution time of all 41 T2 subroutines in the original NWCHEM TCE CCSD execution.

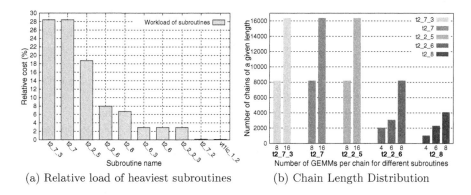

(a) Relative load of heaviest subroutines   (b) Chain Length Distribution

Fig. 3. CCSD statistics for beta-carotene and tilesize = 45.

The scalability tests for the original TCE generated code and the dataflow version of PaRSEC:T2 were performed on the *Cascade* computer system at EMSL/PNNL. Each node has 128 GB of main memory and is a dual-socket Intel Xeon E5-2670 (Sandy Bridge EP) system with a total of 16 cores running at 2.6 GHz. We performed various performance tests utilizing 1, 2, 4, 8, and 16 cores per node. NWCHEM v6.5 was compiled with the Intel 14.0.3 compiler, using the optimized BLAS library MKL 11.1, provided on Cascade.

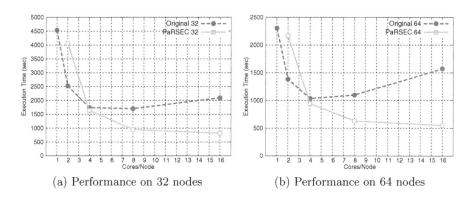

(a) Performance on 32 nodes   (b) Performance on 64 nodes

Fig. 4. Execution time comparison using beta-carotene on EMSL/PNNL Cascade (Color figure online)

3.2 Discussion

Figure 4 shows the execution time of the entire CCSD kernel when the implementation found in the original NWCHEM code is used, and when our PARSEC based dataflow implementation is used for the (earlier mentioned) 29 PaRSEC:T2 subroutines. Each of the experiments were run three times; the variance between the runs, however, is so small that it is not visible in the figures. Also, the correctness of the final computed energies have been verified for each run, and differences occur only in the last digit or two (meaning, the energies match for up to the 14th decimal place). In the graph we depict the behavior of the original code using the dark (green) dashed line and the behavior of the PARSEC implementation using the light (orange) lines. Once again, the execution time of the PARSEC runs does not exclude any steps performed by the modified code.

On a 32 node partition, the PARSEC version of the CCSD code performs best for 16 cores/node while the original code performs best for 8 cores/node. Comparing the two, the PaRSEC execution runs more than twice as fast – to be precise, it executes in 48 % of the best time of the original. If we ignore the PaRSEC run on 16 cores/node – in an effort to compare performance when both versions use 8 cores/node and thus have similar power consumption – we find that PaRSEC still runs 44 % faster than the original.

The results are similar on a 64 node partition: the PaRSEC version of CCSD is fastest (for 16 cores/node) with a 43 % runtime improvement compared to the original code (which on 64 nodes performs best for 4 cores/node). It is also interesting to point out that for 64 nodes, while PaRSEC manages to use an increasing number of cores – all the way up to $64 \times 16 = 1024$ cores – to improve performance, the original code exhibits a slowdown beyond 4 cores/node. This behavior is not surprising since (1) the unit of parallelism of the original code (chain of GEMMs) is much coarser than that of PARSEC (single GEMM), and (2) the original code uses a global atomic variable for load balancing while PARSEC distributes the work in a round robin fashion and avoids any kind of global agreement in the critical path.

4   Related Work

An alternate approach for achieving better load balancing in the TCE CC code is the Inspector-Executor methods [12]. This method applies performance model based cost estimation techniques for the computations to assign tasks to processors. This technique focuses on balancing the computational cost without taking into consideration the data locality.

ACES III [9] is another method that has been used effectively to parallelize CC codes. In this work, the CC algorithms are designed in a domain specific language called the Super Instruction Assembly Language (SIAL) [5]. This serves a similar function as the TCE, but with an even higher level of abstraction to the equations. The SIAL program, in turn, is run by a MPMD parallel virtual machine, the Super Instruction Processor (SIP). SIP has components that coordinate the work by tasks, communicate information between tasks for retrieving data, and then for execution.

The Dynamic Load-balanced Tensor Contractions framework [8] has been designed with the goal to provide dynamic task partitioning for tensor contraction expressions. Each contraction is decomposed into fine-grained units of tasks. Units from independent contractions can be executed in parallel. As in TCE, the tensors are distributed among all processes via global address space. However, since GA does not explicitly manage data redistribution, the communication pattern resulting from one-sided accesses is often irregular [14].

5 Conclusion and Future Work

We have successfully demonstrated the feasibility of converting TCE generated code into a form that can execute in a dataflow-based task scheduling environment, such as PARSEC. Our effort substantiates that utilizing dataflow-based execution for Coupled Cluster methods enables more efficient and scalable computation – as our performance evaluation reveals a performance boost of 2x for the entire CCSD kernel.

As a next step, we will automate the conversion of the entire NWCHEM TCE CC implementation into a dataflow form so that it can be integrated to more software levels of NWChem with minimal human involvement. Ultimately, the generation of a dataflow version will be adopted by the TCE engine.

Acknowledgment. This material is based upon work supported in part by the Air Force Office of Scientific Research under AFOSR Award No. FA9550-12-1-0476, and the DOE Office of Science, Advanced Scientific Computing Research, under award No. DE-SC0006733 "SUPER - Institute for Sustained Performance, Energy and Resilience". A portion of this research was performed using EMSL, a DOE Office of Science User Facility sponsored by the Office of Biological and Environmental Research and located at Pacific Northwest National Laboratory.

References

1. Bartlett, R.J., Musial, M.: Coupled-cluster theory in quantum chemistry. Rev. Mod. Phys. **79**(1), 291–352 (2007)
2. Bosilca, G., Bouteiller, A., Danalis, A., Herault, T., Lemarinier, P., Dongarra, J.: DAGuE: a generic distributed DAG engine for high performance computing. Parallel Comput. **38**(12), 37–51 (2012)
3. Cosnard, M., Loi, M.: Automatic task graph generation techniques. In: Proceedings of the 28th Hawaii International Conference on System Sciences (1995)
4. Danalis, A., Bosilca, G., Bouteiller, A., Herault, T., Dongarra, J.: PTG: an abstraction for unhindered parallelism. In: Proceedings of International Workshop on Domain-Specific Languages and High-Level Frameworks for High Performance Computing (WOLFHPC) (2014)
5. Deumens, E., Lotrich, V.F., Perera, A., Ponton, M.J., Sanders, B.A., Bartlett, R.J.: Software design of ACES III with the super instruction architecture. Wiley Interdisc. Rev. Comput. Mol. Sci. **1**(6), 895–901 (2011)
6. Hirata, S.: Tensor contraction engine: abstraction and automated parallel implementation of configuration-interaction, coupled-cluster, and many-body perturbation theories. J. Phys. Chem. A **107**(46), 9887–9897 (2003)
7. Kowalski, K., Krishnamoorthy, S., Olson, R., Tipparaju, V., Aprà, E.: Scalable implementations of accurate excited-state coupled cluster theories: application of high-level methods to porphyrin-based systems. In: High Performance Computing, Networking, Storage and Analysis (SC), 2011, pp. 1–10 (2011)

8. Lai, P.W., Stock, K., Rajbhandari, S., Krishnamoorthy, S., Sadayappan, P.: A framework for load balancing of tensor contraction expressions via dynamic task partitioning. In: Proceedings of the International Conference on High Performance Computing, Networking, Storage and Analysis (SC), pp. 1–10. ACM (2013)
9. Lotrich, V., Flocke, N., Ponton, M., Yau, A., Perera, A., Deumens, E., Bartlett, R.: Parallel implementation of electronic structure energy, gradient and hessian calculations. J. Chem. Phys. **128**, 194104-1–194104-15 (2008)
10. McCraw, H., Danalis, A., Herault, T., Bosilca, G., Dongarra, J., Kowalski, K., Windus, T.: Utilizing dataflow-based execution for coupled cluster methods. In: Proceedings of IEEE Cluster 2014, pp. 296–297 (2014)
11. Nieplocha, J., Palmer, B., Tipparaju, V., Krishnan, M., Trease, H., Apra, E.: Advances, applications and performance of the global arrays shared memory programming toolkit. Int. J. High Perform. Comput. Appl. **20**(2), 203–231 (2006)
12. Ozog, D., Shende, S., Malony, A., Hammond, J., Dinan, J., Balaji, P.: Inspector/executor load balancing algorithms for block-sparse tensor contractions. In: Proceedings of the 27th International ACM Conference on International Conference on Supercomputing, ICS 2013, pp. 483–484. ACM (2013)
13. Purvis, G., Bartlett, R.: A full coupled-cluster singles and doubles model - the inclusion of disconnected triples. J. Chem. Phys. **76**(4), 1910–1918 (1982)
14. Solomonik, E., Matthews, D., Hammond, J., Demmel, J.: Cyclops tensor framework: reducing communication and eliminating load imbalance in massively parallel contractions. In: 2013 IEEE 27th International Symposium on Parallel Distributed Processing (IPDPS), pp. 813–824 (2013)
15. Valiev, M., Bylaska, E.J., Govind, N., Kowalski, K., Straatsma, T.P., Van Dam, H.J.J., Wang, D., Nieplocha, J., Aprà, E., Windus, T.L., de Jong, W.: NWChem: a comprehensive and scalable open-source solution for large scale molecular simulations. Comput. Phys. Commun. **181**(9), 1477–1489 (2010)

Parallelization and Optimization of a CAD Model Processing Tool from the Automotive Industry to Distributed Memory Parallel Computers

Luis Ayuso[1]([✉]), Juan J. Durillo[1], Bernhard Kornberger[2], Martin Schifko[3], and Thomas Fahringer[1]

[1] Institute of Computer Science, University of Innsbruck, Innsbruck, Austria
{luis,juan,tf}@dps.uibk.ac.at
[2] Geom Softwareentwicklung, Graz, Austria
bkorn@geom.at
[3] ESS - Engineering Software Steyr GmbH, Munich, Germany
martin.schifko@essteyr.com

Abstract. In the early phases of an automotive industry development, design time can be substantially improved by using automated tools that assist engineers to perform repetitive and time consuming tasks, speeding up automotive products development and providing quality guarantees over otherwise error-prone processes. MERGE is such a tool from industry that prepares CAD models for electrophoretic deposition simulation. In this paper we describe the parallelization and optimization of MERGE for distributed memory parallel architectures. For this purpose we create a dynamic tree of tasks at runtime, analyze its load behavior and execute it through a master-worker compute paradigm based on different scheduling policies. Our implementation is based on a hybrid MPI-OpenMP version which results in a considerable improvement of both resource utilization and performance. Empirical performance results are presented for our new approach which achieve a speedup of up to 18 on an SMP cluster architecture.

Keywords: Distributed memory architecture · MPI-OpenMP · CAD · Intree precedence · Task scheduling

1 Introduction

In the late stages of a car design, all the independent parts that need to be manufactured are precisely modeled as independent volumetric triangulated meshes. During the manufacturing process, these parts will be welded together into a single object called *"body in white"*. In the process of preparing the car manufacturing, simulations are executed using the triangulated volumetric mesh of the body in white to prevent errors during manufacturing, i.e. electrophoretic deposition simulation (*ALSIM* software [1]).

ⓒ Springer International Publishing Switzerland 2016
R. Wyrzykowski et al. (Eds.): PPAM 2015, Part I, LNCS 9573, pp. 377–388, 2016.
DOI: 10.1007/978-3-319-32149-3_36

The *MERGE* application [2] is a repair-, connect-, and re-meshing tool for triangular meshes used in the automotive industry. It provides a fully automatic solution for the preparation of 2D and 3D CAD models used in simulations.

The MERGE application exhibits a coarse grain task execution tree with in-tree dependency constraints. This kind of arrangements suffer from a scalability problem in both resource utilization and execution time due to inherent loss of parallelism in the later stages of execution. Scheduling of task with in-tree precedence has been widely studied in the past [3–5]. The lack of shared data between tasks execution, however, exhibits potential for MERGE to be ported onto a distributed memory parallel architecture.

This paper presents a solution which improves both execution time and resource utilization by constructing the task tree in a dynamic manner and exploiting nested parallelism along the critical path tasks based on innovative scheduling policies. We have developed a distributed memory parallel MERGE version based on a hybrid MPI and OpenMP solution to manage and exploit the different levels of parallelism. Hybrid MPI-OpenMP is a popular parallel paradigm widely used in HPC [6,7]. Experimental results will be shown to demonstrate the effectiveness of our approach on a cluster of SMP nodes.

The rest of this paper is organized as follows: Sect. 2 presents the application and the architecture of the used MPI-OpenMP implementation. Section 3 models the MERGE application execution time. Section 4 describes the scheduling policies and the decision mechanism. Experimental results can be found in Sect. 5. Section 6 is the related work, and Sect. 7 summarizes our conclusions.

2   The Merge Application and the Master-Worker Architecture

The MERGE application serves two purposes: firstly merges all input meshes into a single triangulated mesh which defines the boundaries of the body in white volume. Secondly, it also guarantees the quality of the produced triangles.

Although the application is called MERGE, the Merge process is only one of the three phases of the application execution, where each of the phases compromises dozens of algorithms. The three phases will be referred to as *Repair*, *Merge*, and *Export* in the remainder of this paper.

The real world process of welding together car parts is modeled by the *Union* operation of meshes in software, this operation *merges* together pairs of volumetric triangulated meshes. Merging meshes can lead to a degradation of the quality of the triangles that can compromise the robustness of the computational geometry algorithms. Problems may be caused by almost-degenerated triangles and intersections of nearby co-planar triangles. The MERGE application uses a clean triangular representation for each of its algorithms, which guarantees the computation robustness and preserves the look and feel of the CAD designs.

The Repair phase accepts n input meshes and produces n *repaired* output meshes. This phase enforces the constraints to be fulfilled by the input of the Merge phase as described by Schifko et al. [8].

The Merge phase computes the union of the all repaired volumetric meshes, represented by a single volumetric triangulated mesh. The Merge phase carries out a reduction by applying the Union operation in pairs of meshes. For an input of n repaired meshes it will produce one single body in white mesh output.

The MERGE application uses Computational Geometry Algorithms Library (CGAL)[9] for robustness reasons. The Export phase converts arbitrary precision values into values with limited accuracy (floating point coordinates).

2.1 Master-Worker Architecture for MERGE

The distributed memory version of the program was implemented in two nested layers: The inner layer consists of different stand-alone programs, each of which perform the specific phases (Repair, Merge, and Export). Individual executions of these programs are denoted as *tasks* in the reminder of this paper.

Our previous work [10] focused on a shared memory parallelization of the MERGE application based on OpenMP. Although tasks exhibit shared memory parallelism, not all algorithm implementations used in the tasks could be parallelized due to inherent sequential dependencies. Therefore scalability of the shared memory parallelism is limited.

The outermost layer is a MPI program with a master-slave architecture where a master process dispatches tasks to worker processes. One single worker process is spawned shared memory node. It can execute as many tasks concurrently as cores are available in the node. The master process performs the dynamic construction of the task dependency tree and determines the number of cores to be used for each task. The master process can also act as worker process to maximize resource utilization, with the difference that one local core will be exclusively dedicated to the master process.

3   Execution Model of MERGE

To parallelize the MERGE application on a distributed memory architecture, it is necessary to understand the nature of the problem. The distributed memory parallelism is implemented on the coarse grain task arrangement of the program. This arrangement is composed of the three phases described in Sect. 2.

Figure 1 shows an example execution where 3 input meshes (a, b, and c) are *repaired* and stored as intermediate results a', b', and c'; these three repaired meshes are then merged into a single one by applying the binary operator Union. The task execution tree has in-tree precedence constraints, since tasks are connected by edges (meshes) and every task has one or no successor task.

The Union operation used in the Merge phase is associative and commutative, which allows for different combinations of the dependency tree to be constructed dynamically during runtime. We will refer to this feature as *arrangement flexibility*. The example illustrated in Fig. 1 shows one execution of the three possible combinations: $(a' \bigcup b') \bigcup c'$, $(a' \bigcup c') \bigcup b'$, and $(b' \bigcup c') \bigcup a'$. This feature will be exploited to improve our parallel MERGE implementation performance.

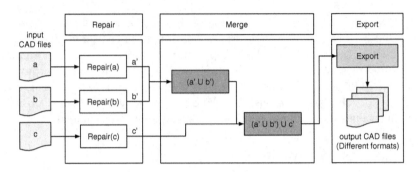

Fig. 1. Execution example of Merge with 3 input meshes

For any given task arrangement exists a critical path through the tree which dominates the overall execution time.

3.1   Load Imbalance at Repair Phase

As described in Sect. 2, the Repair phase enforces that the meshes fulfill the geometrical requirements of the Merge phase. The Repair phase is vulnerable to load imbalance for two reasons:

1. Geometrical properties: meshes containing specific geometrical properties will trigger algorithms which may not be executed for other input meshes, i.e. self-intersection corrections will not be performed if no self-intersections are found.
2. Number of triangles: although the various algorithms have different computational complexities, the execution time of the algorithms correlates with the number of triangles. Unfortunately it is expensive to estimate which set of algorithms will be used during a specific Repair task as the geometrical traits that trigger the algorithms may not be computable in an acceptable time.

Let M be the set of input meshes to be repaired, then the execution time for the Repair phase is modeled as:

$$t(Repair) = max\{t_r(m)\} \quad \forall m \in M. \tag{1}$$

Equation (1) shows the impact of long-lasting tasks during Repair phase. Executing Merge phase operations with meshes that have completed the Repair phase can speed up the overall program execution.

3.2   Load During Merge Stage

The Merge phase carries out a reduction operation on the set of repaired meshes by applying the Union operator. A simple mechanism to solve the Merge phase is to use an execution arrangement of the shape of a balanced binary tree: for n inputs, $\lceil log_2(n) \rceil$ tree levels will be required to conduct $n - 1$ Union operations.

Let $M_t(T, i)$ be the set of meshes to be merged in the i level of the reduction tree T, $t_u(n, m)$ the execution time of the Union operation on the meshes n and m. Assuming that we have an infinite number of cores and a balanced binary tree arrangement for the Merge phase, the minimum execution time of this phase is given by:

$$t(Merge) = \sum_{i=0}^{\lceil log_2(n) \rceil} max\{t_u(n, m)\} \quad n, m \in M_t(T, i) \tag{2}$$

Equation (2) shows that the execution time of the Merge phase is the sum of the execution time for all the $\lceil log_2(n) \rceil$ levels of the tree. The execution time of each level is equal to the execution time of the task with longest execution time of level i.

At this point, it is important to consider that the task arrangement of the Merge phase is flexible. This feature has two consequences:

1. For each level i in the tree, let m be the number of meshes in the set $M_t(T, i)$. Since the Union operator is commutative, there are $\binom{m}{2}$ possible combinations to merge all pairs of meshes in a level. The number of combinations to solve the complete Merge phase is large, even for moderate input mesh sizes. An exhaustive evaluation of all possible combinations is unfeasible.
2. The arrangement flexibility allows unbalanced trees, where branches compute the Union operation over a subset of M, leading to an arrangement with number of levels greater than $\lceil log_2(n) \rceil$. Although Eq. (2) shows that the execution time of the Merge application depends on the number of tree levels, execution times can benefit from an appropriate usage of this property.

3.3   Outlier Analysis

Equations (1) and (2) are dominated by long executing tasks, as a consequence of processing a certain set of meshes. The meshes processed by the long-lasting tasks exhibit properties with a severe impact on performance. Load imbalance during the Merge phase delays the Export phase execution.

Each level of the tree will exhibit a set of input meshes:

$$M_o(T, i) \subset M_t(T, i) \quad \forall m, n \in M_o(T, i) \mid t_u(m, n) > threshold. \tag{3}$$

Let $M_o(T, i)$ be the set of outlier meshes in level i, therefore the outlier meshes are the meshes utilized as input by the most time consuming tasks in level i. The definition of the outlier meshes is dependent on the execution time of the tasks that operate on them, although the time is not known before execution and therefore the threshold can not be known in advance.

Considering an environment with p cores and $p > n$, where n is the number of input meshes, the minimum execution time of the Merge phase in presence of o outlier meshes is given by:

$$t(Merge) = \sum_{i=0}^{\lceil log_2(o+1) \rceil} max\{t_u(n, m)\} \quad n, m \in M_o(T, i) \tag{4}$$

Equation (4) shows that the execution time of the Merge phase depends exclusively on the number of outlier inputs. The reduction tree is partitioned into two subtrees: the critical tree of long-lasting tasks and the tree of non critical tasks that can be sequentialized and executed utilizing idle times of cores.

Identifying the cause of load imbalance during execution may be difficult and computationally expensive, since the algorithms used by Repair and Merge tasks are triggered on demand depending on geometrical traits present in the input meshes. Nevertheless, the complexity of these algorithms is related to the number of triangles present in each input mesh. The size of a mesh is its number of triangles. The meshes used by any task during the MERGE application execution can be sorted according to their size. This order provides a fast mechanism for prioritizing tasks. A trivial outlier analysis is conducted over the input meshes of each Merge level: We considered a task to be an outlier when the sum of the input mesh sizes is greater than a threshold defined by $Q3 + (Q3 - Q1)$, where $Q3$ and $Q1$ are 25th and 75th percentiles respectively.

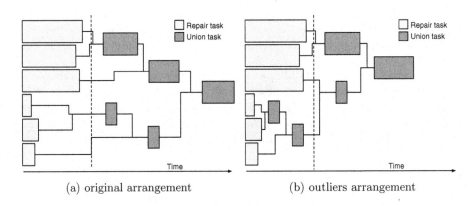

(a) original arrangement   (b) outliers arrangement

Fig. 2. Arrangement flexibility

As a result of load imbalance, some of the cores may remain idle. Our approach can use these idle time to execute any available task. The dynamic build of the task execution tree improves resource utilization. Figure 2 outlines the order of execution of two tasks arrangements for the same input meshes. Figure 2(a) shows the execution of the Repair and Merge phases when using a binary balanced tree as modeled in Eq. (2). Figure 2(b) shows the arrangement utilizing core idle time as modeled by Eq. (4). While the leftmost figure performs the Merge phase in three levels, the rightmost figure can achieve a valid output using only two levels in the critical path.

Although arrangement flexibility was utilized in our previous work to cope long-lasting tasks execution time, the outliers analysis provides another source to improve the MERGE application performance: tasks involving outlier meshes can be executed with shared memory parallelism. The combination of arrangement flexibility and outliers analysis can improve execution times by reducing the number of tasks in the critical path and improving the execution time of the remaining tasks in the critical path, which will be demonstrated through experiments in Sect. 5.

4 Scheduling

The execution time of the Merge application is bounded by the nature of the in-tree precedence constraints between tasks: this task arrangement causes a progressive loss of parallelism, with one single task to be executed in its last phase. Nevertheless, dynamic task generation and in-task parallelism can lead to performance improvement.

Although in-task shared memory parallelism can reduce the execution time of tasks, the in-task parallelism does not scale for increasing number of cores. The scheduler identifies tasks in the critical path to enable in-task parallelism while non-critical tasks are executed by a single core.

4.1 Scheduling Policies

According to the classification of Pinedo et al. [11], our scheduler uses a *Non-preemptive Dynamic Policy*. This kind of scheduler will utilize all the information available at the time it is invoked. This information will be dynamically updated until all task tasks are finished. Since all the Repair tasks are known at time zero, the algorithm behaves as *Non-preemptive Static List Policy Scheduler* during the Repair phase. However, as soon as the first Union task is available, the behavior of the scheduler changes.

We use the term *promotion* to refer to the activity of increasing the number of threads to be used to execute a task. The promotion mechanism maximizes the resource utilization by providing the maximum number of available cores to execute the tasks, although real resource utilization depends on the in-task degree of parallelization. In addition to the classical dynamic queue policy, promotion is utilized to improve performance with the following two policies:

1. Promote Unions: The Union phase is a tree reduction that reduces the number of repaired meshes n to one last volumetric mesh, by applying the Union operator. During the late steps of the Merge phase, the number of remaining meshes is less than the number of available cores and not enough tasks are generated to utilize all the cores. This will happen eventually on every target parallel hardware, with a more severe impact on larger systems where an increased number of cores runs out of tasks to execute in the earlier tree levels. Task promotion enables in-task shared memory parallelism for tasks in the later levels of the task execution tree, maximizing the resource utilization. This policy will not avoid the inherent loss of parallelism due to the in-tree dependency tree, however, it will exhibit the resource degradation per node instead of per core.

Algorithm 1. Scheduler: get_tasks

Input: Q ... list of queues, *available_cores*
Output: list of tuples: $(task, number\_cores)$
1: **function** *get_tasks*
2:     $tasks := []$
3:     $used\_cores := 0$
4:     **for all** q **in** Q **do**
5:         $tasks \leftarrow get\_outliers(\text{q}, available\_cores)$
6:         $used\_cores := count\_used\_cores(tasks)$
7:         **for** $i := 0 \ldots min(len(\text{ q }), available\_cores - used\_cores)$ **do**
8:             $tasks \leftarrow (pop(q), 1)$
9:             $used\_cores := used\_cores + 1$
10:     **if** $used\_cores < available\_cores$ **then**
11:         $tasks := promote(tasks, available\_cores - used\_cores)$
        return $tasks$

---

2. Promote outliers: As described in Sect. 3, the task dependency tree has in-tree precedence and the execution time is bounded by the execution time of the critical path. A reduction in the execution time of the critical tasks will reduce the application execution time. This can be achieved by increasing the number of cores utilized to execute the tasks operating on outliers.

4.2   Scheduler decision algorithm

Algorithm 1 shows the scheduler decision mechanism. When the master process is informed that a worker has available cores, the scheduler is queried for tasks. The master thread will invoke the *get_tasks* routine with the number of available cores in the worker. The objective of this function is to return a list of tuples (t, n), where n is the number of cores to use for the execution of task t. The normal procedure of the scheduler can be seen in the loop at line 7, one task at a time is extracted from the queue. Line 5 shows the mechanism that prioritizes tasks based on the outliers analysis described in Sect. 3. Line 10 shows the mechanism to improve resource utilization using task promotion.

5   Experiments

The experiments described in this Section have been executed on a cluster of 10 nodes equipped with two Xeon X5650 6-core CPUs each, running at 2.7 GHz with hyper-threading capabilities disabled. The nodes are equipped with 24 GB of RAM and are connected with 4x QDR Infiniband. The implementation uses Mpich 3.1.3 and the source code has been compiled with GCC 4.8.2 with optimization level -O3. The system runs Red Hat Enterprise Linux Server release 6.3.

Table 1 describes the datasets utilized in the experiments. Each of the presented experiments has as input a set of meshes called *dataset*. The datasets used in this Section present different features: number of mesh files, total number of

Table 1. Input datasets for MERGE

Dataset	Meshes	Triangles	Outliers	% outliers	SEM
Spheres	1555	1,461,120	259	16.66 %	89.49
Cubes	125	40,500	0	0.00 %	0.0
Dist-0.5	259	114,124	27	10.42 %	940.65
Dist-1.0	259	303,652	35	13.51 %	277.32
Dist-1.5	259	1,440,304	42	16.21 %	1,752.86

triangles, the number of input files where the triangle count exceeds the outliers threshold defined in Sect. 3, and standard error of the mean number of triangles (SEM) for each dataset.

As described in Sect. 2, the original task in-tree arrangement can benefit from a large number of inputs. The dataset Spheres contains 1555 input files to demonstrate this behavior. The dataset Cubes presents a uniform distribution of mesh sizes and avoids the use of outliers promotion. This dataset shows the worst case scenario for our scheduler, although real datasets do not present uniform sizes. The impact of outliers on the execution time is shown with the datasets Dist-0.5, Dist-1.0, and Dist-1.5. These datasets are a subset of the Spheres dataset where the number of triangles per mesh has been altered by a random factor α, where α has a uniform distribution with $\mu = 0$ and $\sigma = 0.5$, 0.1 and 1.5 respectively.

The experiments have been conducted for increasing the number of nodes and repeated 15 times. The target system has 12 cores per node and each node is managed by one worker process.

Figure 3 shows the mean speedup for each dataset with different scheduling policies and number of cores used.

- **No promote**: This experiment is the baseline of our development. Tasks are queued by priority order and in-task parallelism is not allowed. The critical path dominates the execution time and almost no improvement is achieved by increasing the number of cores. No Promote policy exhibits best performance for the dataset with a larger number of meshes (Spheres). This experiment also shows that for the same number of meshes, the number of outliers have a negative impact on execution time. A significant trait of this configuration is the lack of scalability: The Merge phase execution time is dominated by the critical path tasks, which need to be executed sequentially. In the case of the Cubes dataset, exist a path involving intersecting meshes, while other meshes do not share any volume, making the merge phase computationally cheap.
- **Promote Unions**: Union tasks are promoted to use all available cores in a compute node, as the number of remaining tasks becomes smaller than the total number of available cores. This scheduling policy improves the execution time except for Dist-1.0 and Dist-1.5 datasets. These datasets exhibit no performance improvement because of load imbalance produced by the number of

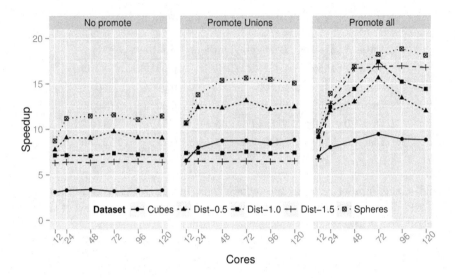

Fig. 3. Merge speedup with the three scheduling policies

outliers and the difference in size between them. Notice that Spheres does not exhibit this behavior since the mesh sizes present less standard error, which leads to a moderate imbalance that can still be compensated by the large number of meshes.

- **Promote all**: This policy extends Promote Unions by identifying and promoting task outliers. Highly unbalanced datasets benefit from our outliers analysis. The two datasets that exhibit no speedup improvement with the previous scheduling polices present a performance boost with the Promote all policy. The Cubes dataset has no improvement compared to the previous scheduling policy since the dataset contains no mesh outliers that can be promoted.

The experiments in this Section show the effectiveness of our scheduling policies for large in-tree tasks arrangements in the presence of load imbalance. The promote all scheduling policy improves execution time for all datasets compared to the no promote scheduling policy, achieving a speedup of 18.8 for the best scenario.

Although our scheduling policies in combination with task execution tree do not achieve good efficiency (96 cores for a speed-up of 18.8), the overall execution time reduction is still of great interest for a real-world industry application. In order to improve on the efficiency, we would need to replace some of the underlying algorithms that implement the MERGE tasks which was beyond the scope of our research.

6   Related Work

The work presented in this paper is closely related to MPI-OpenMP scheduling techniques and scheduling for stochastic unbalanced executions.

In the work of Dong et al. [12], a mechanism to provide non-homogeneous number of threads for the OpenMP parallelism is presented. Although their problem exhibit different execution times for the fine grain tasks, the outermost parallelism implemented in MPI exhibits flat parallelism. MERGE requires a more sophisticated solution to schedule the graph execution tree.

In [3], Frostig proposes a scheduling policy which minimizes makespan in scenarios of imbalance. The proposed technique HLF:LHR (highest level first, lowest hazard rate) requires the distribution of the tasks execution time to be known. Since the execution time of the MERGE application tasks depends on large number of variables (i.e. specific geometrical features), no assumption can be made on the distribution of the tasks execution time.

The complete MERGE application task arrangement can be modeled in functional programming by: $Export(reduce(Union, map(Repair, M)))$, where M is the set of input meshes. Although Map-Reduce is a popular paradigm for parallel programming [13–15], we believe that the parallelization of MERGE is not well suited for the Map-Reduce paradigm because the tasks dependency tree utilized by MERGE has one single output while the current Map-Reduce solutions expose a typical output size much larger than the number of available cores, relying on data-level parallelism to achieve balanced executions.

7 Conclusion

In this paper, we described the problem of parallelizing and optimizing an important tool from the automotive industry called MERGE which generates meshes for car parts to be used for simulation processes. Previous work [10] targeted a shared memory parallelization whereas in this work, we created a new version of MERGE with the goal to execute it on distributed memory parallel architectures.

We started our work with an analysis demonstrating various limitations of the underlying task execution tree of MERGE. We researched various scheduling policies for this application exploring both shared and distributed memory parallelism. A particular challenge of our research was to deal with load imbalance implied by outliers (task with exceptional large execution times). Furthermore, flexible arrangement of tasks of different MERGE phases exhibited promising potential to improve the overall performance which we exploited as part of our scheduling policies. Nested shared memory parallelism as part of a hybrid MPI-OpenMP version resulted in reducing the execution time along the critical path. Our scheduling also improved resource utilization by using only those resources (cores) required by the execution of MERGE.

Experiments on a cluster of SMPs demonstrate a speed-up of up to 18 without changing numerical algorithms of the underlying MERGE tasks. Future work will comprise the study of different numerical methods for the MERGE tasks that will increase the potential for resource efficiency.

Acknowledgments. This research has been funded by the Austrian Research Promotion Agency under contract 834307 (AutoCore), and supported by the Austrian

Ministry of Science BMWF as part of the UniInfrastrukturprogramm of the Focal Point Scientific Computing at the University of Innsbruck.

References

1. Strodthoff, B., Schifko, M., Jttler, B.: Horizontal decomposition of triangulated solids for the simulation of dip-coating processes. CAD Comput. Aided Des. **43**(12), 1891–1901 (2011)
2. Engineering Software Steyr GmbHd: Merge software (2015). http://www.essteyr.com/en/products/merge.html
3. Frostig, E.: A Stochastic Scheduling Problem with Intree Precedence Constraints on JSTOR (1988)
4. Kulkarni, V.G., Chimento, P.F.: Optimal scheduling of exponential tasks with intree precedence constraints on two parallel processors subject to failure and repair. Oper. Res. **40**(3-supplement-2), S263–S271 (1992)
5. Diakité, S., Nicod, J.M., Philippe, L., Toch, L.: Assessing new approaches to schedule a batch of identical intree-shaped workflows on a heterogeneous platform. Int. J. Parallel Emergent Distrib. Syst. **27**(1), 79–107 (2012)
6. Drosinos, N., Koziris, N.: Performance comparison of pure MPI vs hybrid MPI-OpenMP parallelization models on SMP clusters. In: 18th International Parallel and Distributed Processing Symposium, 2004, Proceedings, pp. 15–24. IEEE (2004)
7. Xuan, H., Tong, W., Gong, Z., Lan, Y.: Implementation and performance analysis of hybrid MPI+OpenMP programming for parallel MLFMA on SMP cluster. In: 2012 Third International Conference on Intelligent Control and Information Processing, pp. 744–748. IEEE, July 2012
8. Schifko, M., Jüttler, B., Kornberger, B.: Industrial application of exact boolean operations for meshes. In: Proceedings of the 26th Spring Conference on Computer Graphics, SCCG 2010, pp. 165–172. ACM, New York (2010)
9. CGAL, Computational Geometry Algorithms Library. http://www.cgal.org
10. Ayuso, L., Jordan, H., Fahringer, T., Kornberger, B., Schifko, M., Höckner, B., Moosbrugger, S., Verma, K.: Parallelizing a CAD model processing tool from the automotive industry. In: Lopes, L., et al. (eds.) Euro-Par 2014, Part I. LNCS, vol. 8805, pp. 24–35. Springer, Heidelberg (2014)
11. Pinedo, M.L.: Scheduling: Theory, Algorithms, and Systems. Springer Science & Business Media, New York (2012)
12. Dong, S., Karniadakis, G.E.: Dual-level parallelism for deterministic and stochastic CFD problems. Supercomputing, ACM/IEEE, pp. 1–17 (2002)
13. Dean, J., Ghemawat, S.: Mapreduce: simplified data processing on large clusters. Commun. ACM **51**(1), 107–113 (2008)
14. Mercier, G., Clet-Ortega, J.: Towards an efficient process placement policy for MPI applications in multicore environments. In: Ropo, M., Westerholm, J., Dongarra, J. (eds.) PVM/MPI. LNCS, vol. 5759, pp. 104–115. Springer, Heidelberg (2009)
15. Gates, A.F., Natkovich, O., Chopra, S., Kamath, P., Narayanamurthy, S.M., Olston, C., Benjaminn, R., Srinavasan, S., Srivastava, U.: Building a high-level dataflow system on top of map-reduce: the pig experience. In: VLDB 2009, pp. 1–12 (2009)

GPU Accelerated Simulations of Magnetic Resonance Imaging of Vascular Structures

Krzysztof Jurczuk[1(✉)], Dariusz Murawski[1], Marek Kretowski[1],
and Johanne Bezy-Wendling[2,3]

[1] Faculty of Computer Science, Bialystok University of Technology, Wiejska 45a,
15-351 Bialystok, Poland
{k.jurczuk,m.kretowski}@pb.edu.pl
[2] INSERM, U1099, 35000 Rennes, France
[3] University of Rennes 1, LTSI, 35000 Rennes, France

Abstract. Computer simulations of magnetic resonance imaging (MRI) are important tools in improving this imaging modality and developing new imaging techniques. However, MRI models often have to be simplified to enable simulations to be carried out in a reasonable time. The computational complexity associated with the tracking of magnetic fields' perturbations with high spatial and temporal resolutions is very high and, thus, it calls for using parallel computing environments. In this paper, we present a GPU-based parallel approach to simulate MRI of vascular structures. The magnetization calculation in different spatial coordinates is spread over GPU cores. We apply CUDA framework and take advantage of GPU memory hierarchy to efficiently exploit GPU computational power. Experimental results with different GPUs and various images show that the proposed algorithm substantially speedups the simulation. The proposed GPU-based approach may be easily adopted in modeling of other flow related phenomena like perfusion or diffusion.

Keywords: Computer simulation · Graphics processing unit (GPU) · Magnetic resonance imaging (MRI) · Parallel computing · Vascular structures

1 Introduction

Computational modeling and computer simulations play a very important role in contemporary science and industry. In medical field, they provide an additional insight to increase the comprehension of complex life processes that can be obscured during both *in vitro* and *in vivo* experiments [1]. Computer simulations are also usually cheaper and less time consuming than experimental investigations. Moreover, they can be performed with no need for patients to participate and thus are not limited by the examination duration.

In our research, we focus on modeling of magnetic resonance imaging (MRI) that is one of the most important diagnostic tools in modern medicine [2]. Although MRI is known as a highly detailed 3D imaging modality, there is still

© Springer International Publishing Switzerland 2016
R. Wyrzykowski et al. (Eds.): PPAM 2015, Part I, LNCS 9573, pp. 389–398, 2016.
DOI: 10.1007/978-3-319-32149-3_37

a lot of difficulties in magnetic resonance (MR) image formation and interpretation, especially in the area of blood flow. Imaging of such areas is crucial since vascular diseases are the cause of large mortality rate in the populations of developed countries [3]. Thus, we concentrate mainly on MRI of vascular structures.

The modeling of MR flow imaging is not a trivial task since it requires the integration of many processes/phenomenon (like physiology, hemodynamics, anatomy, imaging technology) in one computational model [4]. The challenge is also the decision about level of model detail (closeness to reality) as well as mathematical modeling itself. Another issue is high computational complexity of the model and, thus, the need for high performance computing to face up to such simulations. Many computations are required to simulate each physical phenomenon itself as well as interactions between them. The computational needs grow fast with the size of vascular structures. Large vascular simulations were shown to be vital to correctly investigate internal processes in human bodies [5].

There have been proposed many approaches in modeling of MR flow imaging [6–8] to name a few. The long simulation time was always one of the main factors limiting the extension of these models to a 3D version or to study complex vascular networks. Thus, it seems that further progress in computational modeling of MRI does not only depend on sophisticated equations, but also on the development of parallel architectures and algorithms. In our recent study [4], we successfully applied cluster computing in modeling of MR flow imaging. It allowed complex vascular structures to be investigated.

In this paper, we propose a GPU-based approach in modeling of MR flow imaging. Our motivation is to bring the possibility to perform MRI simulations fast on a single workstation. This way, complex vascular structure simulations can become independent of computer clusters that might be expensive, maintenance demanding and are not always accessible. In addition, modern GPUs are more often able to provide higher price/performance factor than computer clusters. What is also important that the proposed parallel approach may be easily applied in modeling of other flow related imaging like perfusion or diffusion MRI.

As far as we know, the proposed GPU-based algorithm is the first approach in modeling of MRI of vascular structures. Although, a GPU-based MRI simulator [9] was published recently, it allows only the stationary magnetization (without blood flow) to be investigated. The same research group also extended their simulator to model various motions [10]. However, their extended solution does not still enable vascular structures to be taken into account since the magnetization transport model (based blood flow) is significantly simplified.

In the next section, the model of MR flow imaging is described. In Sect. 3 the GPU-based approach is proposed. Section 4 presents the evaluation of the GPU-based algorithm. Section 5 provides conclusion and plans for future research.

2   Computational Model Description

In our previous research, we developed a three-component model of MR flow imaging [4] (Fig. 1). The first component is used for vascular structures generation based on physiological and hemodynamic parameters [11]. The second one

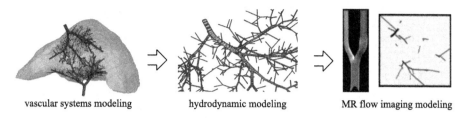

vascular systems modeling   hydrodynamic modeling   MR flow imaging modeling

Fig. 1. Three-component MRI model overview. The model allows us to generate vascular structures, then simulate flow behaviour and finally reproduce MRI processes.

allows flow simulations to be performed in the generated vascular structures [8]. The last component makes use of the generated vascular structures and flow characteristics to simulate MRI of vascular structures. Since, in this paper, the algorithm of imaging is parallelized, only the last model component is described.

The 3D imaged area (object) is divided into cubic elements [8]. To each cubic element basic MR parameters (proton density, relaxation times), determined by the represented part of a tissue, are assigned. In addition, each cubic element contains hydrodynamic parameters, i.e. mean flow velocity and direction of the fluid filling it (generated by the flow model). For stationary tissue structures (e.g. bones, vessel walls), the flow velocity equals zero.

Imaging simulation is divided into small time portions called time steps Δt. After each time step, local magnetizations of all cubic elements are modified taking into account both the flow influence ($\Delta \mathbf{M}_F$) and MRI processes ($\mathbf{A}_{MRI}$):

$$\mathbf{M}(\mathbf{r}, t + \Delta t) = \mathbf{A}_{MRI}(\mathbf{r}, \Delta t)[\mathbf{M}(\mathbf{r}, t) + \Delta \mathbf{M}_F(\mathbf{r}, \Delta t)] , \quad (1)$$

where $\mathbf{M}$ is the magnetization of the cubic element at spatial position $\mathbf{r}$.

First, the flow influence is computed (see left part of Fig. 2). In each cubic element the magnetization fractions are propagated to the neighboring cubic elements (see black rectangles labeled by "a", "b" and "c") based on the flow velocity and direction ($\mathbf{u} = u_x\hat{\mathbf{i}} + u_y\hat{\mathbf{j}} + u_z\hat{\mathbf{k}}$, where $\hat{\mathbf{i}}$, $\hat{\mathbf{j}}$, $\hat{\mathbf{k}}$ are unit vectors in x, y, z directions). This way, parts of magnetization can leave some cubic elements. At the same time, the magnetization fractions that leave some cubic elements enter to neighboring cubic elements. The mean magnetization changes during a time step Δt for a cubic element at $\mathbf{r}$ position are modeled as follows $\Delta \mathbf{M}_F(\mathbf{r}, \Delta t) = \Delta \mathbf{M}_{IN}(\mathbf{r}, \Delta t) - \Delta \mathbf{M}_{OUT}(\mathbf{r}, \Delta t)$ where:

$$\Delta \mathbf{M}_{IN}(\mathbf{r}, \Delta t) = \mathbf{M}\left(\mathbf{r} - \Delta d \frac{u_x(\mathbf{r})}{|u_x(\mathbf{r})|}\hat{\mathbf{i}} - \Delta d \frac{u_y(\mathbf{r})}{|u_y(\mathbf{r})|}\hat{\mathbf{j}}, t\right) |u_x(\mathbf{r})||u_y(\mathbf{r})|+$$

$$\mathbf{M}\left(\mathbf{r} - \Delta d \frac{u_x(\mathbf{r})}{|u_x(\mathbf{r})|}\hat{\mathbf{i}}, t\right) |u_x(\mathbf{r})|(1 - |u_y(\mathbf{r})|)+ \quad (2)$$

$$\mathbf{M}\left(\mathbf{r} - \Delta d \frac{u_y(\mathbf{r})}{|u_y(\mathbf{r})|}\hat{\mathbf{j}}, t\right) (1 - |u_x(\mathbf{r})|)|u_y(\mathbf{r})|,$$

$$\Delta \mathbf{M}_{OUT}(\mathbf{r}, \Delta t) = \mathbf{M}(\mathbf{r}, t)[|u_x(\mathbf{r})||u_y(\mathbf{r})|+$$
$$|u_x(\mathbf{r})|(1 - |u_y(\mathbf{r})|) + (1 - |u_x(\mathbf{r})|)|u_y(\mathbf{r})|]. \quad (3)$$

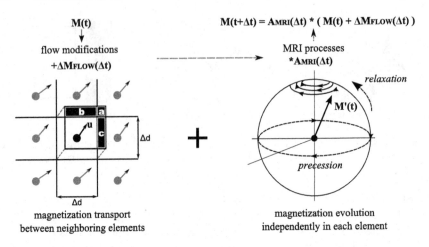

Fig. 2. The connection of MRI and magnetization transport algorithms.

In 3D modeling, each sum component in Eqs. 2 and 3 is made up of two more cases, accordingly, with $|u_z(\mathbf{r})|$ or $(1 - |u_z(\mathbf{r})|)$ term.

Later, the MRI influence is computed (right part of Fig. 2). It is modeled by the Bloch equation. We apply its discrete solution in the form of rotation matrices and exponential scaling [12]. Such an approach allows us to track the magnetization changes ($\mathbf{A}_{MRI}$) induced by MRI events without any integration:

$$\mathbf{A}_{MRI}(\mathbf{r}, \Delta t) = \mathbf{E}_{RELAX}(\mathbf{r}, \Delta t)\mathbf{R}_z(\Theta)\mathbf{R}_{RF}(\mathbf{r}, \Delta t) , \qquad (4)$$

where $\mathbf{E}_{RELAX}$ is responsible for the relaxation phenomena, $\mathbf{R}_z$ is the rotation matrix about the z-axis through angle Θ used to model the influence of spatial encoding gradients and magnetic field inhomogeneities, while $\mathbf{R}_{RF}$ reproduces the excitation process [8].

Based on Faraday's law of an electromagnetic induction, the MR signal coming from the imaged object at a time t is expressed as a sum of the transverse magnetizations of all cubic elements. In MRI, such a signal is acquired many times during successive sequence repetitions, with different phase encoding gradients. These different gradients are used to encode spatial positions of particular signals that compose the total received signal. Within a single repetition period, the acquired signal fills one k-space (readout) matrix row. The MR image is created by applying the fast Fourier transform (FFT) to the fully filled matrix.

This coupled MR flow imaging model closely follows the physical process of MRI. In addition, it does not need additional mechanisms to consider the blood flow influence during most of MRI events, in contrast to prior works [6,7].

3   GPU-based Parallelization of MR Flow Imaging

The most time consuming part of the algorithm is the calculation of new magnetization values ($\mathbf{M}(\mathbf{r}, t + \Delta t)$). It has to be done many times after each time step (Δt) that ought to be small enough (usually of the order of tens to hundreds of microseconds). Second, in each time step, both the blood flow and MRI influences have to be taken into account in each cubic element (which size is usually of the order of tens to hundreds of micrometers). Thus, data decomposition strategy is naturally applied (see Fig. 3).

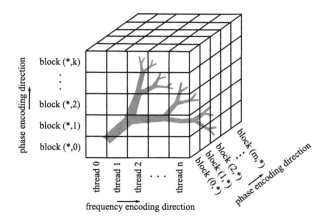

Fig. 3. Data decomposition strategy: 2D grid of blocks along the phase encoding directions and 1D block of threads along the frequency encoding direction.

We decide to use 2D grid of blocks along the directions of phase encoding gradients. The cubic elements along the direction of frequency encoding gradient are, in turn, spread over threads. By default, the number of blocks and threads are set to the number of cubic elements (object size) in the considered directions. If the object size excesses the maximum number of threads/blocks (which is hardware dependent), then the maximum value is used and a single thread/block processes more than one cubic element.

Figure 4 presents the idea behind the GPU-based algorithm of MR flow imaging. At the beginning, at host (CPU), the 3D object and the experiment parameters are initialized. Then, they are sent to the device (GPU) and saved in the allocated space in global memory. This CPU to GPU data transfer is performed only once before MRI procedure and this data is kept there during the whole MRI experiment. This way, each thread has an access to this data. Both the object (flow, MR parameters) and tracked magnetization values are organized as 3D matrices of floating point values arranged in one-dimensional arrays. The results (k-space matrix that is also arranged in a one-dimensional array) are transferred to CPU when all MRI repetitions are finished.

After the initialization, the MR image formation is started and here the GPU-based parallelization is applied. In each excitation/repetition, various MRI events are simulated and finally a single line of the MR signal is acquired and saved in the k-space matrix. Each MRI event (excitation, spatial encoding, relaxation, ...)

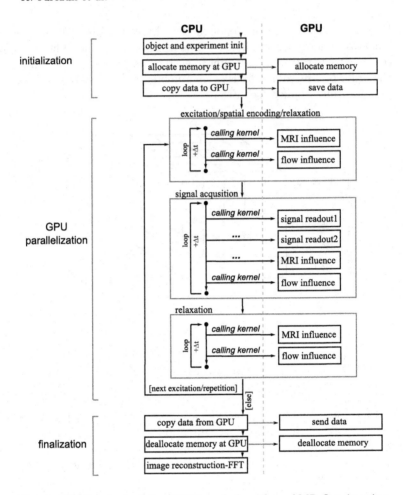

Fig. 4. The flowchart of the GPU-based algorithm of MR flow imaging

is performed in consecutive time steps (Δt). The magnetization of a cubic element in the next time step depends on its current value (time dependency), the MRI influence and magnetization of neighboring cubic elements (spatial dependency), accordingly with Eq. 1. Thus, in each time step, two kernel functions are called. The first one is responsible for the MRI influence, while the second one for the magnetization transport induced by blood flow. Such an approach provides the synchronization for all threads (both inside and between blocks) after each time step. Thereby, the time and spatial dependency between magnetizations of cubic elements is provided. During the MRI influence calculations, each thread accesses only the local magnetization values, while the magnetization transport phase needs to reach into the neighboring cubic elements.

The signal acquisition phase differs a little from other MRI events. Here, not only the MRI and flow influences are simulated, but also the signal from all cubic elements has to be read and saved in the k-space matrix. This operation is carried out by two additional kernel functions (*signal readout1* and *signal readout2*, see Fig. 4). These two functions, in a single time step, provide the sum of all magnetizations (from all cubic elements). The first kernel performs the reduction of magnetization values for cubic elements inside blocks. It uses the mechanism of shared memory inside thread blocks. The second kernel finishes the reduction with the use of one block and a table in global memory.

After the signal acquisition, the relaxation process is simulated and if the next repetition is needed, the algorithm starts again from the excitation. Otherwise, the k-space matrix is transferred from the device to the host. Finally, the MR image is created by the application of FFT to the received matrix at CPU.

In order to efficiently use hardware resources, we take advantage of memory hierarchy of modern GPUs. The following algorithm improvements, among others, are applied: *(i)* Data stored in global memory and frequently used (e.g. cubic element magnetization) is transferred to local variables at the beginning of kernel functions. At the end of the kernel, the data is transferred back to global memory. *(ii)* In the serial algorithm, the results of some repeated calculations (e.g. partial computation of the magnetization increase during relaxation after Δt time step) are saved in auxiliary tables before the MRI procedure and used when needed. In the GPU-based algorithm, we do not use such global auxiliary tables. Results of frequently repeated calculations are saved locally in a kernel function when they are done for the first time. Such a mechanism increases the number of arithmetic operations but, at the same time, reduces redundant loads from GPU global memory. *(iii)* *float4* data type is used to store magnetization description in GPU global memory, despite the fact that the magnetization is a 3D vector and we only need three float values. This way, coalesced memory access may be provided, resulting in efficient memory requests and transfers [13]. As regards local thread variables, temporary magnetizations are stored in *float3* variables since in threads register space is an important issue.

4   Experimental Results

All experiments were performed on a workstation equipped with a quad-core processor (Intel Core i7-870, 8M Cache, 2.93 GHz), 32 GB RAM and a single graphics card. We tested three different Kepler-based NVIDIA graphics cards: *(i)* GeForce GTX 780 (3 GB memory, 2304(12×192) CUDA cores), *(ii)* Quadro K5000 (4 GB memory, 1536(8×192) CUDA cores) and *(iii)* GeForce GTX Titan Black (6 GB memory, 2880(15×192) CUDA cores).

We used 64-bit Ubuntu Linux 14.04.02 LTS as an operating system. The sequential algorithm was implemented in C++, compiled with the use of gcc version 4.8.2 and optimized many times since of long simulation time. The GPU-based parallelization was implemented in CUDA-C and compiled by nvcc CUDA 7.0 [14]. Single precision calculations were analyzed.

We present results for four various vascular structures investigated in our previous papers [4,8], where detailed flow and MRI settings were reported.

Figure 5 shows examples of simulated MR flow images representing more and more complicated vascular structure. The gradient echo (GE) imaging sequence was used to obtained the images. Other basic imaging parameters are reported next to the MR images.

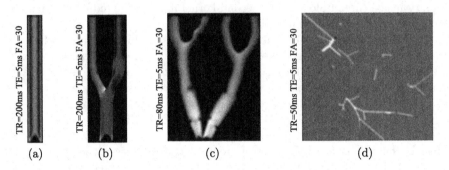

Fig. 5. Simulated MR images: (a) a single vessel, (b) a single bifurcation, (c) multiple bifurcations, (d) slice through the liver volume with many vascular structures.

Figure 6 shows the obtained mean speedup (over different imaging parameters) for four tested vascular structures, relative to the sequential implementation run on a single core. The speedup obtained with the use of four CPU cores by an OpenMP parallelization is also presented. It is clearly visible that the GPU-based parallelization provides a significant decrease in computation time. All GPUs are able to obtain a speedup at least one order higher than the OpenMP parallelization. Currently, the time needed to simulate e.g. MR image in Fig. 5(d) equals about 1.5 h with the use of Titan Black GPU, instead of about 10 days by a single core CPU or 5 days using OpenMP parallelization and four CPU cores. Moreover, the achieved speedup is comparable (images (a–c)) or even higher (image (d)) than the one obtained by a computer cluster of 16 nodes each equipped with 2 quad-core CPUs (Xeon 2.66 GHz) and 16 GB RAM [4].

In Fig. 6 we can also see that the tested GPUs provide different speedups. Concerning GTX Titan Black and GTX 780 GPUs, the difference is quite small and it could be explained by a slightly faster memory (7.0 vs 6.0 Gbps) and a stronger computational unit (2880 cores of 980 MHz vs. 2304 cores of 900 MHz) of the former one. As regards Quadro K5000 GPU, it achieves much smaller acceleration than other tested GPUs. Quadro family GPUs are known to be designed to especially accelerate applications to design, rendering and 3D visual modeling (like CAD software) at the expense of lower computational performance in games and GPGPU. They are also usually more expensive than GTX series GPUs and offer lower power consumption. Anyway, the obtained results show that such a engineering workstation equipped with a Quadro family GPU can also be successfully used in fast MR flow imaging simulations. Considering the price/performance factor, the GTX 780 GPU wins since, currently, it is about 3 times cheaper than Titan Black and 5 times than Quadro K5000.

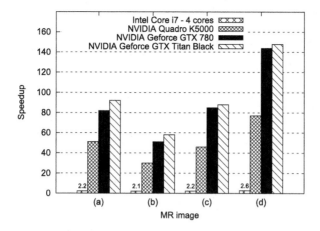

Fig. 6. Speedup of the GPU (as well as multi-core) accelerated algorithms relative to the serial implementation run on a single core of Intel Core I7 CPU. Images shown in Fig. 5 and various GPUs are tested.

There is also a difference in speedup between tested objects. The first reason could be the fraction of cubic elements that represent the object/image background where no MRI is simulated (black areas in images). The second one is the size of the object/image that does not always allow computational resources to be exploited efficiently. The object size and fraction of background elements were as follows: (a) 106×739, 0.37, (b) 167×543, 0.6, (c) 572×740, 0.7, (d) 1000×1000×70, 0. The simulation box can differ from the object size due to additional background cubic elements added to objects to apply FFT easily [8]. The worst results were obtained for the single bifurcation object where there was a lot of background elements and the horizontal object (167) as well as simulation box size (320) was too small to provide load balancing. The best speedup, in turn, was achieved for image (d) since there was no background elements and object size was big enough. We verified execution times with respect to the object size for the straight vessel geometry (Fig. 5(a)) (72×705, 144×1410 and 288×2820 original/flow grid size). The obtained speedup grew with the increase of object size, from about 80 to 120, respectively. The thorough evaluation of the scaling properties of the algorithm was left for future investigation.

5   Conclusion

In this paper we propose a GPU-based parallel approach to simulate MRI of vascular structures. The approach was tested at various GPUs and vascular structures. The results clearly show that it provides a significant speedup relative to the CPU implementation. Thus, it opens a perspective to simulate more and more complex (that is more realistic) vascular structures. Moreover, the proposed GPU-based approach may be easily adapted in other algorithms concerning related phenomena like diffusion or perfusion.

In the future, we want to deal with multi-GPU approaches, both at a single workstation as well as on computing clusters. We also plan to use the presented parallelization in the modeling of contrast agent propagation and dynamic contrast-enhanced MRI [15].

Acknowledgments. This work was supported by the grants W/WI/2/2014 and S/WI/2/2013 from Bialystok University of Technology.

References

1. Hoppensteadt, F.C., Peskin, C.S.: Modeling and Simulation in Medicine and the Life Sciences. Springer-Verlang, New York (2004)
2. Westbrook, C., Roth, C.K., Talbot, J.T.: MRI in Practice. Wiley-Blackwell, Malden (2011)
3. Aiyagari, V., Gorelick, P.B.: Hypertension and Stroke: Pathophysiology and Management. Humana Press, Totowa (2011)
4. Jurczuk, K., Kretowski, M., Eliat, P.-A., Saint-Jalmes, H., Bezy-Wendling, J.: In silico modeling of magnetic resonance flow imaging in complex vascular networks. IEEE Trans. Med. Imaging **33**(11), 2191–2209 (2014)
5. Grinberg, L., Cheever, E., Anor, T., Madsen, J.R., Karniadakis, G.E.: Modeling blood flow circulation in intracranial arterial networks: A comparative 3-D/1-D simulation study. Ann. Biomed. Eng. **39**(1), 297–309 (2011)
6. Lorthois, S., Stroud-Rossman, J., Berger, S., Jou, L.D., Saloner, D.: Numerical simulation of magnetic resonance angiographies of an anatomically realistic stenotic carotid bifurcation. Ann. Biomed. Eng. **33**(3), 270–283 (2005)
7. Marshall, I.: Computational simulations and experimental studies of 3D phase-contrast imaging of fluid flow in carotid bifurcation geometries. J. Magn. Reson. Imaging **31**(4), 928–934 (2010)
8. Jurczuk, K., Kretowski, M., Bellanger, J.-J., Eliat, P.-A., Saint-Jalmes, H., Bezy-Wendling, J.: Computational modeling of MR flow imaging by the lattice Boltzmann method and Bloch equation. Magn. Reson. Imaging **31**(7), 1163–1173 (2013)
9. Xanthis, C.G., Venetis, I.E., Chalkias, A.V., Aletras, A.H.: MRISIMUL: A GPU-based parallel approach to MRI simulations. IEEE Trans. Med. Imaging **33**(3), 607–617 (2014)
10. Xanthis, C.G., Venetis, I.E., Aletras, A.H.: High performance MRI simulations of motion on multi-GPU systems. J. Cardiovasc. Mag. Reson. **16**, 48 (2014)
11. Kretowski, M., Rolland, Y., Bezy-Wendling, J., Coatrieux, J.-L.: Physiologically based modeling for medical image analysis: application to 3D vascular networks and CT scan angiography. IEEE Trans. Med. Imaging **22**(2), 248–257 (2003)
12. Bittoun, J., Taquin, J., Sauzade, M.: A computer algorithm for the simulation of any nuclear magnetic resonance (NMR) imaging method. Magn. Reson. Imaging **2**(2), 113–120 (1984)
13. NVIDIA, CUDA C Best Practices Guide in CUDA Toolkit Documentation v7.0. http://docs.nvidia.com/cuda/cuda-c-best-practices-guide/
14. Wilt, N.: Cuda Handbook: A Comprehensive Guide to GPU Programming. Addison-Wesley, Boston (2013)
15. Mescam, M., Kretowski, M., Bezy-Wendling, J.: Multiscale model of liver DCE-MRI towards a better understanding of tumor complexity. IEEE Trans. Med. Imaging **29**(3), 699–707 (2010)

Parallel Algorithms for Wireless LAN Planning

Andrzej Gnatowski[1(✉)] and Jan Kwiatkowski[2]

[1] Department of Automatics, Mechatronics and Control Systems,
Faculty of Electronics, Wroclaw University of Technology,
Janiszewskiego 11-17, 50-372 Wroclaw, Poland
andrzej.gnatowski@pwr.edu.pl

[2] Department of Informatics, Faculty of Computer Science and Management,
Wroclaw University of Technology, Wybrzeże Wyspianskiego 27,
50-370 Wroclaw, Poland
jan.kwiatkowski@pwr.edu.pl

Abstract. Nowadays wireless services are a part of the everyday life of billions of people around the world. To decrease not only their design and maintenance costs but also to enhance user comfort there is a high demand for specialized software that helps during theirs designing. Unfortunately commonly available and used software is demanding in terms of computing power in order for receiving the results in the reasonable time. The aim of the presented research is verification whether using GPU it is possible accelerate computation in reasonable way. The paper presents a study of possible parallelization of algorithms used for signal power simulation as well as algorithms that optimize placement of routers. The main goal of the paper is to show the impact of GPU usage on different stages of wireless network planning process. The first received results are very promising, the speedup received is even up to 46 for simulation algorithm and up to 15 for optimization algorithms.

Keywords: Wireless LAN · Network planning · Parallel execution at GPU

1 Introduction

Wireless communication is an essential part of everyday life and can be implemented with using a number of technologies, one of them is a wireless local area network (WLAN). These networks are widespread, their used range is very wide from private houses to public facilities and companies headquarters. The key components constituting these networks are devices used directly for wireless communication, equipped with transmitting antennas - wireless routers and access points (AP). Although they differ in design and functionality, the same principles should be applied during theirs placement in the network. Therefore in the paper these terms are used interchangeably. The location of wireless routers has a direct impact on the quality of wireless networks. There are several parameters that can be taken into consideration while evaluating the quality of WLAN,

© Springer International Publishing Switzerland 2016
R. Wyrzykowski et al. (Eds.): PPAM 2015, Part I, LNCS 9573, pp. 399–410, 2016.
DOI: 10.1007/978-3-319-32149-3_38

for example [1–3]: signal power, signal to noise ratio, network load, theoretical average speed of data transmission, average delays and others.

Hence, the ability to estimate the network parameters already at their planning stage is very valuable. For this aim dedicated algorithms for simulating the propagation of electromagnetic waves in the indoor environment are used. Due to the number of possible locations of routers can be very large, to minimize the amount of time needed to predict network parameters, algorithms, which optimize the locations of routers are used. Two different optimization scenarios are possible: the first one (*) that maximizes network quality by determining the locations for the predefined number of routers r and the second one (**) that determines minimal number of routers (and their locations), which meets minimal network quality requirements. The first of above scenarios can be easily transformed into second one, therefore in the paper only the first scenario is considered. Moreover during network planning the FAP (Frequency Assignment Problem) is not taken into consideration. We set up as in [4,5] that the network planning can be done in two steps, when in the first step only QoS is taking into consideration and the second step is related to FAP problem.

In practice, there might to be too many possible routers configurations when using optimization algorithms because of too high computation demand. Therefore, it is necessary to use commercially available specialized software that support network planning process, for example [6,7]. The use of more sophisticated methods of simulation and optimization leads to longer computation time, especially for larger networks. As alternative approach a parallel processing can be used to receive results in the reasonable time [8]. The aim of presented research is to examine whether significant reduction of computation time related to planning of WLAN using the GPU can be achieved.

The structure of the paper is as follows. Section 2 introduces wireless router placement problem. Section 3 presents algorithms used in solving the problem, and the way how they are parallelized. In Sect. 4, results of performed computational experiments are presented. Finally, Sect. 5 summarized presented work and discuss the future plans.

2   Wireless Router Placement Problem

Indoor WLANs are often installed in multistorey buildings. Wireless routers from all floors can interfere with one another. Therefore there is a need to model network in an entire building. There is given set of N_f floors $\mathcal{F} = \{1, 2, \ldots, N_f\}$. Let $\mathcal{C} = \{c_1, c_2, \ldots, c_{N_c}\}$ be the set of possible wireless routers locations. Each location is defined by three parameters: $c_i = (c_{i,x}, c_{i,y}, c_{i,f})$, where $c_{i,x}$ and $c_{i,y}$ describe position of location i on the floor $c_{i,f} \in \mathcal{F}$. Router i from the set $\mathcal{R} = \{1, 2, \ldots, N_r\}$ is described by its location $p_i \in \mathcal{P} = \{p_1, p_2, \ldots, p_{N_r}\} \subset \mathcal{C}$ and signal power s_i (it is assumed that routers has omnidirectional antennas). Since signal power of wireless routers is regulated by law, it is assumed that $s_1 = s_2 = \cdots = s_r = s$. In each location at most one wireless router can be placed:

$$\bigvee_{c_i \in \mathcal{C}} 0 \leq |\{p_i \in \mathcal{P} : p_i = c_i\}| \leq 1 . \tag{1}$$

The problem (*), mentioned in the introduction, can be formally defined as to *find such feasible solution $\mathcal{P}^* \in P$, where P is a set of all possible routers positions configurations, that minimizes the objective function $f : P \to \mathbb{R}$*:

$$\underset{\mathcal{P} \in P}{\forall} \, f(\mathcal{P}) \leq f(\mathcal{P}^*) \,, \tag{2}$$

where solution is defined by elements of set $\mathcal{P}$. The problem (*) can be easily modified to (**): *finding minimal router number N_r^* providing sufficient network quality k (evaluated by the objective function f)*:

$$N_r^* = \min_{N_r \in \mathbb{N}} \{f(\mathcal{P}^*) \geq k\} \,. \tag{3}$$

There are several factors that should be taken into consideration while designing objective function. One of the possible approaches to cope with this multi-criterial problem is to use weighted arithmetic mean:

$$f(\mathcal{P}) = \frac{\sum\limits_{i=1}^{N_F} a_i f_i(\mathcal{P})}{\sum\limits_{i=0}^{N_F} a_i} \,, \tag{4}$$

where $f(\mathcal{P})$ - objective function, N_F - number of components of objective function, $f_i(\mathcal{P})$ - component i of objective function, a_i - weight of component i, $\mathcal{P}$ - locations of wireless routers. Similar approach was described in [2]. In this paper, signal power and network load have been taken into consideration.

Let $\mathcal{M} = \{m_1, m_2, \ldots, m_{N_m}\}$ be the set of locations where certain signal power is demanded. Desired signal power in location m_i is denoted as d_i [dBm]. Signal power component of objective function is based on [2] and is defined as follows:

$$f_1(\mathcal{P}) = \frac{1}{N_m} \sum_{i=1}^{N_m} \max \left\{ d_i - \max_{j \in \mathcal{R}} \{q_{m_i}^j\}, 0 \right\} \,, \tag{5}$$

where $q_{m_i}^j$ - simulated signal power of router j in measurement point m_i [dBm]. Empirical narrow-band propagation model, *multi-wall model* (MWM) [9] is used to simulate signal power of a router at given location. The model can be expressed in form:

$$L = L_{FS} + L_c + \sum_{i=1}^{I} k_i \cdot L_i + k_d^{\frac{k_f+2}{k_f+1}-b} \cdot L_f \,, \tag{6}$$

where: L - suppression between antennas [dB], L_{FS} - free space loss between transmitter and receiver [dB], L_c - fixed loss [dB], k_i - number of penetrated walls of type i, L_i - loss of wall type i [dB], k_f - number of floors between transmitter and receiver, L_f - loss of ceiling [dB], b - empirical parameter, I - number of wall types. The free space loss takes into account both radio wave frequency and distance between transmitter and receiver. It can be expressed by equation [9]:

$$L_{FS} = -27.55 + 20 \log_{10}(f) + 20 \log_{10}(d) \,, \tag{7}$$

where: L_{FS} - free space loss between antennas [dB], f - signal frequency [MHz], d - distance between transmitter and receiver [m]. The loss factors from Eqs. 6 and 7 are not physical wall loses, but rather optimized coefficients determined by empirical measurements. Exemplary model coefficients for 1.8 [GHz] band can be found in [9].

Network load component of objective function is designed to increase with decrease in uniformity of router allocation. It is assumed that user $i \in \{1, 2, \ldots N_u\}$ in location $u_i = (u_{i,x}, u_{i,y}, u_{i,f}) \in \mathcal{U} = \{u_1, u_2, \ldots, u_{N_u}\}$ is assigned to the router which has the highest signal power in u_i. Signal power can be evaluated with Eq. 6. Objective function component is proposed as follows:

$$f_2 = \frac{1}{N_r} \sum_{i=1}^{N_r} \frac{\left|U_i - \frac{N_u}{N_r}\right|}{\frac{N_u}{N_r}} = \frac{1}{N_u} \sum_{i=1}^{N_r} \left|U_i - \frac{N_u}{N_r}\right|, \tag{8}$$

where U_i - number of users assigned to router i.

3   The GPU Algorithms

There are many different optimization algorithms used for wireless routers placement problem. In literature, several algorithms have proven to be effective e.g.: tabu search [1], genetic algorithm [1], pruning [1,2], simulated annealing [1,2,10], neighborhood search [1,2]. In the paper, pruning, simulated annealing (SA) and neighborhood search (NS) have been considered. As the most time-consuming part of the algorithms, calculation of the objective function was only parallelized.

3.1   Objective Function Parallelization

The objective function requires determining signal powers of each of wireless routers in each location of interest from set $\mathcal{L} = \{l_1, l_2, \cdots, l_{N_l}\} = \mathcal{U} \cup \mathcal{M}$ (*multi-wall model* is used). Pseudocode representing parallelized signal power estimation algorithm using MWM is presented in Algorithm 1. Signal power values of routers in each location are stored in $\mathcal{Q} = \{q_{l_1}^1, q_{l_2}^1, \ldots, q_{l_{N_l}}^1, q_{l_1}^2, \cdots, q_{l_{N_l}}^{N_c}\}$. In parallelized loop in line 2, for each location from $\mathcal{L}$ signal power of each router is evaluated. In order to determine value of k_i from Eq. 6, intersections check is done in line 5. Intersection occurs when wall w_j intersects straight line connecting router position p_r and point of signal power measurement l_i. In line 12, number of floors between router r and point of interest i is computed.

Lemma 1. *Using Algorithm 1, signal powers of routers placed in N_c possible locations, measured in N_l locations, with N_w walls present, can be determined in time $O(\log N_w)$ on $O(N_l N_c N_w)$-processor CREW PRAMs.*

Proof. Operations in lines 4–10 can be performed in $O(1)$ time on $O(N_w)$ processors, because intersection check takes $O(1)$ time on a single processor. It is a known fact that sum or maximum of n elements can be evaluated in $O(\log n)$ time

Algorithm 1. Signal power estimation using MWM

1: **procedure** MWM($\mathcal{W}, \mathcal{L}, \mathcal{P}, freq$)
2:     **par for** each pair $i = 1, 2, \ldots, N_l,\ r = 1, 2, \ldots, N_c$ **do**
3:         $V \leftarrow \{v_1, v_2, \ldots, v_{N_w}\}$ ▷ temporary variable
4:         **par for** $i = 1, 2, \ldots, N_{N_w}$ **do**
5:             **if** INTERSECTS(p_r, l_i, w_j) **then**
6:                 $v_i \leftarrow L_{w_j}$
7:             **else**
8:                 $v_i \leftarrow 0$
9:             **end if**
10:        **end par for**
11:        $q_{l_i}^r \leftarrow \mathbf{par} \sum_{k=1}^{N_w} v_k$
12:        $k_f \leftarrow |p_{r,f} - l_{i,f}|$
13:        $q_{l_i}^r \leftarrow q_{l_i}^r + L_f \cdot k_d^{\frac{k_f+2}{k_f+1}-b}$
14:        $q_{l_i}^r \leftarrow q_{l_i}^r + (-27.55 + 20\log_{10}(freq) + 20\log_{10}(d(p_r, l_i))) + s_r$
15:     **end par for**
16:     **return** $\mathcal{Q}$
17: **end procedure**

---

on $O(\frac{n}{\log n})$ processors. Therefore, sum in line 11 can be evaluated in $O(N_w)$ time on $O(\frac{N_w}{\log N_w})$ processors. Therefore, whole algorithm has computational complexity of $O(\log N_w)$ on $O(N_l N_c N_w) + O(N_l N_c \frac{N_w}{\log N_w}) = O(N_l N_c N_w)$ processors. ∎

In Algorithm 2, it is shown how objective function components are calculated. Procedure f_1 corresponds to component of the signal power evaluation, whereas procedure f_2 - network load. Signal power simulation needs to be performed only once for all possible router locations at the beginning an of optimization process.

Lemma 2. *For N_r fixed router locations and N_l known signal power values, using Algorithm 2, objective function can be determined in time $O(\log N_m N_r N_u)$ on $O\left(\frac{N_r N_u + N_m N_r}{\log N_r} + \frac{N_r N_u}{\log N_u}\right)$-processor CREW PRAMs.*

Proof. It is known, that the sum and maximum of n elements can be evaluated in $O(\log n)$ time on $O(\frac{n}{\log n})$ processors. Finding number of occurrences problem in line 14 can be solved in $O(log N_u)$ time on $O(\frac{N_u}{\log N_u})$ processors. Therefore, procedure f_1 can be performed in $O(\log N_m) + O(\log N_r) = O(\log N_m N_r)$ time on:

$$O\left(N_m \frac{N_r}{\log N_r}\right) + O\left(\frac{N_m}{\log N_m}\right) = O\left(\frac{N_m N_r}{\log N_r}\right) \tag{9}$$

processors and procedure f_2 in $2O(\log N_r) + O(\log N_u) = O(\log N_r N_u)$ time on:

$$O\left(N_u \frac{N_r}{\log N_r}\right) + O\left(N_r \frac{N_u}{\log N_u}\right) + O\left(\frac{N_r}{\log N_r}\right)$$
$$= O\left(\frac{N_r N_u}{\log N_r}\right) + O\left(\frac{N_r N_u}{\log N_u}\right) \tag{10}$$

Algorithm 2. Calculation of objective function components

```
1: procedure f₁(Q,D)
2:     V ← {v₁, v₂, ..., v_{N_m}}                              ▷ temporary variable
3:     par for i = 1, 2, ..., N_m do
4:         v_i ← max {d_i − par max_{r∈R} {q^r_{m_i}}, 0}
5:     end par for
6:     return  (1/N_m) par ∑_{i=1}^{N_m} v_i
7: end procedure
8: procedure f₂(Q,U)
9:     V ← {v₁, v₂, ..., v_{N_u}}                              ▷ temporary variable
10:    par for i = 1, 2, ..., N_u do
11:        v_i ← par arg max_{r∈R} {q^r_{u_i}}
12:    end par for
13:    par for i = 1, 2, ..., N_r do
14:        U_i ← numberOfUsers(V, i)
15:    end par for
16:    return  (1/N_u) par ∑_{i=1}^{N_r} |U_i − N_u/N_r|
17: end procedure
```

processors. Finally, the objective function can be determined in $O(\log N_m N_r) + O(\log N_r N_u) = O(\log N_m N_r N_u)$ time with the use of $O\left(\frac{N_r N_u + N_m N_r}{\log N_r} + \frac{N_r N_u}{\log N_u}\right)$ processors. ∎

3.2 Optimization Algorithms Outline

Pruning is a relatively simple greedy approximate algorithm used, for instance, in: [1,2]. In the first phase N_c routers are placed in each of N_c possible locations. Then, in each iteration there is selected the router which removal will cause the value of the objective function of remaining routers to be the lowest. This router is removed from the set of routers. Consequently, the next step of $N_c - 1$ routers is initiated. The algorithm terminates when the number of routers reaches the desired N_r routers. In the pruning algorithm the objective function value is calculated for:

$$A_p(N_c, N_r) = \frac{(N_c - N_r + 1)(N_c - N_r - 1)}{2} = \frac{N_c^2 - (N_r + 1)^2}{2} \quad (11)$$

routers configurations. It should be noted, however, that with the increase of N_r, $A_P(N_c, N_r)$ decreases. Hence, for N_r large in comparison to N_c, pruning is a choice worth considering.

Neighborhood Search is a heuristic algorithm used to place routers, among others, in: [1,2]. The algorithm starts with the initial arrangement of routers $\mathcal{P}_0$. The placement can be drawn or it can be the result of a different algorithm. NS algorithm searches similar placements of routers, trying to move router r in the neighborhood of $\mathcal{P}_1 = \aleph_r(\mathcal{P}_0)$. If the obtained arrengement of routers $\mathcal{P}_1$ has lower value of the objective function than the current one, the state is saved. The algorithm terminates when it can no longer lower the value of the objective function for the state of the system. NS does not have built-in mechanisms for recovery from local minima.

Simulated Annealing is a heuristic algorithm which searches the neighbourhood of states of the system to find the lowest energy state (objective function value). It was used to optimize the distribution of WLAN routers, among others, in: [1,2,10]. In this study, the traditional algorithm of simulated annealing was modified by addition of storage of the best solution found so far. The algorithm starts its operation with an initial placement of routers $\mathcal{P}_0$ for a fixed T_0 initial temperature and the maximum number of steps k_{max}. In each step a new deployment of routers is generated, depending both on the current placement and the temperature. In general, new states should be more similar to the old states, the lower the temperature. The similarity function between states is here the evaluation function. A similar model was adopted as in [2] - several zones of routers' similarity. Router r' located in n zone of r router if:

$$(|c_{r',x} - c_{r,x}| \leq n) \wedge (|c_{r',y} - c_{r,y}| \leq n) \wedge (f_h|c_{r',f} - c_{r,f}| \leq n), \qquad (12)$$

where f_h denotes the height of floors. Size of n zone depends on the temperature. The lower the temperature, the smaller the n zone. After drawing the new state, there is evaluation function value for it calculated. If the new state is better, it is stored (if it is better than the best previously under consideration it is stored as the best state). Otherwise, there is only a slight probability of transition into the state of lower value of the evaluation function. The smaller the difference between the energies and the higher the temperature, the probability is greater (based on [11] p. 550):

$$P(E_{old}, E_{new}, T) = \min\left\{1, \exp\left(\frac{E_{old} - E_{new}}{T}\right)\right\}. \qquad (13)$$

The temperature is modified at the end of each step of the algorithm. In this work, the following form cooling scheme was adopted:

$$T = T_0 \cdot \left(1 - \frac{k}{k_{max}}\right). \qquad (14)$$

The algorithm terminates its actions when the maximum number of steps k_{max} is executed.

4   Computational Experiments

Testing environment was PC equipped with an Intel Core 2 Duo E8500 processor clocked at 4085 MHz, 8GB of RAM and a graphics card Gigabyte GeForce GTX

660 with 960 CUDA cores (980 MHz) and 2 GB of RAM. The operating system was Windows 8.1 Pro, whereas the programming language was CUDA C/C++ ver. 6.0 compiled by Visual Studio Ultimate 2012. Parameters used in the test instances were tuned in the following manner: suppression of randomly-spaced walls was set at $-10[\text{dB}]$, the transmission power of routers – at the level of 20[dBm], evenly distributed possible placement of routers and users (one user for every 64 locations of signal strength measurement, $N_u = N_m/64$, users placed in measurement points from $\mathcal{M}$). The data may not have practical interpretation since the spatial configuration of the walls, routers, and the users have no effect on computation time of the algorithm. Locations where the signal strength of routers were measured $\mathcal{M}$, were decomposed at equal intervals on the grid (hereinafter referred to as the map). The size of the map can be denoted by $\sqrt{N_m} \times \sqrt{N_m}$. Speedup defined as GPU computation time in comparison to single threaded CPU was measured. Two computational experiments were conducted: first one, on the signal power simulation algorithm and the second one, on the optimization algorithms.

4.1   Implementation Details

Due to limited number of cores available, not all of the parallelization mechanisms shown in the Sect. 3 were used. In Algorithm 1, computations of walls intersection checks (lines 4–11) are performed sequentially. In Algorithm 2 sums and maxima in the lines 4, 6, 11, 16 were not parallelized. The coordinates of the location, in which the signal strength was measured, were mapped in dimensions x and y of CUDA block, whereas possible router placement locations in dimension z.

Routers are numbered in consecutive numerical order, as has been shown in the mathematical model. This collection, numbers 1 to N_r is mapped to the z dimension of the block, it means that the router 1 corresponds to $z = 1$. There is not a direct link between the position of the router in space, and z coordinate with. When the size of the problem exceed the available number of threads, it was divided into smaller sub-problems.

An important characteristic of proposed implementation is that the simulated signal power was computed and stored in memory of GPU without any need for large data transfers to CPU. Such result has been achieved due to the fact that both objective function computation and signal power simulation were performed on GPU. Since data transfers are usually bottlenecks of parallel GPU algorithms, this approach facilitates attaining larger speedups. On the other hand, memory size has become a bottleneck of its own, limiting problem size.

4.2   Results of Experiments on Simulation Algorithm

First, the influence of the number of measurement points N_m on the obtained speedup was tested. Map size varied from 64×64 ($N_m = 4096$) to 1152×1152 ($N_m = 1327104$), $N_c = 81$, $N_w = 30$. The results are shown in Fig. 1a. As it is visible, the speedup increases with increased size of the map, setting out at around 46. The absence of further growth of speedup is probably due to the low number of CUDA cores. Similar tests were carried out for a smaller number of possible locations of routers and walls (Fig. 1b) givig similar results.

Then, the impact of the number of walls N_w on the obtained speedup was tested. The size of the map was established at 128×128 ($N_m = 16384$), whereas the number of possible routers' locations - $N_c = 64$. The number of walls varied from 0 to 930. The results of the experiment are shown in Figure 1c. As in the previous experiment, the speedup grew with increasing size of the problem, ultimately reaching approximately 43.5.

Finally, the impact of the number of locations, where routers could be placed N_c, on the obtained speedup was tested. The size of the map was established at 128×128 ($N_m = 16384$), whereas the number of walls at $N_w = 20$. The number

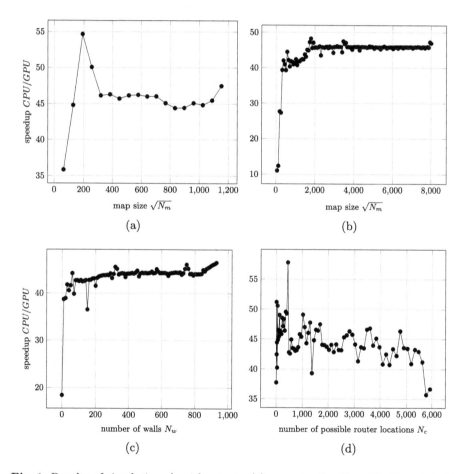

Fig. 1. Results of simulation algorithm tests: (a) map size for $N_c = 81$, $N_w = 30$; (b) map size for $N_c = 4$, $N_w = 5$; (c) number of walls for $N_m = 16384$ (128×128), $N_c = 64$; (d) number of possible router locations for $N_m = 16384$ (128×128), $N_w = 20$.

of possible locations of routers varied from 4 to 5929. The results were shown in Fig. 1d. In contrast to previous tests, the speedup immediately amounted to approximately 45 and decreased with the increase in the number of routers. The initial high speedup can be explained with relatively large initial size of the problem. Similar values can be seen in Fig. 1a, for the size of the map 128×128.

4.3   Results of Experiments on Optimization Algorithms

For optimization algorithms, similarly as in previous experiments, the impact of the parameters: N_c, N_w, N_r on the obtained speedup was tested. SA and

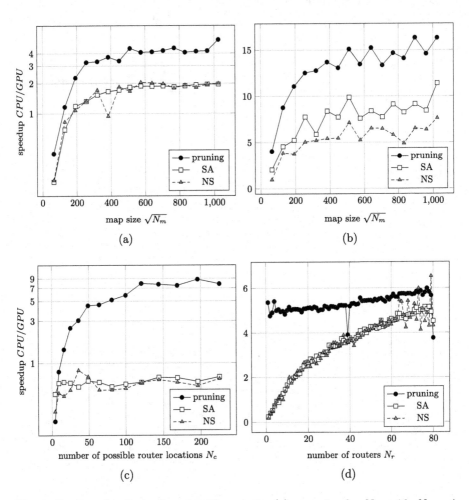

Fig. 2. Results of optimization algorithms tests: (a) map size for $N_c = 16$, $N_r = 4$, $N_w = 10$; (b) map size for $N_c = 64$, $N_r = 20$, $N_w = 10$; (c) number of possible router locations for $N_m = 16384$ (128×128), $N_r = 3$, $N_w = 10$; (d) number of routers for $N_m = 16384$ (128×128), $N_c = 81$, $N_w = 10$.

NA algorithms require appropriate selection of certain parameters in order to obtain good results [2]. However, the aim of this research was not to evaluate the quality of the results obtained by the algorithms. Hence, these parameters were chosen (experimentally) so that all three algorithms obtained the best solutions of similar objective function value. For SA and NS algorithms 1000 iterations were performed, in each case, (whereas for pruning - this number was instance-dependent).

First, the effect of the number of signal strength measurements N_m on the obtained speedup was checked. The study was conducted for two variants: $N_c = 16$, $N_r = 4$, $N_w = 10$ (Fig. 2a) and $N_c = 64$, $N_r = 20$, $N_w = 10$ (Fig. 2b). In the second instance, due to the larger size of the problem higher speedup was observed.

Secondly, the impact of the number of possible routers locations N_c (Fig. 2c) was tested. The size of the map was established at 128×128 ($N_m = 16384$), $N_r = 3$, $N_w = 10$. Pruning algorithm, as the only one, reached the speedup greater than unity. This is due to its specific structure, in which for large N_c and small N_r, there are particularly big number of computations carried out (see formula 11). Such low speedup of SA and NS algorithms was due to small number of routers N_r.

Ultimately, the relationship between the number of routers N_r and the speedup (Fig. 2d) was tested. The size of the map was established at 128×128 ($N_m = 16384$), $N_c = 81$, $N_w = 10$. The target number of routers N_r varied from 1 to 81. Comparing the results of this experiment with the graph in Fig. 2b shows that the achieved maximum speedup for SA and NS algorithms are similar.

Among the tested algorithms, the largest speedups were obtained for pruning algorithm. This is due to the fact, that this is an algorithm which requires computing the value of the objective function for the largest size of the problem. In its initial iterations, the number of routers is close to the number of possible locations where they can be placed. This leads to extending computation time of lines 4 and 11 of Algorithm 2.

5   Conclusions and Future Work

In the paper parallel algorithms for solving the problem of wireless routers placement are proposed. The results are very promising and indicate that the use of GPUs in this field has great potential. In the test configuration, to simulate the signal power, it was possible to obtain speedup up to 46. GPU outperformed CPU even for smaller problem sizes. In case of optimization algorithms, the speedups were smaller but still satisfactory. The best results among the three tested algorithms were obtained for pruning algorithm. Smaller speedups were caused most probably by too simple implemented parallelization of evaluation function. Reducing the amount of sequential computations would probably shorten the execution times of the algorithms. Further work will be related to various propagation models, due to the high practical significance and achieved speedups in the area. The use of methods based on ray-tracing instead of empirical models are taken into consideration.

References

1. Gracia, M., Fernandez-Duran, A., Alonso, J.: Automatic planning tool for deployment of indoor wireless local area networks. In: Proceedings of the 2009 International Conference on Wireless Communications and Mobile Computing: Connecting the World Wirelessly, pp. 1428–1432 (2009)
2. Kamenetsky, M., Unbehaun, M.: Coverage planning for outdoor wireless LAN systems. In: Proceedings of Broadband Communications, Zurich Seminar (2002)
3. Gummandi, R., Wetherall, D., Greenstein, B., Seshan, S.: Understanding and mitigating the impact of RF interference on 802.11 networks. ACM SIGCOMM Comput. Commun. Rev. **37**(4), 385–396 (2007)
4. Lu, J., Runser, K., Gorce, J., F. Valois, F.: Indoor wLAN Planning with a QoS constraint based on a markovian performance evaluation model. In: IEEE 9th International Conference on Wireless and Mobile Computing, Networking and Communications (WiMob), pp. 152–158. IEEE Computer Society (2006)
5. Riihijrvi, J., Mahonen, P., Petrova, M.: Automatic channel allocation for small wireless local area networks using graph colouring algorithm approach. In: Proceedings of the IEEE 15th International Symposium on Personal, Indoor and Mobile Radio Communications, PIMRC 2004, pp. 536–539 (2004)
6. AWE-Communications. http://www.awe-communications.com/
7. iBwave. http://www.ibwave.com/Products/iBwaveDesignEnterprise.aspx
8. Kwiatkowski, J., Gnatowski, A.: Parallelization of wireless routers placement algorithms for the GPU. In: Proceedings of the ISAT 2014–35th International Conference, Information Systems Architecture and Technology - Selected Aspects of Communication and Computational Systems, pp. 87–96 (2014). ISBN 978-83-7493-856-3
9. COST Action 231: Digital Mobile Radio Towards Future Generation Systems: Final Report. European Commission, pp. 175–190 (1999)
10. Xhafa, F., Barolli, A., Sánchez, C., Barolli, L.: A simulated annealing algorithm for router nodes placement problem in Wireless Mesh Networks. Simul. Model. Pract. Theory **19**(10), 2276–2284 (2011)
11. Press, W.H.: Numerical Recipes. The Art of Scientific Computing, 3rd edn. Cambridge University Press, New York (2007)

Toward Parallel Modeling of Solidification Based on the Generalized Finite Difference Method Using Intel Xeon Phi

Lukasz Szustak[1(✉)], Kamil Halbiniak[1], Adam Kulawik[1], Joanna Wrobel[1], and Pawel Gepner[2]

[1] Czestochowa University of Technology, Dabrowskiego 69,
42-201 Czestochowa, Poland
{lszustak,khalbiniak,adam.kulawik,joanna.wrobel}@icis.pcz.pl
[2] Intel Corporation, Swindon, UK
pawel.gepner@intel.com

Abstract. Modern heterogeneous computing platforms have become powerful HPC solutions, which could be applied for a wide range of applications. In particular, the hybrid platforms equipped with Intel Xeon Phi coprocessors offers performance advantages over conventional homogeneous solutions based on CPUs, while supporting practically the same parallel programming model. However, there is still an open issue how scientific applications can utilize efficiently the hybrid platforms equipped with Intel coprocessors.

In this paper we propose a method for porting a real-life scientific application to computing platforms with Intel Xeon Phi. We focus on the parallel implementation of a numerical model of solidification, which is based on the generalized finite difference method. We develop a sequence of steps that are necessary for porting this application to platforms with accelerators, assuming no significant modifications of the code. The proposed method considers not only efficient data transfers that allow for overlapping computations with data movements, but also takes into account an adequate utilization of cores/threads and vector units. The developed approach allows us to execute the whole application 3.45 times faster than the original parallel version running on two CPUs.

Keywords: Intel Xeon Phi · Numerical model of solidification · Application porting · Optimization of data movements

1 Introduction

In the last years, it becomes evident [7,16] that future designs of microprocessors and HPC systems will be hybrid and heterogeneous in nature. An example of this trend are hybrid platforms equipped with the Intel Xeon Phi coprocessors [9,10]. These heterogeneous solutions rely on integration of two major types of components in various proportion to speed up computation intensive applications: (i) multicores CPU technology, as well as (ii) special-purpose hardware and massively parallel accelerators.

© Springer International Publishing Switzerland 2016
R. Wyrzykowski et al. (Eds.): PPAM 2015, Part I, LNCS 9573, pp. 411–422, 2016.
DOI: 10.1007/978-3-319-32149-3_39

The Intel Xeon Phi coprocessor [4,9] is the first product based on Intel Many Integrated Core (Intel MIC) architecture. It includes a large number of cores with wide vector processing units. It offers notable performance advantages over conventional homogeneous solutions based on CPUs, and supports practically the same parallel programming model. Although this architecture is designed for massively parallel applications, there is still an open issue how scientific applications can utilize efficiently the hybrid platforms equipped with Intel coprocessors.

In this study, we present an example of solving this problem by proposing a method for porting a real-life scientific application to computing platforms with Intel Xeon Phi. We focus on the parallel implementation of a numerical model of the dendritic solidification process in the isothermal conditions [1,14]. In this model, the growth of microstructure during the solidification process is determined by solving a system of two partial differential equations (PDEs). These equations define the phase content, and concentration of components in an alloy. The solutions of PDEs is obtained using the Meshless Finite Difference Method (with 2D geometry) and an explicit scheme of calculations.

In this paper, we present a sequence of steps that are necessary for porting application to platforms with accelerators, assuming no significant modifications of the code. In the proposed adaptation, the coprocessor is responsible for executing the major parallel workloads, while the CPU host is used only to execute the remaining part of the application, that do not require massively parallel resources. The main challenges here include not only providing efficient data transfers that overlap computations with data movements, but also ensuring an adequate utilization of cores/threads and vector units. The proposed method allows us to execute computations 3.45 faster than the original parallel code running on two CPUs.

This paper is organized as follows. Section 2 overviews the Intel Xeon Phi architecture, while Sect. 3 shows details of a target platform. Section 4 introduces the numerical model of solidification, which is based on the generalized finite difference method. Section 5 describes the idea of adaptation of the solidification algorithm to computing platforms with Intel Xeon Phi accelerators. Section 6 outlines a sequence of steps that are necessary for porting the main workloads of the application to the Intel Xeon Phi coprocessor, following the idea proposed in the previous section. Section 7 presents performance results obtained for the proposed method, while Sect. 8 concludes the paper.

2   Intel Xeon Phi Coprocessor Overview

The Intel Xeon Phi coprocessor is the first product based on the Intel Many Integrated Core Architecture (Intel MIC Architecture). Tt targets a variety of HPC segments [9,10,17] such as scientific research, physics, chemistry, biology, and climate simulation [12,13,17].

The coprocessor is equipped with more than 50 cores, caches, memory controllers, and PCIe client logic [4,5,11]. All these components are connected together by the bidirectional ring interconnect. Cores are clocked at about 1 GHz, and allow for running up to 4 hardware threads per each core. An integral part of every core is the vector processing unit, that supports a new 512-bit SIMD instruction set called Intel Initial Many-Core Instructions. Each core has 128 vector registers 512-bit wide, and comes complete with a private L1 and L2 caches that are kept fully coherent by the ring interconnect. The coprocessor has over 6 GB of own on-board GDDR5 main memory (maximum 16 GB). The access to the main memory is realized by 6 or 8 memory controllers, that are evenly distributed on the bidirectional bus. The Intel Xeon Phi coprocessors are delivered in form factor of a PCIe additional device.

The Intel MIC architecture provides a general-purpose programming environment similar to that provided for Intel CPUs [9]. It supports the source-code portability between CPU and coprocessor platforms, that gives possibility to run the same code using different devices: Intel CPU or Intel Xeon Phi. Programmers can write source code using most popular programming languages like C, C++ and Fortran. This architecture supports also traditional parallel programming standards such as OpenMP, Intel Thread Building Blocks, Intel Cilk Plus, C++11 threads and MPI.

One of the basic methods to utilize Intel Xeon Phi computing resources is programming in the native mode [9]. In this mode, a source code is compiled on the host using the cross compiler to generate binary for the MIC architecture. Then, the executable application file can be copied and run directly on the coprocessor. Coprocessors support also the heterogeneous programming model known as the offload mode [9]. In this mode, the programmer select code sections to run on the Intel Xeon Phi. This model uses simple pragmas/directives to specify code sections and data to be offloaded to the target device. The application starts on the CPU side, while selected regions are automatically transferred and run on the target device. If for some reason, the Intel Xeon Phi is unavailable, the code regions are executed on the CPU.

3   Target Platform

In this study, we use the platform [10] equipped with two Intel Xeon E5-2695 v2 processors (Ivy Bridge EP architecture), and the top-of-the-line Intel Xeon Phi 7120P coprocessor. Every processor consists of 12 cores clocked at 2.4 GHz, and 128 GB DDR3-1866 main memory. The coprocessor contains 61 cores clocked at 1.238 GHz, and 16 GB of on-board memory. Two CPUs offer 2×230.4 Gflop/s of theoretical peak performance totally, assuming double precision floating-point operations, while a single coprocessor gives 1.2 Tflop/s. The values of peak performance are given taking into account the usage of SIMD vectorization (words 256- or 512-bit wide for CPU or coprocessor, respectively). Table 1 presents a summary of this platform.

Table 1. Specification of test platform [10]

	2 × Intel Xeon E5-2695 V2	Intel Xeon Phi 7120P
Number of cores	2 × 12	61
Number of threads	2 × 24	244
SIMD length [bits]	256	512
Freq. [GHz]	2.4	1.2
Peak for DP [Gflop/s]	2 × 230.4	1208
L1/L2 cache [KB]	64/256	64/512
LLC[a] [MB]	2 × 30	28.5
Memory size [GB]	128	16
Memory bandwidth [GB/s]	2 × 59.7	352

[a]LLC (Last Level Cache) corresponds to either L3 cache for CPU, or aggregated L2 caches for Intel Xeon Phi.

4   Introduction to Numerical Model of Solidification

In the analyzed numerical examples, a binary alloy of Ni-Cu is considered as a system of the ideal metal mixture in the liquid and solid state. The numerical model [1,14] refers to the dendritic solidification process in the isothermal conditions with constant diffusivity coefficients for the liquid and solid phases. It allows to use the field phase model defined by Warren and Boettinger [14]. In this model, the growth of microstructure during the solidification process is determined by solving a system of two PDEs [1,8,14]. The first equations defines the phase content ϕ:

$$\frac{1}{M_\phi}\frac{\partial\phi}{\partial t} = \bar{\varepsilon}^2\left[\nabla\cdot\left(\eta^2\nabla\phi\right) + \frac{\partial}{\partial y}\left(\eta\eta'\frac{\partial\phi}{\partial x}\right) + \frac{\partial}{\partial x}\left(\eta\eta'\frac{\partial\phi}{\partial y}\right)\right] +$$
$$-cH_B - (1-c)H_A - cor, \tag{1}$$

where: M_ϕ is defined as the solid/liquid interface mobility, ε is a parameter related to the interface width, η is the anisotropy factor, H_A and H_B denotes the free energy of both components, cor is the stochastic factor which models thermodynamic fluctuations near the dendrite tip.

The second equation defines the concentration c of the alloy dopant, which is one of components of the alloy:

$$\frac{\partial c}{\partial t} = \nabla\cdot D_c\left[\nabla c + \frac{V_m}{R}c\left(1-c\right)\left(H_B\left(\phi,T\right) - H_A\left(\phi,T\right)\right)\nabla\phi\right], \tag{2}$$

where: D_c is the diffusion coefficient, V_m is the specific volume, R is the gas constant. In this model, the Meshless Finite Difference Method [2,6] is used to obtain the values of partial derivatives with respect to dimensions x and y, that occur in Eqs. 1 and 2.

In order to parallelize computations with a desired accuracy, the explicit scheme is applied with a small value of the time step $\Delta t = 1e - 7s$:

$$\phi_i^{t+1} = \phi_i^t + \Delta t M_\phi \left[\varepsilon^2 \eta^2 \left(\frac{\partial^2 \phi_i^t}{\partial x^2} + \frac{\partial^2 \phi_i^t}{\partial y^2} \right) - cH_B - (1-c)\, H_A - cor + \right.$$

$$\left. + 2\varepsilon^2 \eta \eta' \left(\frac{\partial \phi_i^t}{\partial x} \frac{\partial \theta}{\partial x} + \frac{\partial \phi_i^t}{\partial y} \frac{\partial \theta}{\partial y} \right) + \varepsilon^2 \left(\eta'^2 + \eta \eta'' \right) \left(\frac{\partial \phi_i^t}{\partial x} \frac{\partial \theta}{\partial y} - \frac{\partial \phi_i^t}{\partial y} \frac{\partial \theta}{\partial x} \right) \right], \tag{3}$$

$$c_i^{t+1} = c_i^t + \Delta t \left\{ \frac{\partial D_c}{\partial x} \left[\frac{\partial c}{\partial x} + \frac{V_m}{R} c(1-c)(H_B - H_A) \frac{\partial \phi}{\partial x} \right] + \right.$$

$$+ D_c \left[\frac{\partial^2 c}{\partial x^2} + \frac{V_m}{R} \left[(1-2c) \frac{\partial c}{\partial x} (H_B - H_A) \frac{\partial \phi}{\partial x} + \right. \right.$$

$$\left. + c(1-c) \frac{\partial (H_B - H_A)}{\partial x} \frac{\partial \phi}{\partial x} + c(1-c)(H_B - H_A) \frac{\partial^2 \phi}{\partial x^2} \right] \right] +$$

$$+ \frac{\partial D_c}{\partial y} \left[\frac{\partial c}{\partial y} + \frac{V_m}{R} c(1-c)(H_B - H_A) \frac{\partial \phi}{\partial y} \right] + \tag{4}$$

$$+ D_c \left[\frac{\partial^2 c}{\partial y^2} + \frac{V_m}{R} \left[(1-2c) \frac{\partial c}{\partial y} (H_B - H_A) \frac{\partial \phi}{\partial y} + \right. \right.$$

$$\left. + c(1-c) \frac{\partial (H_B - H_A)}{\partial y} \frac{\partial \phi}{\partial y} + c(1-c)(H_B - H_A) \frac{\partial^2 \phi}{\partial y^2} \right] \right] \right\},$$

where θ is a function of derivatives of ϕ-variable.

These computations correspond to the forward-in-time algorithms [16]. The application code consists of two main blocks of computations, which are responsible for determining either the phase content ϕ and the dopant concentration c. In the model, the values of ϕ and c are calculated for nodes uniformly distributed across the square region, where 751 nodes along every dimension are chosen as sufficient for providing a required accuracy in the example examined in this paper. The values of derivatives at all nodes (Eqs. 3 and 4) are determined at every time step of calculations. All these computations are the main workload for the resulting numerical algorithm.

5   Idea of Adapting the Solidification Application to Computing Platforms with Intel Xeon Phi

In this section, we consider the idea of adaptation of the solidification application to computing platforms with Intel Xeon Phi accelerators. The main goal of our study is to accelerate calculations using Intel Xeon Phi. We propose to employ the coprocessor to perform the major parallel workloads, and use CPU only to execute the rest of application that do not require massively parallel resources.

As a result, the computations that correspond to Eqs. 3 and 4 are performed using the Intel Xeon Phi coprocessor.

In the studied application, computations are interleaved with writing partial results to a file. In the original version, parallel computations are executed for subsequent time steps, while writing results to the file is performed after the first time step, and then after every 100 time steps. The CPU does not perform any computations during writing results to the file. Figure 1 illustrates execution of the computational core of the studied application using CPU only.

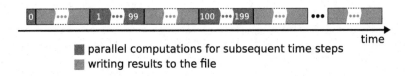

Fig. 1. Implementation of the original version of solidification algorithm

We propose to use Intel Xeon Phi for executing all operations associated with the computational core, and make the CPU responsible for writing partial results to the file. This requires to perform adequate data transfers between processor and coprocessor. Since the studied application belongs to the forward-in-time algorithms, where subsequent time steps depend on the previous ones, all the input data required for the first time step are transferred to the coprocessor, while others input data necessary for subsequent time steps are computed by the coprocessor. So after finishing computations for the first time step, and then for every package of 100 time steps, the coprocessor transfers its outcomes back to the host processor that is responsible for writing these results to the file.

A critical performance challenge here is to overlap workload performed by Intel Xeon Phi with data movements. To reach this goal, we propose to perform simultaneously data transfers between the CPU and Intel Xeon Phi, writing data to the file, as well as computations. This idea is illustrated in Fig. 2.

After finishing computations for the 1st time step, both transfers of the partial results from the coprocessor to CPU and computations for the next package

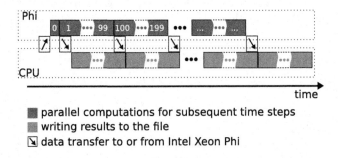

Fig. 2. Idea of parallelization of numerical modeling of solidification using both Intel CPU and Intel Xeon Phi

of 100 time steps are started in the same time. Then during computations for the package of 100 time steps, and after transferring data, the CPU writes required data to the file. Such a scheme repeats for every package of 100 time steps, and gives possibility for overlapping of computations and data movements only when the time of computations will be not shorter than the time of data movements.

6 Porting Application to Intel Xeon Phi

This section presents a sequence of steps that are necessary for porting the main workloads of the application to the Intel Xeon Phi coprocessor, following the idea proposed in the previous section. The first challenge includes overlapping the following operations: (i) data writing to the file on the CPU side, (ii) computations on the coprocessor side, (iii) data transfers between the coprocessor and CPU. The second challenge concerns an adequate utilization of Intel Xeon Phi computing resources. At the same time, we assume no significant modifications of the code, especially for the computing kernels. In this study, we do not provide any improvements for data writing to the file. Furthermore, only the massively parallel regions of the application are transferred to the coprocessor, while the remaining parts are executed on the CPU side. Such a distribution allows for an appropriate utilization of both the coprocessor designed for parallel computations, and the CPU that features a more general usage.

6.1 Optimization of Data Movements

Determining a set of data that have to be transferred between the coprocessor and CPU is the first step for porting the studied application to Intel Xeon Phi. Since the coprocessor is used in the offload mode, the data transfers through the PCIe bus are crucial for the overall performance [4,5]. Generally speaking, the total amount of data transfers between the coprocessor and processor has to be maximally reduced.

This aim is addressed by the proposed idea of adaptation presented in Sect. 5. Following this idea, the required input data are transferred to the coprocessor only once before computations. Then, the appropriate portion of data has to be exchanged after every package of 100 time steps (see Fig. 2).

Selecting an appropriate method for providing efficient data transfers is important for the overall performance. A basic solution to provide the desired efficiency is to ensure a linear (or continuous) access for the required data. It is achieved by choosing an appropriate data structure. Typically, there are two major possibilities for laying out memory [3]: array of structures (AoS) and structure of arrays (SoA). The original version of the studied application utilizes the AoS option. In this case, a periodic access to the required data and/or copying some unnecessary data are necessary for transferring data to and from the coprocessor. To avoid these overheads in the proposed approach, we migrate to the SoA option in order to guarantee both the linear access, and transferring only the necessary data.

```
for(/*...*/)
{
    // Parallel computation for subsequent time steps on Intel MIC side
    #pragma offload target(mic) signal(&offload)
    /*...*/

    // Asynchronous receiving data from coprocessor
    #pragma offload_transfer target(mic)
    /*...*/

    // Asynchronous writing data to the file on CPU side
    /*...*/

    // Waiting for computation finishing
    #pragma offload_wait target(mic) wait(&offload)
    /*...*/
}
```

Fig. 3. Asynchronous data transfers between CPU and coprocessor

To overlap computations and data movements, the asynchronous transfers between the CPU and coprocessor is utilized. The offload mode supports such a solution be applying an adequate sequence of pragmas, as shown in Fig. 3.

The next step is associated with exploration of multiple buffering techniques [15]. To provide simultaneous computations and communications, it is enough to apply two buffers on the coprocessor side, that are responsible for keeping results of subsequent packages of 100 time steps. When one buffer is used for saving results during computation of a given package of 100 time steps, another one is employed for transferring results of the previous package from the coprocessor to CPU. This is achieved at the cost of some extra memory space. Moreover, the double buffering technique is also deployed on the CPU side, which is necessary to overlap data writing to the file with data transfers from the coprocessor to CPU. When data writing to the file utilizes one buffer, the results are transferred to the next one, and so on. In order to improve the overall performance, the memory regions for the buffers are allocated only once for the coprocessor side, and then reused multiple times. It allows us to reduce the number of memory allocations in the offload mode that usually generates a significant performance overhead [4,5].

6.2 Multithreading and Vectorization Optimizations

The previous optimization steps give possibility to overlap computation, communication, as well as data writing to the file. Now the main constraint is to make the time required by computation for a package of 100 time steps no longer then the time of data writing (see Fig. 2). To reach this goal, both cores/threads and vector units should be successfully utilized.

The original version of the studied application employs the OpenMP parallel programming standard to utilize cores/threads. This version uses the basic work-sharing construction `#pragma omp parallel for` to assign work to all the available threads. Since the Intel Xeon Phi coprocessors supports the OpenMP standard, the application code can be successfully executed without any modifications. As a result, all the available threads of the Intel Xeon Phi coprocessor can be utilized for the joint problem solving. To ensure the best overall performance assuming no significant modifications of the code, different setups for the scheduling clauses are evaluated, including `static`, `dynamic`, and `guided`.

The next step required for porting the application to Intel Xeon Phi is associated with applying the vectorization of computations. The compiler-based automatic vectorization [9] seems to be the most convenient method for achieving this goal. The automatic vectorization is provided by the Intel Compiler that automatically uses SIMD instructions available in the Intel Streaming SIMD Extensions. The compiler detects operations in the program that can be done in parallel, and then converts sequences of operations to parallel vector operations. In practice, the automatic vectorization usually occurs when the Intel compiler generates packed SIMD instructions through unrolling the innermost loop.

However, in the studied case the innermost loop can not be vectorized safely because of complexity of computations as well as data dependencies. In fact, calculating a single output element in the innermost loop requires a set of input elements with dynamically determined indexes. An example of such a situation is presented in Fig. 4a. In this case, the automatic vectorization of computations fails because of an irregular data access, unpredictable during compilation. To solve this problem, we propose to change slightly the code by adding temporary vectors responsible for loading the necessary data from the irregular memory region. It is enough to provide SIMD computations. This idea is illustrated in Fig. 4b.

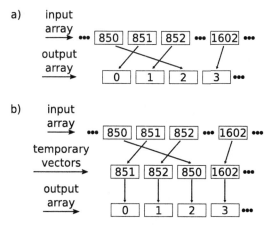

Fig. 4. Idea of vectorization: (a) scalar computation based on irregular data access, (b) vectorization of computation using temporary vectors

Additionally, appropriate keywords and directives should be provided as compiler hints, in order to improve the efficiency of auto-vectorization. The auto-vectorization is also assisted with applying an appropriate data alignment for the vectorized data. This forces the compiler to create data objects in memory aligned to specific byte boundaries.

7   Performance Results

In this section, we present performance results obtained for the approach described in Sects. 5 and 6, assuming double precision floating point numbers. All the benchmarks are compiled using the Intel icpc compiler (v.15.0.2) with the optimization flag -O3. The resulting code is executed in the offload mode on the platform equipped with the Intel Xeon E5-2695 V2 CPU and Intel Xeon Phi 7120P coprocessor, while the original version of the solidification application uses parallel resources of two CPUs. All the tests are performed for 2626 time steps, and the 564 001 nodes (751 nodes along dimensions x and y).

In our benchmarks, we evaluate different loop scheduling options (static, dynamic, auto and guided) with different configuration for the size of chunks. For all the performance tests, we achieve similar performance results with differences below 5 %. The best performance corresponds to the static scheduling with equal-sized chunks of loop assigned to threads.

We check also the impact of the auto-vectorization for the overall performance. For the vectorized regions of the code, the computation are accelerated of about 2 times. This relatively small performance gain is mainly caused by overheads required for providing the irregular data access (see Fig. 4). The total performance gain of vectorization decreases when not vectorizable regions are taken into account. In consequence, the final speedup of the code with vectorization is 1.6 against the scalar code.

Table 2 presents the performance results for modeling solidification obtained for the original and proposed (optimized) codes. The total execution of the basic version takes 335 s. It includes the aggregated time of computations (244 s) and the aggregated time of data writing (94 s). In this case, all computations are not overlapped with writing output data to the file (see Fig. 1).

As compared with the original version, the optimized code performs all the workloads 3.45 times faster. Its execution takes 97 s, and includes computations, data writing, and transfers from and to the coprocessor. The aggregate time required to write data to the file is the same as in the case of the original version (94 s), while the aggregate time of computations (80 s) is shorter about 3 times. The transfer of all the input data takes 2.41 s, and occurs only at the beginning of the application execution, while transfers of partial results from the coprocessor takes totally 0.07 s only.

For about 81.8 % of the application execution time, data movements are overlapped with computations. Moreover, almost all the computations performed by the Intel Xeon Phi device are overlapped with data movements, and this allows us to hide 99.26 % of computations behind data writing.

Table 2. Performance results for modeling solidification using original and optimized codes

	Original version	Optimized version
Total Time [s]	335	97
Aggregate time of computations [s]	242	80
Aggregate time of data writing [s]	94	94
Time of input data transfers from CPU to Intel Xeon Phi [s]	-	2.41
Aggregate time of output data transfers from Intel Xeon Phi to CPU [s]	-	0.07

8   Conclusions

Using computing platforms with Intel Xeon Phi is a promising direction for improving the efficiency of parallel computations in the numerical modeling of solidification. Porting applications to the Intel MIC architecture requires to use the native or offload mode, where the latter proved to be a better solution for the studied application. In this paper, we propose the method for porting and optimizing the application for modeling alloy solidification on computing platforms with Intel Xeon Phi. In comparison with the original parallel version of the code, the optimized version performs all the workloads 3.45 times faster.

The offload mode is an efficient solution for porting large codes that consist of both massively parallel and sequential regions to platforms with Intel MIC. Using the coprocessor to perform major parallel workloads, and employing the CPU only to execute the rest of an application give a strong possibility to accelerate the whole application. Such a workload distribution allows for an appropriate utilization of both the coprocessor designed for massively parallel computations, and the CPU that is designed for the general usage.

The Intel MIC architecture provides a general-purpose programming environment that allows for quick and easy code porting. However, the utilization of the offload mode requires to implement data transfers between the processor and coprocessor efficiently. The solution of this issue is provided by the proposed method of the adaptation of the application for modeling alloy solidification.

This method allows us to overlap (i) data writing to the file on the CPU side, (ii) computations on the coprocessor side, and (iii) data transfers between the coprocessor and CPU. As a result, for about 81.8 % of the application execution time, data movements are overlapped with computations. When considering only the time of computations, the proposed adaptation allows us to hide 99.26 % of computations behind data movements. Such a high degree of overlapping is achieved by applying multithreading and vectorization optimizations. This allows us to reduce the time of executing the parallel workload by the coprocessor, and makes it no longer then the time of data movements.

Acknowledgments. The authors are grateful to the Czestochowa University of Technology for granting access to Intel CPU and Xeon Phi platforms providing by the MICLAB project No. POIG.02.03.00.24-093/13.

References

1. Adrian, H., Spiradek-Hahn, K.: The simulation of dendritic growth in Ni-Cu alloy using the phase field model. Arch. Mater. Sci. Eng. **40**(2), 89–93 (2009)
2. Benito, J.J., Ureñ, F., Gavete, L.: The generalized finite difference method. In: Àlvarez, M.P. (ed.) Leading-Edge Applied Mathematical Modeling Research, pp. 251–293. Nova Science Publishers Inc. (2008)
3. Hager, G., Wellein, G.: Introduction to High Performance Computing for Scientists and Engineers. CRC Press, Boca Raton (2011)
4. Intel Xeon Phi Coprocessor System Software Developers Guide. Intel Corporation (2013)
5. Jeffers, J., Reinders, J.: Intel Xeon Phi Coprocessor High-Performance Programming. Elsevier Inc., Waltham (2013)
6. Kulawik, A.: The modeling of the phenomena of the heat treatment of the medium carbon steel. Wydawnictwo Politechnki Czestochowskiej, Monografia, no. 281 (2013) (in Polish)
7. Kurzak, J., Bader, D., Dongarra, J. (eds.): Scientific Computing with Multicore and Accelerators. CRC Press, Boca Raton (2011)
8. Longinova, T., Amberg, G., Ågren, J.: Phase-field simulations of non-isothermal binary alloy solidification. Acta Mater. **49**(4), 573–581 (2001)
9. Parallel Programming and Optimization with Intel Xeon Phi Coprocessors, Handbook on the Development and Optimization of Parallel Applications for Intel Xeon Processors and Intel Xeon Phi Coprocessors. Colfax International, Sunnyvale, CA (2013)
10. Pilot Laboratory of Massively Parallel Systems (MICLAB). http://miclab.pl
11. Rahman, R.: Intel Xeon Phi Coprocessor Architecture and Tools: The Guide for Application Developers. APress, Berkeley (2013)
12. Szustak, L., Rojek, K., Gepner, P.: Using Intel Xeon Phi coprocessor to accelerate computations in MPDATA algorithm. In: Wyrzykowski, R., Dongarra, J., Karczewski, K., Waśniewski, J. (eds.) PPAM 2013, Part I. LNCS, vol. 8384, pp. 582–592. Springer, Heidelberg (2014)
13. Szustak, L., Rojek, K., Olas, T., Kuczynski, L., Halbiniak, K., Gepner, P.: Adaptation of MPDATA heterogeneous stencil computation to Intel Xeon Phi coprocessor. Sci. Program. **2015**, 14 (2015)
14. Warren, J.A., Boettinger, W.J.: Prediction of dendritic growth and microsegregation patterns in a binary alloy using the phase-field method. Acta Metall. Mater. **43**(2), 689–703 (1995)
15. Wyrzykowski, R., Rojek, K., Szustak, L.: Model-driven adaptation of double-precision matrix multiplication to the Cell processor architecture. Parallel Comput. **38**(4–5), 260–276 (2012)
16. Wyrzykowski, R., Szustak, L., Rojek, K.: Parallelization of 2D MPDATA EULAG algorithm on hybrid architectures with GPU accelerators. Parallel Comput. **40**(8), 425–447 (2014)
17. Xue, W., Yang, C., Fu, H., Wang, X., Xu, Y., Liao, J., Gan, L., Lu, Y., Ranjan, R., Wang, L.: Ultra-scalable CPU-MIC acceleration of mesoscale atmospheric modeling on Tianhe-2. IEEE Trans. Comput. **64**(8), 2382–2393 (2015)

Optimized Parallel Model of Human Detection Based on the Multi-Scale Covariance Descriptor

Nesrine Abid[1]([✉]), Tarek Ouni[1], Kais Loukil[1], A. Chiheb Ammari[2,3], and Mohamed Abid[1]

[1] Laboratory of Computer and Embedded Systems,
National School of Engineering of Sfax, Sfax, Tunisia
nesrineabid88@gmail.com
[2] MMA Laboratory, National Institute of the Applied
Sciences and Technology, Tunis, Tunisia
[3] Renewable Energy Group, Department of Electrical and Computer Engineering,
Faculty of Engineering, King Abdulaziz University, Jeddah 21589, Saudi Arabia

Abstract. Human detection based on the multi-scale covariance descriptor outperforms many other antecedent descriptors. However, it has the disadvantage of being highly time consuming. The complexity of this type of application intensifies the need of multiprocessor architecture (MPSOC) to meet real time constraints. A well-balanced application model is necessary for an efficient implementation into MPSOC architecture. In this paper, a high-level independent target-architecture parallelization approach is used to propose an optimized parallel model of a multi-scale covariance human detection system. The main characteristic of this approach is the exploration of both task and data levels of parallelism. For this end, an initial model is proposed. This model is implemented and validated at a high level interface. The potential bottlenecks of this model are identified using communication and computation workload analysis. Based on this analysis, an optimized parallel model with maximum workload balance is provided. Results reveal that the obtained parallel model has more than four times lower execution time in comparison with the sequential model.

Keywords: Multi-scale covariance descriptor · Human detection · MP-SOC · Kahn processor network · Parallel model

1 Introduction

Human detection is a crucial task in video surveillance, robotic and in other fields. It represents a hard issue due to the large variability of the target of interest, which is subjected to occlusions, variations in pose, shape, and appearance. For all these reasons, many solutions of features extraction and classifier construction have been proposed. The typical approaches are characterized by a classifier that operates by means of a sliding window over the image. The classifier may be fed with extracted features for example Haar-like features [1], wavelet,

© Springer International Publishing Switzerland 2016
R. Wyrzykowski et al. (Eds.): PPAM 2015, Part I, LNCS 9573, pp. 423–433, 2016.
DOI: 10.1007/978-3-319-32149-3_40

SIFT [2], Histograms of Oriented Gradients (HOG) [3], or covariance descriptors [4]. The experiments in [3] show the high accuracy of the HOG descriptor in the detection applications. In [4] researchers proposed a region covariance descriptor technique for human detection that outperforms the previous approaches even outperforms the HOG based solution. In [5] authors proposed a multi-scale covariance descriptor (MSCOV). The MSCOV descriptor combines the quadtree integral image structure with the covariance matrix to optimize the trade-off between local and global object description, to discard noisy image information and capture relevant ones and to speed up the computation. In fact, in [6] authors show that the MSCOV outperforms the COV descriptor and other well known human detection descriptors. It has much smaller dimensionality than other representations based on histograms. And even better, they can be calculated super fast. For all these reasons, we propose to implement a human detection application based on multi-scale covariance descriptor for an embedded System-on-Chip. The software implementation of this application using a mono processor architecture running on a stratixII Altera FPGA has taken 41 s to detect human from a 150×99 image. Although beside the detection accuracy, detection speed is an important measure in human detection. In order to accelerate detection speed, many researchers adopted the hardware acceleration [7]. In this context, authors in [8] propose a software/hardware human detection architecture implemented on an ATOM processor and a FPGA-based accelerator. This design could achieve nearly 20x speedup for the critical function block compared with the processor-only solution but the total detection execution time still slow about 11.453 s. In [9] authors propose a novel human detection architecture using ZedBoard platform. The detection application was implemented on ARM Coretex-A9 processor connected to hardware accelerator. The Whole detection system had been achieved 3.22 times speed up but about 2.68 % loss of the recognition accuracy. In addition, hardware acceleration based approach show fast developing time which is not suitable for application requiring ultra low cost. So it is costly in delay and area size. The most used hardware description languages is VHDL and Verilog but it requires a great deal of programming effort. In the other hand, the high level language modified C syntax such as SystemC and HandelC which aim to raise FPGA programming from gate-level to a high-level have some limitations for example HandelC does not support pointer and standard floating point. Therefore, to achieve faster processing needed for real time detection without sacrificing quality, we motivate the use of a multiprocessor approach. The two mains advantages of using the multiprocessor implementation is first the high flexibility in that developers can directly modify instructions in codes. Second, the used programming language is based on C syntax that is very adapted in image processing. Prior to any multiprocessing implementation, a parallel specification of the application is required. In this paper, we propose to use a high-level parallelization approach of the MSCOV human detection application. This approach is based on the use of Kahn Process Network (KPN) [10] parallel models implemented by Y-chart Applications Programmers Interface (YAPI) C++ library [11]. The key characteristic of this approach is the simultaneous use of data and task levels of parallelism for the optimal balance of the parallel model. In this context, an initial KPN parallel model of the MSCOV human detection system is proposed. This model is implemented through YAPI environment. A computation and communication workload analysis of the obtained parallel model is considered to identify the potential bottlenecks and to provide a global guidance when optimizing concurrency between processes. Task level splitting, merging and data-level partitioning are then simultaneously explored to get an optimized parallel YAPI/KPN model. The goal of this optimization is to get a parallel model of a MSCOV human detection with the best computation and communication workload balance. The paper is organized as follows: Sect. 2 presents the human detection algorithm based on MSCOV descriptor. In Sect. 3 we present the used parallelization approach and then we discuss the different steps used to get an optimized parallel model. Section 4 concludes the paper.

2 Multi-scale Covariance Descriptor for Human Detection

The human detection system that we are targeting is based on a MSCOV descriptor [5] followed by a Support Vector Machine (SVM) classification [12]. The SVM algorithm operates under a sliding window. The window is a rectangular region that scans the entire image with a vertical and horizontal constant stride. First, a MSCOV descriptor is calculated for each window. Afterwards, the descriptor is classified by the SVM classifier. The block diagram of the MSCOV human detection application is shown in Fig. 1.

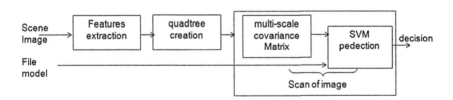

Fig. 1. MSCOV based human detection

As shown in Fig. 1, the studied application is composed of four modules. The first module extracts features. The second module creates the quadtree and stores extracted features in the quadtree nodes. In the third module the quadtree is scanned to collect image quadrants which are inside scan widow as shown in Fig. 2. Finally, the fourth module performs the classification.

2.1 Features Extraction

Ten features Fc are used in the MSCOV descriptor which are pixel location "x" and "y", quadtree level "l", intensity "Ng", color components values "Cr" and

"Cb",firstorder partial derivative in x "xsobel", firstorder partial derivative in y "ysobel", magnitude "mag" and edge orientation "d". The "x, y, l" features are extracted when creating the quadtree. The "Cb, Cr, Ng" features can be computed in parallel. Once the "Ng" is extracted, the "xsobel" and "ysobel" can be computed in parallel. Using "xsobel" and "ysobel" the "grad" and "mag" can also be computed in parallel.

2.2 Quadtree Creation

The Quadtree is a multi-scale image representation quadtree that divide recursively the image into four equal size quadrant (node) as shown in Fig. 2. Each node in the quadtree stores the extracted feature vector (Fci), the sum of each feature Pc(k) and the multiplication of each two features Qc(k, l). Where P and Q of nodes whose child are leafs are computed by

$$P_c(k) = \sum_{i=0}^{3} F_{ci}(k) \tag{1}$$

$$Q_c(k, l) = \sum_{i=0}^{3} F_{ci}(k) * F_{ci}(l) \tag{2}$$

And P and Q of parent internal nodes are computed from child ones:

$$P_c(k) = \sum_{i=0}^{3} P_{ci}(k) \tag{3}$$

$$Q_c(k, l) = \sum_{i=0}^{3} q_{ci}(k, l) \tag{4}$$

Where Ci is child node of node c. Finally, the multi-scale covariance matrices are calculated for each scan window by the following relation:

$$C_c(k) = \frac{1}{N_C}[Q_C(k, l) - \frac{1}{N_C}P_C(k)P_C(l)] \tag{5}$$

where Nc is the number of descendants nodes of C. We refer readers to [5] for more details about the multi-scale covariance descriptor.

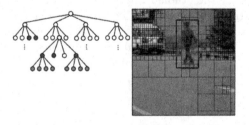

Fig. 2. Schematic of the quadtree in MSCOV human detection application.

2.3   The Classifier

The classifier classifies unlabeled descriptors based on their similarity with descriptors in their training sets. For classification of the MSCOV descriptors we are using a SVM algorithm [9]. SVM is a classification method, which has proved to be very efficient in such case of high dimensional data. Two classes considered one for human that equals 1 and the other class is for no human objects and equals 0.

3   Parallelization Approach

To achieve faster processing for real time detection using MSCOV descriptors, a multiprocessing approach is motivated. Prior to any multiprocessor implementation, a parallel specification is required to functionally describe the studied application as a set of processes exchanging data according to an appropriate model of computation. The KPN [10] is the most used for multimedia application. Several multiprocessor designing frameworks [13,14] are based on the KPN model. To execute the KPN in a parallel fashion, several implementations are provided. Since the most language chosen for writing image processing programs is the C/C++, we choose to implement the KPN process using the Y-chart Applications Programmers Interface (YAPI) [11]. Using YAPI, each process is modeled as a light-weight thread that communicates data with other processes via unbounded FIFO channels.

3.1   The Initial Parallel Model

First, only task-level parallelism is exploited to extract the maximum parallelism from blocks diagram. The application is decomposed in separate blocks. Each block defines one single process that runs a separate stage of an algorithm. Next, the inter-process communication is established using a message passing KPN primitives. Going through this procedure, the Initial proposed model shown in Fig. 3 is obtained and then implemented using the YAPI environment. This model performs as follows: the Ng, Cb and Cr processes collect image data from the input file. The Xsobel and Ysobel processes calculate the first order derivative. The mag and grad processes calculate the second order derivative. The outputs of these processes represent the features. Each feature is stored in the quadtree nodes by "quadtree" process. The obtained quadtree is forwarded to Compcov process to calculate covariance matrix. Finally each obtained descriptor is classified by "SVM" process.

The KPN/YAPI implementation is validated by high level functional simulation. The correctness of the parallel code is proved by comparing both execution results of sequential and parallel code using the same input. For performance evaluation of the proposed parallel model, two important functional properties are generated: the communication workload and the computation workload. The communication workload for a particular process network FIFO communication

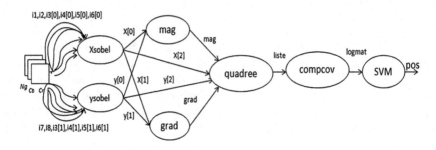

Fig. 3. Initial parallel model of the MSCOV human detection application

channel represents the amount of data exchange over this channel. The computation workload represents the processing time. These computation and communication characteristics define the concurrency properties of the model and measure the efficiency of the computation division over the different processes. Thereby a parallel model with good concurrency proprieties should have a balanced computation workload for all the network processes together with a balanced communication workload over its different FIFO communication channels.

The Communication Workload. The obtained communication workload results for an image of 640×422 resolution are shown in Fig. 4. Rtoken and Wtokens present respectively the total number of Read tokens and Write tokens. The average amount of data communicated per call between two processes is represented by Tsize.

As shown in Fig. 4., the Initial model has 21 FIFO channels transmitting 270080 tokens of 4 bytes size. So the total number of bytes communicated over each channel is equal to 270080*4*1 bytes. This big transmitted amount of data

	size	Tsize	Wtokens	Wcalls	T/W	Rtokens	Rcalls	T/R
mscovdetect.i	270080	4	270080	270080	1	270080	1	270080
mscovdetect.i1	270080	4	270080	270080	1	270080	270080	1
mscovdetect.i2	270080	4	270080	270080	1	270080	270080	1
mscovdetect.i3[0]	270080	4	270080	270080	1	270080	270080	1
mscovdetect.i3[1]	270080	4	270080	270080	1	270080	270080	1
mscovdetect.i4[0]	270080	4	270080	270080	1	270080	270080	1
mscovdetect.i4[1]	270080	4	270080	270080	1	270080	270080	1
mscovdetect.i5[0]	270080	4	270080	270080	1	270080	270080	1
mscovdetect.i5[1]	270080	4	270080	270080	1	270080	270080	1
mscovdetect.i6[0]	270080	4	270080	270080	1	270080	270080	1
mscovdetect.i6[1]	270080	4	270080	270080	1	270080	270080	1
mscovdetect.i7	270080	4	270080	270080	1	270080	270080	1
mscovdetect.i8	270080	4	270080	270080	1	270080	270080	1
mscovdetect.X[0]	270080	4	270080	270080	1	270080	270080	1
mscovdetect.X[1]	270080	4	270080	270080	1	270080	270080	1
mscovdetect.X[2]	270080	4	270080	270080	1	270080	1	270080
mscovdetect.Y[0]	270080	4	270080	270080	1	270080	270080	1
mscovdetect.Y[1]	270080	4	270080	270080	1	270080	270080	1
mscovdetect.Y[2]	270080	4	270080	270080	1	270080	1	270080
mscovdetect.mag	270080	4	270080	270080	1	270080	1	270080
mscovdetect.grad	270080	4	270080	270080	1	270080	1	270080
mscovdetect.liste	1000	4	1	1	1	1	1	1
mscovdetect.o	1000	4	77700	777	100	77700	777	100
mscovdetect.logmat	1000	4	77700	777	100	77700	777	100
mscovdetect.pos	128	4	1554	777	2	1554	777	2

Fig. 4. Communication workload of the initial parallel model

will require the connected processes to spend a lot of time dealing with communication. In addition, four other channels are exchanging fewer amounts of data. All this make the communication workload of the initial model unbalanced. For a better communication behavior, we should be looking for using data and task level parallelism or merging techniques.

Computation Workload Analysis. The computational workload is also necessary for analyzing the concurrency properties. The obtained results in terms of CPU percentage time spent in the execution of each process are presented in Fig. 5.

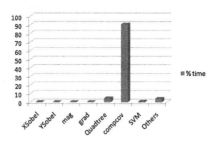

Fig. 5. Parallel computational profiling of the initial model

As shown in Fig. 5, the Initial model has too much unbalanced computational workload. All processes have a negligible load compared to "compcov" process which has the higher load. The "compcov" process is very computational-expensive with more than 90 % of the total computation time complexity. Therefore it is clear, using the obtained results of communication and computation workload, that the Initial parallel model of MSCOV based human detection has poor concurrency properties. This stimulates its optimization for a better communication and computation workload balance. Typically, to get a better communication behavior, we should use data level parallelism and task level splitting or merging. Data level parallelism consists in splitting the data exchanged over selected channels thus duplicating the associated processes of the model. Task level merging consists in merging the processes that are exchanging large data structures. Using task level splitting the available task-level parallelism is extracted by further splitting the computing nodes. The decision on using data splitting and task merging or splitting depends on the computational workload of the studied network. Generally, the processes, that have low computation with high communication loads, are merged while data splitting is applied for those with high computational workload.

3.2   The Optimized Parallel Model

This section presents the different steps that have been used to derive in a structured way a parallel implementation of the MSCOV human detection application with a balanced workload and good communication behavior. Using the profiling results of Figs. 4 and 5, it is clear that the processes "xsobel", "yso-bel", "grad" and "mag" have negligible computations loads and transmit a very big data amounts to "quadtree" process. So we propose to merge it into only one "sob_quad" process. In this case "xsobel", "ysobel", "grad" and "mag" fea-tures are calculated when the quadtree is created. So the associated channels transmitting very large tokens structures are removed. For a better concurrency optimization, the data splitting of the most computational-expensive "comp-cov" process is proposed to distribute the computation of this process in four processes "compcov1", "compcov2", "compcov3" and "compcov4". For this end a data dependence analyses is performed.

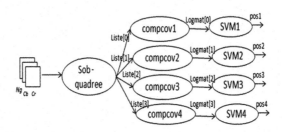

Fig. 6. Proposed optimized parallel KPN model of the MSCOV human detection application.

Concurrency Results of the Optimized Model. The obtained opti-mized parallel model of the MSCOV human detection application is given in Fig. 6. This figure shows the task-merging of the "sob_quad" process and the data-partitioning "compcov1", "compcov2", "compcov3" and "compcov4". This model has been implemented and validated at YAPI system level. The obtained communication and computational workload results are shown respectively in Figs. 7 and 8 for the 640×422 image. It is clear from Fig. 7 that the optimized proposed model has better communication behavior compared to the Initial model. The number of channels and the total number of communicated tokens are reduced. In addition, as indicated in Fig. 8, merging tasks decrease the time processes spent in communication and the data spilling distribute the computa-tion over processes. This final proposed model of the MSCOV human detection has obviously better communication and computational behavior compared to the Initial model.

3.3   Results Discussion

To evaluate the obtained model, we compare the proposed model first with the initial model and the sequential model of the MSCOV human detection application then with some works in literature. Table 1 presents the execution time performance of each model. The results prove that our obtained model outperforms the other previous models.

	size	Tsize	Wtokens	Wcalls	T/W	Rtokens	Rcalls	T/R
mscovdetect.liste[0]	1000	4	1	1	1	1	1	1
mscovdetect.liste[1]	1000	4	1	1	1	1	1	1
mscovdetect.liste[2]	1000	4	1	1	1	1	1	1
mscovdetect.liste[3]	1000	4	1	1	1	1	1	1
mscovdetect.logmat1	1000	4	18900	189	100	18900	189	100
mscovdetect.logmat2	1000	4	18900	189	100	18900	189	100
mscovdetect.logmat3	1000	4	18900	189	100	18900	189	100
mscovdetect.logmat4	1000	4	18900	189	100	18900	189	100
mscovdetect.posR1	1000	4	378	189	2	378	189	2
mscovdetect.posR2	1000	4	378	189	2	378	189	2
mscovdetect.posR3	1000	4	378	189	2	378	189	2
mscovdetect.posR4	1000	4	378	189	2	378	189	2

Fig. 7. Communication workload of the optimized parallel model

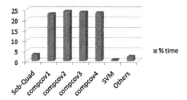

Fig. 8. Parallel computational profiling of the final model

The initial model is parallelized only with tasks partitioning parallelism. The use of this form of parallelism improves execution time performance. The total processing time has decreased from 14.875 s for the sequential code to 10.094 s for the initial parallel model. Depending on the computational and communication workload the initial model may have to spend more time dealing with communication rather than computation, which is completely unacceptable. To achieve a better performance and get an optimized parallel model for MSCOV based human detection, we have used a high level parallelization approach based on task and data level splitting and merging. As given in Table 1, the obtained model achieves a better execution time performance in comparison with the initial model. It accelerates the execution time three times as much as the initial. Table 2 shows the comparison of the proposed model with some works in literature of accelerating human detection application. It presents the number of detection window per seconds, the execution time performance and the total speed-up of each work. All results are given when computing over the image size 640×422 pixels.

Table 1. Comparison of the three models

	Execution time s	Speed-up
Sequential	14.875	-
Initial model	10.094	1/3
Optimized model	3.25	3/4

Table 2. Comparison with previous works

	[8]	[9]	Our model
Windows/second	-	195.133	239.07
Execution time (s)	11.453	3.981	3.25
Total speed-up	2.128	3.22	4.577
Accuracy deterioration	0 %	2.7 %	0 %

As shown in Table 2, The total execution time in [9] archives about x3.22 speedup but the accuracy is degreed by 2.7 %. In [8] the total execution time archives about x2.12 speedup without accuracy deterioration but with a high time. However, our proposed optimized model accelerates the execution time x4.57 without any accuracy deterioration and with 99,99 % detection accuracy on MIT CBCL dataset.

4   Conclusion

In this paper an optimized parallel model of a MSCOV based human detection system is developed using a high-level independent target-architecture parallelization approach. This approach is based on the use of KPN parallel programming models. It is characterized by the exploration of task and data levels of parallelism and merging to ensure for the optimal balance of the model computation and communication behavior. In this context, we proposed an Initial parallel model that exhibits the maximum task level parallelism. This model is then implemented and validated using the YAPI environment. The communication and computation workload analysis of the KPN/YAPI Initial model has showed very poor concurrency properties. To improve the model concurrency properties, data level parallelism and task level splitting or merging are applied. At the end, an optimized parallel model of the MSCOV human detection system is obtained. This model gives considerable computation and inters process communication workload balance without sacrificing quality.

References

1. Lienhart, R., Maydt, J.: An extended set of Haar-like features for rapid object detection. In: ICIP 2002, pp. 900–903. IEEE Press (2002)
2. David, G.: Distinctive image features from scale-invariant keypoints. J. Comput. Vis. **147**(60), 91–110 (2004)
3. Dalal, N., Triggs, B.: Histograms of oriented gradients for human detection. In: Computer Vision and Pattern Recognition, pp. 886–893. IEEE Computer Society, Compute (2005)
4. Tuzel, O., Porikli, F., Meer, P.: Pedestrian detection via classification on Riemannian manifolds. Trans. Pattern Anal. Mach. Intell. **30**, 1713–1727 (2008)

5. Walid, A., Hichem, S., Mohamed, A.: A fast multi-scale covariance descriptor for object re-identification. Pattern Recogn. Lett. **33**, 1902–1907 (2011)
6. Walid, A., Hichem, S., Smach, F., Mohamed, A.: The multi-scale covariance descriptor. In: IEEE Workshop on Biometric Measurements and Systems for Security and Medical Applications, Performances Analysis in Human Detection, pp. 1–5 (2012)
7. Abid, N., Ayedi, W., Ammari, A.: SW/HW implementation of Image covariance descriptor for person detection system. In: ATSIP, pp. 115–119. IEEE (2014)
8. Zhu, Y., Liu, Y., Zhang, D., Li, S., Zhang, P.: Acceleration of pedestrian detection algorithm in novel C2RTL HW/SW codesign platform. In: International Conference Green Circuits and Systems, pp. 615–620 (2010)
9. Mao, H., Takaaki, M., Hideharu, A.: Data reduction and parallelisation for human detection system. In: The 19th Workshop on Synthesis And System Integration of Mixed Information Technologies, pp. 134–139 (2015)
10. Kahn, G.: The semantics of a simple language for parallel programming. In: Proceedings of IFIP 1974 (1974)
11. Kock, E., Essink, G., Smits, W., Wolf, P., Brunel, J.Y., Kruijtzer, W.M., Lieverse, P., Vissers, K.A.: YAPI: application modeling for signal processing system. In: IEEE Procceeding Design Automation Conference, pp. 402–405 (2000)
12. Fradkin, D., Muchnik, I.: Support vector machines for classification. Mathematics Subject Classification (2000)
13. Corre, Y., Diguet, J.-P., Lagadec, L., Heller, D., Blouin, D.: Fast template-based heterogeneous MPSoC synthesis on FPGA. In: Brisk, P., Figueiredo Coutinho, J.G., Diniz, P.C. (eds.) ARC 2013. LNCS, vol. 7806, pp. 154–166. Springer, Heidelberg (2013)
14. Lavagno, L., Lazarescu, M., Walters, J., Kienhuis, B.: HEAP: a highly efficient adaptive multi-processor framework. In: European Projects in Embedded System Design: EPESD, pp. 1050–1062 (2012)

Neural Networks, Evolutionary Computing and Metaheuristics

Parallel Extremal Optimization with Guided Search and Crossover Applied to Load Balancing

Eryk Laskowski[1]([⊠]), Marek Tudruj[1,4], Ivanoe De Falco[2], Umberto Scafuri[2], Ernesto Tarantino[2], and Richard Olejnik[3]

[1] Institute of Computer Science, Polish Academy of Sciences, Warsaw, Poland
{laskowsk,tudruj}@ipipan.waw.pl
[2] Institute of High Performance Computing and Networking, CNR, Naples, Italy
{ivanoe.defalco,umberto.scafuri,ernesto.tarantino}@na.icar.cnr.it
[3] Computer Science Laboratory, University of Science and Technology of Lille,
Villeneuve-d'Ascq, France
richard.olejnik@lifl.fr
[4] Polish-Japanese Academy of Information Technology, Warsaw, Poland

Abstract. Extremal Optimization is a nature-inspired optimization method which features small computational and memory complexity. Due to these features it can be efficiently used as an engine for processor load balancing. The paper presents how improved Extremal Optimization algorithms can be applied to processor load balancing. Extremal Optimization detects the best strategy of tasks migration leading to balanced application execution and reduction in execution time. The proposed algorithm improvements cover several aspects. One is algorithms parallelization in a multithreaded environment. The second one is adding some problem knowledge to improve the convergence of the algorithms. The third aspect is the enrichment of the parallel algorithms by inclusion of some elements of genetic algorithms – namely the crossover operation. The load balancing based on improved Extremal Optimization aim at better convergence of the algorithm, smaller number of task migrations to be done and reduced execution time of applications. The quality of the proposed algorithms is assessed by experiments with simulated parallelized load balancing of distributed program graphs.

Keywords: Nature inspired optimization · Load balancing · Extremal optimization · Distributed computing

1 Introduction

Processor load balancing is one of the most important problems in the methodology of parallel and distributed systems. Good surveys and classifications of general load balancing methods are presented in [1,6]. In our paper we address application of a specific nature inspired algorithm to load balancing. Nature inspired algorithms applied to load balancing, including genetic algorithms, simulated annealing, swarm intelligence methods, ant colonies and similar have

© Springer International Publishing Switzerland 2016
R. Wyrzykowski et al. (Eds.): PPAM 2015, Part I, LNCS 9573, pp. 437–447, 2016.
DOI: 10.1007/978-3-319-32149-3_41

received attention in many papers, a good survey of this subject can be found in [7]. Among relevant papers enumerated in the survey we have not spotted any reports on research on application of Extremal Optimization (EO) to load balancing [2]. In our earlier papers [3–5] we have shown that Extremal Optimization is a nature inspired method which can be applied with success in processor load balancing. The common feature of load balancing supported by EO is that EO has a very appealing computational and memory complexities. Using this approach, the load balancing decisions can be worked out in small number of iterations while working on a single solution. Among other nature inspired methods EO very well matches optimization problems where solutions are represented by integers, true for load balancing algorithms.

The algorithms presented in this paper are parallelized versions of the EO–based load balancing algorithms [3–5], in which EO has been used for load balancing of processors in execution of distributed programs. In the previous paper, we have modified the EO algorithm to replace the fully random processor selection by the stochastic selection with the probability guided by some knowledge of the problem (the EO–GS algorithm). The guidance is based on a formula which examines how a migrated task fits a given processor in terms of the global computational balance in the system and processor communication load.

Parallelization of EO can be considered in two ways. The first way is to parallelize actions aiming at a possibly stronger improvement of the current EO solution with the introduction of a population-based solution processing. It can be done using a really parallel system or a sequential system for solutions with parallel component improvement. The second way consists in using a multipoint strategy during solution improvement, in which many components of an EO solution are selected and improved in a concurrent way.

Parallel EO implementation has already some modest bibliography. In [8] the authors propose an extended EO model in which a single EO solution is replaced by a set of EO solutions which are processed using the general EO strategy. These solutions are subject to a specific selection and mutation to provide a set of solution vectors to be next processed in parallel. A set of EO improvements including various forms of parallelization was proposed in [10–12] in the context of solving problems in molecular biology. The MEO (Modified EO) approach consists in random generation of several neighbour solutions, from which the single best solution is selected during component improvement. The PMEO method is a Population-based Modified EO which uses a combination of a population-based approach to solution generation with the MEO approach to the selection of the best solution for further improvements. The third approach identified in the papers is the Distributed Modified EO (DMEO) which is a combination of the PMEO approach and the distributed genetic algorithms methodology. The DMEO is based on distribution of a population of solutions into islands. The islands evolve using the PMEO method. There are transfers of best solutions between the islands with back transfers of the replaced ones.

In this paper we propose an EO-based parallel approach which includes crossover operations on selected variants of the global solution configurations.

The modified EO is applied for solving processor load balancing in execution of programs represented as layered graphs of tasks. In the parallel EO–GS, a method similar to PMEO is applied but with an additional fitness function which is a base for the stochastic selection of the best solution state in the neighbourhood of the one chosen for improvement. We have introduced the crossover operation into EO as a way of transferring good features of solutions obtained between series of improvements performed using EO–GS performed in parallel. Three methods of using the crossover were applied. Two of them enable selecting for further parallel improvement a single solution by choosing the best solution crossed with another stochastically selected or choosing the best crossover product in a set of the parallel crossover results. The third enables periodic generation for further improvement of a set of solutions obtained by crossover of a best solution with all others improved in parallel. The solutions with crossover were experimentally positively assessed against parallel EO without best solution exchange and non-parallel EO–GS, which justified the use of the crossover operations.

The paper is organized as follows. Section 2 explains how the guided state changes were included into Extremal Optimization. Section 3 shows the overall theoretical principles of applied load balancing method. Section 4 presents the considered versions of the parallel EO algorithms. Section 5 describes the experiments performed with the proposed versions of the algorithm.

2   Guided State Changes in Extremal Optimization

Extremal Optimization is an optimization technique based on self-organized criticality [9]. In Extremal Optimization we use iterative updates of a single solution S built of a number of components s_i, which are variables of the problem. For each component, a local fitness value ϕ_i is evaluated to select the worst variable s_w in the solution. In a generic EO, S is modified at each iteration step, by randomly updating the worst variable. As a result, a solution S' is created which belongs to the neighbourhood $Neigh(S, s_w)$ of S. For S' the global fitness function $\Phi(S)$ is evaluated which assess the quality of S'. The new solution S' replaces S if its global fitness is better than that of S. We can avoid staying in a local optimum in such EO, by using a probabilistic version τ–EO, [2]. It is based on a user-defined parameter τ, used in stochastic selection of the updated component. In a minimization problem solved by τ–EO , the solution components are first assigned ranks k, $1 \leq k \leq n$, where n is the number of the components, consistently with the increasing order of their local fitness values. It is done by a permutation π of the component labels i such that: $\phi_\pi(1) \leq \phi_\pi(2) \leq \ldots \phi_\pi(n)$. The worst component s_i is of rank 1, while the best one is of rank n. Then, the component selection probability over the ranks k is defined as follows: $p_k \sim k^{-\tau}$, for a given value of the parameter τ. At each iteration, a component rank k is selected in the current solution S according to p_k. Next, the respective component s_j with $j = \pi(k)$ randomly changes its state and S moves to a neighboring solution, $S' \in Neigh(S, s_j)$, unconditionally. The parameters of the τ–EO are: the total number of iterations $\mathcal{N}_{\text{iter}}$ and the probabilistic selection parameter τ.

Algorithm 1. EO algorithm with Guided State Changes (EO–GS)

initialize configuration S at will
$S_{\text{best}} \leftarrow S$
while total number of iterations $\mathcal{N}_{\text{iter}}$ not reached **do**
  evaluate ϕ_i for each variable s_i of the current solution S
  rank the variables s_i based on their local fitness ϕ_i
  choose the rank k according to $k^{-\tau}$ so that the variable s_j with $j = \pi(k)$ is selected
  evaluate ω_s for each neighbour $S_v \in Neigh(S, s_j)$, generated by s_j change of the
  current solution S
  rank neighbours $S_v \in Neigh(S, s_j)$ based on the target function ω_s
  choose $S' \in Neigh(S, s_j)$ according to the exponential distribution
  accept $S \leftarrow S'$ unconditionally
  if $\Phi(S) < \Phi(S_{\text{best}})$ **then**
    $S_{\text{best}} \leftarrow S$
  end if
end while
return S_{best} and $\Phi(S_{\text{best}})$

τ–EO with guided state changes (EO–GS) has been proposed to improve the convergence speed of EO optimization. For this, some knowledge of the problem properties is used for next solution selection in consecutive EO iterations with the help of an additional local target function ω_s. This function is evaluated for all neighbour solutions existing in $Neigh(S, s_{\pi(k)})$ for the selected rank k. Then, the neighbour solutions are sorted and assigned GS-ranks g with the use of the function ω_s. The new state $S' \in Neigh(S, s_{\pi(k)})$ is selected in a stochastic way using the exponential distribution with the selection probability $p \sim \text{Exp}(g, \lambda) = \lambda e^{-\lambda g}$. Due to this, better neighbour solutions are more probable to be selected. The bias to better neighbours is controlled by the λ parameter. The general scheme of the EO–GS is shown as Algorithm 1.

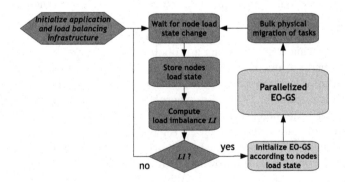

Fig. 1. The general scheme of load balancing based on parallel EO–GS.

3   Load Balancing Based on the EO Approach

The proposed load balancing method is meant for a clusters of multicore processors interconnected by a message passing network. The load balancing algorithms dynamically control assignment of program tasks (threads) t_k, $k \in \{1 \ldots |T|\}$ to processors (computing nodes) n, $n \in [0, |N| - 1]$, where T and N are the sets of all the tasks and the computing nodes, respectively. The goal is the minimal total program execution time, achieved by task migration between processors. The load balancing method is based on a series of steps in which detection and correction of processor load imbalance is done, Fig. 1. The imbalance detection relies on some run-time infrastructure which observes the state of the executive computer system and the execution states of application programs. Processors (computing nodes) periodically report their current loads to the load balancing monitor. When load imbalance is discovered, a version of EO or EO–GS is launched which identify tasks which are to be migrated among processors to improve the general balance of processor loads. Then, the physical task migrations are performed and the system returns to the load imbalance detection.

To evaluate the load of the system two indicators are used. The first is the computing power of a node n: $Ind_{power}(n)$, which is the sum of potential computing powers of all the active cores on the node. The second is the percentage of the CPU power available for application threads on the node n: $Time^{\%}_{CPU}(n)$, periodically estimated on computing nodes. System load imbalance LI is a boolean defined based on the difference of the CPU availability between the currently most heavily and the least heavily loaded computing nodes:

$$LI = \max_{n \in N}(Time^{\%}_{CPU}(n)) - \min_{n \in N}(Time^{\%}_{CPU}(n)) \geq \alpha$$

The load imbalance equal true requires a load correction. The value of α is set using an experimental approach (we set it between 25 % and 75 %).

An application is characterized by two programmer-supplied parameters, based on the volume of computations and communications tasks: $COM(t_s, t_d)$ is the communication metrics between tasks t_s and t_d, $WP(t)$ is the load weight metrics introduced by a task t. $COM(t_s, t_d)$ and $WP(t)$ metrics can provide exact values, e.g. for well-defined tasks sizes and inter-task communication in regular parallel applications, or only some predictions e.g. when the execution time depends on the processed data.

A task mapping solution S is represented by a vector $\mu = (\mu_1, \ldots, \mu_{|T|})$ of $|T|$ integers ranging in the interval $[0, |N| - 1]$. $\mu_i = j$ means that the solution S under consideration maps the i–th task t_i onto the computing node j.

Complete theoretical foundations for the proposed EO-based load balancing have been already presented in [4,5], here we will only recall basic definitions of the global and local fitness functions $\Phi(S)$ and $\phi(t)$, and the target node selection procedure in EO–GS. The global fitness function $\Phi(S)$ is defined as follows:

$$\Phi(S) = attrExtTotal(S) * \Delta_1 + migration(S) * \Delta_2 \\ + imbalance(S) * [1 - (\Delta_1 + \Delta_2)] \tag{1}$$

where $1 > \Delta_1 \geq 0$, $1 > \Delta_2 \geq 0$ and $\Delta_1 + \Delta_2 < 1$ hold.

The function $attrExtTotal(S)$ represents the impact of the total external communication (i.e. between tasks placed on different nodes) on the quality of a given mapping S. This function is normalized in the range $[0,1]$: when all tasks are placed on the same node $attrExtTotal(S) = 0$, when tasks are placed in the way that all communication is external $attrExtTotal(S) = 1$. The function $migration(S)$ is a migration costs metrics. Its value is in the range $[0,1]$, i.e. it is equal to 0 when there is no migration, $migration(S) = 1$ when all tasks have to be migrated, otherwise $0 \leq migration(S) \leq 1$. The function $imbalance(S)$ represents the numerical load imbalance metrics in the solution S. It is equal to 1 when in S there exists at least one unloaded computing node, otherwise it is equal to the normalized average absolute load deviation of tasks in S.

The local fitness function of a task $\phi(t)$ forces moving tasks away from overloaded nodes, while preserving low external (inter-node) communication. The γ parameter $(0 < \gamma < 1)$ allows tuning the weight of load metrics.

$$\phi(t) = \gamma * load(\mu_t) + (1 - \gamma) * rank(t) \tag{2}$$

The function $load(n)$ indicates how much the load of node n, which executes t, exceeds the average load of all nodes. It is normalized versus the heaviest load among all the nodes. The $rank(t)$ function governs the selection of best candidates for migration. Chance for migration have tasks, with low communication with their current node and low load deviation from the average load.

In the EO–GS algorithm migration target node selection is based on additional "biased" stochastic approach, to favour some solutions over others. In our case, the valid solution state neighbourhood includes the use of all system nodes. Therefore, at each update of rank k, all nodes $n \in N$ are sorted using the $\omega(n1, n2)$ function, $n1, n2 \in N$, with the assignment of GS-ranks g to them. Then, one of the nodes is selected using the exponential distribution $Exp(g, \lambda) = \lambda e^{-\lambda g}$. The proposed $\omega(n1, n2)$ function used for the sorting algorithm is based on a pairwise ordering of computing nodes $n1, n2$ as targets for migration of task j. It takes into account normalized load deviation of nodes $n1, n2$ ($relload(n)$) and attraction of task j to each of these nodes ($attrext(j, n)$):

$$\omega(n1, n2) = \begin{cases} sgn(relload(n1) - relload(n2)) & \text{when } relload(n1) \neq relload(n2) \\ sgn(attrext(j, n2) - attrext(j, n1)) & \text{otherwise.} \end{cases} \tag{3}$$

4   Parallel Extremal Optimization Algorithms

The general scheme of four parallel versions of the EO algorithm applied in this paper to load balancing of distributed programs is shown in Fig. 2. All the versions are based on a parallelized execution of the EO–GS Algorithm 1 presented in Sect. 2. The scheme starts with an initialization of the EO based on the current loads of all computing nodes. Next, a set of EO–GS algorithm iterations is executed in parallel on P system resources available for load balancing control. The population-type parallel EO algorithm is applied here, which replaces the single improvement of the EO solution by a parallel search for improvements performed on a population of EO solutions.

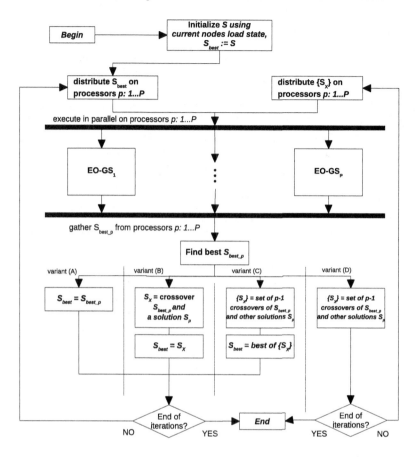

Fig. 2. The general scheme of the parallel version of the EO algorithm.

When iterations in all P parallel EO–GS branches are completed, the best solution S_{best_p} is selected out of best solutions in the branches. Next, the decision is taken which solutions will be improved in subsequent series of iterations of EO–GS in P parallel branches of the algorithm. We have 4 variants (A, B, C, D) of the rules which govern these decisions. Three of them (B, C, D) include a single point crossover on so-far solutions in stochastically selected points. In all crossovers, two off-spring solutions are generated out of which this one (S_x) is selected for which $\Phi(S_x)$ is better. Variant A (without crossover) selects only one globally best solution S_{best_p} which is next distributed to P parallel branches for parallel improvement. Variant B finds and selects the crossover of S_{best_p} with a stochastically selected other final solution from the preceding iterations. Variant C finds crossovers of S_{best_p} with $P-1$ other solutions from the preceding iterations and selects the one for which the global fitness function is the best. Variant D finds crossovers of S_{best_p} with $P-1$ other final solutions from the preceding iterations but it selects for further improvement S_{best_p} and $P-1$ best solutions in all pairs of generated crossovers in respect to values of $\Phi(S)$.

5   Experimental Assessment

The goal of the experiments was to compare the presented variants of parallel EO–GS to classic (sequential) EO and EO–GS. The experiments were performed using simulated execution of application programs in a message-passing distributed system. The DEVS-based system simulator [13] was used and parallel EO algorithms were run in a cluster of Intel i7-based, 8–core workstations.

We used a set of 5 synthetic exemplary applications, which were randomly generated in such way that program tasks are set in parallel layers. Tasks in a layer could communicate. At the boundaries between layers there was also a global exchange of data. The number of tasks in an application varied from 128 to 576 and the communication/computation ratio was in the range $[0.10, 0.20]$. The first four exemplary programs were irregular applications in which the execution time of tasks was different and depended on the processed data. Such programs exhibit unpredictable both execution time of tasks and the communication schemes, so that serious load imbalance can occur in computing nodes. The last program type was a regular parallel application in which load imbalance can appear due to non–optimized task placement or runtime conditions change.

All parallel EO–GS variants and sequential EO and EO–GS used the same local and global fitness functions. The following parameters for load balancing control were used: $\alpha = 0.5, \beta = 0.5, \gamma = 0.75, \Delta_1 = 0.13, \Delta_2 = 0.17, \tau = 1.5$, and for EO–GS $\lambda = 0.5$, the number of computing nodes: 2, 4, 8, 16, or 32. Each experiment was repeated 10 times, for each run a random, unoptimized initial task placement was used, then the results were averaged. Other settings of control parameters, together with the broader selection of exemplary applicatins were presented in [3, 4], the comparison of the execution time of EO-based load balancing methods was shown in [5].

We tested performance of standard and parallelized EO in load balancing for a fixed search space in all algorithms. For sequential EO algorithms, we set the number of iterations to 500, which is sufficient for the size of used exemplary programs according to our earlier experiments [3]. For population-based parallel versions, the number of iterations depended on the number of parallel branches in the EO algorithm (250 iterations for 2 branches, 125 iterations for 4 branches and 63 iterations for 8 branches). Thus, all load balancing algorithms performed the solution space search equivalent to 500 global fitness calculations.

Figure 3(a) and (b) show the average irregular application speedup and speedup improvements with load balancing based on sequential EO for the number of computing nodes set to 2, 4, 8, 16 and 32. The reference for the speedup improvement was the speedup with the standard, sequential EO. We achieved the speedup improvement up to 12 %, with better speedup improvements for higher number of computing nodes. The improvements with parallel EO–GS over sequential EO–GS is over 4 % for 32 computing nodes. PEO algorithm, which is not population-based, obtained almost no speedup improvement in this experiment.

We investigated also changes of migrations number for different load balancing algorithms in irregular applications, Fig. 4. Population-based parallel

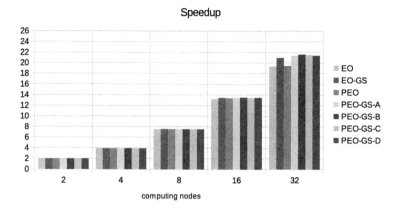

(a)

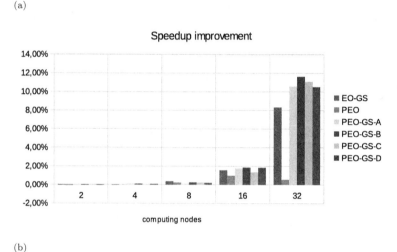

(b)

Fig. 3. Irregular application average speedup (a) and average speedup improvement (b) with parallelized EO–GS against sequential EO versus computing nodes number.

EO–GS algorithms (PEO-GS-A, -B, -C, -D), achieved big reductions of migrations number. The reduction in the range 20 %–40 % is higher for all cases than reduction for sequential EO–GS and PEO. We see that parallelization is able to substantially reduce time overhead of EO-based load balancing.

The results of experiments with the regular application are not shown since they have given speedup improvements within the range of 2 %, for the same setup of parameters of EO and load balancing control. The migrations number reduction for regular applications was up to 60 %.

Our experiments revealed that the population-based parallelization (algorithms PEO-GS-A, -B, -C, -D) gives improvement due to repeated exchange of the best solutions among parallel branches of the algorithm. The improvement needs no additional computations, since all algorithms perform the same search work.

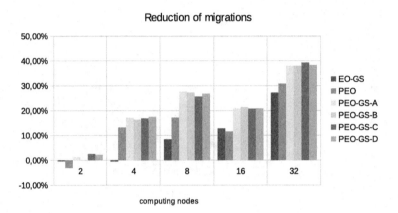

Fig. 4. Average migrations number change with parallelized EO–GS against sequential EO for irregular applications versus computing nodes number.

So, an increase of iterations number is replaced by widening of search area using parallel algorithm branches. Population-type parallelized EO–GS method is able to find load balancing solutions of high quality for irregular applications, both in the terms of applications speedup and the migrations number. This positive result is consistent for all tested irregular applications.

6   Conclusions

The paper has presented parallel algorithms for dynamic processor load balancing in execution of distributed programs. The proposed algorithms are based on EO–GS method, which is an improved version of Extremal Optimization, a nature-inspired optimization technique exploiting self-organized criticality. The load balancing procedure consists of cyclic execution of two actions: the load imbalance detection and correction based on the parallelized EO–GS algorithms. The purpose of the parallel EO–GS algorithms is to determine candidates for task migrations to correct the global imbalance in processor loads.

The contribution of the paper is the proposal of population-based, parallelized EO–GS methods, including four variants of stochastic exchange of best solutions between parallel branches of the algorithm. Three of them are based of crossover operator, used in genetic algorithms.

The experiments with simulated load balancing based on the presented algorithms have shown that the proposed variants of parallel EO–GS algorithms are efficient in processor load balancing for irregular applications. The parallel EO–GS with crossover enabled better quality of load balancing compared to other tested methods. Additional profit of using many parallel state changes in EO is a reduction in the number of task migrations needed to balance the system.

References

1. Barker, K., Chrisochoides, N.: An evaluation of a framework for the dynamic load balancing of highly adaptive and irregular parallel applications. In: Proceedings of the ACM/IEEE Conference on Supercomputing. ACM Press, Phoenix (2003)
2. Boettcher, S., Percus, A.G.: Extremal optimization: methods derived from coevolution. In: Proceedings of the Genetic and Evolutionary Computation Conference(GECCO 1999), pp. 825–832. Morgan Kaufmann, San Francisco (1999)
3. De Falco, I., Laskowski, E., Olejnik, R., Scafuri, U., Tarantino, E., Tudruj, M.: Load balancing in distributed applications based on extremal optimization. In: Esparcia-Alcázar, A.I. (ed.) EvoApplications 2013. LNCS, vol. 7835, pp. 52–61. Springer, Heidelberg (2013)
4. De Falco, I., Laskowski, E., Olejnik, R., Scafuri, U., Tarantino, E., Tudruj, M.: Improving extremal optimization in loadbalancing by local search. In: Esparcia-Alcázar, A.I., Mora, A.M. (eds.) EvoApplications 2014. LNCS, vol. 8602, pp. 51–62. Springer, Heidelberg (2014)
5. De Falco, I., Laskowski, E., Olejnik, R., Scafuri, U., Tarantino, E., Tudruj, M.: Extremal optimization applied to load balancing in execution of distributed programs. Appl. Soft Comput. 30(5), 501–513 (2015)
6. Khan, R.Z., Ali, J.: Classification of task partitioning and load balancing strategies in distributed parallel computing systems. Int. J. Comput. Appl. 60(17), 48–53 (2012)
7. Mishra, M., Agarwal, S., Mishra, P., Singh, S.: Comparative analysis of various evolutionary techniques of load balancing: a review. Int. J. Comput. Appl. 63(15), 8–13 (2013)
8. Randall, M., Lewis, A.: An extended extremal optimisation model for parallel architectures. In: 2nd IEEE International Conference on e-Science and Grid Computing, e-Science 2006, p. 114 (2006)
9. Sneppen, K., et al.: Evolution as a self-organized critical phenomenon. Proc. Natl. Acad. Sci. 92, 5209–5213 (1995)
10. Tamura, K., Kitakami, H., Nakada, A.: Reducing crossovers in reconciliation graphs with extremal optimization (in japanese). Trans. Inf. Process. Soc. Jpn. 49(4) (TOM 20), 105–116 (2008)
11. Tamura, K., Kitakami, H., Nakada, A.: Island-model-based distributed modified extremal optimization for reducing crossovers in reconciliation graph. Transactions on Engineering Technologies. LNCS, vol. 275. Springer, New York (2013)
12. Tamura, K., Kitakami, H., Nakada, A.: Distributed modified extremal optimization using island model for reducing crossovers in reconciliation graph. Eng. Lett. 21(2), EL_21_2_05, 81–88 (2013)
13. Zeigler, B.: Hierarchical, modular discrete-event modelling in an object-oriented environment. Simulation 49(5), 219–230 (1987)

Parallel Differential Evolution
in the PGAS Programming Model
Implemented with PCJ Java Library

Łukasz Górski[1]([✉]), Franciszek Rakowski[2], and Piotr Bała[2]([✉])

[1] Faculty of Mathematics and Computer Science,
Nicolaus Copernicus University, Toruń, Poland
`lgorski@mat.umk.pl`
[2] Interdisciplinary Centre for Mathematical and Computational Modelling,
University of Warsaw, Warsaw, Poland
`{rakowski,bala}@icm.edu.pl`

Abstract. New ways to exploit parallelism of large scientific codes are still researched on. In this paper we present parallelization of the differential evolution algorithm. The simulations are implemented in Java programming language using PGAS programing paradigm enabled by the PCJ library. The developed solution has been used to test differential evolution on a number of mathematical function as well as to fine-tune the parameters of nematode's *C. Elegans* connectome model. The results have shown that a good scalability and performance was achieved with relatively simple and easy to develop code.

Keywords: Parallel processing · Differential evolution · Parallel genetic algorithm · PGAS · Java

1 Introduction

Concurrent and distributed programming remains a vibrant area of research. Even though at least a decade has passed since *concurrency revolution* was hailed [1], we are still on the lookout for new methods and paradigms that would allow to exploit parallelism and mitigate difficulties programmers encounter when designing and reasoning about concurrent programs. Moreover, as the heterogeneous computing systems are now routinely deployed, the interconnection throughput becomes the limiting factor for the scalability of well-written concurrent algorithms [2]. Therefore, performance wise implementations that reduce inter-thread communications can be deemed more advantageous.

In this paper we investigate the feasibility of employing the differential evolution algorithm in the parallel environment. The algorithm is evaluated against standard mathematical functions (Rosenbrock saddle and quartic function) as well as a real-life application. In particular, the proposed algorithm was used to fine-tune the parameters of the connectome model of nematode *C. Elegans*. The last application allowed to address some differential evolution's per-formance

© Springer International Publishing Switzerland 2016
R. Wyrzykowski et al. (Eds.): PPAM 2015, Part I, LNCS 9573, pp. 448–458, 2016.
DOI: 10.1007/978-3-319-32149-3_42

characteristics that would not be exhibited with the use of ordinary mathematical functions. The implementation presented hereinafter will use the Java programming language coupled with actively developed PCJ library [3] which enables the use of PGAS programming paradigm with the JVM languages.

2   Related Work

2.1   PGAS Languages

Our work uses Java-based PCJ library which implements PGAS programmwing paradigm. The PGAS model shares some characteristics with both the message-passing (MPI) and shared memory programming (OpenMP) models. In the PGAS, by default, a private storage area affiliated with particular thread is assumed. A programmer may decide to publish some of the private variables and thus make them accessible to all the threads of execution. The popular implementations of PGAS paradigm include: UPC [4], Fortran (PGAS programming primitives were incorporated in its latest standard [5]; further enhancements to Fortran capabilities are available within Coarray Fortran 2.0 project [6]). Additionally, languages like Chapel [7] or X10 [8] extend PGAS model with capability to run asynchronous remote tasks (so-called APGAS model).

2.2   Evolutionary Computation

While the paper focuses on the (parallel) differential evolution algorithm, it is not the only nature-inspired metaheurisitc for function optimization [9]. Subject matter study dates back to the sixties and includes a selection of methods. Fogel [10] and Rechenberg [11] together with Schwefel [12] pioneered the use of then called evolution programming and evolution strategies, respectively [9]. Holland was the first to use the term genetic algorithm [9,13]. Differential evolution (DE) was described firstly by Storn and Price [14], and since then was found to be a great contester for solving real-value optimization problems [9]. It has won or secured high ranking places in a number of competitions. A number of DE variants exist, including self-adaptive ones [9]. Differential evolution algorithm should be thought as a type of traditional evolutionary algorithm, in a sense that it employs the same computational steps. The place, where it differs from standard method is that the solution space is further explored by using the difference of candidate vectors (cf. Sect. 4 for details). It is a simple, yet powerful algorithm, that outperforms genetic algorithms on many numerical singleobjective problems [19]. The feasibility of PGAS parallel differential evolution algorithm has already been put under scrutiny with the use of UPC and benchmarked against a set of well known mathematical functions [15]. Whilst the performance results were promising, a further investigation based on real-life applications is fully warranted by the aforementioned study and has been undertaken in the current paper. While we focus on the parallelization of differential evolution algorithm in the PGAS model, a study of partial parallelizations was also conducted in other works. In this respect, Hadoop and MapReduce was used for the fitness function parallelization [20].

3 PCJ Library and PGAS Programming Paradigm

PCJ library is a self-contained solution (i.e. single .jar file without any external dependencies) that enables PGAS programming in the JVM. Its simplicity has already been recognized in high performance computing community, receiving accolades in the High Performance Computing Benchmark Challenge (class 2 prize for code elegance) [3]. The library adheres to standard PGAS programming conventions. Each thread of execution is represented by class implementing the StartPoint interface, with void start() method being the entry point of every thread. By default all class' fields are accessible only to the particular thread; they may be annotated with the @Shared annotation and thus made accessible to other threads of execution (shared variable support requires the extension of Storage class). The library supports also fully automatic deployment of concurrent code to local as well as remote computing nodes, with the use of deploy method. Its first argument represents the class implementing StartPoint interface, second one - the Storage class that holds shared variables; finally a list of computing nodes is given.

The basic primitives of PGAS programming offered by the PCJ library are as follows and may be executed over all the threads of execution or by a subset forming a group:

1. get(int nodeId, String name) - read a shared variable (tagged by name) published by another thread identified with nodeId); both synchronous and asynchronous read with FutureObject is supported.
2. put(int nodeId, String name, T newValue) - dual to get, put writes to a shared variable (tagged by name) owned by a thread identified with textttnodeId; the operation is non-blocking and may return before target variable is updated
3. barrier() - blocks the threads until all pass the synchronization point in the program; a two-point version of barrier that synchronizes only the pair of threads is supported as well
4. broadcast(String name, T newValue) - broadcasts the newValue and writes it to each thread's shared variable tagged by name
5. waitfor(String name) - due to the asynchronity of communication primitives a measure that allows one thread to block until another changes one of its shared variables (tagged with name) was introduced.

Use of the PCJ library does not require any additional libraries or tools and is transparent to any Java programmer.

4 Differential Evolution Algorithm

4.1 Introduction

The pseudocode for standard differential evolution algorithm is given as Algorithm 1 [15]. Parallel version follows the same generic principles, and adds (coarse-grained) parallelism by affiliating a set population with every thread of execution. Underlined code fragments present how the sequential version can be updated for parallel execution.

Input: Objective function f
Output: Candidate vector v_{best} that minimizes f
population $\leftarrow \{v_1, v_2, \dots, v_M\}$ // random candidate vectors
while *termination criteria not satisfied* **do**
    foreach $v \in population$ **do**
        $v_{tmp} \leftarrow$ MUTATE (*population*)   // eq. (3)
        adjust v_{tmp} coordinates so that they fit problem's domain
        $v_{tmp} \leftarrow$ CROSSOVER(v, v_{tmp})   // eqs. (1) & (2)
        if $f(v_{tmp}) < f(v)$ **then**
          | population $\leftarrow$ population $\setminus \{v\} \cup \{v_{tmp}\}$
        end
        $v_{best} \leftarrow \underset{u \in population}{\arg\min} \ f(u)$
        MIGRATE(v_{best})   // sect. 4.2
    end
end
MIN-REDUCE(v_{best})
Algorithm 1. Pseudocode of (parallel) differential evolution algorithm

---

To find the minimum of a given function the following steps are employed: (*i*) generation of candidate vectors (genomes); (*ii*) mutation, crossover and selection of the best candidate vector, based on its evaluation against a given function. The algorithm is defined by the following parameters: M (number of candidate vectors), N (arity of the objective function), C (crossover probability; set to 0.8 according to the experimental data and good practices), F (scaling factor; set to 0.9), φ (migration factor, set to 0.9).

The strength of differential evolution algorithm lies in its crossover and mutation functions, which allows for a better exploration of the solution parameter space when compared to other implementations [9]. In current implementation we follow [15] and define the crossover as follows (Eqs. (1) and (2)). Crossover replaces parameters from vector v by the parameters from vector x with $1 - C$ probability.

$$l = rand(0, N - 1) \tag{1}$$

$$v[m] = \begin{cases} v[m] & if(rand(0,1) < C) \ or \ m = l \\ x[m] & otherwise \end{cases} \tag{2}$$

Differential mutation is given by the Eq. (3). A new vector v is created based on three different vectors from the population (chosen randomly).

$$v = v_{r1} + F \cdot (v_{r2} - v_{r3}) \tag{3}$$

Thus defined, differential evolution is an embarrassingly parallel algorithm. By affiliating a population (i.e. set of candidate vectors) with a particular thread it is possible to perform differential evolution on every thread irrespectively of the others.

4.2   Migration

To further improve the accuracy of the results it is a common practice to introduce migration. In current implementation we allow a best candidate vector to be copied to the next population, based on a ring topology (island model).

We have tested three migrations policies. Please consult Fig. 1 for a graphical representation of strategies discussed herein. Firstly, the best candidate vector was transferred on every iteration of the algorithm and a proper synchronization measures were introduced (Strategy I). Secondly, the same algorithm was used, but the synchronization was dropped (Strategy II). Lastly, to mitigate latency introduced by the synchronization, a specific migration policy was introduced (Strategy III). A migration was performed every 30 generations of the thread. 10 generations later a target thread used the migrated data, irrespective of whether a successful migration was performed or not. Such choice does not heavily impact algorithm's accuracy and offers a good tradeoff between correctness and performance (cf. Sect. 4.2). In fact, in other context, it has already been suggested that in order to leverage the advantages of parallelism one may have to sacrifice the full correctness and settle on algorithms that are not fully deterministic, but accurate enough for the task at hand [16].

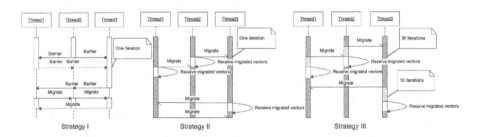

Fig. 1. Migration strategies

The performance experiments were performed on the cluster system with the following node configuration: $2 \times$ Intel Xeon CPU E5-2697 v3 @ 2.60 GHz (28 cores per node), 128 GiB RAM. Java 8 from Oracle has been used. Each experiment was repeated at least five times and figures depict the averaged performance results, together with standard deviation values.

As exploration of solution space for *C. Elegans'* connectome model is a time consuming task, a simpler cases were chosen to test migration strategies performance. In particular, two standard function were chosen: Rosenbrock saddle and quartic function.

Rosenbrock saddle is a well known function routinely used for optimization strategies testing. It is usually defined as $r(x, y) = (1-x)^2 + 100 \cdot (y - x^2)^2$ (though a different formulations are possible) and has a global minimum at $(x, y) = (1, 1)$, which is located in a long, parabolic valley. Generally, it is hard for a computer program to obtain that global minimum numerically. Even though, usually, for testing purposes it is assumed that $x, y \in [-2.048, 2.048]$, we have decided to increase the problem's difficulty by enlarging the solution space and set the parameters' domain to $x, y \in [-20, 20]$.

Quartic function definition is as follows: $q(\overrightarrow{x}) = \sum_{i=1}^{30}(i \cdot x_i^4 + \eta)$, $\overrightarrow{x} = [x_1, \ldots, x_{30}]$, where η is a random variable with uniform distribution in the range $[0, 1]$. Functions' global minimum is equal to $q(\overrightarrow{0}) = 30 \cdot E\eta = 15$. Quartic function is a simple unimodal function padded with noise: its evaluation on the same point never returns the same value. Therefore, a good performance on this test indicates that algorithm is well-suited for a noisy data analysis [17]. Accordingly, to increase the problem's difficulty we have chosen that $x_i \in [-20, 20]$ over the conventional range $x_i \in [-1.28, 1.28]$, thus enlarging the solution space for exploration.

To exhibit performance characteristics, a significant dataset was chosen. 40 000 iterations of differential evolution were executed. Each thread was affiliated with 300 candidate vectors. Tests were performed using $2^i, 0 \le i \le 7$ threads. Accurate results could have been calculated with much smaller dataset, however the migration strategies performance was in focus here; the running times of about 1 s in case of smaller datasets would not allow the properly exhibit timing results. Figure 2 shows the achieved execution time for both benchmarked functions and all migration strategies. In both cases Strategy II and Strategy III were much more efficient than Strategy I. Strategy III was chosen to be used with further experiments, as it reduces network congestion to the greatest extent.

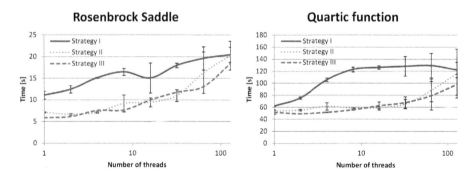

Fig. 2. Migration strategies execution time for Rosenbrock saddle and quartic function benchmark

5   Practical Application - *C. Elegans'* Connectome Model

Use of Java PCJ library allowed to perform the calculations of the large model in a distributed environment. Apart from abovementioned functions, the algorithm was also examined in the more real-life scenario.

5.1   Introduction

Nematode *C. Elegans* is a model organism whose neuronal development has been studied extensively. It remains the only organism whose connectome is fully known.

The problem the differential evolution algorithm was applied to was defined as follows: groups of nematode's motoric interneurons were ablated using laser, distinctively altering its movement patterns. The mathematical model of those interneurons (that accounts for ablations) is defined by a set of ordinary differential equations, with eight parameters. The genetic algorithm was used to explore the solution space and fit model's parameters so that its predictions are in line with the empirical data. Therefore a function we optimized took eight arguments - model's parameters and returned an error value, a difference between model's prediction and the measured nematode's behavior.

The concrete values of model's parameters, while promising, are out of the scope of this discussion and will be published in a more focused paper. Hereinafter the scalability and encountered performance bottlenecks will be the subject of analysis.

5.2   Scalability

Figure 3 shows the performance characteristic - expressed as a number of tested parameter configurations per second. The experimental dataset amounted to a population of 35 candidate vectors affiliated with each thread (weak scalability) that were evaluated through 200 iterations, $2^i, 0 \le i \le 8$ threads were used.

A scaling close to the ideal was achieved. Incidentally, the increase in the number of threads allowed to consistently achieve a better quality results (as proven by obtained energy values, i.e. the minima of fitness function evaluations).

5.3   Performance Analysis and Tuning

The connectome model is composed of several differential equations, that have to be solved for every algorithm step and every set of candidate vectors. For this purpose We have used the 5(4) Dormand Prince integrator from Apache Commons library [18]. While the evaluation of simple mathematical functions (like those analyzed hereinbefore) takes roughly constant time, irrespective of its arguments, the single invocation times for differential integrator varied from 1 to 10 s. As the equation are non-continuous and nonlinear it was impossible to predict the running time of the integrator. To overcome the relatively long calculation time, a caching was introduced so that the target function is

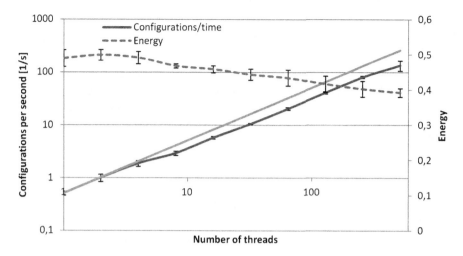

Fig. 3. Scalability results in terms of tested configurations per second. The ideal scaling is presented for reference. The energy (obtained minimum of fitness function) of the final configuration is shown

not reevaluated unnecessarily. In simpler cases (simple mathematical functions) such reevaluation was performed, for example, when the candidate vector was migrated to a new population, as the cost was insignificant.

The non-homogeneity of a single step of calculation had an impact on the total calculation time as well. Figure 4 shows a typical behavior: while most of the threads in a single calculation took about 6 h to finish, some of them finished as soon as after 3 h and others slowed the program down to a total of 9 h execution time. In all cases the differential integration was affecting the total running time to the largest extent (taking more than 99 % of the total running time).

Figure 5 shows the quantified measure of how the number of threads used impacts the variability of calculation times. The deviation from the ideal load balancing δ (we talk about ideal load balancing when all threads take the same amount of time to run; for example, trivially, when there is only a single thread, an ideal load balancing is achieved) was calculated according to the Eq. (4). Thread count increase effects in larger deviation from perfect load balancing.

$$\delta = \frac{longest\_running\_thread\_time - shortest\_running\_thread\_time}{average\_running\_time} \tag{4}$$

Correlation analysis reveals that the fastest-running thread does not necessarily yield most optimal results. Figure 6 presents the results averaged from ten runs of the algorithm, with 200 iterations and 35 candidate vectors per thread. Best energy value for a genome set affiliated with particular thread was sampled every ten algorithm's iterations. The diagram represents the obtained deciles of the threads as sorted by their completion time (which varied from about 4 h to 9 h). For completeness, the minimum and maximum of obtained values is plotted as well.

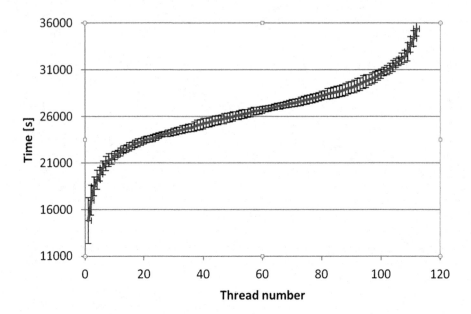

Fig. 4. Histogram of threads' execution time

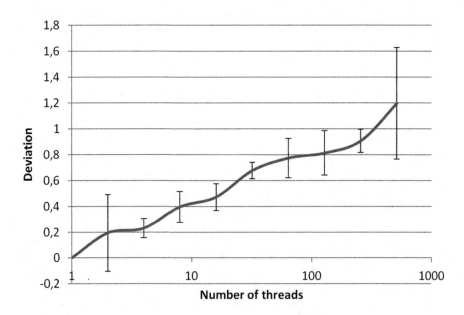

Fig. 5. Deviation from ideal thread balancing per number of threads

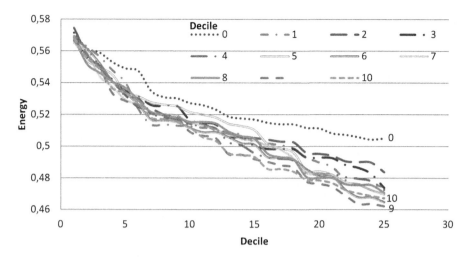

Fig. 6. Energy per PCJ thread. Number indicates decile of the thread (with 0 = min, 10 = max) as sorted by their completion time

6   Conclusion and Future Work

The PGAS paradigm is a very potent and actively research programming paradigm. The study has once again proven its usability, this time in the context of genetic programming. The PCJ library supplements core Java capabilities with a robust and port-able implementation of the PGAS programming model.

We plan to extend our study in two areas: firstly, new application of PCJ library that would allow to exhibit its performance and features; secondly, a further fine-tuning of differential evolution algorithm is fully warranted. Overcoming performance bottlenecks connected with sub-optimal load balancing of computing nodes when calculating *C. Elegans'* connectome model parameters is of immediate interest in this respect. Whereas evolutionary algorithms usability has already been put under scrutiny when used to load balance computing systems, the load balancing of the evolutionary algorithm itself has been a neglected field of study.

Acknowledgement. This work has been performed using the PL-Grid infrastructure. Partial support from CHIST-ERA consortium is acknowledged.

References

1. Sutter, H.: The free lunch is over. a fundamental turn toward concurrency in software. Dr. Dobbs J. **30**(3), 202–210 (2005)
2. Tasoulis, D.K., Pavlidis, N.G., Plagianakos, V.P., Vrahatis, M.N.: Parallel differential evolution. In: IEEE Congress on Evolutionary Computation (CEC) (2004)
3. Parallel Computing in Java. Homepage: http://pcj.icm.edu.pl/. Accessed 6 November 2015

4. Berkeley UPC. Homepage: http://upc.lbl.gov/. Accessed 6 November 2015
5. Information technology - Programming languages - Fortran. ISO, Language standard ISO/IEC: 1539–1 (2010)
6. Rice University: Coarray Fortran 2.0. Homepage: http://caf.rice.edu/. Accessed 6 November 2015
7. Chapel Programming Language. Homepage: http://chapel.cray.com/. Accessed 6 November 2015
8. X10 Programming Language. http://x10-lang.org/. Accessed 6 November 2015
9. Das, S., Suganthan, P.N.: Differential evolution: a survey of the state-of-the-art. IEEE Trans. Evol. Comput. **15**(1), 4–32 (2010)
10. Fogel, L.J., Owens, A.J., Walsh, M.J.: Artificial Intelligence through Simulated Evolution. Wiley, New York (1966)
11. Rechenberg, I.: Evolutionsstrategie - optimierung technischer systeme nach prinzipien der biologischen evolution, Ph.D. thesis (1971)
12. Schwafel, H.-P.: Numerische optimierung von computer-modellen. Ph.D. thesis (1974)
13. Holland, J.H.: Adaptation in Natural and Artificial Systems. University of Michigan Press, Ann Arbor (1975)
14. Storn, R., Price, K.V.: Differential evolution. a simple and eficient adaptive scheme for global optimization over continuous spaces: ICSI, TR-95-012 (1995). http://icsi.berkeley.edu/storn/litera.html. Accessed 6 November 2015
15. Kromer, P., Platos, J., Snasel, V.: Parallel differential evolution in unified parallel C. In: IEEE Congress on Evolutionary Computation (CEC), pp. 642–649. Cancun (2013)
16. Ungar, D.: Everything you know (about parallel programming) is wrong!. IBM Research Technical report, A Wild Screed About the Future (2011)
17. Feoktisov, V.: Differential Evolution. In Search of Solutions. Springer, New York (2007)
18. Apache Commons. homepage: https://commons.apache.org/. Accessed 2 November 2015
19. Tušar, T., Filipič, B.: Differential evolution versus genetic algorithms in multiobjective optimization. In: Obayashi, S., Deb, K., Poloni, C., Hiroyasu, T., Murata, T. (eds.) EMO 2007. LNCS, vol. 4403, pp. 257–271. Springer, Heidelberg (2007)
20. Zhou, C.: Fast parallelization of differential evolution algorithm using mapreduce. In: Proceedings of the 12th Annual Conference on Genetic and Evolutionary Computation, Portland, Oregon, USA, pp. 1113–1114 (2010)

Adaptation of Deep Belief Networks to Modern Multicore Architectures

Tomasz Olas[1]([✉]), Wojciech K. Mleczko[2], Robert K. Nowicki[2],
and Roman Wyrzykowski[1]

[1] Institute of Computer and Information Sciences, Czestochowa University
of Technology, Dabrowskiego 69, 42-201 Czestochowa, Poland
{olas,roman}@icis.pcz.pl
[2] Institute of Computational Intelligence, Czestochowa University of Technology,
Armii Krajowej 36, 42-201 Czestochowa, Poland
{wojciech.mleczko,robert.nowicki}@iisi.pcz.pl
http://www.icis.pcz.pl
http://www.iisi.pcz.pl

Abstract. In our previous paper [17], the parallel realization of
Restricted Boltzman Machines (RBMs) was discussed. This research confirmed a potential usefulness of Intel MIC parallel architecture for implementation of RBMs.

In this work, we investigate how the Intel MIC and Intel CPU architectures can be applied to implement the complete learning process using
Deep Belief Networks (DBNs), which layers correspond to RBMs. The
learning procedure is based on the matrix approach, where learning samples are grouped into packages, and represented as matrices. This approach is now applied for both the initial learning, and fine-tuning stages
of learning. The influence of the package size on the accuracy of learning, as well as on the performance of computations are studied using
conventional CPU and Intel Xeon Phi architectures.

Keywords: Deep belief network · Restricted Boltzman machine · Parallel programming · Multicore architectures · OpenMP · Vectorization ·
Intel Xeon Phi

1 Introduction

The Intel Xeon Phi coprocessor is the first generation product of the Intel MIC
(Many Integrated Core) architecture. It combines many Intel CPU cores onto
a single chip [8]. The Intel Xeon Phi architecture is targeted for highly parallel
HPC workloads, and offers a high peak performance (more than 1 TFlops at
double-precision) with a high memory bandwidth (more than 300 GB/s). A key
attribute of this architecture is that unlike GPU accelerators [20, 29] Intel Xeon
Phi coprocessors can execute applications compiled from the same C/C++ or
Fortran code as conventional Intel Xeon CPUs.

© Springer International Publishing Switzerland 2016
R. Wyrzykowski et al. (Eds.): PPAM 2015, Part I, LNCS 9573, pp. 459–472, 2016.
DOI: 10.1007/978-3-319-32149-3_43

Neural networks possess a natural parallelism that can be implemented on various physical architectures [1–6,18,28]. One of sophisticated types of neural networks is the Restricted Boltzmann Machine (RBM) [11,23], which can processes the probability distribution, and is applied to filtering, image recognition, and modelling [9]. RBMs are extensively applied for implementing Deep Belief Networks (DBNs) [7,13], which are a powerfull model used in the machine learning.

In our previous paper [17], the parallel realization of Restricted Boltzman Machines was proposed. The implementation intends to use multicore architectures of modern CPUs and Intel Xeon Phi coprocessor. The learning procedure is based on the matrix description of RBM, where the learning samples are grouped into packages, and represented as matrices. The influence of the package size on the convergence of learning, as well as on the performance of computation, are studied for various number of threads, using conventional CPU and Intel Xeon Phi architectures. Our research confirms a potential usefulness of MIC parallel architecture for implementation of RBMs. In this work, we investigate how the Intel MIC and Intel CPU architectures can be applied to implement the complete learning process using Deep Belief Networks, which layers corresponds to RBMs.

The material of the paper is organized as follows. Section 2 outlines the Intel MIC architecture, while Sect. 3 provides introduction to architecture of Deep Belief Networks. Section 4 presents our approach to the adaptation of the DBN learning to multicore architectures using learning with accumulation. Details of parallel implementation of the resulting algorithms are described in Sect. 5, as well as performance results of experiments performed both on Intel Xeon Phi coprocessors and Intel Xeon CPUs. Section 6 presents conclusions and future works.

2   Intel MIC Architecture

The Intel Xeon Phi coprocessor comprises of up to 61 processor cores connected by a high performance 512-bit bidirectional ring. Each core is capable of 4-way hyper-threading, which gives up to 244 logical cores [25]. An important component of the coprocessors core is its vector processing unit (VPU), which features 512-bit wide registers. Thus, the VPU can execute 16 single-precision or 8 double-precision instructions per cycle. Each instruction can be a floating point multiply-add.

The Intel Xeon Phi coprocessor implements a very high bandwidth memory subsystem. The memory controllers and the PCIe client logic provide a direct interface to GDDR5 memory on the coprocessor and the PCIe bus, respectively. The coprocessor has over 6 GB of on-board memory (maximum 16 GB). Each core contains a 32 KB, 8-way set associative L1 cache, and 512 KB, 8-way L2 cache. The high-speed ring connects together all the cores, caches, memory controllers and PCIe client logic of Intel Xeon Phi coprocessors. As a result, caches are fully coherent and implement the x86 memory model. The effective utilization of caches is key to achieving high performance on Intel Xeon Phi coprocessors.

The main advantage of Xeon Phi accelerators is that it is built to deliver a general-purpose programming environment similar to that provided for Intel CPUs. The coprocessor is supported by a rich development environment that includes compilers from C/C++ and Fortran languages, numerous libraries such as threading libraries (OpenMP, Cilk Plus, etc.) and high performance math libraries (e.q., MKL library), performance characterizing and tuning tools (e.g., Intel VTune Amplifier), and debuggers.

In principle, programming applications for Intel Xeon Phi coprocessors is not significantly different from programing for conventional Intel x86 processors. However, after empirical performance and programmability studies performed by many researchers [10,22,26,27] it is clear that to achieve high performance, Intel Xeon Phi still needs help from programmers, and that merely relying on compilers with traditional programming models is still far from reality. In fact, high degree of parallelism of Xeon Phi accelerators is best suited to applications that are structured to use the parallelism. Almost all codes would gain from some tuning beyond the initial base performance to achieve higher performance. The hidden benefit [19] is that this "transforming-and-tuning" approach doubles advantages of programming investments for Intel Xeon Phi coprocessors that generally apply directly to any general-purpose processor as well, offering more forward scaling to future computing architectures.

3 Introduction to Deep Belief Network Architecture

Deep Belief Networks can be viewed as a composition of a visible input layer, a few hidden layers, and output layer. Input data of the visible layer are normalized in the range [0,1], while states of the hidden layers are binary values, activated by the sigmoid function.

The Restricted Boltzmann Machine is the basic block of Deep Belief Networks (see Fig. 1); it is trained by a learning algorithm called Contrastive Divergence [13], which uses the Gibbs sampling and the reconstruction error to train the weights of RBM. The energy function of RBM is defined by [24]:

$$E^{(l)}(\mathbf{v}, \mathbf{h}) = \exp\left(-\sum_{i \epsilon visible} b_{vi}^{(l)} v_i^{(l)} - \sum_{j \epsilon hidden} b_{hj}^{(l)} h_j^{(l)} - \sum_{i,j} v_i^{(l)} h_j^{(l)} w_{ij}^{(l)}\right), \quad (1)$$

where v_i, h_j are binary states of visible unit i and hidden unit j respectively, b_{vi} b_{hj} are their biases, and w_{ij} is the weight characterizing their connection. The network assigns a probability to every possible pair of visible and hidden vectors via the following energy function:

$$p^{(l)}(\mathbf{v}, \mathbf{h}) = \frac{1}{Z} e^{-E^{(l)}(\mathbf{v}, \mathbf{h})}, \quad (2)$$

where the partition function Z is given by summing over all the possible pairs of visible and hidden vectors:

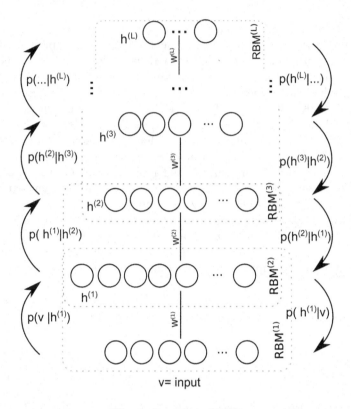

Fig. 1. Learning of DBN

$$Z^{(l)} = \sum_{\mathbf{v},\mathbf{h}} e^{-E^{(l)}(\mathbf{v},\mathbf{h})}. \tag{3}$$

The probability assigned by the network to a visible vector $\mathbf{v}$ is given by summing over all the possible hidden vectors:

$$p^{(l)}(\mathbf{v}) = \frac{1}{Z^{(l)}} \sum_{\mathbf{h}} e^{-E^{(l)}(\mathbf{v},\mathbf{h})}, \tag{4}$$

Given a random input configuration $\mathbf{v}$, the state of the hidden unit j is set to 1 with probability:

$$P^{(l)}\left(h_j^{(l)} = 1 | \mathbf{v}^{(l)}\right) = \sigma\left(b_{hj}^{(l)} + \sum_i v_i^{(l)} w_{ij}^{(l)}\right), \tag{5}$$

where $\sigma(x)$ is the logistic sigmoid function $\frac{1}{1+\exp(-x)}$. Similarly, given a random hidden vector, the state of the visible unit i can be set to 1 with probability:

$$P^{(l)}\left(v_i^{(l)} = 1 | \mathbf{h}^{(l)}\right) = \sigma\left(b_{vi}^{(l)} + \sum_i h_j^{(l)} w_{ij}\right). \tag{6}$$

The probability assigned by the network to the training image can be raised by adjusting the weights and biases to lower the energy of that image, as well as to raise the energy of other images, especially those that have low energies and therefore make a big contribution to the partition function. The derivative of the log probability of a training vector with respect to a weight is surprisingly simple:

$$\frac{\partial \log p^{(l)}(\mathbf{v})}{\partial w_{ij}^{(l)}} = \langle v_i^{(l)} h_j^{(l)} \rangle_0 - \langle v_i^{(l)} h_j^{(l)} \rangle_\infty, \tag{7}$$

where $\langle \cdot \rangle_0$ denotes the initial values, while $\langle \cdot \rangle_\infty$ denotes the values after alternating Gibbs sampling [12]. It could be processed many times.

One iteration of alternating Gibbs sampling consists of updating all the hidden units in parallel, using Eq. (6), followed by updating all the visible units in parallel using Eq. (6) [11]. To solve this problem, Hinton proposed a much faster learning procedure called the Contrastive Divergence algorithm [11,12]. This procedure can be applied in order to correct the weights and biases of the network:

$$\Delta w_{ij}^{(l)} = \eta \left(\langle v_i^{(l)} h_j^{(l)} \rangle_0 - \langle v_i^{(l)} h_j^{(l)} \rangle_\infty \right), \tag{8}$$

$$\Delta b_{vi}^{(l)} = \eta (v_{i0}^{(l)} - v_{i\infty}^{(l)}), \tag{9}$$

$$\Delta b_{hj}^{(l)} = \eta (h_{j0}^{(l)} - h_{j\infty}^{(l)}). \tag{10}$$

Every next layer is stacked on top of the previous RBM as shown in Fig. 1. The training process is performed in an unsupervised manner allowing the system to learn complex functions by mapping the input to the output directly from data. All weights of DBN must be pre-trained layer by layer, through the RBM training.

After pre-training, the weights of DBNs are fine-tuned by the standard back-propagation algorithm and the steepest descent algorithm, as assumed by the Multi-Layer Perceptron. For this purpose, we create an additional layer using the output from the last RBM layer, which represent a model of logistic regression in form of a probabilistic classifier. An additional layer is a bilayer network with output units defined as follows:

$$y_j^{(L)} = softmax_j(w_{ij}^{(L)} h_i^{(L)} + b_j^{(L)}), \tag{11}$$

where L is an additional layer, $y_j^{(L)}$ is output of the network. $w_{ij}^{(L)}$ and $b_j^{(L)}$ are respectively weight and bias of the extra layer, their initial value are set to 0. Values of $h_i^{(L)}$ are obtained from the last RBM layer using a scholar DBN. The softmax function is calculated as follows:

$$softmax_j(w_{ij}^{(L)} h_i^{(L)} + b_j^{(L)}) = \frac{e^{w_{ij}^{(L)} h_i^{(L)} + b_j^{(L)}}}{\sum_j e^{w_{ij}^{(L)} h_i^{(L)} + b_j^{(L)}}}. \tag{12}$$

4 Adaptation of DBN Learning to Multicore Architectures Using Learning with Accumulation

To adapt DBN to multicore architecture, we apply the procedures described in Sect. 3 to a package of learning samples. As a consequence, the weight adjustment is calculated and applied once for all the samples in the package.

The size of a package is denoted by u, and it should be correlated with the number of threads in the available multicore architecture. The resulting number of packages is specified as $\tau_{\max}$. If the number of samples is less than $u\tau_{\max}$, the last package is smaller, but for the sake of clarity this case will be omitted.

This idea could be implemented in various ways. In our approach, the values of all the samples assigned to the τ–th package are represented by the single matrix $\mathbf{V}_0(\tau)$ ([16,21]):

$$\mathbf{V}_0(\tau) = \begin{bmatrix} \mathbf{v}_0\,(\tau u - u + 1) \\ \vdots \\ \mathbf{v}_0\,(\tau u) \end{bmatrix} = \begin{bmatrix} v_{01}\,(\tau u - u + 1) & \cdots & v_{0M}\,(\tau u - u + 1) \\ \vdots & \ddots & \vdots \\ v_{01}\,(\tau u) & \cdots & v_{0M}\,(\tau u) \end{bmatrix}. \quad (13)$$

This matrix contains the input values for the first layer of DBN, i.e.

$$\mathbf{V}_0^{(1)}\,(\tau) = \mathbf{V}_0\,(\tau). \quad (14)$$

Then, according to Eqs. (6) and (5) matrices $\mathbf{H}_0^{(l)}(\tau)$ and $\mathbf{V}_0^{(l)}(\tau)$ are calculated for $l = 1, \ldots, L$, where:

$$\mathbf{H}_0^{(l)}(\tau) = \begin{bmatrix} \mathbf{h}_0^{(l)}\,(\tau u - u + 1) \\ \vdots \\ \mathbf{h}_0^{(l)}\,(\tau u) \end{bmatrix} = \begin{bmatrix} h_{01}^{(l)}\,(\tau u - u + 1) & \cdots & h_{0N}^{(l)}\,(\tau u - u + 1) \\ \vdots & \ddots & \vdots \\ h_{01}^{(l)}\,(\tau u) & \cdots & h_{0N}^{(l)}\,(\tau u) \end{bmatrix}. \quad (15)$$

The index 0 refers to the initial values, in contrast to the final values denoted by the index ∞. The connection between the subsequent RBMs is realized as follows:

$$\mathbf{V}_0^{(l)}\,(\tau) = \mathbf{H}_\infty^{(l-1)}\,(\tau). \quad (16)$$

These computations are illustrated in Fig. 2. The subsequent computation steps related to processing the successive RBMs are performed in the same way. By that means, the required learning with accumulation is implemented.

The DBN learning is composed of two stages: initial learning and fine-tuning stage. In this paper, it is proposed to apply the package approach (see Fig. 2) not only to the initial learning, as it was assumed in our previous paper [17], but also to the fine-tuning stage. These stages should be repeated many times in an iterative way, where each iteration is usually called "epoch" in the literature [13].

4.1 Initial Learning

The weights of connections between visible and hidden layers are updated once after processing the whole package of samples, so the size of the package has

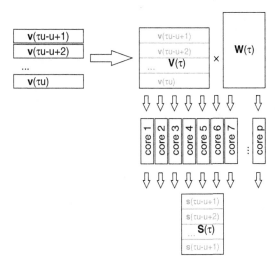

Fig. 2. Parallel implementation of RBM based on package approach

direct influence on the frequency of an update. The matrix of updates derived for the package τ can be written as:

$$\Delta \mathbf{W}^{(l)}(\tau) = \eta \left(\mathbf{V}_0^{(l)T}(\tau) \mathbf{H}_0^{(l)}(\tau) - \mathbf{V}_\infty^{(l)T}(\tau) \mathbf{H}_\infty^{(l)}(\tau) \right) \tag{17}$$

or adding a momentum component $\alpha \Delta \mathbf{W}^{(l)}(\tau - 1)$, assuming $\Delta w_{ij}^{(l)}(0) = 0$. The biases are updated in the same way:

$$\Delta \mathbf{B}_{\mathrm{v}}^{(l)}(\tau) = \eta \left(\mathbf{V}_0^{(l)}(\tau) - \mathbf{V}_\infty^{(l)}(\tau) \right) \tag{18}$$

and

$$\Delta \mathbf{B}_{\mathrm{h}}^{(l)}(\tau) = \eta \left(\mathbf{H}_0^{(l)}(\tau) - \mathbf{H}_\infty^{(l)}(\tau) \right). \tag{19}$$

Obviously, the momentum components of corrections $\alpha \Delta \mathbf{B}_{\mathrm{v}}^{(l)}(\tau - 1)$ and $\alpha \Delta \mathbf{B}_{\mathrm{h}}^{(l)}(\tau - 1)$ can be added.

4.2   Fine-Tuning of DBN

After the initial learning, the fine-tuning is performed. During this phase, the data are propagated between the subsequent RBMs as follows:

$$\mathbf{V}_0^{(l)}(\tau) = \mathbf{H}_0^{(l-1)}(\tau), \tag{20}$$

where the values of the matrix $\mathbf{H}^{(l)}(\tau)$ are calculated, according to Eq. (5):

$$h_j^{(l)}(\tau) = \begin{cases} 1 & \xi(j,t) < y_{\mathrm{h}j}^{(l)}(\tau) \\ 0 & \xi(j,t) \geq y_{\mathrm{h}j}^{(l)}(\tau) \end{cases}. \tag{21}$$

The values $\mathbf{Y}_0^{(l)}(\tau)$ are calculated as in the classic feedforward neural networks with the following activation function:

$$f(x) = \frac{e^x}{\sum e^x}. \qquad (22)$$

The correction of the weights is realized using the common supervised iterative learning procedure, where desired values are defined for the package τ by the model matrix $\mathbf{D}(\tau)$.

5   Parallel Implementation and Experimental Results

5.1   Details of Parallel Implementation

The proposed algorithms are fully implemented in the C++ language using the OpenMP standard for parallelizing computation. The initial learning stage is based on the RBM algorithm already implemented in the previous paper [17]. However, in comparison with this implementation, several modifications are introduced in this paper. The BLAS library is now used only to implement the matrix-matrix multiplication. The rest of computing kernels was rewritten aiming at optimizing the reusage of cache memories. Moreover, the efficient use of vector processing is of vital importance for utilizing the computing power of modern multicore architectures. Vectorization is even more important for the MIC architecture than for Intel Xeon CPUs.

For this aim, we use the auto-vectorization option available in the Intel compiler, which is additionally supported by applying suitable directives of the compiler. It is illustrated in Fig. 3, which shows a fragment of code corresponding to a part of the fine-tuning stage.

A suitable data alignment is crucial for the vectorization efficiency. The data alignment is a method to force the compiler to create data objects in memory on specific byte boundaries. This is done to increase efficiency of data loads and stores to and from the processor. For example, for the Intel MIC architecture, memory movements are optimal when the data starting address lies on 64 byte boundaries. Usually it is achieved using the _mm_malloc function. However, not always it is enough for the compiler to vectorize a given loop. In this case, it is necessary to force the compiler by using, e.g., the **vector aligned** directive.

Our code is compiled using the Intel C++ compiler available in the Intel Parallel Studio XE 2015 environment, where the native mode is assumed for Intel Xeon Phi coprocessors. Additionally, the MKL library is utilized for the efficient implementation of BLAS routines, as well as for generating pseudorandom numbers. In particular, the Fast Mersenne Twister pseudorandom number generator VSL_BRNG_SFMT19937 is utilized. In the code shown in Fig. 3, the corresponding MKL routine vdRngUniform is invoked by every thread. All the experiments are performed in double precision, with an extensive usage of vectorization (256-bit AVX2 standard for Intel CPUs, and 512-bit AVX-512 vector extension for Intel Xeon Phi).

```
void HiddenLayer::sample_h_given_v(double* input,
             double* sample, double u) {
    const int id = omp_get_thread_num();
    const int _nIn = (nIn / ALIGN_DOUBLE) * ALIGN_DOUBLE;

#pragma omp for
    for (int ib = 0; ib < u; ++ib) {
        const int nb_ = ib * nOut_;
        const int nn_ = ib * nIn_;

        vdRngUniform(VSL_RNG_METHOD_UNIFORM_STD,
            randomStreams[id], nOut, randoms[id], 0.0, 1.0);

        for (int i = 0; i < nOut; ++i) {
            const int ni_ = i * nIn_;
            double linear_output = 0.0;
#pragma vector aligned
#pragma simd
            for (int j = 0; j < _nIn; ++j)
                linear_output += W[ni_ + j] * input[nn_ + j];
            for (int j = _nIn; j < nIn; ++j)
                linear_output += W[ni_ + j] * input[nn_ + j];

            linear_output += b[i];
            const double value = sigmoid(linear_output);
            sample[nb_ + i] = binomial(value, randoms[id][i]);
        }
    }
}
```

Fig. 3. Fragment of code illustrating auto-vectorization supported by compiler directives

5.2 Experimental Results

In this section, we investigate experimentally the performance of the proposed approach to parallelizing the DBN learning on multicore architectures. For test purposes, we use the MNIST database [15], which contains samples of handwritten digits. In our case, the training set consists of 5000 examples, while the test set contains 1000 examples. For test purposes, the fixed number of epochs is adopted as the stopping criterion. We assume 50 epochs for both the initial learning and fine-tuning stage. These tests are performed for two RBM layers, where sizes of hidden layers are set as 640 and 320.

All the experiments are performed both on Intel Xeon Phi and Intel Xeon CPU platforms managed by the MICLAB project [14]. A single node of the testing platform consists of two 12-core Intel Xeon E5-2695 v2 CPUs (2.40GHz, 30 MB L3 cache), equipped with 128 GB ECC RAM memory (1866MHz) providing 2×59.7 GB/s memory bandwidth. Each node is also eqipped with two Intel

Xeon Phi 7120P coprocessors, 61 cores each. Each coprocessor provides 16 GB on-board memory with 352 GB/s memory bandwidth. The peak performance of this platform is given by 1208.3 GFlop/s for a single coprocessor, and 480.8 GFlop/s for two CPUs.

First of all, we test the influence of the package size u on the accuracy of the learning process, where u is the same for both the initial learning, and fine-tuning stage. The results presented in Fig. 4 show that accuracy depends significantly on the package size, and there is a range of sizes which allows us to achieve an acceptable accuracy.

At the same time, the package size affects strongly the execution time of the learning process. This effect is demonstrated in Fig. 5, which shows the influence of the package size on the execution time for two CPUs and a single Intel Xeon Phi. It is assumed that all the computing cores are utilized – 24 threads on two CPUs, and 240 threads on Intel Xeon Phi (4 threads per core). It can be concluded that for Intel Xeon Phi the execution time decreases exponentially with increasing the package size. Up to a certain size u, this tendency is preserved for CPU as well. In consequence, for 24 CPU cores/threads, already $u = 100$ allows us to reduce the execution time significantly. At the same time, the efficient usage of 240 threads in the coprocessor requires increasing the package size 5 times, up to $u = 500$.

These tests allows for the preliminary selection of the package size $u = 500$ for our testing problem running on both architectures. Table 1 compares the

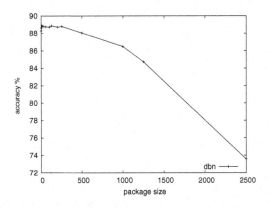

Fig. 4. Influence of package size on accuracy of learning

Table 1. Execution times (in seconds) for DBN learning executed both on Intel Xeon CPUs and Intel Xeon Phi coprocessor

package size	$u = 1$	$u = 500$	
version	sequential	sequential	parallel
2 × Intel Xeon	481.12	232.48	17.01
Intel Xeon Phi	4609.73	1085.52	16.99

execution times achieved both for a single node with two CPUs, and a single coprocessor. It can be concluded that both platforms provide practically the same execution time, for our testing problem.

Figure 6 presents the speedup of the DBN learning for the initial learning and fine-tuning phases depending on the number of threads, separately for CPUs and coprocessor. In our experiments, the number of threads is up to 24 for two CPUs, and up to 240 (1, 2, 3 or 4 threads per core) for Intel Xeon Phi. For both architectures, when calculating the speedup $S_p = T_s/T_p$, the parallel execution time T_p is compared with the sequential execution time T_s on a single CPU core. The conclusion is that the speedup increases with increasing the number of threads, both for CPUs and coprocessor. At the same time, running 4 threads per core instead of 3 threads increases the speedup slightly. The highest speedup for the initial learning is achieved on the coprocessor $(S_p = 14.16)$, while for the fine-tuning stage the highest speedup is obtained for CPUs $(S_p = 16.14)$.

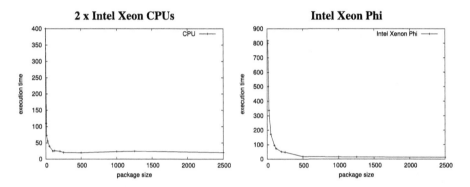

Fig. 5. Execution time for DBN learing depending on package size u for CPUs and Intel Xeon Phi ($u \in< 10, 2500 >$)

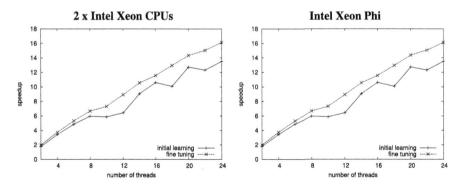

Fig. 6. Speedup of DBN learning stages (initial training and fine-tuning) depending on number of threads, for CPUs and coprocessor

6   Conclusions and Future Work

In the paper, we propose the method for parallelizing the complete learning process using Deep Belief Networks, which layers corresponds to Restricted Boltzman Machines. The learning procedure is based on the matrix approach, where the learning samples are grouped into packages, and represented as matrices. This approach is now applied for both the initial learning, and fine-tuning stages of learning. In consequence, the proposed method allows us to use efficiently multicore architectures of modern general-purpose processors and Intel Xeon Phi coprocessors.

The proposed algorithms are fully implemented in the C++ language using the OpenMP standard for parallelizing computation. Additionally we utilize the optimized matrix-matrix multiplication routine from the Intel MKL library, as well as the loop-level vectorization. The efficient usage of vectorization requires the compiler to be assisted by a programmer providing a suitable modifications of code.

In our experiments, we firstly investigate the influence of the package size on the accuracy of learning. They show that there is a range of sizes which allows for achieving an acceptable accuracy. At the same time, the package size affects strongly the execution time of the learning process. Also, we compare the performance of a single Intel Xeon Phi 7120P coprocessor against two 12-core Intel Xeon E5-2695 v2 CPUs. The experiments show practical the same performance on both platforms, for the testing problems. At the same time, we expect the advantage of coprocessor against CPUs for larger problems.

In future works, we plan an exhaustive investigation of behaviour of both platforms depending on different values of such parameters as: package size, number of epochs, number of layers and their sizes, etc. Other directions of our research in near future include integrating the energy consumption of the application as an important optimization criterion, and possibility to exploit the full computation power provided by the hybrid platform consisting of CPUs and coprocessors.

Acknowledgements. This project was supported by the National Centre for Research and Development under MICLAB project No. POIG.02.03.00.24-093/13, and by the Polish Ministry of Science and Education under Grant No. BS-1-112-304/99/S, as well as by the Polish National Science Centre under grant No. DEC-2012/05/B/ST6/03620.

The authors are grateful to the Czestochowa University of Technology for granting access to Intel CPU and Xeon Phi platforms providing by the MICLAB project.

References

1. Bilski, J., Litwiński, S., Smoląg, J.: Parallel realisation of qr algorithm for neural networks learning. In: Rutkowski, L., Siekmann, J.H., Tadeusiewicz, R., Zadeh, L.A. (eds.) ICAISC 2004. LNCS (LNAI), vol. 3070, pp. 158–165. Springer, Heidelberg (2004)

2. Bilski, J., Smolag, J.: Parallel realisation of the recurrent rtrn neural network learning. In: Rutkowski, L., Tadeusiewicz, R., Zadeh, L.A., Zurada, J.M. (eds.) ICAISC 2008. LNCS (LNAI), vol. 5097, pp. 11–16. Springer, Heidelberg (2008)

3. Bilski, J., Smolag, J.: Parallel realisation of the recurrent elman neural network learning. In: Rutkowski, L., Scherer, R., Tadeusiewicz, R., Zadeh, L.A., Zurada, J.M. (eds.) ICAISC 2010, Part II. LNCS, vol. 6114, pp. 19–25. Springer, Heidelberg (2010)

4. Bilski, J., Smolag, J.: Parallel realisation of the recurrent multi layer perceptron learning. In: Rutkowski, L., Korytkowski, M., Scherer, R., Tadeusiewicz, R., Zadeh, L.A., Zurada, J.M. (eds.) ICAISC 2012, Part I. LNCS, vol. 7267, pp. 12–20. Springer, Heidelberg (2012)

5. Bilski, J., Smolag, J.: Parallel approach to learning of the recurrent jordan neural network. In: Rutkowski, L., Korytkowski, M., Scherer, R., Tadeusiewicz, R., Zadeh, L.A., Zurada, J.M. (eds.) ICAISC 2013, Part I. LNCS, vol. 7894, pp. 32–40. Springer, Heidelberg (2013)

6. Bilski, J., Smolag, J., Galushkin, A.I.: The parallel approach to the conjugate gradient learning algorithm for the feedforward neural networks. In: Rutkowski, L., Korytkowski, M., Scherer, R., Tadeusiewicz, R., Zadeh, L.A., Zurada, J.M. (eds.) ICAISC 2014, Part I. LNCS, vol. 8467, pp. 12–21. Springer, Heidelberg (2014)

7. Chu, J.L., Krzyzak, A.: The recognition of partially occluded objects with support vector machines and convolutional neural networks and deep belief networks. J. Artif. Intell. Soft Comput. Res. 4(1), 5–19 (2014)

8. Corporation, I.: Intel Xeon Phi Coprocessor System Software Developer's Guide. The Intel Corporation, Technical report, June 2013

9. Dourlens, S., Ramdane-Cherif, A.: Modeling & understanding environment using semantic agents. J. Artif. Intell. Soft Comput. Res. 1(4), 301–314 (2011)

10. Fang, J., Varbanescu, A.L., Sips, H.: Benchmarking Intel Xeon Phi to Guide Kernel Design.Delft University of Technology Parallel and Distributed Systems Report Series, No. PDS-2013-005, pp. 1–22 (2013)

11. Hinton, G.: Training products of experts by minimizing contrastive divergence. Neural Comput. 14(8), 1771–1800 (2002)

12. Hinton, G.: A practical guide to training restricted Boltzmann machines. Momentum 9(1), 926 (2010)

13. Hinton, G., Osindero, S., Teh, Y.W.: A fast learning algorithm for deep belief nets. Neural Comput. 18(7), 1527–1554 (2006)

14. http://miclab.pl: MICLAB pilot laboratory of massively parallel systems. Web Page (2015)

15. http://yann.lecun.com/exdb/mnist/: The mnist database of handwritten digits

16. Karpathy, A., Toderici, G., Shetty, S., Leung, T., Sukthankar, R., Fei-Fei, L.: Large-scale video classification with convolutional neural networks. In: 2014 IEEE Conference on Computer Vision and Pattern Recognition (CVPR), pp. 1725–1732, June 2014

17. Olas, T., Mleczko, W.K., Nowicki, R.K., Wyrzykowski, R., Krzyzak, A.: Artificial intelligence and soft computing. In: Proceedings of the 14th International Conference, ICAISC 2015, Zakopane, Poland, 14–18 June 2015, Part I, chap. Adaptation of RBM Learning for Intel MIC Architecture, pp. 90–101. Springer International Publishing, Cham (2015)

18. Patan, K., Patan, M.: Optimal training strategies for locally recurrent neural networks. J. Artif. Intell. Soft Comput. Res. 1(2), 103–114 (2011)

19. Reinders, J.: An Overview of Programming for Intel Xeon Processors and Intel Xeon Phi Coprocessors. Technical report (2012)

20. Rojek, K.A., Ciznicki, M., Rosa, B., Kopta, P., Kulczewski, M., Kurowski, K., Piotrowski, Z.P., Szustak, L., Wojcik, D.K., Wyrzykowski, R.: Adaptation of fluid model eulag to graphics processing unit architecture. Concurr. Comput.: Pract. Exper. **27**(4), 937–957 (2015)

21. Russakovsky, O., Deng, J., Su, H., Krause, J., Satheesh, S., Ma, S., Huang, Z., Karpathy, A., Khosla, A., Bernstein, M., et al.: Imagenet large scale visual recognition challenge. arXiv preprint (2014). arXiv:1409.0575

22. Saule, E., Kaya, K., Catalyurek, U., Saule, E., Kaya, K., Catalyurek, U.: Performance evaluation of sparse matrix multiplication kernels on intel xeon phi. In: Wyrzykowski, R., Dongarra, J., Karczewski, K., Waśniewski, J. (eds.) Parallel Processing and Applied Mathematics. LNCS, vol. 8384, pp. 559–570. Springer, Heidelberg (2014)

23. Smolensky, P.: Information processing in dynamical systems: foundations of harmony theory. In: Rumelhart, D.E., McLelland, J.L. (eds.) Parallel Distributed Processing: Explorations in the Microstructure of Cognition, vol. 1 Fundations, chapter 6, pp. 194–281. MIT (1986)

24. Smolensky, P.: Information processing in dynamical systems: foundations of harmony theory (1986)

25. Staff, C.I., Reinders, J.: Parallel Programming and Optimization with Intel® Xeon PhiTM Coprocessors: Handbook on the Development and Optimization of Parallel Aplications for Intel® Xeon Coprocessors and Intel® Xeon PhiTM Coprocessors. Colfax International (2013)

26. Szustak, L., Rojek, K., Gepner, P.: Using intel xeon phi coprocessor to accelerate computations in MPDATA algorithm. In: Wyrzykowski, R., Dongarra, J., Karczewski, K., Waśniewski, J. (eds.) PPAM 2013, Part I. LNCS, vol. 8384, pp. 582–592. Springer, Heidelberg (2014)

27. Szustak, L., Rojek, K., Olas, T., Kuczynski, L., Halbiniak, K., Gepner, P.: Adaptation of MPDATA heterogeneous stencil computation to intel xeon phi coprocessor. Sci. Program **2015**, 14 (2015)

28. Tambouratzis, T., Chernikova, D., Pázsit, I.: Pulse shape discrimination of neutrons and gamma rays using Kohonen artificial neural networks. J. Artif. Intell. Soft Comput. Res. **3**(2), 77–88 (2013)

29. Wyrzykowski, R., Szustak, L., Rojek, K.: Parallelization of 2d MPDATA EULAG algorithm on hybrid architectures with GPU accelerators. Parallel Comput. **40**(8), 425–447 (2014)

Implementing Deep Learning Algorithms on Graphics Processor Units

Karol Grzegorczyk, Marcin Kurdziel$^{(\boxtimes)}$, and Piotr Iwo Wójcik

Department of Computer Science, Faculty of Computer Science,
Electronics and Telecommunications, AGH University of Science and Technology,
al. A. Mickiewicza 30, 30-059 Krakow, Poland
{kgr,kurdziel,pwojcik}@agh.edu.pl

Abstract. Deep learning has recently become a subject of vigorous research in academia and is seeing increasing use in industry. It is often considered a major advance in machine learning. However, deep learning is computationally demanding and therefore requires highly optimized software and high performance computing hardware. In this paper we share our experience from implementing core deep learning algorithms on a contemporary general-purpose graphics processor units. We describe in details the design decisions and considerations made during the implementation and show that it significantly outperforms high-level matrix-algebra implementations on a typical deep learning task. Finally, we outline a few use cases and research directions that we carried out with this software.

Keywords: Deep learning · Graphics processor units · Neural networks

1 Introduction

The fundamental idea behind deep learning is to automatically obtain multiple layers of features instead of hand-engineering them. Unfortunately, the back-propagation algorithm [17] – the most common approach to training neural networks – was for many years unsuited for learning models having more than one hidden layer. Deep learning took off in the mid 2000s, mainly due to the significant improvement in computing hardware. Prior to that time, algorithms and theoretical models which constitute the core of deep learning were unfeasible in many practical applications, due to their computational cost. Fortunately, those algorithms and methods can be efficiently parallelized. Since general purpose computing frameworks for graphics processor units (GPUs) have become available, highly parallel scientific computations are often run on those units instead of CPUs. In this article we present our experience from implementing essential deep learning algorithms on general-purpose GPUs.

Probably the most straightforward way to implement deep learning algorithms is by using a high-level matrix algebra software, like e.g. MATLAB. Implementations written in such software packages are concise and easy-to-maintain,

© Springer International Publishing Switzerland 2016
R. Wyrzykowski et al. (Eds.): PPAM 2015, Part I, LNCS 9573, pp. 473–482, 2016.
DOI: 10.1007/978-3-319-32149-3_44

but not as fast as their low-level GPU-accelerated counterparts. In experimental evaluation we demonstrate what level of performance improvement can be achieved by implementing deep learning on GPUs instead of using a high-level matrix algebra environments.

Due to the lack of space, in this work we do not describe deep learning algorithms in details. An introductory material on this subject can be found in [2,7].

2   Related Work

In recent years, a number of deep learning libraries and toolkits have been created. In many cases they are maintained by leading research teams in the field. Caffe [9] is a framework maintained by BVLC group at the Berkeley University, that focuses mainly on convolutional neural networks. It exposes Python and MATLAB bindings and can be run on either CPU or GPU. One of the advantages of Caffe is that it is shipped with a collection of already trained models, which are ready for use. Torch [3] is another deep learning framework maintained primarily by the Facebook AI Research team and the Google DeepMind. Experiments for Torch are written in Lua scripting language and, as in the case of Caffe, can be run on either CPU or GPU. Researchers from the Machine Learning Laboratory of the University of Montreal, led by Yoshua Bengio, are maintaining Pylearn2 [4] library. It is built on top of Theano [1], a scientific computing library for Python. Models in Theano are described as computational graphs and are compiled to sets of optimized low-level intrinsics.

3   Implementation Details

Our implementation was written in C++ and CUDA and has an easy-to-use Python API. Currently it provides several supervised and unsupervised neural network models, which can be used for a variety of tasks. Specifically, three main types of deep learning models are supported: Deep Belief Networks (DBN), deep autoencoders and multilayer perceptrons (MLP). Recurrent neural networks are currently in implementation. We deliberately decided not to support convolutional neural networks since these models are already well-supported by several high-quality frameworks, like those mentioned in the previous section.

Neural networks can use a number of different neuron types. In addition to the standard sigmoid units, we also implement linear units with optional Gaussian noise, rectified linear units [14] (often abbreviated as ReLU), Softmax units in the output layer and the Constrained Poisson model [19] for the bag-of-words data. The two main training algorithms that we support are: backpropagation for feedforward networks and Contrastive Divergence (CD) [6] for energy-based models. Contrastive Divergence is implemented in two variants, namely classical CD (a.k.a. CD-k) and Persistent Contrastive Divergence [22]. Our network can employ three important regularization mechanisms, namely weight decay (in L_1 and L_2 penalty variants), sparsity, which encourages sparse activations in hidden units [7], and dropout, which trains a large ensemble of neural networks with shared weights [20]. The only supported optimization algorithm is stochastic gradient descent with the momentum method. Both classical momentum [16] and Nesterov's Accelerated Gradient [15] interpreted as a momentum method [21] are supported. All these algorithms can be run in either single or double floating-point precision.

3.1   Data Layout and Implementation Strategy

The vast majority of operations in the stochastic gradient descent algorithm involve vector-matrix operations. For example, in order to compute hidden activations we need to multiply an input vector by the weight matrix. However, on GPU performing one matrix-matrix multiplication is significantly faster than an equivalent number of vector-matrix multiplications. Having this in mind, we implemented the so-called mini-batch variant of the stochastic gradient descent. In this variant, instead of training the network with a single example one takes a number of examples (i.e., a mini-batch) and calculates their mean training gradient using mainly matrix-matrix operations. We adopted a convention that each training example is represented as a row in the mini-batch matrix. The same layout is used in the hidden activations matrix. Weight vectors for the hidden units are stored as columns of the weight matrix. Because some of our models, e.g. Restricted Boltzmann Machines (RBM), are undirected we need to maintain biases for both visible and hidden units. In order to reduce the number of multiplications during the training process, the biases are added as an additional row and column to the weight matrix. This also requires that an additional column filled with ones is appended to each mini-batch matrix. This column is never modified.

Most of the neural network training algorithms map well to the level 3 Basic Linear Algebra Subprograms (BLAS), which is a standard for high performance linear algebra calculations. We use the Nvidia CUDA BLAS (cuBLAS) library, a GPU-accelerated version of the BLAS standard. Because cuBLAS uses column-major storage, all matrices in our software are stored in this manner. In the training process we did our best to employ as few BLAS operations as possible. For example, consider the Contrastive Divergence approximation to the RBM gradient [6]:

$$\Delta w_{ij} = \varepsilon(\langle v_i h_j \rangle - \langle v_i h_j \rangle_R), \tag{1}$$

where ε is a learning rate, $\langle v_i h_i \rangle$ is the expected product of visible and hidden activations given a training example $\mathbf{v}$ and $\langle v_i h_i \rangle_R$ is their expected product under the distribution of samples 'fantasized' by the RBM. When vectorized, this approximation to the gradient can be calculated with just two BLAS GEMM operations, one for the training mini-batch and the other for the RBM fantasy:

$$\mathbf{W} = \mathbf{W} + \varepsilon \mathbf{v}^T \mathbf{h}, \tag{2}$$

$$\mathbf{W} = \mathbf{W} - \varepsilon \mathbf{v}_R^T \mathbf{h}_R. \tag{3}$$

As part of GEMMs, the gradient is also divided by the mini-batch size.

For a limited number of operations that are not supported by cuBLAS we created a handful of custom CUDA kernels (e.g. for L_2 penalty, stochastic activations, cost functions, etc.). Due to the fact that a substantial portion of custom kernels compute in-place operations on activation matrices, we designed kernels in such a way that each column of the matrix is processed by a separate one-dimensional thread block (group of CUDA threads). Therefore GPU threads for our custom kernels are always organized in flat grid of flat thread blocks. We employed this convention because mini-batch and activation matrices are in column-major order and usually have significantly more columns than rows. Kernels on a GPU are executed in groups of 32 threads (i.e., warps) and therefore block size should be a multiple of 32. While the maximum number of threads per block varies with the version of CUDA, and is either 512 or 1024, in our implementation block size is limited mainly by the size of the mini-batch. In particular, block size should match the mini-batch size. Otherwise some of the threads in a block would have no elements to process. On the other hand, according to [7] mini-batch size should be between 10 and 100 cases. Taking this into consideration, we adopted a block size of 128 threads. Therefore, this is also the most efficient size of a mini-batch in our software. If a kernel operates on a matrix that has more than 128 rows (e.g. weight matrix) then each thread strides through the column and processes multiple matrix elements.

Finally, it is worth to note that all operations on the GPU (cuBLAS calls, custom kernels calls, random number generation and memory transfers) share one common CUDA stream and are executed asynchronously with respect to the host.

3.2   Memory Management

During training all of the model parameters and network activations are stored in the GPU's device memory. For example, in multilayer networks data fed through the network is not transferred via the host memory but an output of one layer is directly an input to the next layer. Furthermore, when a network is initialized from an existing model (e.g. MLP initialized from DBN or MLP created from the encoder part of a deep autoencoder) the weight matrices are not copied but 'borrowed'. That is, the new network reuses the memory space that was previously allocated for the original model. No additional memory allocations or memory transfers are performed. Note that in case of a deep autoencoder we must untie the encoding and decoding part. Thus, while the encoder part is borrowed from the original model, the decoder part is its transposed copy. We use these optimizations to save GPU memory, which allows us to build larger models. All objects and data structures which are shared among multiple models are referenced using C++11 smart pointers. This way we do not need to manually deallocate their GPU device memory. Moreover, since GPU memory allocation is a synchronous operation, in order to speed-up the training process all memory buffers are allocated only once and reused throughout the training. Device memory for all the matrices is allocated with an additional padding added to each column. This 'pitched' allocation ensures that matrix columns begin at 512-byte aligned addresses. The memory allocated for padding is never used, but alignment improves the performance of cuBLAS and our custom kernels.

Large training sets will usually not fit in the GPU memory. On the other hand, transferring only one mini-batch at a time might be inefficient. We therefore split the dataset into chunks consisting of several mini-batches and store these chunks in the GPU memory.

3.3   Random Numbers

Certain deep learning algorithms are stochastic in nature, and consequently rely heavily on random numbers. For this purpose we utilize the Nvidia CUDA Random Number Generation library (cuRAND), which is able to efficiently produce pseudo-random numbers drawn from a variety of distributions. We use the cuRAND device API. That is, we allocate a matrix of random number generator (RNG) states in the GPU memory, and when we need to populate some other matrix or array with random numbers we do it in parallel entirely on the GPU. Device API is significantly faster than the host API. For the two distributions that are most often used when training deep networks, i.e., Gaussian and uniform distribution, device API generates as robust random numbers as the host API. In one type of network layers, namely the Constrained Poisson model, we also use Poisson distributed random numbers. In this case device API has slightly worse statistical properties than the host API. Therefor to validate the implementation, we implemented this model in MATLAB, using its Poisson random number generator. We found that these two implementations yielded very similar results when used in deep networks.

While drawing random numbers, each CUDA thread uses its own RNG state. However, we deliberately run less threads than the number of elements in the destination matrix. The reason for this is that we want each thread to generate at least a few random numbers. This offsets the time spent in reading and writing the global RNG state. If the number of threads matched the size of the needed random matrix, the time spent in fetching and storing of the global RNG state would negatively impact the performance.

We use deterministic seeds when initializing matrix of RNG states. In general, our goal was to provide binary compatibility among experiments. If two identical experiments are run with the same seed they receive the same sequence of random numbers. Note, however, that there are some external factors that can violate binary compatibility. For example, if two experiments are run using different version of CUDA, or on different GPU models, there can be numerical differences in their outcomes.

3.4   Bindings

In order to simplify the experiments we implemented an extensive set of Python bindings, using the Boost Python library. Therefore, the high-level experiments can be created entirely in Python. Bindings expose all of the main library functionalities and perform memory management for the underlying objects.

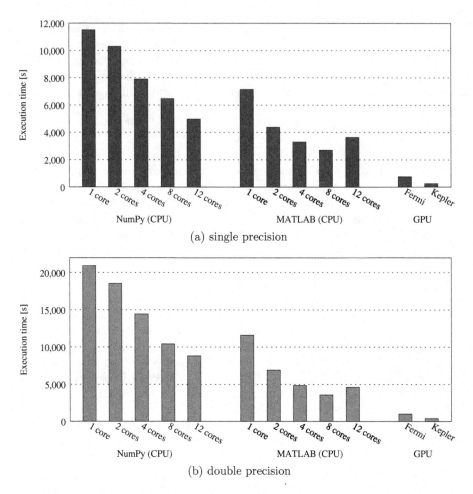

(a) single precision

(b) double precision

Fig. 1. Performance comparison between NumPy/OpenBLAS, MATLAB and GPU implementations. Execution times are reported for a 100-epoch pretraining task on the MNIST training set. Network used was a 784-1000-500-250-30 DBN.

Exporting C++ objects via Boost allows us to extend them in Python. This is useful for features that are not easily supported by Boost, like serialization/deserialization via the Python cPickle module. Moreover, bindings contain several utility functions, such as allocation of NumPy arrays as page-locked ('pinned') memory, which improves their host-device transfer throughput.

4   Experimental Evaluation

In order to validate the correctness of our implementation we reproduced experiments from key publications in the field. For example, we implemented MNIST [11] classification task as described in the seminal deep learning

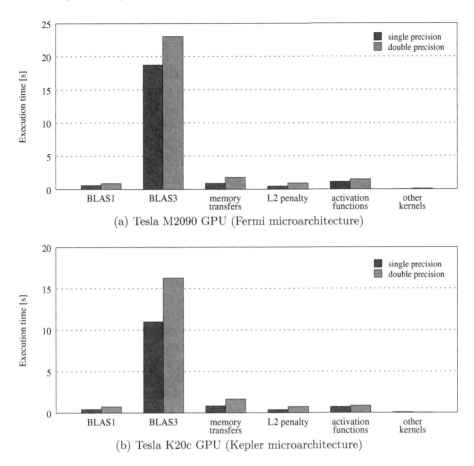

Fig. 2. Kernel execution times on GPU. Times are reported for 5-epoch of the pre-training task from Fig. 1

paper [8] and replicated their results. We also constructed a 2D visualization of this dataset with an autoencoder, and got equivalent spatial separation of digits. In classification of NORB [12] and CIFAR-10 [10] images we obtained results very close to those reported in [10,14], respectively. Using networks with dropout for the MNIST dataset we were able to match the performance reported in [20]. We also experimented with document retrieval task on the 20-newsgroups corpus[1], using deep autoencoder networks similar to those described in [18,19]. We obtained comparable precision-recall numbers.

To evaluate the performance of our implementation we conducted comparative experiments on a dimensionality reduction task. For the test dataset we have chosen MNIST. As a benchmark for comparison we used the original MATLAB source code from [8] as well as a NumPy/OpenBLAS-based implementation in

---

[1] Available at http://qwone.com/~jason/20Newsgroups.

Python. From the networks used in [8] we chose a DBN with 784-1000-500-250-30 architecture. All layers except the last one used sigmoid activation function. Last hidden layer was Gaussian. The network was pretrained with CD-1, using stochastic gradient descent with 100-element mini-batches and momentum. We built the same network architecture using our library and ran it on two GPU microarchitectures developed by Nvidia, namely Fermi (Tesla M2090) and Kepler (Tesla K20c). MATLAB and OpenBLAS have built-in support for multi-threading. Therefore we executed the reference experiments on 1, 2, 4, 8 and 12 cores. The tests were carried out on a dual-socket Intel Xeon E5645 machine. We used MATLAB version 2013b, OpenBLAS version 0.2.6, CUDA version 5.5, GCC version 4.8.2, Python version 2.7.5, NumPy version 1.9.2 and Boost version 1.55.

Execution times for 100-epoch pretraining task are presented in Fig. 1. We report times for both single and double floating point precision. Note, that the GPU implementation run on Kepler microarchitecture is approximately 9–10 times faster than the fastest multicore CPU implementation.

We also analyzed the above experiment with the nvprof profiling tool from the CUDA Toolkit. Figure 2 shows kernel execution times for a 5-epoch pretraining task, grouped by the kernel type. The kernels are grouped into: vector-vector operations (BLAS1), matrix-matrix operations (BLAS3), memory transfers (host to device, device to device and device to host) and custom kernels (mainly L_2 penalty and activation functions). As can be seen, the overwhelming fraction of the time is spent in matrix-matrix operations, which is expected when learning a neural network. Note that cuBLAS matrix-matrix operations are extremely well optimized and it is usually reasonable to use them instead of custom kernels.

5   Conclusions

In this paper we shared our experience from efficient implementation of important deep learning algorithms on modern graphics processor units. In particular, we described our main design decisions, which we believe could be useful for other researchers in the field. Experimental evaluation shows that the most important factor affecting the performance of network training are efficient matrix-matrix operations, and in particular matrix multiplication. In this respect we described optimized memory layout for network matrices. We also presented design considerations regarding random number generation and memory management. Overall, our GPU-based software notably outperforms a high-level, CPU-based matrix-algebra implementations. Experiments show an up to several-fold decrease in execution time.

We have carried out a number of different experiments using software described in this article. For example, we conducted research aimed at investigating effects of the Sparse Initialization technique [13] in pretrained networks. Results of this work are reported in [5]. Other directions of research that we currently pursue focus on regularization techniques in unsupervised models and on recurrent neural networks.

Acknowledgments. This research is supported by the Polish National Science Centre grant no. DEC-2013/09/B/ST6/01549 "Interactive Visual Text Analytics (IVTA): Development of novel, user-driven text mining and visualization methods for large text corpora exploration."

Special thanks are due to (partial) financial supported by the Polish Ministry of Science and Higher Education under AGH University of Science and Technology grant 11.11.230.124 (statutory project).

This research was carried out with the support of the "HPC Infrastructure for Grand Challenges of Science and Engineering" Project, co-financed by the European Regional Development Fund under the Innovative Economy Operational Programme.

This research was supported in part by PL-Grid Infrastructure.

References

1. Bastien, F., Lamblin, P., Pascanu, R., Bergstra, J., Goodfellow, I., Bergeron, A., Bouchard, N., Warde-Farley, D., Bengio, Y.: Theano: new features and speed improvements. arXiv preprint (2012). arxiv:1211.5590

2. Bengio, Y.: Learning deep architectures for ai. Found. Trends® Mach. Learn. **2**(1), 1–127 (2009)

3. Collobert, R., Kavukcuoglu, K., Farabet, C.: Torch7: a matlab-like environment for machine learning. In: BigLearn, NIPS Workshop. No. EPFL-CONF-192376 (2011)

4. Goodfellow, I.J., Warde-Farley, D., Lamblin, P., Dumoulin, V., Mirza, M., Pascanu, R., Bergstra, J., Bastien, F., Bengio, Y.: Pylearn2: a machine learning research library. arXiv preprint (2013). arxiv:1308.4214

5. Grzegorczyk, K., Kurdziel, M., Wójcik, P.I.: Effects of sparse initialization in deep belief networks. Comput. Sci. **16**(4), 313–327 (2015)

6. Hinton, G.E.: Training products of experts by minimizing contrastive divergence. Neural Comput. **14**(8), 1771–1800 (2002)

7. Hinton, G.E.: A Practical guide to training restricted boltzmann machines. In: Montavon, G., Orr, G.B., Müller, K.-R. (eds.) Neural Networks: Tricks of the Trade, 2nd edn. LNCS, vol. 7700, pp. 599–619. Springer, Heidelberg (2012)

8. Hinton, G.E., Salakhutdinov, R.R.: Reducing the dimensionality of data with neural networks. Science **313**(5786), 504–507 (2006)

9. Jia, Y., Shelhamer, E., Donahue, J., Karayev, S., Long, J., Girshick, R., Guadarrama, S., Darrell, T.: Caffe: convolutional architecture for fast feature embedding. arXiv preprint (2014). arxiv:1408.5093

10. Krizhevsky, A.: Learning multiple layers of features from tiny images. Master's thesis, University of Toronto (2009)

11. LeCun, Y., Bottou, L., Bengio, Y., Haffner, P.: Gradient-based learning applied to document recognition. Proc. IEEE **86**(11), 2278–2324 (1998)

12. LeCun, Y., Huang, F.J., Bottou, L.: Learning methods for generic object recognition with invariance to pose and lighting. In: Proceedings of the 2004 IEEE Computer Society Conference on Computer Vision and Pattern Recognition (CVPR 2004), vol. 2, pp. 97–104. IEEE (2004)

13. Martens, J.: Deep learning via hessian-free optimization. In: Proceedings of the 27th International Conference on Machine Learning (ICML 2010), pp. 735–742 (2010)

14. Nair, V., Hinton, G.E.: Rectified linear units improve restricted boltzmann machines. In: Proceedings of the 27th International Conference on Machine Learning (ICML 2010), pp. 807–814 (2010)

15. Nesterov, Y.: A method of solving a convex programming problem with convergence rate $O(1/k^2)$. Sov. Math. Dokl. **27**(2), 372–376 (1983)
16. Polyak, B.T.: Some methods of speeding up the convergence of iteration methods. USSR Comput. Math. Math. Phys. **4**(5), 1–17 (1964)
17. Rumelhart, D.E., Hinton, G.E., Williams, R.J.: Learning representations by back-propagating errors. Nature **323**(6088), 533–536 (1986)
18. Salakhutdinov, R.: Learning deep generative models. Ph.D. thesis, University of Toronto (2009)
19. Salakhutdinov, R., Hinton, G.: Semantic hashing. Int. J. Approximate Reasoning **50**(7), 969–978 (2009)
20. Srivastava, N., Hinton, G., Krizhevsky, A., Sutskever, I., Salakhutdinov, R.: Dropout: a simple way to prevent neural networks from overfitting. J. Mach. Learn. Res. **15**, 1929–1958 (2014)
21. Sutskever, I., Martens, J., Dahl, G., Hinton, G.: On the importance of initialization and momentum in deep learning. In: Proceedings of the 30th International Conference on Machine Learning (ICML 2013), pp. 1139–1147 (2013)
22. Tieleman, T.: Training restricted boltzmann machines using approximations to the likelihood gradient. In: Proceedings of the 25th International Conference on Machine Learning, pp. 1064–1071. ACM (2008)

Fuzzy Transducers as a Tool for Translating Noisy Data in Electrical Load Forecast System

Mariusz Flasiński[✉], Janusz Jurek, and Tomasz Peszek

Information Technology Systems Department, Jagiellonian University in Cracow,
ul. prof. St. Lojasiewicza 4, 30-348 Cracow, Poland
`mariusz.flasinski@uj.edu.pl`

Abstract. Transforming noisy data into symbolic information is one of the crucial problems in syntactic pattern recognition/forecasting systems. A formal model of such transformation is defined in the paper. Firstly, the architecture of the model, which is based on the use of a transducer and probabilistic neural networks is described. Then, the experimental results of the application of the model in electrical load forecast system are presented. A parallelization of the model and possible other areas of applications are discussed.

Keywords: Fuzzy transducer · Noisy data · Syntactic pattern recognition · Electrical load forecast

1 Introduction

A syntactic/structural analysis is one of two main approaches to pattern recognition [3,6,8]. This approach is based on the mathematical linguistics model, namely theory of formal languages, grammars and automata. The recognition is performed in the following two phases.

(1) In the first phase a structural representation of a pattern is generated. Firstly, a pattern is segmented in order to identify elementary patterns, called *primitives*. Secondly, the symbolic representation of the pattern in the form of a string (or other structure), which consists of symbols representing primitives is defined. This representation is treated as a word belonging to a formal language.

(2) During the next phase such a word is analyzed by a formal automaton, strictly speaking by its implementation i.e. a parser, which is constructed on the basis of a formal grammar generating the corresponding language. As a result a derivation of the analyzed word is obtained. The derivation is used by the syntactic pattern recognition system for describing structural features of the pattern and for recognizing (classifying) it.

A distortion (fuzziness/noisiness) of an analyzed pattern is one of the main practical problems of syntactic pattern recognition. If a pattern is distorted and a result of the first phase is improper, then the second phase usually does not

© Springer International Publishing Switzerland 2016
R. Wyrzykowski et al. (Eds.): PPAM 2015, Part I, LNCS 9573, pp. 483–492, 2016.
DOI: 10.1007/978-3-319-32149-3_45

bring a satisfactory result either. Let us stress, that even one misrecognized primitive could lead to the failure of the syntax analysis of the whole pattern.

There are several methods supporting a recognition in such a case known in the literature (e.g. [5,8]), but they are based on some improvements in the second phase only (e.g. error-correcting parsing, stochastic parsing).

In the paper we present a model, in which the uncertainty factor (fuzziness/distortion) is taken into account in the whole process of a recognition, i.e. in the phase of a primitive identification as well as in the phase of a syntax analysis. The main idea of the model consists in using a fuzzy transducer for translating noisy data into symbolic information. The transducer is based on a collection of GDPLL(k) parsers, which are able to efficiently analyze a large class of formal languages (even many context-sensitive languages).

Moreover, an implementation of the model may be parallelized. In case of the problem of a recognition of distorted patterns, the parallelization may improve significantly the performance of the recognition process.

The model has been successfully verified in a system for electrical load forecasting in Tauron Polish Energy Distribution Division in Gliwice, Poland. The application has required the analysis of noisy data (e.g. an air temperature forecast or an insolation forecast) and an identification of structural dependencies in a description of a day, for which an electrical load prediction should be generated.

The paper is organized as follows. In the next section we present a general idea of the model and its architecture. In the third section definitions concerning a fuzzy transducer and a GDPLL(k) parser, which is its main part are introduced. The application of the transducer in a system for electrical load forecasting is described in the fourth section. Conclusions are included in the final section.

2   Model of the Recognition of Fuzzy/Distorted Patterns

As we have mentioned it in the introduction, one of the main practical problems with applications of syntactic pattern recognition concerns an uncertainty factor (distortions/fuzziness/noise) in analyzed patterns.

Usually, numeric data are used as an input for a recognition system. According to the syntactic approach, such data have to be translated into the symbolic information. If the translation is inaccurate, then the syntactic recognition will end with a failure in most cases.

The model proposed is delivers a solution for the problem. The general idea of the model of the syntactic recognition process of noisy/distorted patterns is shown in Fig. 1.

The preprocessing module is responsible for the identification of so-called noisy primitives. Let us define such primitives.

Definition 1. Let $\Sigma = \{a_{i_1}, ..., a_{i_s}\}$ is a set of symbols. A *noisy primitive* is a vector:

$$(a_{i_1} p_{i_1}, ..., a_{i_s} p_{i_s}), \text{ where:}$$

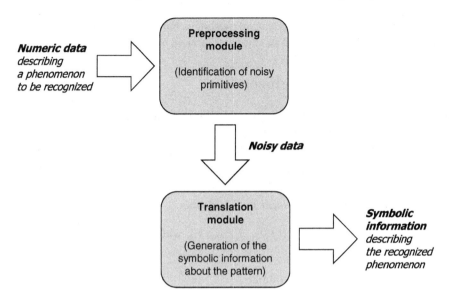

Fig. 1. The model of the recognition process of noisy/distorted patterns

$a_{i_1}, ..., a_{i_s} \in \Sigma$ are different symbols,

$p_{i_1}, ..., p_{i_s}$ are probabilistic measures corresponding to each symbol such that:
- $p_{i_k} > 0, \ k = 1, ..., s,$
- $\Sigma_{k=1}^{s} p_{i_k} \leq 1,$
- $p_{i_1} \geq \cdots \geq p_{i_s}.$

Each pair (a_{i_k}, p_{i_k}) will be called *a unit* of a noisy primitive, and each symbol a_{i_k} will be called *an element* of a noisy primitive. The *dimension* of a noisy primitive is the number of its units. The set of noisy primitives over Σ will be denoted as NP_Σ. □

The preprocessor module is implemented with the help of probabilistic neural networks (PNN) introduced by Specht in [13] at the beginning of the 1990s. Apart from a classification, the networks deliver some additional information on how far an unknown object is related to other classes (not only to the winner one). In our case PNN recognizes pre-defined symbols with their probabilities, i.e. noisy primitives are the output of PNN.

The translation module is responsible for a generation of symbolic information about the *whole* pattern on the basis of a sequence of noisy primitives delivered by the preprocessing module. Such information is called a *symbolic (variant) identification* of a pattern. Let us introduce the following definition.

Definition 2. Let Σ is a set of (terminal) symbols. A *symbolic (variant) identification* of a pattern is a set of pairs: (w_i, p_i), where: $w_i \in \Sigma^*$ is a recognized word (one of the most probable patterns), p_i is a probability of the recognition of word w_i. Each pair (w_i, p_i) is an *unit of a symbolic (variant) identification*. The set of symbolic identifications of a pattern over Σ will be denoted as SI_Σ. □

The construction of the translation module is based on a formalism of a fuzzy transducer, which is described in the next section.

3   FGDPLL(k) Transducer

A formalism of a transducer has been well-known in the language translation area and syntactic pattern recognition for many years [8]. A finite transducer is a finite-state automaton with an output. Thus, it is a recognizer, which generates an output string each time a transition is made. Let us present the following definition.

Definition 3. A *finite transducer* is a six-tuple:

$$T = (Q, \Sigma, \Delta, \delta, q_0, F), \text{ where:}$$

Q is a finite set of states;
Σ is a finite set of input symbols;
Δ is a finite set of output symbols;
δ is a finite subset of $Q \times \Sigma \times Q \times \Delta^*$;
$q_0 \in Q$ is the initial state;
$F \subseteq Q$ is the set of final states. □

If $(q, a, q', \alpha) \in \delta$ then the transducer in a state q, after reading an input $a \in \Sigma$, makes a transition to a state q', and generates an output $\alpha \in \Delta^*$.

A finite transducer is a standard (simple) translator. In our model, we use much more sophisticated tool, the so-called FGDPLL(k) transducer as a basis for the translation module. This model has been defined as a formalism for constructing an electrical load forecast system. An identification of structural dependencies in data corresponding to an electrical load is the main task of the system. Such dependencies cannot be described with the use of regular grammars or even context free-grammars. Therefore, we use a special class of quasi-context sensitive grammars, GDPLL(k) grammars [4,9–12], to describe patterns of an electrical load, and GDPLL(k) parsers (automata) to analyze them.

Before we define an FGDPLL(k) transducer, let us introduce a definition of a GDPLL(k) automaton [11]. A GDPLL(k) automaton is a system of two cooperating elementary automata: LLA(k) and OIA.

An LLA(k) automaton is a pushdown automaton, which is constructed in a similar way to the well-known LL(k) automaton [1]. The only difference is that the LLA(k) automaton can propose *more than one* possible productions at a derivation step. (It means that the LLA(k) automaton is not a deterministic automaton). Let us present the following definition [11].

Definition 4. A *LLA(k) automaton* is a seven-tuple

$$LLA(k) = (\Sigma, \Gamma, \Pi, \delta_{LLA_1}, \delta_{LLA_2}, Z_0, p_0), \text{ where:}$$

Σ is a finite input alphabet;

Γ is a finite alphabet of a stack;

Π is a finite set of productions' indices;

δ_{LLA_1} and δ_{LLA_2} are transition functions defined as follows:

$\delta_{LLA_1} : \Gamma \times \Sigma^{*k} \longrightarrow \Pi^* \cup \{err, acc, rem\}$, where Σ^{*k} is a set of terminal sentences of the length equal to or less than k,

$\delta_{LLA_2} : \Gamma \times \Sigma^{*k} \times \Pi \longrightarrow \Gamma^*$;

$Z_0 \in \Gamma$ is the starting symbol of a stack;

$p_0 \in \Pi$ is the starting symbol of a productions' tape. □

To achieve a determinism of the GDPLL(k) automaton, its second automaton, called an Operation Interpreter Automaton (OIA), chooses the proper production out of possible productions (determined by δ_{LLA_1}). Let us introduce its formal definition [11].

Definition 5. An *OIA automaton* is a five-tuple

$$OIA = (\Pi, M, \delta_{OIA_1}, \delta_{OIA_2}, p_0), \text{ where:}$$

Π is a finite set of productions' indices;

M is a memory;

δ_{OIA_1} and δ_{OIA_2} are transition functions defined as follows:

$\delta_{OIA_1} : \Pi \times M \longrightarrow \{TRUE, FALSE\}$,

$\delta_{OIA_2} : \Pi \times M \longrightarrow M$;

$p_0 \in \Pi$ is the starting symbol of a productions' tape. □

The production to be applied is chosen via checking the predicates of applicability of possible productions on the base of the current state of the memory M. The first transition function δ_{OIA_1} is responsible for this task. After the production is selected, the OIA automaton changes the contents of M by performing operations corresponding to the chosen production (it is made by δ_{OIA_2}). Then the LLA(k) changes the contents of the stack (accordingly to the δ_{LLA_2}).

Now, we can present the definition of GDPLL(k) automaton.

Definition 6. A *GDPLL(k) automaton*, $A_{GDPLL(k)}$, is a pair: $A_{GDPLL(k)} = $ (LLA(k),OIA). □

Detailed characteristics GDPLL(k) automata can be found in [4,10,11].

A general scheme of a transducer for translating noisy primitives into a symbolic identification, based on GDPLL(k) parsers is shown in Fig. 2.

The input for an FGDPLL(k) transducer is of the form of a string of noisy primitives. The FGDPLL(k) transducer maintains a pool of GDPLL(k) parsers, which are needed to perform simultaneously several derivation processes for each possible value of an input primitive. The control of the transducer is responsible for verifying, which of the derivation processes should be continued. For this purpose it uses an auxiliary memory, which contains computed probability values for each derivation process. The output of the transducer is of the form of a symbolic (variant) identification of the pattern. A formal definition of a fuzzy FGDPLL(k) transducer is given below.

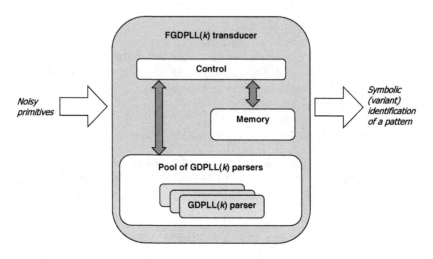

Fig. 2. The scheme of FGDPLL(k) transducer

Definition 7. An *FGDPLL(k) transducer* (a fuzzy GDPLL(k)-based transducer) is a seven-tuple:

$$T = (Q, \Sigma, M_A, \delta, q_0, F, S_{GDPLL(k)}), \text{ where:}$$

Q is a finite set of states;
Σ is a finite set of input symbols;
M_A is an auxiliary memory;
δ is a finite subset of $Q \times NP_\Sigma \times M_A \times Q \times M_A \times (SI_\Sigma \cup \emptyset)$, where NP_Σ is the set of noisy primitives, SI_Σ is the set of symbolic identifications of a pattern;
$q_0 \in Q$ is the initial state;
$F \subseteq Q$ is the set of final states;
$S_{GDPLL(k)} = \{A_{GDPLL(k)} : A_{GDPLL(k)} \text{ is a GDPLL}(k) \text{ automaton } \}$.   □

The main differences between a definition of a common transducer and a definition of FGDPLL(k) transducer are as follows

- An FGDPLL(k) transducer is equipped with an additional memory structure M_A which contains computed probability values for each derivation process performed by GDPLL(k) automata.
- δ, which determines the transducer transitions, reads a noisy primitive instead of an input symbol, and generates a symbolic identification of a pattern. If $(q, n, m, q', m', w) \in \delta$ then the transducer in a state q, after reading an input $n \in NP_\Sigma$ and the current content of the auxiliary memory m, makes a transition to a state q', changes the memory to m' and generates $w \in (SI_\Sigma \cup \emptyset)$.
- An FGDPLL(k) transducer is a complex automaton based on GDPLL(k) parsers, therefore its definition contains the set of GDPLL(k) automata $S_{GDPLL(k)}$.

4   Application of FGDPLL(k) Transducer

Accurate electrical load forecasting is very important for a cost-efficient operation of a power distribution company. Better forecasts mean better trade profits for a distributor. Although a lot of methods and software tools have been developed for this purpose [2,14], there is still a need for constructing more accurate methods and systems.

The FGDPLL(k) transducer has been successfully tested in a system for electrical load forecasting in Tauron Polish Energy Distribution Division in Gliwice, Poland. A prediction of an electrical demand of customers one day ahead is prepared daily at 8 am. The prediction concerns the 24-hour period, and the demand is defined for every hour of this period. On the basis of this prediction Tauron buys a proper amount of energy on the Polish energy market.

There are many methods of forecasting applied by Tauron, for example neural networks, using such parameters as: a time, a type of a day (a weekday, Saturday, Sunday or a holiday), a weather (e.g. temperature or insolation); the autoregressive integrated moving average (ARIMA) method based on historic data; methods based on expert reports prepared by experienced specialists in the field of electrical load forecasting. These methods are combined together to obtain the prediction.

The FGDPLL(k) transducer has been applied in Tauron to correct predictions, and to make it more precise on the basis of *structural* information. A scheme of the transducer-based part of the forecasting system is presented in Fig. 3.

The input consists of numeric data (mostly), corresponding to the electrical load forecast for a given day. The data could be divided in the following two groups.

(1) An area identifier; a type of the day (a working day or a non-working day); a day of a week (Monday–Sunday); a season (spring, summer, autumn, or winter); an amount of non-working days before a given day; an amount of non-working days after a given day.

(2) A load forecast for each hour in a two-days period prepared in a "traditional" way by Tauron; an air temperature forecast for each hour in a two-days period; an insolation forecast for each hour in two days ahead period;

Input data are preprocessed into the form of fuzzy primitives. It is performed with the help of two probabilistic neural networks [13]: PNN_1 and PNN_2.

The data from the first group are translated into one fuzzy primitive describing the classification of the day. It is done with the help of the PNN_1 network. Its output layer neurons correspond to classes to which the presented input data can be assigned. The higher output value of a particular output neuron, the higher probability that the presented data instance belongs to an associated class. Let us notice that classes have been identified on the base of the historic data in the neural network learning phase. The classes are marked with subsequent symbols from the set $\{a, b, c, d, e, f, a_1, ..., f_1, ..., a_9, ..., f_9\}$. Class symbols related to neurons with highest output values become the elements of the first

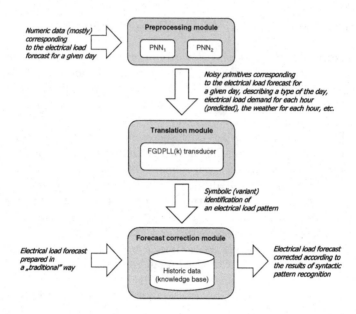

Fig. 3. The scheme of the transducer-based part of the electrical load forecasting system

fuzzy primitive. Relevant probabilities are calculated by normalizing the values of appropriate output layer neurons.

The data from the second group are translated into 24 fuzzy primitives describing conditions (forecasted demand, forecasted air temperature, forecasted insolation) for each hour of the day. It is done with the help of PNN_2 network in a similar way as it is described above. The classes are marked here by the symbols from the set $\{g, ..., z, g_1, ..., z_1, ..., g_9, ..., z_9\}$.

All fuzzy primitives (1+24) are concatenated and submitted to FGDPLL(k) transducer analysis. The FGDPLL(k) transducer translate them into a symbolic (variant) identification of an electrical load pattern. The probability of each unit of symbolic (variant) identification is calculated as the product of probabilities of elements taken from subsequent input fuzzy primitives.

The forecast correction module computes a proper correction to the prediction prepared in a "traditional" way in Tauron. This task is performed with the use of the knowledge base, according to historic data corresponding to current (real) weather circumstances and a real electrical load. Let a result delivered by the FGDPLL(k) transducer be $(w_1, p_1), ..., (w_n, p_n)$. Then, for each $i = 1, ..., n$ the knowledge base is searched for cases represented by w_i. The correction module uses a pattern matching technique to find adequate cases. A correction of the forecast is calculated as the weighted sum of corrections based on historic data (where the weight is the probability p_i of a given pattern). This way structural information about an electrical load pattern is included in the forecasting process.

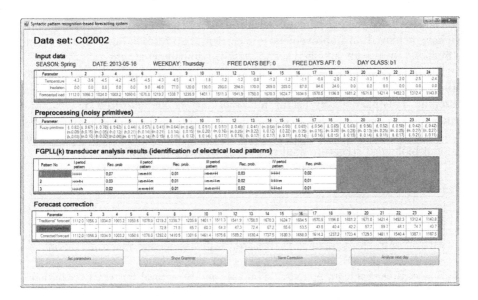

Fig. 4. The syntactic pattern recognition-based forecasting system.

One of the windows of the syntactic pattern recognition-based forecasting system is presented in Fig. 4. It contains information about parameters for the forecast for a given day (numeric data), the transducer output (symbolic information), and a correction of the forecast.

The system has been tested on data concerning 2013. $MAPE$ (mean absolute percentage error) has been used for an evaluation of an accuracy of the forecast. The corrections of forecasts based on FGDPLL(k) transducer improve $MAPE$ up to 0, 3. The improvement can result in significant profits, if we consider energy purchasing on the Polish energy market.

At the end of this section let us notice that the transducer can be implemented as a parallel application. GDPLL(k) parsers being the base elements of the transducer are computationally efficient (the linear computation time). Nevertheless, in case of fuzzy data, the parsing time increases remarkably because of the need of checking many different derivation paths. Hence, a parallelization is of a big importance, especially in real-time environments. A method of a parallelization of derivation processes has been presented in [7]. The same method can be used in case of the transducer. The experiments have shown that the use of a parallel model reduces processing time considerably.

5 Concluding Remarks

A method of improving electrical load forecasting with the help of a FGDPLL(k) transducer has been presented in the paper. The model allows one to translate fuzzy numeric data into symbolic information needed to correct the forecast. Symbolic information is a result of a syntactic analysis of input patterns.

The results of the experiments with the use of the transducer are promising. We have achieved an improvement of electrical load forecasting that can result in financial profits for an electrical energy distribution company.

The model can be applied in many application areas, especially in such pattern recognition/forecating systems that use symbolic descriptions of patterns for reasoning. It is particularly useful in case of patters characterized by *structural* dependencies. Therefore, for structurally-oriented recognition problems such as: character recognition, scene analysis, speech recognition, texture analysis, chemical and biological structure analysis, the model can be applied.

References

1. Aho, A.V., Ullman, J.D.: The Theory of Parsing, Translation, and Compiling. Prentice-Hall, Englewood Cliffs, NJ (1972)
2. Alfares, H.K., Nazeeruddin, M.: Electric load forecasting: literature survey and classifcation of methods. Int. J. Syst. Sci. **33**, 23–34 (2002)
3. Bunke, H.O., Sanfeliu, A. (eds.): Syntactic and Structural Pattern Recognition – Theory and Applications. World Scientific, Singapore (1990)
4. Flasiński, M., Jurek, J.: Dynamically programmed automata for quasi context sensitive languages as a tool for inference support in pattern recognition-based real-time control expert systems. Pattern Recogn. **32**, 671–690 (1999)
5. Flasiński, M., Jurek, J.: On the analysis of fuzzy string patterns with the help of extended and stochastic GDPLL(k) grammars. Fundam. Informaticae **71**, 1–14 (2006). IOS Press, Amsterdam-Berlin-Oxford
6. Flasiński, M., Jurek, J.: Fundamental methodological issues of syntactic pattern recognition. Pattern Anal. Appl. **17**, 465–480 (2014). Springer-Verlag, Berlin Heidelberg, Germany
7. Flasiński, M., Jurek, J., Peszek, T.: Parallel Processing Model for Syntactic Pattern Recognition-Based Electrical Load Forecast. In: Wyrzykowski, R., Dongarra, J., Karczewski, K., Waśniewski, J. (eds.) PPAM 2013, Part I. LNCS, vol. 8384, pp. 338–347. Springer, Heidelberg (2014)
8. Fu, K.S.: Syntactic Pattern Recognition and Applications. Prentice Hall, Chicago (1982)
9. Jurek, J.: Towards grammatical inferencing of GDPLL(k) grammars for applications in syntactic pattern recognition-based expert systems. In: Rutkowski, L., Siekmann, J.H., Tadeusiewicz, R., Zadeh, L.A. (eds.) ICAISC 2004. LNCS (LNAI), vol. 3070, pp. 604–609. Springer, Heidelberg (2004)
10. Jurek, J.: Recent developments of the syntactic pattern recognition model based on quasi-context sensitive languages. Pattern Recogn. Lett. **26**, 1011–1018 (2005). Elsevier, Amsterdam, The Netherlands
11. Jurek, J.: Syntactic Pattern Recognition with the Help of GDPLL(k) Grammars (in Polish). Jagiellonian University Publishers, Cracow (2005)
12. Jurek, J.: Grammatical inference as a tool for constructing self-learning syntactic pattern recognition-based agents. In: Bubak, M., Albada, G.D., Dongarra, J., Sloot, P.M.A. (eds.) ICCS 2008, Part III. LNCS, vol. 5103, pp. 712–721. Springer, Heidelberg (2008)
13. Specht, D.F.: Probabilistic neural networks. Neural Netw. **3**, 109–118 (1990)
14. Taylor, J., McSharry, P.: Short-term load forecasting methods: an evaluation based on european data. IEEE Trans. Power Syst. **22**, 2213–2219 (2008)

Towards a Scalable Distributed Fitness Evaluation Service

Włodzimierz Funika[1,2(✉)] and Paweł Koperek[2]

[1] ACC CYFRONET AGH, AGH, ul. Nawojki 11, 30-950 Kraków, Poland
funika@agh.edu.pl
[2] Faculty of Computer Science, Electronics and Telecommunication,
Department of Computer Science, AGH, al. Mickiewicza 30, 30-059 Kraków, Poland

Abstract. Organizations across the globe gather more and more data. Large datasets require new approaches to analysis and processing, which include methods based on machine learning. In particular, the *symbolic regression* can provide many useful insights. Unfortunately, due to high resource requirements, the use of this method for large datasets might be unfeasible. In this paper we analyze a bottleneck in an open-source implementation of this method, we call *hubert*. We identify that the evaluation of individuals is the most costly operation. As a solution to this problem, we propose a new evaluation service based on the Apache Spark framework, which attempts to speed up computations by distributing them on a cluster of machines. We compare the performance of the service by analyzing the execution time for a number of samples with use of both implementations. Then we discuss how the computation time improves with increased amount of resources. Finally we draw conclusions and outline plans for further research.

Keywords: Distributed systems · Evolutionary programming · Symbolic regression · Scaling · Apache spark

1 Introduction

Nowadays many organizations around the world gather more and more data. Nearly unlimited storage and computing resources only encourage collecting all available data. Unfortunately mining the ever-growing datasets is a very challenging task. It requires excellent domain knowledge and understanding of what data is being analyzed. Furthermore, scalable algorithms and technologies need to be used to provide results in reasonable time.

Due to its complexity, until recently such work was mostly a domain for humans. Today as computer techniques become more and more advanced, automated analysis gains more and more attention. According to [5], computers are now used at multiple stages of research process: from gathering knowledge about related work and similar experiments, through automatic data analysis [15], up to complete automatic systems capable of creating and verifying new hypotheses on

© Springer International Publishing Switzerland 2016
R. Wyrzykowski et al. (Eds.): PPAM 2015, Part I, LNCS 9573, pp. 493–502, 2016.
DOI: 10.1007/978-3-319-32149-3_46

their own [10]. Completely autonomous data mining is not yet attainable, however there are many ways to help scientists and business intelligence specialists in their daily jobs.

Eureqa [14] and hubert [7,8] are examples of a new kind of tools, which were designed to help in the identification of meaningful relationships in available data. They both use a symbolic regression method to automatically search for relationships between variables. A result of such an analysis is provided as a mathematical formula describing discovered connections. Symbolic regression is based on evolutionary programming. It attempts to solve problems defined by the user through generating and improving a population of possible solutions. Unfortunately, implementing such an approach imposes limits on the amount of analyzed information. Evaluation usually comprises applying a solution to the whole data set which requires reading it from storage. Because of a very high number of evaluations, in case of large, multimillion row data series, the time required to obtain a satisfactory solution becomes unacceptably high.

The problem of scalability can be addressed by use of technologies related to Big Data: Apache Hadoop MapReduce [2] or Apache Spark [17]. Both tools were designed to allow efficient distributed processing of large datasets. Their major advantage is the ability to horizontally scale to large numbers of nodes: there are successful deployments of production clusters with thousands of nodes and petabytes of storage [3,12]. Although those frameworks have a lot in common, they are significantly different w.r.t. the basic concepts they are based on. The former provides an abstraction for a processing algorithm where each part of computations is independent. Passing information between jobs requires persisting information to a storage system, e.g., HDFS. Apache Spark on the other hand exploits in-memory processing techniques. This allows implementation of iterative algorithms, applying different computations to the same data.

In this paper we outline an implementation of fitness function evaluation based on the Apache Spark framework. First, we discuss the technical details of some related tools and frameworks and the symbolic regression itself. Then we describe the architecture of the complete system and discuss a comparison of sample datasets processed with use of Apache Spark and hubert. Later we analyze how increasing the amount of resources influences the processing time. Finally, we draw conclusions from the conducted experiments and discuss directions for further research.

2   Background and Related Work

In this section we present information about the tools and concepts used in our research.

2.1   Symbolic Regression

Symbolic regression [11] is a method of regression analysis which focuses on identifying the mathematical expression, in its symbolic form, which would be a very good fit within a given dataset. The parameter space and functional form of equations are being searched in the same time. The method relies on genetic programming. At first, a set of individuals — mathematical expressions, is randomly generated. Each expression is built from specified primitive elements such as algebraic operations $(+, -, *, /)$, variables $(x, y, ...)$, constants $(3.1415, 2.71...)$, etc. Although initially they don't fit the input at all, they get gradually improved with an evolutionary process.

Owing to such a general definition, this method can be applied to solve a magnitude of different problems. Unfortunately, achieving meaningful results is challenging. The key issue lies in choosing proper learning parameters, problem description and most importantly the correct cost function. Very good results can be obtained with the method proposed in [14], which forces the algorithm to discover *implicit relationships*. An implicit relationship is a function of form $f(x, y) = 0$ whereas the explicit function is represented as $y = f(x)$.

2.2 Parallel Implementations of Evolutionary Algorithms

Improving the processing time of evolutionary algorithms and genetic programming in particular, receives much attention from researchers. The major work in this area includes creating implementations utilizing parallel computing paradigms: MPI [13] or MapReduce [4]. These tools focus on parallelizing all steps of the algorithm at once. The solution space is divided into many small populations and all steps of the algorithm (population generation, fitness evaluation, mutation and crossing-over) for a particular set of individuals are processed by a separate physical CPU. This so-called *island* model can be tuned in various ways. First, a straight-forward approach is to start the evolution with different parameters or starting conditions for each population. This broadens the searched solution space but might lead to creating many local solutions (*niching*). Usually obtaining a single, best global individual is preferred. In such a case migrating specimens between populations can be conducted as an additional step executed before starting a new evolution iteration. Evolutionary computations can be parallelized with focus on specific steps of the algorithm, e.g., individual evaluation or mutation only. Another approach is to limit the computation time by limiting the evaluation with the size of used data, e.g., by splitting the original dataset between populations.

2.3 Apache Spark

Apache Spark is a framework for parallel processing of big data sets in a fault-tolerant manner. It is based on a new concept of distributed memory abstraction - Resilient Distributed Datasets (RDD). RDDs are motivated by the limitation of current computing frameworks: poor support for iterative algorithms and interactive data mining tools. They provide a shared memory model, which instead of fine-grained updates prefers coarse-grained transformations like *map*, *filter* or *join*. Such operations can be applied at once to many data items. Fault-tolerance is achieved by logging all transformations used to create a dataset. In case of any error, only some required operations need to be computed again. RDDs are immutable and can only be created by reading from a data source or as an effect of transformations of an existing dataset. The processing is lazy: actual computations are only triggered by actions which require access to the output. RDDs can be cached in memory or persisted on hard-drive for further reuse.

2.4 Hubert

Hubert is a result of our prior work on the use of symbolic regression to the monitoring of computer systems. It provides an open source implementation of the ideas described in [14, 16]. The goal of the project is to discover hidden relationships in monitoring-data streams in order to help gain deeper insight into the behaviour of complex computer systems. Such relationships are described with use of precise mathematical expressions which are individuals from the genetic algorithm perspective. Each individual is represented as an expression tree built from primitive blocks ($+$, $-$, $\times$, $/$, sin, cos, variables defined in an input dataset, constants).

The process can be used to analyze any series of numerical data. In case of hubert we focused on the data coming from the monitoring of computer systems. In this case the best discovered equations can describe dependencies between the components and model complex behaviour of the system. Such information can be used to improve the architecture of the system or to tune its parameters to improve its performance. The iterative nature of such an analysis makes the model evolve over time and keeps it up to date.

The tool was written in Java language thus it can be easily integrated with other technologies which use Java Virtual Machine. It is an open-source project - we encourage the reader to use and extend it and adapt to users' specific needs [8].

3 Overview of the Evaluation Service Concept

The evaluation of individuals is the most resource-consuming part of evolutionary algorithms. It requires reading the input data set and evaluating the value of the assessed specimen over all its data points. This often takes over 99 % of the whole processing time. In case of more complex individuals and more data this value gets even closer to 100 %. Since the same set is used in every iteration multiple times, the best way to speed up the computations is to cache it in memory. Unfortunately in case of the datasets whose sizes exceed the memory of a single machine, this solution is not possible to use. In such a case, the time of evaluation increases drastically because of heavy I/O usage. This renders the symbolic regression algorithms not feasible to use for larger volumes of data.

To address this limitation on the input size both in hubert and in symbolic regression in general, we made use of the Apache Spark framework to implement a new fitness evaluation service. First, the user needs to register datasets he/she wishes to use. They can be stored in HDFS or on local hard-drives of computing nodes. We prefer the former solution. In this case the Spark framework can easily split computations in such a way that each node processes only the part of data which is stored on its local hard-drive. Later, when the client (e.g., hubert) needs to evaluate a specific individual, it sends a query to the service instead of performing the computations itself. The evaluation query contains the individual - the mathematical expression, information about which dataset to use and specifies which formula should be used for numerical differentiation (forward, backward and central finite differences).

At the registering step the dataset is being preprocessed and results are cached in memory. The preprocessing includes generating all pairs of variables and computing numerically derivatives for each pair. This creates $C_n^2 = \binom{n}{2}$ data series (n is the number of variables), nearly of the same size as the input data set. All of them need to be stored at the same time in memory - they are reused each time an individual evaluation occurs. For a sample dataset of currency quotations containing 4 variables (EUR/USD, EUR/GBP, CHF/PLN, USD/CHF), each holding a 2,5 GB data series (the data cover the period from 05.2005 to 06.2014), this means that $\binom{4}{2} = \frac{4!}{2!2!} = 6$ series of 2,5 GB each need to be cached. This number rapidly grows with the growth of input dimensions. Unfortunately, this imposes limits on the tool's capabilities and has to be kept in mind when estimating how much resources are required to ensure low request response times. If the user does not manually trigger the preprocessing before submitting an evaluation request, it is automatically run upon having received the first request.

Implementing the evaluation as a service has several advantages. First, it can be used in tools other than hubert. It also abstracts the Spark API. In case a better processing solution can be applied to speedup computations, such a change won't require any changes on the service users' side. The use of the service allows for handling multiple requests at the same time thus improves cluster usage. Furthermore it is a starting point for migrating hubert as a whole to a microservices architecture, which would enable further improvements, e.g., using populations with a greater number of individuals or evolving solutions for multiple datasets in parallel.

4   Back-End Architecture

The architecture of the solution under discussion is presented in Fig. 1. The system comprises a *front-end node*, which contains an instance of the evaluation service and infrastructure services: *HDFS Name Node* and *Spark Master*. Back-end nodes are running *HDFS Data Nodes* and *Spark Slave*.

Once a request is received, the evaluation service submits a new job to *Spark Master*. *Spark Master* splits the job and sends it as a serialized Java bytecode to its slaves. If the input data is not cached in memory yet, it is delivered by HDFS services. Owing to this, CPUs process the data which is actually stored nearest to them - on their hard-drives. The progress is constantly monitored. If an executor fails it is automatically restarted, the missing computations are rescheduled on other machines. Finally, when all the stages of processing finish, the evaluation service returns a result. The related RDDs are automatically cached in memory.

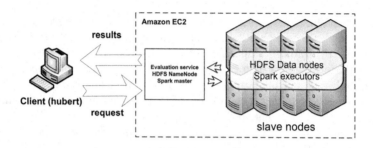

Fig. 1. General architecture diagram

5   Comparison with Hubert

To evaluate the new implementation and examine whether it processes the data faster than the initial one, we compared the execution time of processing for three sample individuals generated by hubert:

- A: $sin(x + y)$
- B: $(x - y + cos(x) - 4.906 + 5.8 + x - y)/(cos(4.56575) + cos(x) + sin(x) * x/y)$
- C: $(((x - y) * x * 1.0951405 * sin(x * y) * cos(cos(y) + 1.0951405/3.01411)) * (sin((x - y) + 2.377/2.817) * cos(x) * (x + y) - (sin(x) - sin(y)))) - (cos(cos(x/y)/((x + y) - (x - 2.3776817))/sin((7.305318 + x) + (x - y))))$

We present them in their non-simplified form - as such they are processed by both tools.

Each specimen was evaluated against three datasets of different sizes: 10 MB, 100 MB, 1024 MB. Each of them contained a different time window of financial data series: price quotes for the euro and American dollar currency pair.

The computations were carried out in Amazon Elastic Computing Cloud [1]. Spark cluster consisted of nine instances of *m1.medium* type virtual machines (one master and eight slaves), each using 1 VCPU, 3.75 GB RAM and 410 GB local hard drive storage.

A comparison of computation times is given in Fig. 2. In all cases the service implementation has shorter execution times than hubert. The bigger the dataset, the bigger speed up was observed. In case of the simplest individual - A, and the biggest dataset - 1024 MB, the service was almost 70 times faster. The complexity of the analyzed individual has an opposite effect: for more complex individuals the advantage of using a cluster is less visible, but still significant - nearly 20 times.

6   Scalability Analysis

To analyze how scalable the new service is, we observed how the execution time changes when more resources are added to the computing cluster. We examined two scenarios:

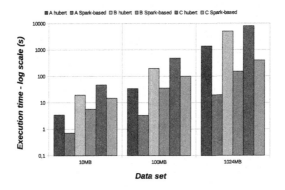

Fig. 2. Hubert and service execution time comparison

- *scaling up* - using the same number of machines but with more cores, RAM etc.
- *scaling out* - using more machines of the same kind.

We analyzed both cases in Amazon Elastic Compute Cloud environment. For *scaling up* we processed the same dataset with use of a cluster of a fixed size but each time with machines with different specification:

- *m1.medium* - 1 vCPU, 3.75 GB RAM
- *m1.large* - 2 vCPU, 7.5 GB RAM
- *m1.xlarge* - 4 vCPU, 15 GB RAM

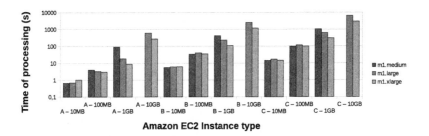

Fig. 3. Scaling up of the Spark-based service

In *scaling out* we created clusters of sizes: 3, 6 and 12, consisting of *m1.medium* machines. Figures 3 and 4 present the results of experiments with scaling up and scaling out.

The results obtained prove that the service is able to effectively use more resources. However to fully take advantage of them, a big enough dataset is required. For smaller datasets (10 MB, 100 MB) the processing time is almost the same. The more data is processed the greater the performance improvement is visible. For 1 GB the speed up varies from 3,5x to 10x after increasing the

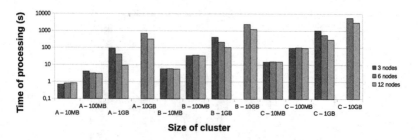

Fig. 4. Scaling out of the Spark-based service

amount of resources by a factor of 4. Unfortunately the smallest (3 node) cluster was unable to process the biggest (10 GB) sample. Observations indicate that the speed up would follow the same pattern as that observed in smaller ones.

The base case for both approaches is the same: 3 node cluster consisting of *m1.medium* instances. In each step the same number of cores and RAM was added to the system. We can therefore compare the results obtained for scaling out and scaling up and notice that both approaches have similar influence on the performance (Tables 1 and 2). Scaling out seems to be a bit more profitable because of greater I/O throughput available across the cluster.

Table 1. Relative speed up of execution in comparison to a base case of 3-node *m1.medium* cluster (*scale up* approach)

	A	A	A	B	B	B	C	C	C
	10 MB	100 MB	1 GB	10 MB	100 MB	1 GB	10 MB	100 MB	1 GB
m1.large	0,96	1,22	**4,89**	0,92	0,85	**1,83**	0,85	0,85	**1,67**
m1.xlarge	0,64	1,34	**10,29**	0,91	0,96	**3,81**	0,98	0,99	**3,43**

Table 2. Relative speed up of execution in comparison to a base case of 3-node *m1.medium* cluster (*scale out* approach)

	A	A	A	B	B	B	C	C	C
	10 MB	100 MB	1 GB	10 MB	100 MB	1 GB	10 MB	100 MB	1 GB
6 nodes	0,84	1,30	**2,24**	0,97	0,93	**1,01**	0,95	0,94	**1,86**
12 nodes	0,81	1,37	**10,11**	1,00	1,02	**4,05**	0,97	1,01	**3,53**

7   Conclusions and Further Work

Symbolic regression is a useful research method. It can be used to create mathematical models based on numerical data series. In particular, such a model can be used to describe the behaviour of a computer system in a given time range.

Unfortunately, in many cases the amount of data acquired by monitoring facilities is large [6,9]. To make the idea of evolutionary computations feasible for use in this scenario, we demonstrated how to speed them up by applying new concepts stemming from distributed computing.

In this paper we presented an implementation of the fitness evaluation service which can be easily integrated with the existing symbolic regression tools. The new service scales very well when increasing the amount of available resources and the speed up is significant enough to enable the processing of large datasets in reasonable time. By using hubert together with the new service, scientists can automatically create mathematical models for relationships within an analyzed dataset, even if this does not fit in a single machine's memory. This makes symbolic regression applicable to a new range of problems for which the size of input prevented from the use of genetic programming-based methods.

The results obtained show that Apache Spark is a viable solution to execute machine learning algorithms. The RDD memory model was simple to use. Unfortunately, we noticed that not all types of computation can be easily represented in such a way. If the algorithm requires combining subsequent values of data series, it is necessary to copy the whole input, change data indices and perform a costly *join* operation over the original and new RDD. Spark applications can be easily tested in a single machine configuration. We encourage the reader to use this facility to assess the usability of the framework, before deciding on exploiting it in production. Otherwise the limitations of that model might induce an unnecessary overhead which will outweigh other benefits.

Hubert proved that it is a robust implementation of symbolic regression. The tool's architecture is also very flexible. We were able to easily refactor the tool's code to a form of service. According to the results obtained, the single processor implementation of fitness evaluation is still very useful. The use of a distributed computing framework in case of small datasets induces too much overhead. Instead of completely migrating to a new evaluation implementation, we plan to allow switching between both of them.

The work on the open-source implementation of symbolic regression is ongoing. We plan to further optimize the execution time of the evaluation service. We believe that enabling low-latency responses of several second duration is possible. Furthermore we see another opportunity to increase scalability in migrating the hubert's architecture from the monolithic code base to a microservices architecture, which allows for independent scaling of different parts of the system and potentially hosting them on a separate clusters.

Acknowledgement. We would like to thank dr. Maciej Malawski for his valuable help with Amazon EC2 experiments. This research is supported by AGH grant no. 11.11.230.124 as well as by the PLGrid Core project.

References

1. Amazon.com Inc: AWS Amazon Elastic Compute Cloud (EC2) - Scalable Cloud Hosting (2014). http://aws.amazon.com/ec2. Accessed 02 April 2015
2. Apache Software Foundation: Welcome to apache TM hadoop! (2014). http://hadoop.apache.org/. Accessed 11 March 2015
3. Baldeschwieler, E.: Yahoo! launches world's largest hadoop production application (2008). https://developer.yahoo.com/blogs/hadoop/yahoo-launches-world-largest-hadoop-production-application-398.html. Accessed 11 March 2015
4. Du, X., Ni, Y., Yao, Z., Xiao, R., Xie, D.: High performance parallel evolutionary algorithm model based on mapreduce framework. Int. J. Comput. Appl. Technol. **46**(3), 290–295 (2013)
5. Evans, J., Rzhetsky, A.: Machine science. Science **329**, 399–400 (2010)
6. Funika, W., Godowski, P., Pegiel, P., Król, D.: Semantic-oriented performance monitoring of distributed applications. Comput. Inf. **31**(2), 427–446 (2012). http://www.cai.sk/ojs/index.php/cai/article/view/948
7. Funika, W., Koperek, P.: Genetic programming in automatic discovery of relationships in computer system monitoring data. In: Wyrzykowski, R., Dongarra, J., Karczewski, K., Waśniewski, J. (eds.) PPAM 2013, Part I. LNCS, vol. 8384, pp. 371–380. Springer, Heidelberg (2014)
8. Funika, W., Koperek, P.: Hubert project source code (2015). https://github.com/pkoperek/hubert. Accessed 15 March 2015
9. Funika, W., Kupisz, M., Koperek, P.: Towards autonomic semantic-based management of distributed applications. Comput. Sci. (AGH) **11**, 51–64 (2010). http://journals.agh.edu.pl/csci/article/view/116
10. King, R.D., et al.: The automation of science. Science **324**, 85–89 (2009)
11. Koza, J.: Genetic Programming: On the Programming of Computers by Means of Natural Selection. MIT Press, Cambridge (1992). http://mitpress.mit.edu/books/genetic-programming
12. Ryan, A.: Under the hood: Hadoop distributed filesystem reliability with namenode and avatarnode (2012). https://www.facebook.com/notes/facebook-engineering/under-the-hood-hadoop-distributed-filesystem-reliability-with-namenode-and-avata/10150088759153920. Accessed 11 April 2015
13. Salhi, A., Glaser, H., De Roure, D.: Parallel implementation of a genetic-programming based tool for symbolic regression. Inf. Process. Lett. **66**(6), 299–307 (1998). http://dx.doi.org/10.1016/S0020-0190(98)00056-8
14. Schmidt, M., Lipson, H.: Distilling free-form natural laws from experimental data. Science **324**(5923), 81–85 (2009)
15. Schmidt, M.D., Lipson, H.: Data-mining dynamical systems: automated symbolic system identification for exploratory analysis. In: ASME Conference Proceedings, vol. 2008(48364), pp. 643–649 (2008). http://dx.doi.org/10.1115/esda2008-59309
16. Schmidt, M., Lipson, H.: Age-fitness pareto optimization. In: Pelikan, M., Branke, J. (eds.) GECCO, pp. 543–544. ACM (2010). http://dblp.uni-trier.de/db/conf/gecco/gecco2010.html#SchmidtL10
17. Zaharia, M., Chowdhury, M., Das, T., Dave, A., Ma, J., McCauley, M., Franklin, M.J., Shenker, S., Stoica, I.: Resilient distributed datasets: a fault-tolerant abstraction for in-memory cluster computing. In: Proceedings of the 9th USENIX Conference on Networked Systems Design and Implementation, p. 2. NSDI 2012, USENIX Association, Berkeley, CA, USA (2012). http://dl.acm.org/citation.cfm?id=2228298.2228301

Minisymposium on GPU Computing

Revisiting the Gauss-Huard Algorithm for the Solution of Linear Systems on Graphics Accelerators

Peter Benner[1], Pablo Ezzatti[2], Enrique S. Quintana-Ortí[3], and Alfredo Remón[1(✉)]

[1] Max Planck Institute for Dynamics of Complex Technical Systems,
39106 Magdeburg, Germany
{benner,remon}@mpi-magdeburg.mpg.de
[2] Instituto de Computación, Universidad de la República,
11300 Montevideo, Uruguay
pezzatti@fing.edu.uy
[3] Dep. de Ingeniería y Ciencia de la Computación, Universidad Jaime I,
12701 Castellón, Spain
quintana@icc.uji.es

Abstract. In 1979, P. Huard presented an efficient variant of the Gauss-Jordan elimination for the solution of linear systems. In particular, this alternative algorithm exhibits the same computational cost as the traditional LU-based solver, and is considerably cheaper than the Gauss-Jordan algorithm, but there exist no recent high performance implementations of the Gauss-Huard (GH) variant that allow a comparison of these approaches. In this paper we present a reliable GH solver for hybrid platforms equipped with conventional multi-core technology and a graphics processing unit (GPU). The experimental results show that the GH algorithm can beat high performance versions of the LU solver, from tuned libraries for CPU-GPU servers such as MAGMA, for problems of small to moderate scale.

Keywords: Dense linear systems · Gauss-Huard algorithm · LU factorization · Multi-core processors · Graphics Processing Units (GPUs) · High performance

1 Introduction

The solution of linear systems $Ax = b$, where $A \in \mathbb{R}^{n \times n}$, $b, x \in \mathbb{R}^n$ and x represents the sought-after solution, is a fundamental problem in many scientific and engineering applications [6]. When the coefficient matrix A is dense, the most popular approach to tackle this problem is based on the LU factorization (i.e., Gaussian elimination), which in practice is complemented by (partial) row pivoting for numerical stability [9].

The Gauss-Jordan elimination (GJE), with row pivoting has been demonstrated to be a reliable and efficient algorithm for matrix inversion on a variety

© Springer International Publishing Switzerland 2016
R. Wyrzykowski et al. (Eds.): PPAM 2015, Part I, LNCS 9573, pp. 505–514, 2016.
DOI: 10.1007/978-3-319-32149-3_47

of current high performance systems, including platforms equipped with accelerators [3,13]. However, when applied to the solution of linear systems, GJE incurs a larger computational cost than traditional approaches based on the LU factorization. Concretely, GJE requires about n^3 floating-point arithmetic operations, or flops, while an LU-based solver requires around $2n^3/3$ flops only. This shortage was addressed by P. Huard in the late 1970s by introducing a variant of GJE to solve linear systems, the Gauss-Huard algorithm (GHA) [12], with a computational cost analogous to that of the LU-based method. When combined with *column* pivoting, GHA offers a numerical stability that is comparable with that of the LU-based method with row pivoting [5].

Hybrid computer servers consisting of general-purpose multicore processors (CPUs) and a many-core graphics processing unit (GPU) have progressed dramatically over the last decade. The success of these platforms is due to their moderate price, high performance, favorable throughput-per-Watt, and the availability of relatively simple application programming interfaces as well as standard parallel programming tools and libraries. Some of the most powerful supercomputers in the world are currently built using hybrid CPU-GPU servers [15]. In addition, this technology has reported remarkable performance for the solution of many scientific and engineering problems with high computational requirements. In the domain of dense linear algebra, many efforts have demonstrated the benefits of GPU-accelerated computing; see, among others, [2,4,18] and the MAGMA library [17].

In this paper we present a reliable and high-performance implementation of the GHA with column pivoting that exploits the intrinsic concurrency of the algorithm and its low communication requirements on hybrid CPU-GPU platforms. This study is relevant because, since its introduction more than 35 years ago, GHA has been analyzed only once in the context of high performance implementations, in 1994, for message-passing systems [11], and with no experimental results provided there. Thus, in view of the recent advances in computer architecture in general, and the surge of graphics accelerators in particular, an experimental evaluation which determines whether this approach can be competitive with conventional methods, based on the LU factorization, is in time. This work complements the evaluations in [7,14], where different hardware architectures are targeted, specifically, the Intel Xeon Phi processor and a platform formed by a low-power GPU and an ARM processor.

The rest of the paper is structured as follows. In Sect. 2, we briefly review the LU-based and GHA approaches to solve linear systems. In Sect. 3, we introduce the new high performance solver based on GHA for CPU-GPU platforms, and we experimentally evaluate the result, in Sect. 4. Finally, we draw a few conclusions and discuss future work in Sect. 5.

2   Solution of Linear Systems

In this section we revisit two alternative approaches for the solution of linear systems: the traditional LU-based algorithm and GHA, both in-place methods with a similar cost in terms of flops and memory requirements.

2.1 Solution of Linear Systems via the LU Factorization

This approach commences by computing the LU factorization of the matrix A. Furthermore, to ensure numerical stability, (partial) row pivoting is employed during this decomposition [9]. In practice the factorization takes the form $PA = LU$, where $P \in \mathbb{R}^{n \times n}$ is a permutation matrix (implicitly stored as a vector), and the factors $L, U \in \mathbb{R}^{n \times n}$ are, respectively, unit lower triangular and upper triangular and overwrite the corresponding parts of A (the diagonal of L is not stored as it consists only of 1s). The original system is then equivalent to $LUx = (Pb) = \hat{b}$, and x can be obtained by solving the triangular linear systems $Ly = \hat{b}$ followed by $Ux = y$.

The LU-based solver presents a few properties with a negative impact on performance. In particular, it is divided into three consecutive stages (LU factorization, upper triangular solve and lower triangular solve), implicitly defining

---

Algorithm: $\left[\hat{A}, p\right] := \text{GaussHuard\_unb}(\hat{A})$

Partition $\hat{A} \to \left(\begin{array}{c|c} \hat{A}_{TL} & \hat{A}_{TR} \\ \hline \hat{A}_{BL} & \hat{A}_{BR} \end{array}\right), p \to \left(\begin{array}{c|c} p_L & p_R \end{array}\right)$

   where $\hat{A}_{TL}$ are 0×0, and p_L has 0 elements

while $m(\hat{A}_{TL}) < m(\hat{A})$ **do**
   Repartition

$\left(\begin{array}{c|c} \hat{A}_{TL} & \hat{A}_{TR} \\ \hline \hat{A}_{BL} & \hat{A}_{BR} \end{array}\right) \to \left(\begin{array}{c|c|c} \hat{A}_{00} & \hat{a}_{01} & \hat{A}_{02} \\ \hline \hat{a}_{10}^T & \hat{\alpha}_{11} & \hat{a}_{12}^T \\ \hline \hat{A}_{20} & \hat{a}_{21} & \hat{A}_{22} \end{array}\right), \left(\begin{array}{c|c} p_L & p_R \end{array}\right) \to \left(\begin{array}{c|c|c} p_0 & \pi_1 & p_2 \end{array}\right)$

   where $\hat{\alpha}_{11}, \pi_1$ are 1×1

$\left[\hat{\alpha}_{11}, \hat{a}_{12}^T\right] := \left[\hat{\alpha}_{11}, \hat{a}_{12}^T\right] - \hat{a}_{10}^T \cdot \left[\hat{a}_{01}, \hat{A}_{02}\right]$; Row elimination

$\pi_1 := \text{PivIndex}\left(\left[\hat{\alpha}_{11}, \hat{a}_{12}^T\right]\right)$

$\left(\begin{array}{c|c} \hline \hat{a}_{01} & \hat{A}_{02} \\ \hline \hat{\alpha}_{11} & \hat{a}_{12}^T \\ \hline \hat{a}_{21} & \hat{A}_{22} \end{array}\right) := \left(\begin{array}{c|c} \hline \hat{a}_{01} & \hat{A}_{02} \\ \hline \hat{\alpha}_{11} & \hat{a}_{12}^T \\ \hline \hat{a}_{21} & \hat{A}_{22} \end{array}\right) P(\pi_1)$; Pivoting

$\left[\hat{\alpha}_{11}, \hat{a}_{12}^T\right] := \left[\hat{\alpha}_{11}, \hat{a}_{12}^T\right] / \hat{\alpha}_{11}$; Diagonalization (row scaling)

$\hat{A}_{02} := \hat{A}_{02} - \hat{a}_{01} \cdot \hat{a}_{12}^T$; Column elimination

---

Continue with

$\left(\begin{array}{c|c} \hat{A}_{TL} & \hat{A}_{TR} \\ \hline \hat{A}_{BL} & \hat{A}_{BR} \end{array}\right) \leftarrow \left(\begin{array}{c|c|c} \hat{A}_{00} & \hat{a}_{01} & \hat{A}_{02} \\ \hline \hat{a}_{10}^T & \hat{\alpha}_{11} & \hat{a}_{12}^T \\ \hline \hat{A}_{20} & \hat{a}_{21} & \hat{A}_{22} \end{array}\right), \left(\begin{array}{c|c} p_L & p_R \end{array}\right) \leftarrow \left(\begin{array}{c|c|c} p_0 & \pi_1 & p_2 \end{array}\right)$

endwhile

Fig. 1. Basic (unblocked) Gauss-Huard algorithm for the solution of $Ax = b$. On entry, $\hat{A} = [A, b]$. Upon completion, due to pivoting, the last column of $\hat{A}$ is overwritten with a permutation of the solution, $P(p)x$. In notation, $m(\cdot)$ stands for the number of rows of its input. Furthermore, $\text{PivIndex}\left([\hat{\alpha}_{11}, \hat{a}_{12}^T]\right)$ returns the index of the entry of largest magnitude in its input (excluding the last element, which corresponds to an entry of b); and $P(\pi_1)$ denotes the corresponding permutation.

two synchronization points. In addition, it requires the solution of two triangular linear systems, an operation that presents a rich collection of data dependencies and more reduced concurrency than other Level-3 BLAS kernels. Furthermore, small triangular linear systems also appear during the factorization stage.

2.2 The Gauss-Huard Algorithm

Figure 1 offers an algorithmic description of GHA for the solution of the linear system $Ax = b$ using the FLAME notation [10]. At a given iteration, the algorithm requires three main operations: diagonalization, row elimination, and column elimination. In theory, as the procedure progresses, block $\hat{A}_{00}$ becomes the identity matrix as the diagonalization sets $\hat{\alpha}_{11} = 1$, while the row and column eliminations set both $[\hat{a}_{10}^T]$ and $[\hat{a}_{01}]$ to 0. In practice $[\hat{a}_{10}^T]$ and $[\hat{a}_{01}]$ will not be accessed later and their updates are not required. Hence, our implementations avoid the updates of $[\hat{a}_{10}^T]$ and $[\hat{a}_{01}]$ (by removing them from the row and column elimination processes). Nevertheless, all the diagonal elements in $\hat{A}_{00}$ equal 1 due to the diagonalization step.

Algorithm: $\boxed{\hat{A},p}$:= GaussHuard_blk($\hat{A}$)

Partition $\hat{A} \rightarrow \left(\begin{array}{c|c} \hat{A}_{TL} & \hat{A}_{TR} \\ \hline \hat{A}_{BL} & \hat{A}_{BR} \end{array} \right), p \rightarrow \left(p_L \,\middle|\, p_R \right)$

 where $\hat{A}_{TL}$ is 0×0, and p_L has 0 elements

while $m(\hat{A}_{TL}) < m(\hat{A})$ **do**

 Determine block size c

 Repartition

$$\left(\begin{array}{c|c} \hat{A}_{TL} & \hat{A}_{TR} \\ \hline \hat{A}_{BL} & \hat{A}_{BR} \end{array} \right) \rightarrow \left(\begin{array}{c|c|c} \hat{A}_{00} & \hat{A}_{01} & \hat{A}_{02} \\ \hline \hat{A}_{10} & \hat{A}_{11} & \hat{A}_{12} \\ \hline \hat{A}_{20} & \hat{A}_{21} & \hat{A}_{22} \end{array} \right), \left(p_L \,\middle|\, p_R \right) \rightarrow \left(p_0 \,\middle|\, p_1 \,\middle|\, p_2 \right)$$

 where $\hat{A}_{11}$ is $c \times c$ and p_1 has c elements

$\left[\hat{A}_{11}, \hat{A}_{12} \right] := \left[\hat{A}_{11}, \hat{A}_{12} \right] - \hat{A}_{10} \cdot \left[\hat{A}_{01}, \hat{A}_{02} \right]$; Block-row elimination

$\left[\hat{A}_{11}, \hat{A}_{12}, p_1 \right] := $ GaussHuard_unb $\left(\left[\hat{A}_{11}, \hat{A}_{12} \right] \right)$; Block diagonalization

$\left(\begin{array}{c|c} & \hat{A}_{01} & \hat{A}_{02} \\ \hline \hline \hat{A}_{21} & \hat{A}_{22} \end{array} \right) := \left(\begin{array}{c|c} & \hat{A}_{01} & \hat{A}_{02} \\ \hline \hline \hat{A}_{21} & \hat{A}_{22} \end{array} \right) P(p_1)$; Pivoting

$\hat{A}_{02} := \hat{A}_{02} - \hat{A}_{01} \cdot \hat{A}_{12}$; Block-column elimination

---

Continue with

$$\left(\begin{array}{c|c} \hat{A}_{TL} & \hat{A}_{TR} \\ \hline \hat{A}_{BL} & \hat{A}_{BR} \end{array} \right) \leftarrow \left(\begin{array}{c|c|c} \hat{A}_{00} & \hat{A}_{01} & \hat{A}_{02} \\ \hline \hat{A}_{10} & \hat{A}_{11} & \hat{A}_{12} \\ \hline \hat{A}_{20} & \hat{A}_{21} & \hat{A}_{22} \end{array} \right), \left(p_L \,\middle|\, p_R \right) \leftarrow \left(p_0 \,\middle|\, p_1 \,\middle|\, p_2 \right)$$

endwhile

Fig. 2. Blocked Gauss-Huard algorithm for the solution of $Ax = b$. On entry, $\hat{A} = [A, b]$. Upon completion, due to pivoting, the last column of $\hat{A}$ is overwritten with a permutation of the solution, $P(p)x$.

A blocked variant of GHA was introduced in [11] for distributed-memory (message-passing) systems. Unfortunately, the authors did not perform any experimental evaluation of the implementation, and simply stated that its performance could be expected to be close to that of an LU-based solver. The authors also introduced a variant of this algorithm that merges the block-column elimination of one iteration with the block-row elimination of the following iteration into a single matrix multiplication.

The blocked version of GHA is illustrated in Fig. 2. The algorithm processes c columns of the matrix per iteration of the loop body, with c often referred to as the algorithmic block size. Again, our implementations do not perform the updates of $[\hat{A}_{10}]$ and $[\hat{A}_{01}]$ to save unnecessary operations.

3   Efficient Hybrid CPU-GPU Implementations

CPU-GPU platforms have reached a relevant position for HPC applications, in particular in the domain of linear algebra, and state-of-the-art scientific libraries, such as MAGMA [17], PETSc [1], ViennaCL [16] and CULA [8] nowadays include routines to solve linear systems that efficiently leverage the manycore architecture of GPUs to deliver considerable speed-ups.

In this context, we present and evaluate a high performance implementation of GHA that targets hybrid CPU-GPU servers. The implementation decomposes the algorithm into a collection of (sub)operations, executes each part in the most appropriate device, and off-loads the computationally intensive operations to the GPU while incurring a low to moderate overhead due to data transfers between the memory address spaces of CPU and GPU. Whenever possible, the implementation also overlaps computations in both architectures.

In more detail, our CPU-GPU implementation of GHA is based on the GaussHuard_blk algorithm. Provided the algorithmic block size is chosen as $c \ll n$, most of the computations in the algorithm are cast in terms of the matrix multiplications performed during the block-column and block-row eliminations. This type of operation exhibits a high level of concurrency that suits the GPU architecture and consequently, they are carried out by the GPU. In contrast, the block diagonalization presents a moderate computational cost and its concurrency is constrained by the pivoting scheme. Thus, this operation is performed in the CPU to attain reasonable performance. Finally, the pivoting stage (except for $[\hat{A}_{11}, \hat{A}_{12}]$ which was already pivoted during the block diagonalization) is performed in the GPU to reduce the number of data movements and to exploit the larger memory bandwidth of the accelerator.

This partitioning of the workload requires, at each step of the algorithm, three transfers of the block $[\hat{A}_{11}, \hat{A}_{12}]$ between the memory spaces of the CPU and the GPU. Specifically, it must be first transferred from the CPU to the GPU before the calculation of the block-row elimination. Later, it is transferred back to the CPU to be updated during the block diagonalization; and finally sent back to the GPU to compute the block-column elimination. Thus, the volume of data transferred at each iteration is proportional to n and c. After the procedure is completed, the solution vector (in the last column of $\hat{A}$) must be retrieved by the CPU.

The efficiency of this initial CPU-GPU implementation strongly depends on the value chosen for c: a small value of this parameter reduces the performance of the matrix multiplications that are off-loaded to the GPU, though it ensures that the efficiency of the algorithm is driven by computation instead of data transfers. Conversely, a large value of this parameter negatively affects the performance of the complete procedure because it shifts a significant part of the algorithm's flops into the diagonalization stage. To tackle this scenario, we propose an implementation that operates with two block sizes, b_c and b_g, for the CPU and GPU operations, respectively. This strategy allows the use of the optimal block size independently for each architecture. Concretely, our implementation carries out the block diagonalization of $[\hat{A}_{11}, \hat{A}_{12}]$ in GAUSSHUARD_BLK using a blocked variant of the same algorithm that exclusively proceeds on the CPU, with an algorithmic block size $b_c < c$. At a higher level, the GAUSSHUARD_BLK algorithm operates with the block size $b_g = c$, which determines one of three dimensions of the matrix multiplications involved in the block-row and block-column eliminations.

The following list comprises a number of additional code optimizations included in our implementation:

- The matrix is initially copied to the GPU, saving one transfer of $[\hat{A}_{11}, \hat{A}_{12}]$ per iteration. The overhead of the initial transfer is mostly hidden by overlapping this copy with the diagonalization of the first panel. Furthermore, a single large transfer is often more efficient than several smaller transfers.
- The block diagonalization in the CPU is overlapped with the application of column swaps due to pivoting in blocks $[\hat{A}_{01}, \hat{A}_{02}]$ and $[\hat{A}_{21}, \hat{A}_{22}]$.
- Whenever possible, the operations are performed via calls to kernels of BLAS (e.g., Intel MKL for the CPU and NVIDIA CUBLAS libraries for the GPU), though a few key scalar and Level-1 BLAS operations are performed via *ad-hoc* kernels, parallelized with OpenMP, which yield higher performance.

4   Experimental Evaluation

This section introduces the experimental setup and reports the results obtained from the evaluation of the CPU-GPU implementation of GHA.

4.1   Test Environment

The target CPU-GPU server contained 2 Intel Xeon(R) CPU E5-2620 v2 (6 cores at 2.1 GHz, Sandy-Bridge architecture) with 128 GB of DDR3 memory, connected to an NVIDIA K40 GPU (2,880 CUDA cores at 745 MHz, Kepler architecture, compute capability 3.5, 12 GB of DDR5 memory) via a PCI-e bus. The operating system was Centos v6.5 Linux. The codes were compiled using gcc v.4.4.7 and the NVIDIA CUDA compiler v.6.5 (both with the -O3 flag enabled). High performance implementations of BLAS were obtained by linking the solver to Intel MKL v11.1 and NVIDIA CUBLAS v6.5.

4.2   Experimental Analysis

We next evaluate the performance of three solvers applied to linear systems of dimension n varying between 1.000 and 10.000 (1K and 10K) with a single right-hand side vector b. Concretely, we compare our CPU-GPU implementation of GHA described in the previous section with two highly-tuned implementations of LU-based solvers, namely routine dgesv from Intel MKL and routine magma_dgesv from MAGMA. Although the Intel MKL solver only runs in the CPU, it is included in the comparison for reference. In contrast, MAGMA can be considered as the state-of-the-art hybrid CPU-GPU solver. Note that the quality of the numerical results obtained by the three are comparable due to the use of pivoting in the GHA-based solver.

The results correspond to the average taken over 5 independent executions. For the CPU-GPU codes (GHA and MAGMA), the execution times include the cost of all CPU-GPU communication. We tested several pairs of algorithmic block sizes (b_c, b_g) and numbers of CPU threads for our implementation of GHA, but for simplicity we only report the results obtained with the best configuration. We also evaluated configurations with different numbers of threads for the Intel MKL and MAGMA solvers, but again we only show the best results.

Table 1 reports the execution times (in seconds) of GHA as well as those obtained with the Intel MKL and MAGMA routines. The results demonstrate that the Intel MKL solver outperforms the hybrid solvers for small problems, due to the overhead intrinsic to CPU-GPU communication. For $n = 2K$, MKL and GHA offer similar performance and, starting from that point, the latter solver outperforms its counterpart in the Intel MKL library. The GHA variant is also more efficient than the MAGMA solver when the dimension of the system is smaller than $n = 6K$. In particular, for $n = 3K$, GHA is about 75 % faster than

Table 1. Execution time (in seconds) and performance (in billions of flops/sec. or GFLOPS) obtained by the three solvers evaluated with problems of varying dimension n.

n	GHA		MAGMA		Intel MKL	
	TIME	GFLOPS	TIME	GFLOPS	TIME	GFLOPS
1K	0.06	12.81	0.14	05.20	0.03	27.59
2K	0.10	56.76	0.17	34.12	0.10	57.33
3K	0.18	106.86	0.32	60.44	0.24	81.61
4K	0.31	148.83	0.42	109.92	0.54	84.89
5K	0.42	210.14	0.44	201.13	1.03	86.50
6K	0.61	253.57	0.59	260.84	1.23	125.75
7K	0.84	293.61	0.77	320.31	2.33	105.85
8K	1.20	304.24	1.13	323.28	2.53	144.90
9K	1.54	338.86	1.41	369.83	3.35	155.77
10K	1.83	391.16	1.71	418.61	4.13	173.32

the MAGMA implementation while, for larger problems, the performance gap between both solvers shrinks. Finally, the routine from MAGMA delivers the highest performance for the solution of the three largest systems. The different behavior of MAGMA and GHA is rooted in that the MAGMA solver is based on the LU factorization and, therefore, operates with triangular matrices. This is specially harmful when the matrix dimension is small to medium, but as the matrix dimension grows, the negative impact of the triangular kernels declines.

The previous experiment showed the high impact that three parameters exerts on the performance of GHA, namely, the number of threads, b_c, and b_g. This motivated a second study to determine optimal or suboptimal values for those parameters. Some initial experiments demonstrated that small values of b_c are required. In particular $b_c = 16$ provided the best execution times, as larger values of this parameter caused a significant degradation of performance reported for the BLAS-1 operations involved in the block diagonalization. This is due to the fact that GHA employs partial column pivoting, asking for row-wise accesses to a matrix that is stored column-wise (following the Fortran/BLAS specification). Figure 3 shows the execution times obtained by GHA in the solution of a system with $n = 4K$, varying the value of b_g and the number of threads. The best performance is obtained with 6 threads and $b_g = 64$. The value of the outer algorithmic block size, $b_g = 64$, which is 4× larger than b_c, exposes enough concurrency during the computation of the block diagonalization. As a result, the best performance is obtained using 6 cores. However, the parallelism is constrained by the low values of both block sizes and, therefore, larger numbers of threads report larger execution times. Besides, lower values of b_g harm the performance obtained by the GPU, while values larger than 64 increase the time dedicated to the block diagonalization and thus, the time the GPU is idle. In summary, the low value of b_c is motivated by the poor performance of the

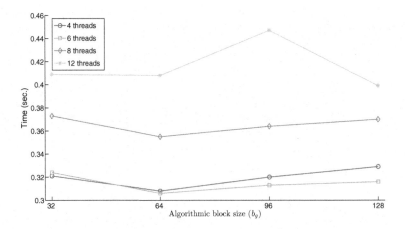

Fig. 3. Execution time for the solution of a system of dimension $n = 4K$ via the GHA solver with different algorithmic block sizes (c) and number of threads.

BLAS-1 during the block diagonalization. In addition, b_g must be initialized to a value that is large enough to attain high performance in the GPU, and allow certain parallelism during the execution of the block diagonalization, but small enough to minimize the time the GPU is idle.

5 Concluding Remarks and Future Work

The Gauss-Huard algorithm (GHA), complemented by column pivoting, is an alternative to solve linear systems, ignored for the last two decades, with the same arithmetic cost and numerical stability as traditional LU-based solvers with row pivoting.

In this work we have revisited GHA to introduce a reliable and high performance implementation of this method for hybrid CPU-GPU platforms. The experimental results demonstrate that GHA is a competitive approach when targeting servers equipped with many-core graphics accelerators. Indeed, our implementation of the solver attains similar or even higher performance compared with that obtained with the tuned LU-based solver in the MAGMA library on the solution of problems of small to moderate dimension. Furthermore, the implementation of GHA clearly outperforms the corresponding kernel in the Intel MKL library for moderate to large problems.

As part of future work, we intend to improve our implementation of GHA by applying more sophisticated optimization techniques. We also plan to target other hardware accelerators such as the Intel Xeon Phi processor.

Acknowledgments. The researcher from the *Universidad Jaime I* was supported by the CICYT projects TIN2011-23283 and TIN2014-53495-R of the *Ministerio de Economía y Competitividad* and FEDER.

References

1. Balay, S., Gropp, W., McInnes, L.C., Smith, B.: PETSc 2.0 users manual. Technical report ANL-95/11, Argonne National Laboratory, October 1996
2. Barrachina, S., Castillo, M., Igual, F.D., Mayo, R., Quintana-Ortí, E.S.: Solving dense linear systems on graphics processors. In: Luque, E., Margalef, T., Benítez, D. (eds.) Euro-Par 2008. LNCS, vol. 5168, pp. 739–748. Springer, Heidelberg (2008)
3. Benner, P., Ezzatti, P., Quintana-Ortí, E.S., Remón, A.: Using hybrid CPU-GPU platforms to accelerate the computation of the matrix sign function. In: Lin, H.-X., Alexander, M., Forsell, M., Knüpfer, A., Prodan, R., Sousa, L., Streit, A. (eds.) Euro-Par 2009. LNCS, vol. 6043, pp. 132–139. Springer, Heidelberg (2010)
4. Benner, P., Ezzatti, P., Quintana-Ortí, E.S., Remón, A.: Matrix inversion on CPU-GPU platforms with applications in control theory. Concur. Comput. Pract. Exp. **25**(8), 1170–1182 (2013)
5. Dekker, T.J., Hoffmann, W., Potma, K.: Stability of the Gauss-Huard algorithm with partial pivoting. Computing **58**, 225–244 (1997)
6. Demmel, J.W.: Applied Numerical Linear Algebra. Society for Industrial and Applied Mathematics, Philadelphia (1997)

7. Dufrechou, E., Ezzatti, P., Quintana-Ortí, E., Remón, A.: Solving linear systems on the Intel Xeon-Phi accelerator via the Gauss-Huard algorithm. In: Osthoff, C., Navaux, P.O.A., Hernandez, C.J.B., Silva Dias, P.L. (eds.) High Performance Computing. CCIS, vol. 565, pp. 107–117. Springer International Publishing, Switzerland (2015)
8. EM Photonics. http://www.culatools.com/
9. Golub, G.H., Van Loan, C.F.: Matrix Computations, 3rd edn. The Johns Hopkins University Press, Baltimore (1996)
10. Gunnels, J.A., Gustavson, F.G., Henry, G.M., van de Geijn, R.A.: FLAME: formal linear algebra methods environment. ACM Trans. Math. Soft. **27**(4), 422–455 (2001)
11. Hoffmann, W., Potma, K., Pronk, G.: Solving dense linear systems by Gauss-Huard's method on a distributed memory system. Future Gener. Comput. Syst. **10**(2–3), 321–325 (1994)
12. Huard, P.: La méthode simplex sans inverse explicite. EDB Bull. Direction Études Rech. Sér. C Math. Inform. **2**, 79–98 (1979)
13. Quintana-Ortí, E.S., Quintana-Ortí, G., Sun, X., van de Geijn, R.A.: A note on parallel matrix inversion. SIAM J. Sci. Comput. **22**, 1762–1771 (2001)
14. Silva, J.P., Dufrechou, E., Quintana-Ortí, E., Remón, A., Benner, P.: Solving dense linear systems with hybrid CPU-GPU platforms. In: Proceedings of the XLI Latin American Computing Conference (2015) (to appear)
15. TOP500.org. http://www.top500.org/
16. TU Wien and FASTMathSciDAC Institute. http://viennacl.sourceforge.net/
17. Univ. of Tennessee. http://icl.cs.utk.edu/magma/
18. Volkov, V., Demmel, J.: LU, QR and Cholesky factorizations using vector capabilities of GPUs. Technical report UCB/EECS-2008-49, EECS Department, University of California, Berkeley, May 2008

Increasing Arithmetic Intensity in Multigrid Methods on GPUs Using Block Smoothers

Matthias Bolten[1(⊠)] and Oliver Letterer[2]

[1] Institut für Mathematik, Universität Kassel, Heinrich-Plett-Straße 40,
34132 Kassel, Germany
bolten@mathematik.uni-kassel.de
[2] Fakultät für Mathematik und Naturwissenschaften,
Bergische Universität Wuppertal, 42097 Wuppertal, Germany

Abstract. Block smoothers that relax small blocks of unknowns provide a way to increase the amount of arithmetic operations needed per smoothing iteration. If the block sizes are small, the variables associated to these blocks fit in fast local memories, thus allowing for a better exploitation of modern computer architectures. At the same time block smoothers are efficient smoothers that allow for more aggressive coarsening resulting in less coarse grids.

We implemented block smoothers in combination with aggressive coarsening in OpenCL, targeting GPUs. Two different data layouts were compared for the smoother. The results show that while the more advanced data layout does not yield a better performance, the introduction of block smoothers in multigrid methods can indeed reduce the time to solution on a GPU.

Keywords: Multigrid methods · High arithmetic intensity · Block smoothers · OpenCL · GPU

1 Introduction

In many simulations in computational science and engineering linear systems

$$Ax = b, A \in \mathbb{R}^{n \times n}, \ x, b \in \mathbb{R}^n,$$

have to be solved, often these linear systems arise from the discretization of partial differential equations (PDEs). Frequently the overall run time of the simulation is governed by the time spent in the solver, therefore there is a huge demand for optimal scalable solvers. If the number of unknowns is large the direct solution of the system is usually not feasible, so iterative methods are used. In many cases, especially when the underlying PDE is elliptic, multigrid methods are known to be optimal iterative solvers, i.e., the convergence rate does not depend on the system dimension. Further, multigrid methods can be parallelized efficiently and scale to huge numbers of processors. The computational complexity that determines the cost of one multigrid cycle is governed by

© Springer International Publishing Switzerland 2016
R. Wyrzykowski et al. (Eds.): PPAM 2015, Part I, LNCS 9573, pp. 515–525, 2016.
DOI: 10.1007/978-3-319-32149-3_48

the number of unknowns per row in the system matrix. In the case of the usually used low-order discretization of a PDE independently of wether a finite difference, finite volume or finite element scheme is used, this number is usually small. As a consequence the number of arithmetic operations that has to be carried out per unknown and thus per memory transfer is low. This limits the achievable performance of multigrid methods in terms of FLOPS compared to the theoretically achievable performance on modern computer architectures. While this limit exists for a long time, in recent years it has become more prominent as the performance of processors increases much faster than the performance of memory does.

On parallel computers the scalability of multigrid methods is limited only through the necessary global information exchange that is inherently necessary to solve the problems at hand due to their global nature. This requirement introduces a logarithmic dependency of the runtime on the number of processors, the degrade in scalability mostly is due to the relatively low amount of work that has to be conducted on the lower levels compared to the communication. One way to hide this effect for larger processor numbers is the use of aggressive coarsening that results in a lower number of levels in the multigrid hierarchy. For multigrid methods employing aggressive coarsening to be as effective as standard multigrid methods more powerful smoothers are needed. This can either be accomplished by carrying out more smoothing steps or by using completely different smoothers, e.g., polynomial smoothers.

In order to overcome both limitations we propose to use block smoothers that result in a higher arithmetic cost than usually employed point smoothers but at the same time reduce high frequency components of the error more effectively. In contrast to line and plane smoothers that are often used in multigrid methods, e.g., when anisotropies are present in the underlying PDE, by block smoothers we generally entitle methods that do not relax the residual at one point at a time but at a set of points concurrently. In the cases considered here these sets will be local subdomains. If the variables are stored in memory appropriately this results in a high locality of the data that is used during a relaxation and thus less memory transfers are needed. Even if the arithmetic cost per unknown is higher than in point smoothers, the combination of a better smoothing factor together with the better use of modern architectures results in an overall reduced time to solution.

In this paper we present an implementation of block smoothers on GPUs using OpenCL to showcase that our proposal is feasible. In fact combining aggressive coarsening with block smoothers on a GPU results in a multigrid method that has a lower convergence rate but a reduced time to solution although more iterations of the multigrid method are necessary.

The rest of the paper is structured as follows: In the next section we will provide a brief introduction into multigrid methods. In Sect. 3 the block smoothers used here are presented and afterwards in the following section the implementation on the GPU is described. Numerical results are shown in Sect. 5 and the paper closes with a conclusion and outlook.

Algorithm 1. Multigrid cycle $x_{n_i} = \mathcal{MG}_i(x_{n_i}, b_{n_i})$

$x_{n_i} \leftarrow \mathcal{S}_i^{\nu_1}(x_{n_i}, b_{n_i})$

$r_{n_i} \leftarrow b_{n_i} - A_i x_{n_i}$

$r_{n_{i+1}} \leftarrow R_i r_{n_i}$

$e_{n_{i+1}} \leftarrow 0$

if $i + 1 = l_{\max}$ **then**

    $e_{n_{l_{\max}}} \leftarrow A_{l_{\max}}^{-1} r_{n_{l_{\max}}}$

else

    for $j = 1, \ldots, \gamma$ **do**

        $e_{n_{i+1}} \leftarrow \mathcal{MG}_{i+1}(e_{n_{i+1}}, r_{n_{i+1}})$

    end for

end if

$e_{n_i} \leftarrow P_i e_{n_{i+1}}$

$x_{n_i} \leftarrow x_{n_i} + e_{n_i}$

$x_{n_i} \leftarrow \tilde{\mathcal{S}}_i^{\nu_2}(x_{n_i}, b_{n_i})$

---

2 Multigrid Methods

Multigrid methods go back to [1,6,7], their use in applications has been promoted in [3,11,12]. In the following we describe geometric multigrid methods that are based on the following observation: When an iterative method like Gauß-Seidel or damped Jacobi is applied to a linear system that arises from the discretization of a simple elliptic PDE like the Poisson equation and when plotting the error on the discretization grid before and after a few steps of the iterative method it is much smoother after the application. As a consequence it will be well-represented on a coarser grid where the problem is less expensive to solve. This idea is applied recursively resulting in the so–called V-cycle, if multigrid is called multiple times recursively, the result are other cycling schemes. In Algorithm 1 the basic multigrid algorithm is provided. In addition to the different grids that are needed to represent the approximation to the solution on the various levels, smoothers and grid transfer operators have to be defined. As smoothers usually point smoothers like the aforementioned Gauß-Seidel method or damped Jacobi are used. Other options include polynomial smoothers, incomplete factorizations or block smoothers like the ones defined in Sect. 3. Based on the observation of the behavior of the error as grid transfer operators methods like linear or higher order interpolation are used to transfer the error from the coarse to the fine level, in the opposite direction the simplest option is injection of the fine level solution, i.e., the current value at a grid point is just copied to the coarse level. Another option that is often used is full-weighting that is the transpose of the linear interpolation, possibly multiplied with a scalar factor. Extensions to multigrid methods that do not require an a priori defined grid hierarchy are known as algebraic multigrid (AMG). An introduction to geometric multigrid methods can be found in [4], more details and an introduction to AMG are found in [16].

The efficient implementation of parallel multigrid has been discussed in different papers, an overview over parallelization of multigrid is provided in [5], here also the aggressive coarsening that is used here is presented. Multigrid methods for GPUs have been presented before, e.g., in [8–10].

A geometric multigrid method starts from a given partial differential equation

$$\mathcal{L}u = f, \qquad \text{in } \Omega$$
$$u = g, \qquad \text{in } \partial\Omega.$$

Other boundary conditions than Dirichlet boundary conditions are possible. The equation is discretized using different discretization parameters h resulting in linear systems of the form

$$L_h u_h = f_h,$$

where the boundary conditions are eliminated to be contained in L_h or they are handled explicitly. Grid transfer operators are defined to transfer quantities between the different levels of discretization in a geometrically motivated manner and simple iterative schemes like Gauß-Seidel are added as described before in Algorithm 1 to obtain a multigrid method.

Here, we limit ourselves to cuboidal domains discretized using regular grids. In the simplest case of the unit cube discretized with n grid points in each direction we end up with a linear system with n^3 unkowns. The coarser levels are obtained by subsequently taking every gth grid point, only. Usually, a coarsening ratio of $g = 2$ is chosen and the grid sizes are chosen such that we end up with 1 unkown, only. In the case of Dirichlet boundary conditions this results in $n = 2^k - 1$ grid points while for periodic boundary conditions we obtain $n = 2^k$ grid points. As interpolation linear interpolation is used, i.e., the value at a fine grid point is taken as the weighted average between neighboring points. The interpolation is taken as full-weighting with the same weights, i.e., the transpose of the restriction operator.

3   Block Smoothers

Point smoothers have a relatively small number of arithmetic operations per memory transfer, resulting in a poor use of modern processor architectures. At the same time, when aggressive coarsening is employed the smoothing factor drops substantially.

In [2] the usage of block smoothers in multigrid methods has been proposed and preliminary results of analyzing block smoothers using local Fourier analysis were given. These block smoothers are using a domain decomposition approach, i.e., the unknowns are partitioned into smaller sets forming a connected subdomain Ω_i of the whole domain under consideration. An introduction to domain decomposition can be found in [14]. The union of the subdomains is the whole domain, i.e.,

$$\bigcup_i \Omega_i = \Omega,$$

the subdomains do not have to be disjoint. One step of the block smoother consists of a loop over the subdomains. Within the loop the residual is calculated, the linear system is being restricted to the variables corresponding to the current subdomain, the restricted system is solved for the restricted residual as right hand side and finally the current guess is updated by adding the result of this small system. This results in a relaxation of a whole subdomain instead of an individual variable. This method is known as block Gauß-Seidel or multiplicative Schwarz. If the residual is not updated within the loop over the domains but rather just once before, the method is a block Jacobi-method or additive Schwarz method. It is known from the underlying theory that multiplicative methods work better than additive, just like in the scalar case [14]. The subdomains can be chosen on each level individually. As this is used as a smoothing procedure it is further not necessary to solve each of the restricted systems exactly, but rather using an iterative method. When small block sizes are chosen even a plain Gauß-Seidel method is well-suited for this task, as its convergence factor depends on the ratio of the grid spacing and the domain size that will be quite large in this case. As in the point relaxation case a lexicographic ordering of the blocks results in a method that is inherently sequential. This is not the case for block Jacobi-type methods, but the smoothing factor of block Jacobi is worse than that of block Gauß–Seidel. By using a multicoloring of the blocks also the Gauß–Seidel variant of the smoother is parallelizable.

As the solution of the restricted linear systems is much more expensive than the relaxation of an individual variable, the overall method is more expensive. On the other hand, the resulting methods are much more efficient as smoothers than point-relaxation methods. If the subdomains are relatively small, the overhead introduced by the method is relatively small, as well. Further, if a memory layout is chosen that keeps all needed variables in the cache, the solution of the linear system and thus the relaxation of one block can be calculated very fast as modern processors can be used more efficiently. This is true for direct solvers that are used to solve the restricted linear system as well as for iterative solvers, as both will benefit from the advantageous memory layout.

As we are dealing with cuboidal domains, we consider cuboidal subdomains, as well. To allow for parallel processing in the smoother a multi-coloring scheme of the blocks is employed.

4   Implementation

A multigrid method for cuboidal domains with equispaced regular grids was implemented in OpenCL to measure the performance gain of the proposed method on GPUs. Parameters like the work-group size were set to be chosen automatically by OpenCL. The multigrid method uses aggressive coarsening to reduce the number of levels that is present, in the following the coarsening factor will be denoted by g. The block smoother uses small cubic blocks with side length equal to the coarsening ratio.

The multigrid method itself uses a simple data layout with lexicographic ordering of the unknowns, the numbering of the grid points includes the boundary values, c.f., Fig. 1, which depicts the two-dimensional analogue of the distribution scheme used. This numbering is neglecting the blocking and it is used

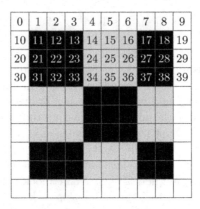

Fig. 1. Numbering of the grid points for the multigrid method

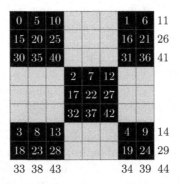

Fig. 2. Numbering of the grid points for the block smoother

for all operations, i.e., calculating the residual, restriction of the residual, and prolongation of the error and correcting the current approximation afterwards.

In order for the block smoother to benefit from the spatial locality of the unknowns of a block the data can be rearranged in memory such that data belonging to one block is stored consecutively in memory. Further, for data access to be as fast as possible on a GPU coalesced access has to be used. Therefore the unknowns of 16 blocks of the same color are interleaved to provide coalesced data access for one half warp. A 2D example for a half warp size of 5 is depicted in Fig. 2. Similar approaches have been used, e.g., in [13].

As the smoother does not need boundary values, the boundaries are not included in this data layout. To accommodate for missing values in subdomains at the boundary padding is used.

The usage of two data layouts results in memory copies before and after a relaxation. Overall we obtain Algorithm 2 for a V-cycle.

We compared this algorithm to an alternative using the previously defined simplified layout, only. To obtain the highest possible performance instead of a red and a black block-sweep two full block Jacobi-sweeps are performed, where the blocks are inverted as in the previous algorithm by a few iterations of Gauß–Seidel. The resulting V-cycle is given by Algorithm 3.

Algorithm 2. V-cycle with block smoother and two data layouts

---

for $\ell = 1, \ldots, \ell_{\max} - 1$ **do**
  for $k = 1, \ldots, \nu_1$ **do**
    $r_\ell = f_\ell - L_\ell u_\ell$
    Copy residual from standard layout to layout for block smoother
    Solve $L_\ell e_\ell = r_\ell$ on red blocks using σ iterations of Gauß-Seidel
    Copy residual from layout for block smoother to standard layout
    $u_\ell = u_\ell + e_\ell$
    $r_\ell = f_\ell - L_\ell u_\ell$
    Copy residual from standard layout to layout for block smoother
    Solve $L_\ell e_\ell = r_\ell$ on black blocks using σ iterations of Gauß-Seidel
    Copy residual from layout for block smoother to standard layout
    $u_\ell = u_\ell + e_\ell$
  end for
  $f_{\ell+1} = R_\ell(f_\ell - L_\ell u_\ell)$
end for
$u_{\ell_{\max}} = L_{\ell_{\max}}^{-1} f_{\ell_{\max}}$
for $\ell = \ell_{\max} - 1, \ldots, 1$ **do**
  $e_\ell = P_\ell u_{\ell+1}$
  $u_\ell = u_\ell + e_\ell$
  for $k = 1, \ldots, \nu_2$ **do**
    Copy residual from standard layout to layout for block smoother
    Solve $L_\ell e_\ell = r_\ell$ on red blocks using σ iterations of Gauß-Seidel
    Copy residual from layout for block smoother to standard layout
    $u_\ell = u_\ell + e_\ell$
    $r_\ell = f_\ell - L_\ell u_\ell$
    Copy residual from standard layout to layout for block smoother
    Solve $L_\ell e_\ell = r_\ell$ on black blocks using σ iterations of Gauß-Seidel
    Copy residual from layout for block smoother to standard layout
    $u_\ell = u_\ell + e_\ell$
  end for
end for

To stop the iteration the 2-norm of the residual is checked after each V-cycle. The calculation of this 2-norm is carried out by squaring all entries and then using a fan-in scheme using two arrays that are used alternately to sum up two variables in one array and storing them in the other until the sum of all squared entries is located in the first entry of one array. This allows to use the GPU for this task, as well.

5 Numerical Results

As test problem we consider the PDE

$$\Delta u(x) = f(x), \qquad \text{for} x \in \Omega = (0,1)^3,$$
$$u(x) = 0, \qquad \text{for} x \in \partial\Omega.$$

Algorithm 3. V-cycle with block smoother one data layout

---
for $\ell = 1, \ldots, \ell_{\max} - 1$ **do**
  for $k = 1, \ldots, 2\nu_1$ **do**
    $r_\ell = f_\ell - L_\ell u_\ell$
    Solve $L_\ell e_\ell = r_\ell$ on all blocks using σ iterations of Gauß-Seidel
    $u_\ell = u_\ell + e_\ell$
  end for
  $f_{\ell+1} = R_\ell(f_\ell - L_\ell u_\ell)$
end for
$u_{\ell_{\max}} = L_{\ell_{\max}}^{-1} f_{\ell_{\max}}$
for $\ell = \ell_{\max} - 1, \ldots, 1$ **do**
  $e_\ell = P_\ell u_{\ell+1}$
  $u_\ell = u_\ell + e_\ell$
  for $k = 1, \ldots, 2\nu_2$ **do**
    Solve $L_\ell e_\ell = r_\ell$ on all blocks using σ iterations of Gauß-Seidel
    $u_\ell = u_\ell + e_\ell$
  end for
end for

The right hand side f was chosen as $3\pi^2 \sin(\pi x_1) \sin(\pi x_2) \sin(\pi x_3)$ such that the analytical solution of the problem is given by

$$u(x) = \sin(\pi x_1) \sin(\pi x_2) \sin(\pi x_3).$$

The problem was discretized using 7-point finite differences.

The implementation was tested in single precision on a NVIDIA Tesla M2050 GPU in the JuDGE cluster at the Jülich Supercomputing Centre. Single precision is sufficient for smaller problems, only, but it can be used as an efficient preconditioner even in the double precision case [15]. The M2050 GPU provides a theoretical peak performance of 1.03 TFLOPS in single precision. In any case, the block size and the coarsening ration were chosen to be the same, resulting in less coarse grids when larger block sizes were chosen.

First, we compare the time for one V-cycle and the performance achieved for Algorithms 2 and 3. The results can be found in Table 1. As expected the necessary copying of the data corrupts the performance a lot, even though Algorithm 3 does twice the amount of operations, the time needed for a V-cycle is smaller.

As expected from a theoretical point of view both algorithms do behave similarly regarding the convergence rate. A plot of the convergence history of both methods can be found in Fig. 3.

As the performance of Algorithm 3 using one data layout, only, was superior, we measured the time to solution with this algorithm, only. We measured the time needed to reduce the error to the discretization error and calculated the obtained performance. In each case in the block Jacobi method the systems belonging to one block were solved approximately with 10 iterations of Gauß–Seidel. The result can be found in Table 2.

Obviously, a block smoother results in a much better performance in terms of GFLOPS when a large block size is chosen. On the other hand, as the coarsening

Table 1. Performance of Algorithms 2 and 3 for $2^5 + 1$, $2^6 + 1$, and $2^7 + 1$ grid points in each direction.

n	$2^5 + 1 = 33$		$2^6 + 1 = 65$		$2^7 + 1 = 129$	
	Time	Performance	Time	Performance	Time	Performance
1 data layout	0.169 s	4.41 GFLOPS	0.302 s	19.76 GFLOPS	1.211 s	39.43 GFLOPS
2 data layouts	0.218 s	1.90 GFLOPS	0.405 s	8.09 GFLOPS	1.263 s	20.72 GFLOPS

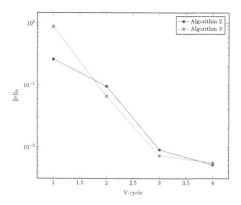

Fig. 3. Convergence history using $6^3 + 1$ grid points in each direction.

Table 2. Time to solution and achieved performance of Algorithm 3 for different grid sizes and coarsening ratios.

n	Block size	Time/iter	#iterations	Performance
$2^5 - 1$	2×2	0.156 s	3	4.76 GFLOPS
$2^6 - 1$	2×2	0.322 s	3	18.50 GFLOPS
$2^7 - 1$	2×2	1.129 s	3	42.25 GFLOPS
$3^5 - 1$	3×3	4.222 s	3	70.28 GFLOPS
$4^3 - 1$	4×4	0.211 s	4	25.11 GFLOPS
$4^4 - 1$	4×4	4.765 s	4	71.22 GFLOPS
$6^3 - 1$	6×6	2.673 s	5	75.43 GFLOPS

ratio is increased as well, the performance in terms of necessary iterations is worsening. Overall, the time to solution is reduced nevertheless, c.f., the time needed for the solution of a system with 63^3 unknowns: When a block size of 2 is used, 3 iterations are needed and each iteration takes 0.322 s, but when a block size of 4 is used, we need one iteration more but each iteration now only takes 0.211 s. The time to solution is 0.966 s in the first case and 0.844 s in the second, so the second approach only takes 87 % of the time of the first.

6   Conclusion and Outlook

Block smoothers provide a way to increase the amount of local arithmetic operations in a way beneficial for multigrid methods. The smoothers allow for a more aggressive coarsening resulting in less coarse grids while at the same time exploiting modern computer architectures. The inclusion of a special data layout for the smoothers, only, does not result in a higher performance, but even a simplistic straightforward approach yields a reduction in time to solution.

We are currently working on analyzing block smoothers theoretically and in incorporating the ideas exploited here in a parallel multigrid method targeting massively parallel computers.

Acknowledgments. We acknowledge the Jülich Supercomputing Centre for providing access to the JuDGE cluster.

References

1. Bakhvalov, N.S.: On the convergence of a relaxation method with natural constraints on the elliptic operator. USSR Comp. Math. Math. Phys. **6**, 101–135 (1966)
2. Bolten, M., Kahl, K.: Using block smoothers in multigrid methods. PAMM **12**, 645–646 (2012)
3. Brandt, A.: Multi-level adaptive solutions to boundary-value problems. Math. Comp. **31**(138), 333–390 (1977)
4. Briggs, W.L., Henson, V.E., McCormick, S.F.: A Multigrid Tutorial. SIAM, Philadelphia (2000)
5. Chow, E., Falgout, R.D., Hu, J.J., Tuminaro, R.S., Yang, U.M.: A survey of parallelization techniques for multigrid solvers. In: Heroux, M.A., Raghavan, P., Simon, H.D. (eds.) Parallel Processing for Scientific Computing, chap. 10, SIAM Series on Software, Environments, and Tools, SIAM, Philadelphia (2006)
6. Fedorenko, R.P.: A relaxation method for solving elliptic difference equations. USSR Comp. Math. Math. Phys. **1**(5), 1092–1096 (1962)
7. Fedorenko, R.P.: The speed of convergence of one iterative process. USSR Comp. Math. Math. Phys. **4**(3), 227–235 (1964)
8. Göddeke, D., Strzodka, R.: Mixed precision GPU-multigrid solvers with strong smoothers. In: Kurzak, J., Bader, D.A., Dongarra, J.J. (eds.) Scientific Computing with Multicore and Accelerators, chap. 7, CRC Press, December 2010
9. Goddeke, D., Strzodka, R., Mohd-Yusof, J., McCormick, P., Wobker, H., Becker, C., Turek, S.: Using gpus to improve multigrid solver performance on a cluster. Int. J. Comput. Sci. Eng. **4**(1), 36–55 (2008)
10. Haase, G., Liebmann, M., Douglas, C.C., Plank, G.: A parallel algebraic multigrid solver on graphics processing units. In: Zhang, W., Chen, Z., Douglas, C.C., Tong, W. (eds.) HPCA 2009. LNCS, vol. 5938, pp. 38–47. Springer, Heidelberg (2010)
11. Hackbusch, W.: On the convergence of a multi-grid iteration applied to finite element equations. Report 77–8, Institute for Applied Mathematics, University of Cologne, West Germany, Cologne (1977)
12. Hackbusch, W.: On the multi-grid method applied to difference equations. Computing **20**, 291–306 (1978)

13. Müthing, S., Ribbrock, D., Göddeke, D.: Integrating multi-threading and accelerators into DUNE-ISTL. In: Abdulle, A., Deparis, S., Kressner, D., Nobile, F., Picasso, M. (eds.) Numerical Mathematics and Advanced Applications - ENUMATH 2013. Lecture Notes in Computational Science and Engineering, vol. 103, pp. 601–609. Springer International Publishing, Switzerland (2015)
14. Smith, B., Bjorstad, P., Gropp, W.: Domain Decomposition: Parallel Multilevel Methods for Elliptic Partial Differential Equations. Cambridge University Press, Cambridge (2004)
15. Tadano, H., Sakurai, T.: On single precision preconditioners for Krylov subspace iterative methods. In: Lirkov, I., Margenov, S., Waśniewski, J. (eds.) LSSC 2007. LNCS, vol. 4818, pp. 721–728. Springer, Heidelberg (2008)
16. Trottenberg, U., Oosterlee, C., Schüller, A.: Multigrid. Academic Press, San Diego (2001)

Optimized CUDA-Based PDE Solver
for Reaction Diffusion Systems
on Arbitrary Surfaces

Samira Michèle Descombes, Daljit Singh Dhillon$^{(\boxtimes)}$, and Matthias Zwicker

Institute of Computer Science, University of Bern, Bern, Switzerland
samira.descombes@students.unibe.ch, djdhillon@gmail.com,
zwicker@inf.unibe.ch
http://www.cgg.unibe.ch/

Abstract. *Partial differential equation* (PDE) solvers are commonly employed to study and characterize the parameter space for *reaction-diffusion* (RD) systems while investigating biological pattern formation. Increasingly, biologists wish to perform such studies with arbitrary surfaces representing 'real' 3D geometries for better insights. In this paper, we present a highly optimized CUDA-based solver for RD equations on triangulated meshes in 3D. We demonstrate our solver using a *chemotactic* model that can be used to study snakeskin pigmentation, for example. We employ a *finite element* based approach to perform *explicit Euler time integrations*. We compare our approach to a naive GPU implementation and provide an in-depth performance analysis, demonstrating the significant speedup afforded by our optimizations. The optimization strategies that we exploit could be generalized to other mesh based processing applications with PDE simulations.

Keywords: CUDA · GPU programming · Reaction-diffusion systems · Nonlinear PDEs · FEM · Explicit time-stepping

1 Introduction

Partial differential equations defined on triangulated meshes in 3D play an important role in many applications. In 3D geometry processing, for example, 3D shapes are smoothed and denoised via curvature flow, or heat diffusion on surfaces can be used to define geometric features. Physical effects such as deformation or cracking can also be modeled using PDEs on surfaces. In this paper we are considering reaction-diffusion equations, which are widely believed to play a crucial role in biological pattern formation. For these applications, efficient PDE solvers on triangulated surfaces are crucial components. To address this need, the goal of this paper is to develop an optimized GPU implementation for one type of surface PDEs, in particular reaction-diffusion equations, and study the improvements that can be achieved over a naive approach by careful profiling.

© Springer International Publishing Switzerland 2016
R. Wyrzykowski et al. (Eds.): PPAM 2015, Part I, LNCS 9573, pp. 526–536, 2016.
DOI: 10.1007/978-3-319-32149-3_49

Reaction-diffusion (RD) models have first been hypothesized by Turing as a mechanism that is involved in biological pattern formation. For biologists it is important to study these equations on 3D geometries to obtain better insights about their behavior in realistic scenarios. In this paper we employ a finite element based approach and perform explicit Euler time integration to solve these equations on triangulated meshes in 3D. The discretization and time integration lead to simple discrete operators that we need to evaluate locally on small neighborhoods of mesh vertices. This offers the possibility of a straightforward parallel implementation on GPUs, where the discrete operators are computed simultaneously on many vertices. We develop a more sophisticated, optimized implementation, however, that demonstrates how techniques such as kernel fusion, partitioning the input data, and effective use of shared memory and device arrays provides significant further performance gains. A contribution of this paper is to show how these techniques could also be leveraged to implement other mesh based PDE solvers efficiently on GPUs. We discuss how the different degrees of freedom that can be exploited to optimize a mesh based PDE solver could be leveraged more generally, for example by exposing them in a domain-specific language that targets this problem. Our discussion is inspired by the recent development and popularity of Halide, a domain-specific language for image processing.

2   Related Work

Partial differential equations based on reaction-diffusion, or Turing [14], models have been widely postulated to be relevant in biological pattern formation, and experimental evidence supporting this hypothesis has been found in various instances. Pioneering work by Kondo and Asai [8] showed for the first time how a simulation of a reaction-diffusion model correctly predicts skin patterns on the marine angelfish, *Pomacanthus*. For an excellent recent survey we refer to the review by Kondo and Miura [9].

There is an abundance of libraries and frameworks for mesh based PDE solvers that target a variety of systems, ranging from GPUs to large-scale clusters. Libraries like PETSc [3] or Sandia's Sierra [2] build on the SPMD model to enable programmers to write code that largely resembles single-threaded programs, while execution may be distributed over a variety of processors. Similarly, Liszt's [5] concept of "for-comprehension" allows the system to choose a parallel implementation, while hiding this from the program code. Other frameworks [4,6] rely on the concept of kernels to express computations that need to be executed on a set of data elements. While some of these systems include back-ends for GPU code generation, they do not easily allow the user to optimize the performance of GPU code. In contrast, Halide [12] is a domain-specific language for image processing that also abstracts away the parallel execution from the specification of an algorithm. A key idea is that it allows the user to provide a so-called schedule in addition to the algorithm, which is a high level description of a desired parallel execution configuration. By adjusting the schedule a user can obtain highly optimized implementations for various back-ends including multicore CPUs and GPUs. Halide is restricted to operate on the rectangular topology of images, however, and it does not consider general meshes. We believe our optimizations could be exposed in a similar high-level language, but targeted at processing data on meshes.

3   Computational Model

We first present a chemotactic RD model that has been proposed for snakeskin pattern formation [11]. We then explain the use of discrete geometric constructs to simulate this PDE based system on an arbitrary mesh. Finally, we give the details of the simulation steps.

Reaction-Diffusion with Chemotaxis. The chemotactic model proposed by Murray et al. [11] is expressed mathematically in a dimensionless form as,

$$\frac{\partial f_n}{\partial t} = D\nabla^2 f_n - \alpha f_n \nabla^2 f_c - \alpha \nabla f_n \cdot \nabla f_c + srf_n(N - f_n),$$

$$\frac{\partial f_c}{\partial t} = \nabla^2 f_c + s\left(\frac{f_n}{1 + f_n} - f_c\right). \tag{1}$$

Here, f_n is a non-negative cell density function, f_c is a non-negative chemoattractant density function, ∇^2 is the Laplacian operator, ∇ is a gradient operator, D a positive cell diffusion rate, α a positive chemotactic rate, r the cell growth rate, s a positive scale factor, and N the maximum cell density capacity. The PDEs operate on a restricted region within the snakeskin surface, subject to zero Neumann boundary conditions. For further discussion of the cell-chemotaxis model see the work by Murray [11] and Winters et al. [1].

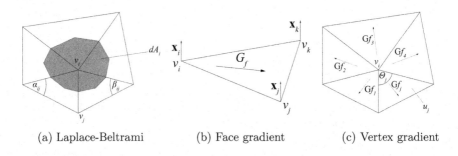

(a) Laplace-Beltrami   (b) Face gradient   (c) Vertex gradient

Fig. 1. Illustrating discrete geometry operations

FEMs and Discrete Geometry. We use discrete geometric operators to simulate Eq. 1 on an arbitrary surface. The Laplacian ∇^2 is replaced by a Laplacian-Beltrami operator that is evaluated for a surface function f around a mesh vertex v_i as,

$$\nabla^2 f(v_i) = \frac{1}{2A_i} \sum_{v_j \in N_1(v_i)} (\cot \alpha_{ij} + \cot \beta_{ij})(f(v_j) - f(v_i)). \tag{2}$$

Here, A_i is the area of the averaging region dA_i around vertex v_i and $(\cot \alpha_{ij} + \cot \beta_{ij})$ is called cotangent weighting. Figure 1a illustrates the calculation. Reuters et al. [13] state that using cotangent weights with the normalizing weights as A_i's is equivalent to a lumped-linear FEM formulation for the Laplace eigenproblem. We thus use these weights as expressed in Eq. 2. Next we compute the function gradients in two steps: (a) computing face-gradients from the function values at vertices, and (b) computing vertex-gradients at the vertices by appropriate weighted averaging over face-gradients. For a face u_i consisting of vertices v_i, v_j and v_k the face-gradient is given by,

$$\nabla f(u_i) = (f(v_j) - f(v_i))\frac{(\mathbf{x}_i - \mathbf{x}_k)^\perp}{2A_T} + (f(v_k) - f(v_i))\frac{(\mathbf{x}_j - \mathbf{x}_i)^\perp}{2A_T}, \quad (3)$$

where $\mathbf{x}$ is the 3D coordinate of vertex v, A_T is the area of the face and $\perp$ is a counterclockwise rotation of a vector by 90°, at its tail, in the triangle plane. The definition is illustrated in Fig. 1b. Next, the gradient for each vertex v_i is computed as,

$$\nabla f(v_i) = \sum_{u_j \in nbr(v_i)} \nabla f(u_j) w_j. \quad (4)$$

Here $nbr(v_i)$ is a set of faces u_j incident on vertex v_i and w_j are the normalized weights for neighborhood-averaging, see Fig. 1c. With an appropriate choice of the weights in Eqs. 3 and 4, these computations approximate first order differentials well [10]. We simply use the incident angles θ_j to generate normalized weights for Eq. 4 since this is appropriate in a discrete geometric sense.

Euler Integration. For our simulation we use explicit time-stepping. For a system state $Y(t)$ defined by (f_n, f_c), we calculate the state at later time $Y(t + \delta t) = Y(t) + \delta t(\partial Y/\partial t)$, where δt is a small time-step. Each simulation begins with: (a) *An initialization* with parameters $D = 0.25$, $r = 1.522$, $\alpha = 12.02$, $s = 1$ and $N = 1$ unless otherwise specified. Also $f_n = N$ and $f_c = N/(1 + N)$ and we add minor random perturbations to both. This is followed by: (b) *A transition loop* where all gradients are computed using discrete operations, as discussed earlier, to evaluate the time-derivatives in Eq. 1 and to update the system state $Y(t)$ through explicit time-stepping. Finally: (c) we terminate the iterations in the transition loop upon reaching convergence or a predetermined number of time-steps. We use time-steps δt of a fixed size that is small enough and $\mathcal{O}(dx^2)$, where dx is the average edge length for the given mesh.

4 GPU Optimizations

In this section we describe in detail our use of various GPU optimization techniques and state their overall impact briefly. Later, in Sect. 5 we provide an in-depth performance analysis for each of the techniques to understand their influence on the speedup.

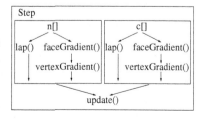

Fig. 2. The *baseline* implementation

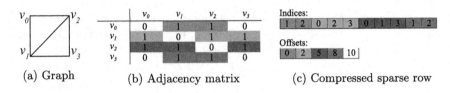

(a) Graph   (b) Adjacency matrix   (c) Compressed sparse row

Fig. 3. Storing neighborhood references

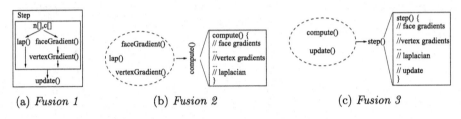

(a) *Fusion 1*   (b) *Fusion 2*   (c) *Fusion 3*

Fig. 4. Implementation outline for *Fusion 1*, *Fusion 2* and *Fusion 3*

Naive Baseline Implementation. We begin with a *baseline* GPU implementation for our iterative transition loop with four CUDA kernels as shown in Fig. 2. Kernel *lap()* computes surface Laplacians for functions f_n and f_c as expressed in Eq. 2. We then compute gradients with kernels *faceGradients()* and *vertexGradients()* implementing Eqs. 3 and 4 respectively. Finally, we compute the time-derivatives and update the system state $Y(t) \equiv (f_n, f_c)$ in kernel *update()* with an explicit time-step. Our *baseline* implementation uses: (a) one thread per vertex, (b) a compressed sparse row (CSR) representation for vertex connectivity (see Fig. 3c), (c) only the memory registers and the global memory. This implementation incurs huge memory overloads due to uncoalesced, repetitive memory accesses for the context data and intermediate results stored in the global memory. Also, it experiences frequent thread stalls due to divergent executions. The naive *baseline* implementation still provides a speedup of about 125× with single-precision operations compared to a single-threaded CPU implementation.

Our Optimized Implementation. Mainly there are two areas that deserve consideration for optimizations, namely: (a) computational flow, and (b) memory accesses. We perform three kernel fusions as depicted in Fig. 4 to optimize the computational flow and use *shared memory* and *device arrays* to optimize memory accesses. With these optimizations, we improve performance by a multiplicative factor of about 4 over the *baseline* version. This gives overall improvements on the order of 500× for the single-precision version and 350× with double-precision operations.

Kernel Fusions. The *baseline* computation flow has two major areas of improvement: (a) shared input data, mainly the function values, are loaded repeatedly for each kernel and (b) intermediate data resides in global memory. We first improve by fusing together the computation kernels for f_n and f_c. This reduces kernel invocations from 7 to 4 (see Fig. 4a). Then, we merge the *computing kernels* together (see Fig. 4b) and finally, we merge the remaining two kernels into one, as shown in Fig. 4c. *Fusion 1* results in 1.5× performance gains but *Fusion 2 & 3* does not give immediate gains. However, *Fusion 2* leads to important data reorganization as explained below, which ultimately reduces data access time considerably with memory optimizations.

Mesh Partitioning. Fusion 2 is problematic since kernel *vertexGradient()* depends on the output of kernel *faceGradient()* and this warrants a synchronization. Within a single kernel however, only block-wide synchronization is possible which is insufficient. Consider the computation flow for the blue block in Fig. 5 where only block-wise synchronization leads to erroneous results in vertex-gradients for shared vertices in adjoining orange block. We thus partition the mesh such that each

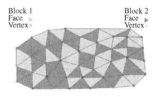

Fig. 5. Vertices and faces arbitrarily assigned to blocks

block is associated to a locally fully-connected *patch*. The process is illustrated by Fig. 6. Partitioning results in a few *halo* faces, shown in Fig. 6c that are added to multiple blocks which add few redundant computations to our computational flow. These overheads are negligible in comparison with later gains due to memory optimizations. We use the METIS library [7] to perform offline multilevel mesh partitioning to: (1) produce partitions of near equal size and (2) minimize the number of halo faces.

For *Fusion 3* we merge kernels *update()* and *compute()* to reduce data transfers for intermediate results. We use double-buffering to avoid data corruption due to concurrent access to function values f_n and f_c while computing the gradients involving *halo* faces. With all three kernel fusions we reduce the computational flow optimally and reorganize data to improve memory accesses as explained next.

Memory Optimizations. There are two main areas of improving memory accesses: (a) The same function values are accessed multiple times during the kernel run and (b) some intermediate results reside in global memory. We first change array indices to point to a fixed size shared memory that stores face-gradients for a given block. Next, we use *shared memory* for vertex-gradients

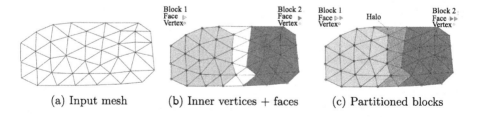

(a) Input mesh   (b) Inner vertices + faces   (c) Partitioned blocks

Fig. 6. Partitioning a mesh into two blocks

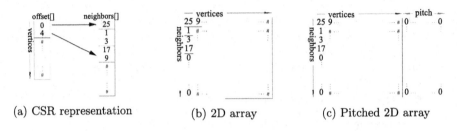

Fig. 7. Reorganizing neighborhood references using *device arrays*

which is slightly complicated. With this new optimization, each thread first calculates the face-gradients and then adds the result to the vertex-gradient stored in the shared memory. Thus, to avoid data corruption due to concurrent accesses during vertex-gradient computations, we make some operations atomic. Using shared memory in conjunction with kernel fusions yields, in general, a multiplicative gain of 2.25× over our *baseline* version.

Next, we use *device arrays* to make memory accesses more efficient. Efficient in this context means coalescing loads and stores. Global memory is divided into segments of 128 bytes. Thus, coalescing global memory accesses needs manual reorganization of the memory. 2D and 3D device arrays support such memory reorganizations. In the following, the process for using device arrays is explained in a simplified manner using the Laplacian computation as an example. Now, the neighbors for each vertex are stored using compressed sparse row representation (CSR), refer Fig. 7a. In each iteration, all threads of a warp access the neighbor with the same neighbor count. This leads to an inefficient access pattern since the array is sorted by vertex and not by neighbor count. Moreover, the first thread of any warp except the first warp accesses a completely random address. For these memory transactions to be fully coalesced two things need to be fulfilled. (1) The first thread of any warps needs to access the first element of a 128 byte segment. (2) All following threads need to access consecutive elements. For this purpose we store the neighbors in a 2D array as shown in Fig. 7b. The vertices are stored into columns where column i stores the neighbors for vertex i. Therefore each row stores the neighbors with the same neighbor count. This representation thus fulfills the second requirement for full coalescing since the 2D array is stored to memory by rows. Next, we add a pitch at the end of each row to ensure that the first entry for the next row begins at an index that is a multiple of 32, see Fig. 7c. This thus satisfies our first criterion for full coalescing. Note however, that 2D device arrays are not sufficient to fully coalesce this access in our case. 3D arrays are needed where the third dimension is the block count. Each slice then represents one block.

Using device arrays in conjunction with previous optimizations yields, in general, a multiplicative gain of 4× in comparison with our *baseline* version. We tried other optimization techniques such as using *vector types*. However, the performance gains for those techniques is not significant and we omit them from our discussion.

Multiprocessors	13
Cores/MP	192
CUDA Cores	2496
Threads per MP	(max) 2048
Blocks per MP	(max) 16
Threads per block	(max) 1024
Global memory	4800 MB
Shared memory per MP	48 KB
Registers per MP	(32 bit) 64K

(a) A rectangular mesh   (b) *Kaa* the virtual lab snake   (c) GPU platform

Fig. 8. Test data and GPU platform details

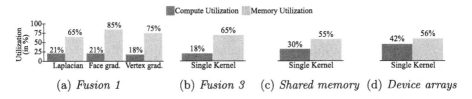

(a) *Fusion 1*   (b) *Fusion 3*   (c) *Shared memory* (d) *Device arrays*

Fig. 9. Compute and memory unit usage for various optimization steps

5   Results and Performance Analysis

In this section we provide an in-depth performance analysis of our implementation with various optimizations. We use an Intel Xeon E5620 processor platform with $4(\times 2)$ cores @2.4 GHz for a naive, multicore reference CPU implementation.[1] Our GPU optimized implementations run on an NVIDIA Tesla K20c platform @704 MHz (core) and @2.6 GHz (memory), see Fig. 8c for hardware details. For evaluations we use a set of 50 highly regular rectangular meshes ranging in resolution from 20K to 1M vertices, see Fig. 8a. We also use a virtual lab snake (*Kaa*) with arbitrary surface geometry as shown Fig. 8b. *Kaa* has 47867 vertices and 94088 triangular faces and a high irregularity in connectivity (neighborhood variance of 3.6 vertices). We also use NVIDIA's graphical profiling tool that offers a collection of different metrics to analyze GPU-accelerated applications.

Performance Evaluation. We first present *utilization metrics* profiled for our different optimization steps using our rectangular mesh with 1M vertices over individual kernel executions. Ideally one would expect 100 % utilization for both *compute units* and *memory bandwidth*, but a benchmark of 60 % simultaneous utilization is often considered very good in practice.

Figure 9a depicts utilization metrics with a barchart for the three main computing kernels for our optimizations in *Fusion 1*. An average compute unit utilization of around 20 % implies that the GPU cores spend most of their time waiting for the data to arrive. Figures 9b, c, and d show that we progressively improve the compute unit utilization with each additional optimization step. For our final optimized implementation (Fig. 9d) we have achieved a considerable improvement

---

[1] Our CPU implementation is not explicitly optimized for the said platform.

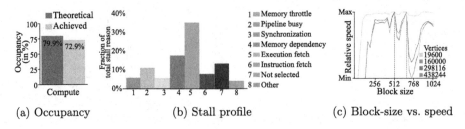

Fig. 10. Various performance profiles: (a) Occupancy metrics for the *Fusion 2* step with the best launch configuration, (b) relative stall profile for the final, most optimized implementation, and (c) impact of the block-size in the launch profile on performance

at 42 % compute unit utilization while compromising marginally on the memory bandwidth utilization at 56 %. Importantly, using *device arrays* increased global load efficiency from 34.6 % to 88.9 % (not shown in the charts) and decreased global load transactions by over 60 % (not shown). Thus, for a data intensive application with relatively limited computational operations, this is a very good trade-off in terms of utilization of the compute and the memory units.

Next, we present an examination of the occupancy metrics. Figure 10a shows the theoretical and achieved occupancies for the best launch configuration for our *Fusion 2* implementation. A considerably low theoretical occupancy (79.9 %) in this case is compounded with even lower achieved occupancy (72.9 %). This simply means that increasing occupancy with a different configuration does not increase performance and indicates that Fusion 2 has latency issues. Some of these issues are resolved with shared memory and especially device array. Our final optimized implementation has theoretical occupancy of 98.4 % and achieved occupancy of 81.5 % for single-precision and 78.2 % for double-precision operations.

We also examine the system latencies for our final optimized implementation with single-precision operations. Figure 10b shows the breakdown of stall reasons averaged over an entire kernel execution. The largest share for warp latencies are due to execution dependencies (34.7 %) and memory dependencies (17.3 %). Memory dependencies are addressed by improving data access patterns. At this point the only inefficient global memory accesses are those to the function value arrays. There are a few common techniques to deal with irregular patterns such as use of: (a) L1 Cache, (b) texture memory, and (c) increased use of shared memory. We tried these options with insignificant or no performance gains. Also, execution dependencies can be mitigated with the use of instruction-level parallelism (IPL). In our case this doesn't work either due to: (a) shared memory overuse, (b) register overuse, or (c) thread overburden. Thus the stall profile show that any further gains in performance may only be achieved with severely diminishing returns on significant programming efforts.

For optimal performance we need to choose a suitable launch configuration, which we found in an empirical manner. Figure 10c shows plots for relative speeds versus block-size for a given mesh resolution. The block-size is the most important factor in determining a launch configuration and for most of the resolutions

Name	Speed up factor			Techniques used
	R1: 19,600	R2: 1,000,000	Kaa	
Baseline	101	128	98	-
Fusion 1	151 (1.5)	189 (1.47)	122 (1.25)	Kernel Fusion
Fusion 2	148 (0.98)	173 (0.92)	125 (1.02)	Kernel Fusion, Partitioning
Fusion 3	160 (1.08)	182 (1.05)	132 (1.06)	Kernel Fusion, Double-Buffer
Shared	234 (1.47)	293 (1.61)	201 (1.52)	Shared Memory, Atomic Operations
Device	366 (1.56)	506 (1.73)	275 (1.37)	Device Array (2D and 3D)
Vector	379 (1.04)	535 (1.06)	317 (1.15)	Vector Type
Double	269 (0.71)	357 (0.67)	224 (0.71)	Double-Precision

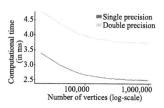

Fig. 11. Speed up factor for each optimization with respect to a naive multicore CPU and with respect to the previous optimization (in brackets). **R1** and **R2** are rectangular meshes of specified resolutions

Fig. 12. Performance comparison of single and double-precision. Time per iteration per 1000 vertices of the mesh

a block-size of 512 or 672 gave best performance. We found that choosing a target block-size of 512 with an assumption of 20 % halo faces yields near optimal launch configuration (i.e., atleast 99 % of the best configuration performance), in general.

Speedups. Finally, we summarize the performance gains for our various optimizations in Fig. 11. We also plot the speed of our final optimization in absolute time in Fig. 12 for the single and double-precision implementations for a rectangular mesh with different resolutions.

6   Conclusions

We have described an optimized GPU implementation for solving non-linear reaction-diffusion (RD) PDEs, which play a crucial role in biological pattern formation, on triangulated meshes in 3D. Optimized tools like our approach will be useful for biologists who study RD models in realistic scenarios with 3D geometries, since they allow them to explore the parameter space of these systems more efficiently. Our works shows that with careful optimizations that take into account parallelism, locality, and recomputation, we can gain a significant speedup over a naive GPU implementation. Therefore, we believe it would be fruitful to allow programmers to explore these degrees of freedom more easily, without resorting to low-level CUDA implementation as we did. Unfortunately, current domain-specific languages for mesh based PDEs do not support this type of optimization. While the Halide language [12] successfully allows the programmer to optimize parallel code execution at an abstract level, and separately from algorithm specification, this approach is tightly connected to the rectangular topology of images (or higher dimensional regular grids). In particular, it exploits the possibility to specify the order in which the dimensions of the grid should be traversed, and whether they should be traversed sequentially or in parallel. It is an interesting challenge for future work to generalize these concepts to arbitrary meshes, such that they can easily be exploited by a programmer for code optimization by providing high-level specifications.

References

1. Winters, K.H., Myerscough, M.R., Maini, P.K., Murray, J.D.: Tracking bifurcating solutions of a model biological pattern generator. IMPACT Comput. Sci. Eng. 2(4), 355–371 (1990)
2. Stewart, J.R., Edwards, H.C.: A framework approach for developing parallel adaptive multiphysics applications. Finite Elem. Anal. Des. 40(12), 1599–1617 (2004). the Fifteenth Annual Robert J. Melosh Competition
3. Balay, S., Gropp, W., McInnes, L., Smith, B.: Efficient management of parallelism in object-oriented numerical software libraries. In: Arge, E., Bruaset, A., Langtangen, H. (eds.) Modern Software Tools for Scientific Computing, pp. 163–202. Birkhäuser, Boston (1997)
4. Brandvik, T., Pullan, G.: Sblock: a framework for efficient stencil-based pde solvers on multi-core platforms. In: Proceedings of the 2010 10th IEEE International Conference on Computer and Information Technology, CIT 2010, pp. 1181–1188. IEEE Computer Society, Washington, DC (2010)
5. DeVito, Z., Joubert, N., Palacios, F., Oakley, S., Medina, M., Barrientos, M., Elsen, E., Ham, F., Aiken, A., Duraisamy, K., Darve, E., Alonso, J., Hanrahan, P.: Liszt: A domain specific language for building portable mesh-based pde solvers. In: Proceedings of 2011 International Conference for High Performance Computing, Networking, Storage and Analysis, SC 2011, pp. 9: 1–9: 12. ACM, New York (2011)
6. Giles, M.B., Mudalige, G.R., Sharif, Z., Markall, G., Kelly, P.H.: Performance analysis of the op2 framework on many-core architectures. SIGMETRICS Perform. Eval. Rev. 38(4), 9–15 (2011)
7. Karypis, G.: Metis - serial graph partitioning and fill-reducing matrixordering, March 2013. http://glaros.dtc.umn.edu/gkhome/metis/metis/overview
8. Kondo, S., Arai, R.: A reaction-diffusion wave on the skin of the marine angelfish pomacanthus. Nature 678(376), 765–768 (1995)
9. Kondo, S., Miura, T.: Reaction-diffusion model as a framework for understanding biological pattern formation. Science 329(5999), 1616–1620 (2010)
10. Meyer, M., Desbrun, M., Schröder, P., Barr, A.H.: Discrete differential-geometry operators for triangulated 2-manifolds. In: Hege, H.-C., Polthier, K. (eds.) Visualization and mathematics III, pp. 35–57. Springer, Heidelberg (2003)
11. Murray, J., Myerscough, M.: Pigmentation pattern formation on snakes. J. Theoret. Biol. 149(3), 339–360 (1991)
12. Ragan-Kelley, J., Adams, A., Paris, S., Levoy, M., Amarasinghe, S., Durand, F.: Decoupling algorithms from schedules for easy optimization of image processing pipelines. ACM Trans. Graph. 31(4), 32:1–32:12 (2012)
13. Reuter, M., Biasotti, S., Giorgi, D., Patan, G., Spagnuolo, M.: Discrete laplace-beltrami operators for shape analysis and segmentation. Comput. Graph. 33(3), 381–390 (2009). IEEE International Conference on Shape Modelling and Applications 2009
14. Turing, A.M.: The chemical basis of morphogenesis. Philos. Trans. R. Soc. Lond. B Biol. Sci. 237(641), 37–72 (1952)

Comparing Different Programming Approaches for SpMV-Operations on GPUs

Jan P. Ecker[1]([✉]), Rudolf Berrendorf[1], Javed Razzaq[1], Simon E. Scholl[1], and Florian Mannuss[2]

[1] Computer Science Department, Bonn-Rhein-Sieg University of Applied Sciences, Sankt Augustin, Germany
{jan.ecker,rudolf.berrendorf,javed.razzaq,simon.scholl}@h-brs.de
[2] EXPEC Advanced Research Center, Saudi Arabian Oil Company, Dhahran, Saudi Arabia
florian.mannuss@aramco.com

Abstract. There exist various different high- and low-level approaches for GPU programming. These include the newer directive based OpenACC programming model, Nvidia's programming platform CUDA and existing libraries like cuSPARSE with a fixed functionality. This work compares the attained performance and development effort of different approaches based on the example of implementing the SpMV operation, which is an important and performance critical building block in many application fields. We show that the main differences in development effort using CUDA and OpenACC are related to the memory management and the thread mapping.

Keywords: CUDA · OpenACC · cuSPARSE · GPU · SpMV · HPC

1 Introduction

Various approaches for programming and using GPUs (Graphics Processing Unit) exist. These approaches range from hardware-oriented approaches like PTX [16] to approaches that are high-level oriented including CUDA [17], OpenCL [12], OpenACC [18], as well as black-box functions in highly optimized mathematical libraries [14].

A common hypotheses is, that a directive based approach like OpenACC can offer a higher development comfort compared to lower-level approaches like CUDA. Furthermore it is often assumed that this will come on the cost of a possibly lower performance, as an advanced programmer has more control over performance critical aspects in a lower level programming model, especially in small but compute intensive kernels. This paper investigates these assumptions for the development of the sparse matrix vector multiplication (SpMV) operation. The SpMV operation is an important building block in many application fields, e.g., in linear solvers [5] and can consume a large part of an applications runtime.

© Springer International Publishing Switzerland 2016
R. Wyrzykowski et al. (Eds.): PPAM 2015, Part I, LNCS 9573, pp. 537–547, 2016.
DOI: 10.1007/978-3-319-32149-3_50

Representatives of classes of programming approaches are chosen in our comparison. For a lower-level approach with more direct control on a programming level we used CUDA, for a higher-level approach we used OpenACC, and especially to relate performance we used the highly optimized vendor library cuSPARSE that offers SpMV functionality for a fixed set of (rather simple) storage formats for sparse matrices. We restrict our evaluation to different generations of Nvidia GPUs.

In this report the SpMV operation for sparse matrices is used as the base of investigation. Special storage formats are used for storing sparse matrices [20] depending on several aspects like structure or target platform. The storage formats can have a huge impact on the memory demand of the matrices and the runtime as well as implementation complexity of the SpMV operation. Therefore different formats from rather simple to quite complex are used for the comparison.

The rest of the paper is structured as follows: Sect. 2 describes the relevant programming approaches. Section 3 discusses the comparison methodology and Sect. 4 describes the different SpMV realizations. Section 5 presents the evaluation and related work is discussed in Sect. 6. We conclude in Sect. 7.

2   Programming Approaches for GPUs

In this section available programming approaches for GPUs are discussed. The emphasis is on those approaches that were used in our comparison (CUDA, cuSPARSE library, OpenACC). We give only a brief overview, the cited literature should be used as a reference.

The CUDA programming model [17] is available as a C, C++ and Fortran language extension and was introduced late 2006 by Nvidia. It allows the definition of special kernel functions which can be launched in a host program and will be executed on a CUDA-capable GPU. The programming model allows fine-grained access to different levels of parallelism.

The cuSPARSE [14] library is developed by Nvidia and provides a fixed set of sparse matrix and vector operations for CUDA-enabled devices. It is designed for the use in C and C++ programs and supports various rather simple sparse matrix formats. The cuSPARSE library offers conversion methods into the supported storage formats and does not allow access to the related data structures of the more sophisticated HYB (and indirectly ELL and COO) format. Therefore it is not possible to develop own methods based on these formats.

OpenACC [18] is a directive based programming approach available for C, C++ and Fortran. It allows the offloading of computations from a host CPU to an attached accelerator. The current version of the OpenACC standard supports different accelerator architectures, like Nvidia and AMD based GPUs and Intel's Xeon Phi platform [19]. Only a small number of compilers support the OpenACC standard including PGI [3] and the newest version 5.1 of GNU GCC [9].

There exist various other approaches for programming GPUs. OpenCL [12] is very similar to CUDA but vendor independent. PTX [16] is a assembly like programming approach. ViennaCL [4], Paralution [2] and LAMA [1] are high-level frameworks, offering mathematical operations like SpMV.

One mayor difference between the different approaches is the memory management. CUDA provides explicit and implicit data handling through API functions. The implicit data handling is introduced in CUDA version 6 and called unified or managed memory. Using managed memory, the CUDA runtime ensures a coherent memory space between the device and the host memory. OpenACC supports explicit and implicit data handling as well by providing special compiler directives. While the CUDA managed memory ensures a coherent memory space at all time, OpenACC allows to specify the program places and direction of memory synchronization. The cuSPARSE library does not provide a memory management at all and e.g., CUDA or OpenACC can be used for handling the data.

Another difference between the approaches is the thread mapping, which describes the dimensions of the different levels of parallelism. The thread mapping influences the utilization of the accelerators and therefore can have a huge impact on the performance. CUDA and OpenACC allows the definition of up to three dimensional thread mappings. In CUDA, the definition of a thread mapping is an essential part of the development process while in OpenACC it is optional. If not specified the OpenACC implementation automatically picks a thread mapping. For the cuSPARSE library it is not required and even not possible to define a thread mapping.

3   Comparison Methodology

The programming approaches are compared by developing SpMV kernels for various sparse matrix storage formats. This allows a comparison of the development effort and performance for different problems. The matrix formats are selected regarding different criteria. The most common formats are used to be able to compare the performance of the developed kernels using CUDA and OpenACC with the existing library approaches in cuSPARSE. Additionally, more complex formats are used to compare the approaches for bigger kernels and more complex development scenarios. Table 1 gives an overview of the different implemented formats and approaches. The SpMV operation for the ELL, CSR and COO format [20] is implemented for all approaches. The more complex formats ELL-BRO [23] and SELL-C-σ [13] are not supported by cuSPARSE and are only implemented using CUDA and OpenACC.

For the rather simple ELL and CSR sparse matrix format, we use the common structure as it is described by Garland and Bell [6]. The COO format is additionally ordered by the row index of the entries which allow more efficient parallel implementations. It should be mentioned that the cuSPARSE library does not support ELL and COO directly but indirectly by using the HYB format.

The formats SELL-C-σ [13] and ELL-BRO [23] are both based on the ELL format. The SELL-C-σ format utilizes additional slicing and reordering approaches to further improve the SpMV performance. The ELL-BRO format

Table 1. Overview of all developed SpMV implementations.

Approach	Sparse matrix storage format				
	ELL	CSR	COO	ELL-BRO	SELL-C-σ
CUDA	+	+	+	+	+
OpenACC	+	+	+	+	+
cuSPARSE	via HYB	+	via HYB	−	−

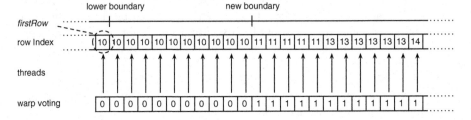

Fig. 1. Process of finding a new lower boundary of a warp in the COO kernel.

uses bit-packing and slicing approaches for reducing the memory required for index structures and therefore the memory bandwidth of the SpMV operation.

4   Realization

The basic ideas for an SpMV implementation of the CSR and ELL format on a GPU were presented by Garland and Bell [6] including the corresponding CUDA kernels. These basic ideas were used for our CUDA and OpenACC implementations. We developed an own SpMV implementation using CUDA for the row-sorted COO format. The concept utilizes multiple active warps working simultaneously on the COO data. In the first step a block distribution is used for calculating the boundaries for each warp. These boundaries are then adjusted by each warp, using CUDA warp voting functions, to match a proper row boundary on the COO data, which ensures exclusive write access for each warp to the elements of the result vector. This process is depicted in Fig. 1. Modified warp local reductions are used for calculating the results of each row of the COO format. The implementation of this concept using OpenACC is not possible, as the programming model does not allow the precise definition of the data distribution in a loop. OpenACC also does not provide access to fine-grained information like e.g., thread IDs which would be required for an alternative implementation. Instead, a simple algorithm using atomic operations is used in OpenACC, which can not utilize the sorting of the matrix elements. The kernels for the ELL-BRO and SELL-C-σ format follow the concepts of the corresponding publications [13,23] for CUDA as well as OpenACC.

The first runtime measurements of the implemented SpMV kernels showed, that the OpenACC implementations performed significantly slower than the

Table 2. Hardware specifications of the used CUDA GPU.

Specification	Tesla K80	Tesla K20m	Tesla M2050
Architecture	Kepler	Kepler	Fermi
Compute capability	3.7	3.5	2.1
Number of CUDA cores	2×2496	2496	448
Clock (MHz)	562 - 875	706	1150
Peak double performance (GFlops)	2×1455	1170	515
Memory bandwidth (GB/s)	2×240	208	148
Memory size (GB)	2×12	6	3
Max. CUDA threads per block	1024	1024	1024
Max. CUDA blocks per grid	$2^{31} - 1$	$2^{31} - 1$	$2^{16} - 1$

CUDA implementation. Therefore, the ELL kernels using CUDA and OpenACC were analyzed using the Nvidia Visual Profiler [15] which reported a high device and memory utilization of the OpenACC kernel, but no further explanation could be found. We moreover compared the generated PTX code of the kernels. The analysis showed that the PGI compiler in version 14.10 uses the debug mode as default. After disabling the debug mode, the generated PTX code of the ELL kernels for CUDA as well as OpenACC look surprisingly identical. It is notable that the PGI compiler made simple low level optimizations like replacing multiplications with bit shifts if possible, which were not done by the CUDA compiler.

5 Evaluation

In this section we first present the evaluation methodology. Afterwards we present the performance numbers and discuss the development effort.

5.1 Evaluation Methodology

The performance of the developed SpMV approaches is determined by measuring the runtime of the SpMV operation. These runtimes include only bare kernel execution and invocation time without data transfer times. Therefore the relevant matrix and vector data is transferred into the device memory beforehand. The measurements are performed on three different CUDA capable devices shown in Table 2. It should be mentioned that only one of the GPUs of the Tesla K80 is used in the benchmarks. For CUDA and cuSPARSE the Nvidia CUDA compiler[1] is used. For compiling the OpenACC approaches the PGI compiler version 14.10 is used. All benchmarks are done by executing the program 20 times and using the median runtime as final result.

---

[1] version 6.5 for Tesla K20m and M2050, version 7.0 for Tesla K80.

Table 3. Set of used test matrices with additional structure information.

#	Matrix	rows / columns	Nnz	Nnz per row	
				μ	σ
1	dielFilterV2real	1157456	48538952	41.93	16.14
2	F1	343791	26837113	78.06	40.80
3	gsm_106857	589446	21758924	36.91	15.63
4	ohne2	181343	11063545	61.00	21.08
5	PR02R	161070	8185136	50.81	19.69
6	RM07R	381689	37464962	98.15	68.68
7	thermal2	1228045	8580313	6.98	0.81
8	TSOPF_FS_b300_c3	84414	13135930	155.61	1181.18
9	TSOPF_RS_b2383_c1	38120	16171169	424.21	484.23
10	matrix_spe10_a	2153544	29192160	13.55	1.10

The benchmarks are done using a set of 10 large matrices. They are shown in Table 3 with additional structural information about number of rows, columns and number of nonzeros as well as the average number of nonzero per row μ and the root mean square deviation σ. They originate from the University of Florida Sparse Matrix Collection [8] and the SPE Comparative Solution Project [21]. These matrices belong to different domains and have different population densities and non-zero structures.

As already discussed in Sect. 2, the thread mapping has a huge impact on the performance. The automatic selection of a near optimal thread mapping is not a trivial task and requires a notable development effort. We used exhaustive parameter runs (taking several days of runtime) to find a near optimal thread configuration for each matrix and used this best time.

Based on the results of these exhaustive parameter runs, we additionally developed trivial heuristics for determining thread mappings for all of the used formats. This was done by incrementally restricting the search space of the parameter runs until one thread mapping remained. The calculated thread mappings are in average 1 % to 10 % slower compared to the best found mappings. This shows that the development of proper heuristics is indeed possible, though this may require a significant development effort. The benchmarks in the next section are based on the best found runtimes.

5.2   Performance Benchmarks

The attained performance of the different approaches is given in the following figures by speedups of the implementations using OpenACC and cuSPARSE relative to CUDA. This allows a common base for all comparisons as the vendor library does not support all formats of investigation. The results for one realization are shown with bars showing minimum, maximum, and average speedup

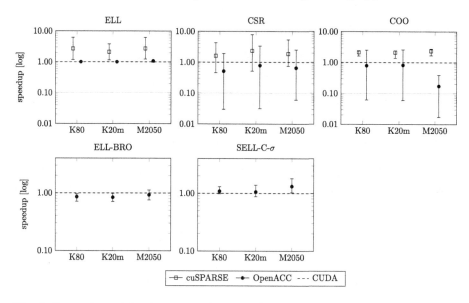

Fig. 2. Error plots with min, max and average best-case speedups of implementations using OpenACC and cuSPARSE relative to CUDA.

over all test matrices on one device. CUDA is also depicted with a value of 1 as it is the reference of the speedup comparison. This allows a very compact notation of all performance data.

In Fig. 2 it can be seen that the cuSPARSE library outperforms the CUDA and OpenACC implementations for most of the matrices with a speedup between 2-4. This is not further surprising as it is a highly optimized library implementation. Moreover it can be seen, that the CUDA and OpenACC implementations perform surprisingly identical for many formats. The ELL and SELL-C-σ kernels deliver very comparable performance on most of the used benchmark systems and matrices. Interestingly OpenACC performs even better, in relation to CUDA, on the older Tesla M2050 than on the newer GPUs. By comparing the PTX code (see Sect. 4) we found out that the compilers generate quite similar low-level code for the (rather simple) ELL format, which explains the very similar runtimes. As the SELL-C-σ kernels are very similar in complexity and structure compared to the ELL kernels, a similar compiler behavior can be expected. The OpenACC ELL-BRO kernel performs slightly worse compared to CUDA with an average speedup of only 0.85.

The OpenACC SpMV implementation for the CSR format performs quite poor compared to CUDA with speedups ranging from 0.5 to 0.8. In CUDA we reached best average performance with a CSR vector approach [6]. We were not able to develop an OpenACC implementation which follows the same concept and a CSR scalar approach [6] is used instead. Further research is required to identify if this is a limitation of the OpenACC specification.

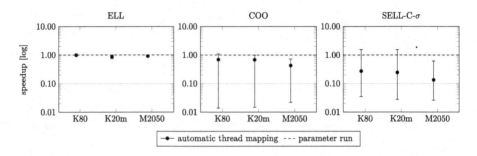

Fig. 3. Error plots with min, max and average OpenACC speedups of the automatic thread mapping relative to the best-case runtimes of the parameter runs.

The fact that the OpenACC implementation of the COO format performs worse than the CUDA implementation is not further surprising as the realized concept does not utilize the additional sorting of the matrix elements (see discussion in Sect. 4). It can be seen that the realization performs much worse on the older Fermi architecture of the M2050. Only around 20 % of the CUDA performance is reached, compared to around 80 % on the K80 and K20m. The much higher performance can be explained by the improved performance of atomic operations on the newer Kepler architecture (7x better).

Finding a proper thread mapping can be very time consuming and can be a significant factor in the development effort. Therefore we evaluated the performance of the automatic thread mapping offered by OpenACC. Figure 3 depicts the speedup of the automatic thread mapping compared to the near optimal runtimes found by the exhaustive parameter runs. The automatic thread mapping works very good for the ELL format. The performance of the CSR and ELL-BRO format is very similar and therefore no plots are given. While the ELL and CSR formats have quite a simple program structure, the ELL-BRO format utilizes quite complex concepts. For the COO format the thread mapping does not work as well and for the SELL-C-σ format it delivers very poor performance. This is itself surprising, as the SELL-C-σ format is very similar to ELL, where a proper thread mapping obviously could be found.

5.3   Development Effort

We evaluated the development effort by comparing the different kernels that were implemented using CUDA and OpenACC. The effort of developing the bare SpMV kernels is very similar for most of the formats. The main difference between the implementations is, that in OpenACC directives are used to accelerate existing code while CUDA requires the definition of additional functions. CUDA also requires a manual data partitioning using different available ID's. As the basic definition of a CUDA function follows a basic scheme most times we anyway expect very comparable development effort. For the ELL, ELL-BRO and SELL-C-σ format the CUDA and OpenACC kernels are nearly identical and therefore it is expected that the development effort is very comparable.

In a direct comparison the CSR kernel implemented using CUDA looks quite a bit more complex compared to using OpenACC. The CUDA kernel uses a warp local reduction which is realized in OpenACC with a simple keyword. However, the development effort is nevertheless similar for both implementations as the warp local reduction is a very basic and an often used building block in CUDA development.

The development effort of the COO SpMV operation using CUDA is much higher compared to the OpenACC implementation. The CUDA implementation utilizes our own, quite complex concept while the OpenACC implementation is based on simple atomic operations as with OpenACC it was not possible to get a eligible work distribution (see Sect. 5.2). A CUDA implementation using the same concept as for OpenACC would again lead to a very comparable kernel and therefore similar development effort.

Beside the effort of implementing the SpMV kernels we identified the memory management and the thread mapping as further important factors of the overall development effort. As we were required to use explicit memory management for all approaches, the effort was comparable. However we expect that the memory management of OpenACC can reduce the development effort in the development process, especially when accelerating existing code. Furthermore we showed that the automatic thread mapping of OpenACC resulted in a quite well performance for many formats. Therefore the use of the automatic thread mapping could be a good alternative, at least for first development versions. This can reduce the development effort in comparison to CUDA significantly.

6   Related Work

The paper from Wienke et al. [24] compares the performance as well as the development effort of OpenACC, PGI Accelerator and OpenCL for two real-world problems. They showed that OpenACC offers a good ratio of performance and development effort. Similarly Sugawara et al. [22] published their work comparing the same programming models but for implementing the matrix multiplication operation. They also found out that OpenACC can reach similar performance compared to OpenCL. In difference to this paper the older OpenACC 1.0 standard is used and OpenCL is used instead of CUDA, which can make a big difference for the performance and development effort comparison.

The papers from Hoshino et al. [11], Herdman et al. [10] and Christgau et al. [7] compare the performance of OpenACC and CUDA for different CFD problems. Hoshino et al. [11] and Christgau et al. [7] state that they were not able to reach good performance using OpenACC compared to highly optimized CUDA kernels. They identified the lack of control over the shared memory and data access patterns as limiting factors of OpenACC. However Herdman et al. [10] reached even better performance using OpenACC compared to CUDA and OpenCL. All three papers focus on the evaluation of the performance and do not discuss the development effort. Additionally the focus of this paper is especially on the SpMV operation.

7   Conclusion

We have implemented SpMV operations for various sparse matrix storage formats using lower-level and high-level programming approaches. By using performance benchmarks it could be shown that the cuSPARSE library delivers the best performance for the supported formats. However the cuSPARSE library is limited to the supported formats and additionally it is not possible to implement own methods based on most of these formats.

Furthermore we could not approve the initial assumption that the use of lower-level approaches like CUDA in general leads to improved performance, at least not with reasonable development effort even for an advanced programmer. We showed that the performance of the OpenACC kernels can be quite comparable to CUDA for the most problems presented in this paper. However, for some problems like implementing the SpMV operation for the COO format limitations of the OpenACC specification are revealed.

It could be shown that the automatic thread mapping of OpenACC can deliver good performance for various different problems which can significantly reduce the development effort. Further research is required to develop proper and fast heuristics for finding proper CUDA thread mappings for the SpMV operation.

Acknowledgements. We would like to thank the CMT team at Saudi Aramco EXPEC ARC for their support and input. Especially we want to thank Ali H. Dogru for making this research project possible.

References

1. LAMA - Library for Accelerated Math Applications. http://www.libama.org/. Accessed 5 August 2015
2. Paralution - The library for iterative sparse methods on CPU and GPU. http://www.paralution.com/. Accessed 5 August 2015
3. PGI Accelerator Compilers with OpenACC Directives. https://www.pgroup.com/resources/accel.htm. Accessed 6 September 2015
4. Vienna Computing Library (ViennaCL). http://viennacl.sourceforge.net/. Accessed 5 August 2015
5. Barrett, R., Berry, M., Chan, T.F., Demmel, J., Donato, J., Dongarra, J., Eijkhout, V., Pozo, R., Romine, C., der Vorst, H.V.: Templates for the Solution of Linear Systems: Building Blocks for Iterative Methods, 2nd edn. SIAM, Philadelphia (1994)
6. Bell, N., Garland, M.: Efficient sparse matrix-vector multiplication on CUDA. Technical report NVR-2008-004, Nvidia Corp., December 2008
7. Christgau, S., Spazier, J., Schnor, B., Hammitsch, M., Babeyko, A., Wächter, J.: A comparison of CUDA and OpenACC: accelerating the tsunami simulation easywave. In: 2014 27th International Conference on Architecture of Computing Systems (ARCS), pp. 1–5. IEEE, February 2014
8. Davis, T.A., Hu, Y.: The university of florida sparse matrix collection. ACM Trans. Math. Softw. **38**(1), 1: 1–1: 25 (2010)

9. GNU GCC: GCC 5 Release Series - Changes, new Features, and Fixes. https://gcc.gnu.org/gcc-5/changes.html. Accessed 5 August 2015
10. Herdman, J., Gaudin, W., McIntosh-Smith, S., Boulton, M., Beckingsale, D., Mallison, A., Jarvis, S.: Accelerating hydrocodes with OpenACC, OpenCL and CUDA. In: 2012 SC Companion on High Performance Computing, Networking, Storage and Analysis (SCC), pp. 465–471. IEEE, November 2012
11. Hoshino, T., Maruyama, N., Matsuoka, S., Takaki, R.: CUDA vs OpenACC: performance case studies with kernel benchmarks and a memory-bound CFD application. In: Proceedings of 2013 13th IEEE/ACM International Symposium on Cluster, Cloud, and Grid Computing, pp. 136–143. IEEE (2013)
12. Khronos OpenCL Working Group: The OpenCL Specification (API Specification), 2 edn. https://www.khronos.org/registry/cl/specs/opencl-2.0.pdf. Accessed 5 August 2015
13. Maggioni, M., Berger-Wolf, T.: An architecture-aware technique for optimizing sparse matrix-vector multiplication on GPUs. Procedia Comput. Sci. **18**, 329–338 (2013). Proceedings of 2013 International Conference on Computational Science. Elsevier B.V
14. Nvidia Corp: Nvidia cuSPARSE. https://developer.nvidia.com/cusparse. Accessed 5 August 2015
15. Nvidia Corp: NVIDIA Visual Profiler. https://developer.nvidia.com/nvidia-visual-profiler. Accessed 5 August 2015
16. Nvidia Corp: Parallel Thread Execution ISA - Application Guide. v4.1st edn., August 2014. http://docs.nvidia.com/cuda/pdf/ptx_isa_4.1.pdf. Accessed 5 August 2015
17. Nvidia Corp: CUDA C Programming Guide. pg-02829-001_v7.0 edn., March 2015. http://docs.nvidia.com/cuda/pdf/CUDA_C_Programming_Guide.pdf. Accessed 06 December 2015
18. OpenACC: OpenACC^{TM} Application Programming Interface, Version 2.0a, August 2013. http://www.openacc-standard.org/. Accessed 5 August 2015
19. Rahman, R.: Intel$^{®}$ Xeon Phi^{TM} Core Micro-architecture, pp. 1–15 (2013)
20. Saad, Y.: Iterative Methods for Sparse Linear Systems, 2nd edn. SIAM, Philadelphia (2003)
21. Society of Petroleum Engineers: SPE Comparative Solution Project. http://www.spe.org/web/csp/
22. Sugawara, M., Hirasawa, S., Komatsu, K., Takizawa, H., Kubayashi, H.: A comparison of performance tunabilities between OpenCL and OpenACC. In: Proceedings of 2013 IEEE 7th International Symposium on Embedded Multicore/Manycore System-on-Chip, pp. 147–152. IEEE (2013)
23. Tang, W., Tan, W., Ray, R., Wong, Y., Chen, W., Kuo, S., Goh, R., Turner, S., Wong, W.: Accelerating sparse matrix-vector multiplication on gpus using bit-representation-optimized schemes. In: Proceedings of SC 2013 Proceedings of the International Conference on High Performance Computing, Networking, Storage and Analysis. No. 26 in Proceedings of ACM/IEEE Supercomputing. ACM (2013)
24. Wienke, S., Springer, P., Terboven, C., an Mey, D.: OpenACC — first experiences with real-world applications. In: Kaklamanis, C., Papatheodorou, T., Spirakis, P.G. (eds.) Euro-Par 2012. LNCS, vol. 7484, pp. 859–870. Springer, Heidelberg (2012)

IVM-Based Work Stealing for Parallel Branch-and-Bound on GPU

Jan Gmys[1]([⊠]), Mohand Mezmaz[1], Nouredine Melab[2], and Daniel Tuyttens[1]

[1] Mathematics and Operational Research Department (MARO), University of Mons, Mons, Belgium
{jan.gmys,mohand.mezmaz,daniel.tuyttens}@umons.ac.be

[2] INRIA Lille Nord Europe, Université Lille 1, CNRS/CRIStAL, Cité Scientifique, 59655 Villeneuve d'Ascq Cedex, France
nouredine.melab@univ-lille1.fr

Abstract. In this paper we present a B&B algorithm entirely based on GPU and propose four work stealing (WS) strategies to balance the workload inside the GPU. Our B&B is based on an Integer-Vector-Matrix (IVM) data structure instead of a pool of permutations, and work units exchanged are intervals of factoradics instead of sets of nodes. To the best of our knowledge, the proposed approach is the pioneering to perform the entire exploration process on GPU. The four WS strategies have been experimented and compared to a multi-core IVM-based approach using standard flow shop scheduling problem instances. The reported results show, on the one hand, that the GPU-based approach is more than 5 times faster than its multi-core counterpart. On the other hand, the best of the four strategies provides a near-optimal load balance while consuming only 2 % of the total execution time of the algorithm.

Keywords: GPU computing · Branch-and-Bound · Combinatorial optimization · Work stealing

1 Introduction

Many permutation-based combinatorial optimization problems, like flow shop, can be solved to optimality using Branch-and-Bound (B&B) algorithms. These algorithms perform an implicit enumeration of all possible solutions by dynamically constructing and exploring a tree. This is done using four operators: branching, bounding, selection and pruning. Execution times of B&B algorithms significantly increase with the size of the problem instance, and often only small or moderately-sized instances can be practically solved. Because of their massive data processing capability and their remarkable cost efficiency, graphics processing units (GPU) are an attractive choice for providing the computing power needed to solve such instances. However, B&B is a highly irregular application, in terms of control flow, memory access patterns and workload distribution. The acceleration of B&B algorithms using GPUs is therefore a challenging task which is addressed by only a few works in the literature [1,2,8]. Most approaches use

© Springer International Publishing Switzerland 2016
R. Wyrzykowski et al. (Eds.): PPAM 2015, Part I, LNCS 9573, pp. 548–558, 2016.
DOI: 10.1007/978-3-319-32149-3_51

pools of subproblems, implemented as linked-lists, stacks or queues to store and manage the B&B tree. The GPU memory imposes limitations on the use of such dynamic data structures, and pool-based approaches perform at least one of the B&B operators on the CPU, requiring costly data transfers between CPU and GPU. The Integer-Vector-Matrix (IVM) data structure [14], used in our GPU-based algorithm, allows to overcome this issue. To the best of our knowledge, the proposed approach is the pioneering to perform the entire exploration process on GPU. Performing all B&B operators on the device raises the issue of balancing the workload inside the GPU. In this paper, we present the GPU-based B&B algorithm using the IVM structure and propose four work stealing (WS) strategies to address the issue of workload imbalance.

2   Parallel B&B and Flow Shop

Serial B&B. B&B algorithms explore the search space of potential solutions by dynamically building a tree whose root node represents the initial problem to be solved. Leaf nodes are possible solutions and internal nodes represent subspaces of the total search space. Subproblems (internal nodes) are stored in a data structure which initially contains only the root node. The best solution found so far is initialized at ∞ and can be improved from one iteration to another. At each iteration, the branching operator partitions a subproblem into several smaller, pairwise disjoint subproblems. For each subproblem the bounding operator is used to compute a lower bound (LB) value of its optimal solution. The pruning operator uses this LB to decide whether a node is eliminated or kept for further decomposition. According to a predefined exploration strategy the selection operator chooses the next node to be processed. For instance, the selection of a node can be based on its depth in the B&B tree, leading to a depth-first search (DFS), which is used in this paper.

Parallel B&B. The pruning mechanism efficiently reduces the size of the search space and thus the computing power needed for its exploration. However, especially for larger instances, the exploration time remains significant and parallel processing is required to speed up the exploration process. A taxonomy of models to parallelize B&B algorithms is presented in [11]. Among the identified models are (1) the parallel tree exploration model, and (2) the parallel evaluation of bounds model. Our GPU-based B&B algorithm uses a combination of both.

Model (1) consists in simultaneously exploring different search subspaces of the initial problem. This means that the four B&B operators are applied in parallel, synchronously or asynchronously, by different B&B processes which explore these subspaces independently. The number of parallel exploration processes is only limited by the capacity to supply them continuously with subproblems to explore, so, the degree of parallelism in model (1) may be important, especially when solving large instances. As the shape of the B&B tree is highly irregular, this work supply strongly depends on the distribution and sharing of the workload. Model (1) can be combined with other parallel B&B models.

Indeed, each independent B&B process may in turn be parallelized, for instance using model (2), which evaluates the LBs of branched subproblems in parallel. This adds a second level second level of parallelism. Model (2) is data-parallel, synchronous and fine-grained (the cost of the evaluation of a bound) and is thus suitable for GPU computing. The GPU-accelerated B&B proposed in [12] is based on model (2) and offloads pools of subproblems for bounding to the GPU. Although good speed-ups are achieved, the data transfers between CPU and GPU constitute a bottleneck for this offload-approach. Our GPU-based B&B uses a 2-level combination of models (1) and (2).

Flow Shop. Permutation flow shop belongs to the category of scheduling problems. It is defined by a set of n jobs to be scheduled in the same order on m machines. The scheduling obeys the chain production principle, i.e. a job cannot be processed on a machine j before it has finished processing on all machines $0, 1, \cdots, j - 1$ located upstream. An operation (the processing of a job) cannot be interrupted, and machines process at most one job at a time. A duration is associated with each operation. The goal is to find a permutation schedule that minimizes the total processing time called makespan. In [3], it is shown that the minimization of the makespan is NP-hard from 3 machines upwards. The effectiveness of B&B strongly depends on the relevance of the used LB for the makespan. The LB proposed by Lageweg *et al.* [6] is used in our bounding operator. This bound is known for its good results and has complexity of $O(m^2 n log(n))$.

In our experiments, the flow shop instances defined by Taillard [15] are used. These instances are divided into 12 groups: 20×5 (i.e. group of instances defined by $n = 20$ jobs and $m = 5$ machines), 20×10, 20×20, 50×5, 50×10, 50×20, 100×5, 100×10, 100×20, 200×10, 200×20, and 500×20. For instances where m is equal to 5 or 10 the used bounding operator gives such good LBs that they can be easily solved using a sequential B&B. Instances where $m = 20$ and $n = 50, 100, 200$, or 500 are very hard to solve. For example, the resolution of $Ta056$ in [13], which is an instance of the 50×20 group, lasted 25 days with an average of 328 processors and a cumulative computation time of about 22 years. Therefore, in our experiments, the validation is performed using the 10 instances where $m = n = 20$.

3   GPU IVM-Based B&B

3.1   Serial IVM-Based B&B

The pool of Fig. 1a is represented as a tree in order to visualize the problem-subproblem relationship between nodes. Figure 1b represents the corresponding Integer-Vector-Matrix (IVM) structure. The notation 23/14 denotes the subproblem where jobs 2 and 3 are scheduled, while 1 and 4 are unscheduled. In the example of Fig. 1 the root node /1234 is decomposed into four nodes, namely 1/234, 2/134, 3/124 and 4/123. In the IVM-approach this corresponds to the first row of the matrix M which contains all job numbers $1, 2, 3, 4$. The example assumes that the algorithm selects and branches the node $2/134$. For the IVM-structure this translates to setting $V(0) = 1$. The so-called *position-vector* V always points to the currently active node at depth I. The decomposition of $2/134$ gives three nodes, $21/34$, $23/14$ and $24/13$. In IVM-terms this consists in copying the elements of row $I = 0$, except the scheduled job $M(0, V(0)) = 2$, to the next row $I = 1$. Also, I is incremented by one when a node is decomposed. The example also assumes that $21/34$ is processed or pruned. The IVM-based selection operator should therefore skip the corresponding cell of M and move rightward to the next cell by incrementing $V(1)$. Cells that correspond to pruned nodes are flagged, multiplying them by -1. Therefore, the algorithm decomposes the second node $23/14$, obtaining two new nodes: $231/4$ and $234/1$. Again, the next-row process performs this decomposition in the IVM.

The depth-first selection, branching and pruning operators can be expressed in terms of actions on the IVM structure. In order to enable the bounding operator to compute the LB of a subproblem encoded by the IVM, a decode operation is required. For example, the solution encoded in Fig. 1b can be directly read by looking (from row $I = 0$ to row $I = 3$) at the values in M pointed by V. With the same V and M, if the current depth is $I = 1$, then the represented subproblem is $23/14$.

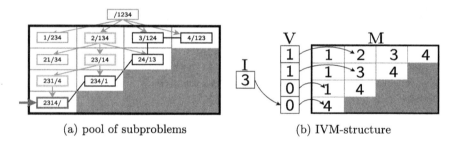

| (a) pool of subproblems | (b) IVM-structure |

Fig. 1. Example of a pool of subproblems and an IVM-structure obtained when solving a permutation problem of size four

3.2   Work Unit: Interval of Factoradics

Throughout the exploration process, the vector V behaves like a counter. In the example of Fig. 1b, V successively takes the values *0000, 0010, 0100, ..., 3200, 3210*. These 24 values correspond to the numbering of the $4!$ solutions using a numbering system in which the weight of the ith position is equal to $i!$ and the digits allowed for the ith position are $0, 1, \ldots, i$. Applied to the numbering of permutations, the French term *numération factorielle* was first used in 1888 [7], while Knuth [5] uses the term *factorial number system*. The term *factoradic*, which seems to be of more recent date is used, for instance, in [10].

Subtrees can be identified with intervals of factoradics. In the example of Fig. 1b, $[0000, 3210[$ is explored by one B&B-process using IVM. It is possible to have two IVMs $R1$, $R2$ such that $R1$ explores $[0000, X[$ and $R2$ explores $[X, 3210[^1$. These factoradic intervals are used as work units, instead of sets of nodes. If $R2$ ends exploring its interval before $R1$ does, then $R2$ steals a portion of $R1$'s interval. Therefore, $R1$ and $R2$ can exchange their interval portions until the exploration of all $[0, N![$. With the exception of rare works such as [13], work units exchanged between processes are sets of nodes. To enable an IVM to explore any interval $[A, B[$ an initialization process is necessary. The correct initialized state of the IVM is such that it would be the same if $V = A$ had been reached through the exploration process. Therefore, initialization differs from the normal B&B process only in the selection operator. Instead of depth-first selecting the next node, initializing IVMs select at each level k the node pointed by $V(k)$ as long as the selected node is promising. If a pruned node is selected, the initialization process is finished and the IVM resumes exploration, searching for the next node to decompose. Thus, the initialization process may last for 1 to N iterations.

3.3   GPU-Based Parallel B&B

IVM is well-adapted to a GPU implementation of the parallel B&B algorithm. Contrary to a linked-list of subproblems, IVM requires no dynamic memory allocation and provides good data locality. A fixed number T of IVM structures is allocated in device memory and the B&B operators, implemented as CUDA-kernels, act on these data structures in parallel. Figure 2 provides an overview of the GPU-based algorithm. Depending on the current IVM-state (*exploring*, *empty* or *initializing*) different actions are performed.

In the `goToNext` kernel, all exploring IVMs try to select the next node to decompose. If this fails the interval is empty and the IVM-state is set accordingly. Initializing IVMs skip this depth-first selection procedure as the next node to decompose is pointed by the new position-vector of the IVM. If the initialization process is terminated, then the state is set to *exploring*. All non-empty IVMs move to the next row in the matrix.

For non-empty IVMs the `decode` kernel reads the IVM structure and builds the father subproblem to be decomposed. The kernel `bound` uses this data and computes the LBs for the children subproblems using one CUDA-thread per LB. The degree of parallelism depends on the number of non-empty IVMs and on the average depth of these explorers in the B&B-tree.

In order to maintain a high device occupancy, the number of empty IVMs should be as low as possible. Also the number of initializing IVMs should be low, because the bounding procedure for these IVMs contributes to computational overhead. As the number of children per non-empty IVM is variable, a `remap` phase precedes the bounding kernel. Using a parallel prefix sum [4] computation,

---

[1] In the rest of this paper the term IVM designates the data structure as well as, by extension, the exploration process associated with a part of the B&B-tree.

the remapping performs a compaction of the mapping of threads onto children subproblems. It also detects the end of the algorithm (⇔ all IVMs empty) and determines global best solution so far using a min-reduce. The **prune** kernel uses the LBs computed in the bounding kernel to decide whether to prune a subproblem or not.

Empty IVMs remain unchanged during an iteration: they try to steal work in the **steal** kernel. A work stealing (WS) operation is determined by a victim selection strategy and a granularity policy, defining how much of a victim's work unit is stolen. In the **selectVictim** phase a one-to-one mapping of empty IVMs onto suitable victim IVMs is built. As in the **steal** kernel all empty IVMs try to acquire work in parallel, two empty IVMs should not select the same victim. Initializing IVMs cannot be selected as WS victims, because this could result in a deadlock situation. Moreover, the victim selection should (1) induce minimal overhead, thus the mapping should be build in parallel, (2) select victim IVMs whose intervals are likely to contain more work than others, (3) serve a maximum of empty IVMs.

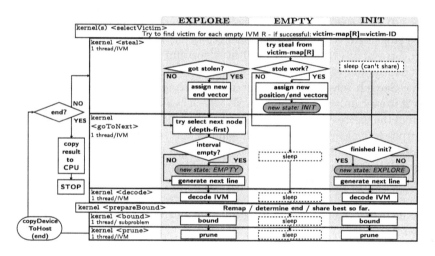

Fig. 2. Flowchart of the CUDA B&B algorithm

4   Work Stealing Strategies

This section explains the IVM-based WS for the GPU-B&B algorithm. In [9] several IVM-based WS strategies have been proposed for asynchronous multi-core B&B algorithms. For instance, the *Ring-1/T* strategy is directly transposable to the synchronous GPU-B&B. The other proposed WS strategies extend this strategy. They are described in Algorithm 1 and correspond to different choices for a set of parameters.

Algorithm 1. Victim selection
Ring-1/T: **B** = 1; **S** = 0; **C** = 0;   Search-1/T: **B** = 1; **S** = 25, **C** = 0;
Large-1/2: **B** = 1; 10 < **S** < T, **C** = 1;   Circle-1/2: **B** = iter%T; 10 < **S** < T; **C** = 1;

```
 1: procedure SELECTVICTIM(B, S, C)
 2:   for (k = B → B+S) do
 3:       beg<<<T threads>>>(k, victim-map, C,...)
 4:   end for
 5: end procedure
 6: procedure <<< >>> BEG(k, victim-map, C,...)
 7:   ivm ← blockIdx.x*blockDim.x + threadIdx.x
 8:   if (state[ivm]=empty) then
 9:       V ← (ivm-k)%T
10:         if (state[V]=exploring AND flag[V]= 0 AND length[V]>C*meanLength) then
11:             victim-map[ivm]← V
12:             flag[V]← 1
13:         end if
14:     end if
15: end procedure
```

The **Ring**-selection policy consists in selecting a victim in round-robin fashion, meaning that an empty IVM $R \in \{1,...,T\}$ tries to steal a portion of work from IVM $(R-1)\%T$. In Algorithm 1 this corresponds to parameters $B = 1$ and $S = 0$. If the state of the selected victim $(R-1)\%T$ is *exploring*, then work can be stolen. No conditions on the length of the victim's interval are imposed, which corresponds to setting $C = 0$ in Algorithm 1. To use this information in the steal-kernel, victim-map[R] is set to $(R-1)\%T$.

The **Search** selection policy extends the *Ring* strategy by checking successively if work can be stolen from IVMs $(R-1)\%T$, $(R-2)\%T$, ... , $(R-S)\%T$. Seaching the entire ring $(S = T)$ results in excessive overhead, so for experimental purposes the length of the search window S is set to 25. The idea behind this strategy is to avoid the following situation: Using the *Ring* policy, if a group of l empty IVMs is queued behind a group of exploring IVMs, then it takes at least l iterations to serve all IVMs in that group. In order to avoid multiple selections of the same victim, a flag-variable is used (Algorithm 1, line 10). The *Search* policy increases the probability that an empty IVM succeeds its WS attempt at a given iteration.

The idea behind the **Large** selection policy is to steal from larger intervals as they are likely to contain more nodes to decompose. This requires computing the length of each interval at each iteration. In order to avoid the costly operation of sorting the IVM-IDs by their corresponding interval-lengths, the mean interval-length is computed prior to the victim selection phase. Having an interval larger than average is added as a criterion for the eligibility of an IVM as a WS victim. In Algorithm 1 this corresponds to setting $C = 1$. However, this length-criterion increases the probability that no victim is found in the search window of fixed length S. The parameter S is therefore allowed to float between 10 and T. If more than 10 % of IVMs are empty at a given iteration, then S is incremented by one, otherwise S is decremented by one. Auto-tuning this parameter requires copying the number of empty IVMs to the host at each iteration.

In the **Circle** selection policy the search window is shifted by one position at each iteration of the algorithm. In Algorithm 1 this corresponds to setting

B = iter%T. Like in the *Large* policy the parameter S is adapted dynamically and half of the victim's interval is stolen. The main idea behind this policy is to reduce the length of the search window as a large value for S induces most overhead.

Two **granularity policies** are used. For *Ring* and *Search*, a thief steals all but the Tth part of its victim's interval. In [9] the authors show that this granularity policy better suits the *Ring* selection strategy than a $\frac{1}{2}$-policy. Stealing all but $1/T^{th}$ of an interval aims at acquiring work not only for itself but also for the following IVMs. Choosing large intervals (*Large* and *Circle*) justifies stealing one half of the victim's interval.

5 Experiments

Hardware/Experimental Protocol. For all experiments an NVIDIA Tesla K20m GPU, and version 5.0.35 of the CUDA Toolkit are used. As explained in Sect. 2, the medium-sized instances *Ta021-Ta030* [15] are used in our experiments. The performances of the GPU-B&B are compared to a multi-core B&B [9] running on 2 Sandy Bridge E5-2650 processors with 32 threads. When an instance is solved twice using a multi-threaded B&B, the number of explored subproblems often differs between the two resolutions. Therefore, we choose to always initialize our B&B by the optimal solution of the instance to be solved. With this initialization, the number of decomposed nodes is the same in each exploration. The number of used IVMs is $T = 768$.

Table 1. Exploration time (in seconds) for solving flowshop instances *Ta021-Ta030*. Comparison of multi-core IVM-based B&B [14] using 32 threads with GPU-IVM B&B using different WS strategies

Inst.	$\times 10^6$ Nodes	Ring(multi-core)	Ring(GPU)		Search(GPU)		Large(GPU)		Circle(GPU)	
		Time	Time	Rate	Time	Rate	Time	Rate	Time	Rate
21	41.4	1371	386	3.6	338	4.1	280	4.9	250	5.5
22	22.1	668	247	2.7	229	2.9	146	4.6	129	5.2
23	140.8	4466	1002	**4.6**	934	**4.8**	915	**5.4**	813	5.5
24	40.1	1142	359	3.2	254	4.5	255	4.5	219	5.2
25	41.4	1422	431	3.3	327	4.3	280	5.1	250	**5.7**
26	71.4	1975	459	4.3	443	4.5	429	4.6	384	5.2
27	57.1	1600	404	3.9	370	4.3	336	4.8	301	5.3
28	8.1	263	120	2.2	79	3.3	59	4.4	52	5.1
29	6.8	216	95	2.3	88	2.5	50	4.4	42	5.1
30	1.6	58	39	1.5	36	1.6	20	2.9	12	4.8
Avg	**43.1**	1318	**354**	**3.7**	**310**	**4.3**	**277**	**4.8**	**245**	**5.4**

Experimental Results. Table 1 shows the exploration time for different WS strategies. It also shows the rate comparing the GPU-based approaches to the multi-core IVM-based B&B presented in [14]. The latter uses 32 (POSIX) threads (= 32 IVMs) and an asynchronous *Ring-1/T* WS strategy. For all instances best performances are achieved with the *Circle* strategy, which allows to complete the exploration on average 5.4 times faster than the multi-core B&B. Comparing the different WS strategies, one can see that *Circle* is less sensitive to varying instance-sizes and shapes. *Circle* provides a relative speedup of $4.8 - 5.7x$ over the multi-core B&B, while the spread $1.5 - 4.6x$ is much larger for the *Ring* strategy. Table 2 shows the *IVM-efficiency* ($= 100\% \times \frac{\#\text{decomposed nodes}}{\#\text{iter} \times \text{T}}$) which provides a measure for the efficiency of the proposed WS strategies. Indeed, an optimal work load balance leads to one node decomposition per IVM per iteration. The results show that the *Circle* strategy is close to the optimal case where *IVM-efficiency* $= 100\%$ – on average only 2.5 % of the $T = 768$ IVMs are either empty or initializing.

Finally, the overhead induced by load balancing is evaluated. Figure 3 shows, for the four WS strategies, the average time spent in different phases of the algorithm. In all cases the bounding phase consumes >85 % of the total execution time. Using the *Ring* strategy, on average 327 seconds are spent in the bounding kernel, against 226 seconds for the *Circle* strategy. This corresponds to the increased efficiency of that kernel (Table 2). Compared to *Ring*, the *Search*-window of length $S = 25$ significantly improves IVM-efficiency – at the cost of spending ≈3 % instead of ≈0.1 % in victim-selection. The *Large* strategy further improves the work load balance, but victim-selection amounts for ≈9 % as the average value for the auto-tuned parameter S increases to 110. Extending *Large*

Table 2. Average percentage of IVMs in *exploring* state (IVM-Efficiency) and average number of iterations performed to complete the exploration

	Ta021	Ta022	Ta023	Ta024	Ta025	Ta026	Ta027	Ta028	Ta029	Ta030	**Avg**	#Iter
Ring	52.0	38.4	74.5	48.2	45.6	71.0	63.9	24.6	29.1	14.8	**46.2**	96781
Search	67.4	46.3	84.2	84.2	73.8	80.4	79.0	53.8	36.7	19.9	**62.5**	74402
Large	93.3	92.9	93.4	93.2	93.3	93.2	93.4	91.5	90.4	76.4	**91.1**	60108
Circle	99.4	98.9	99.8	99.4	99.3	99.7	99.5	96.8	96.8	85.3	**97.5**	56368

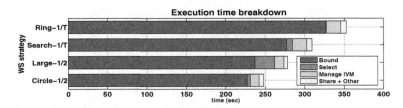

Fig. 3. Average elapsed time for solving instances *Ta021-Ta030* and its repartition among different phases of the algorithm

to *Circle* only slightly improves the load balancing, but, more importantly, as the average S decreases to 15, the cost of victim selection decreases to <2 %.

6   Conclusion and Future Work

We have presented a GPU-based B&B algorithm which is, to the best of our knowledge the pioneering to perform the entire exploration process on the GPU. We focused on balancing the irregular workload and proposed four work stealing (WS) strategies to address this challenge. In these lock-free WS strategies explorers exchange work units in a WS phase introduced in the synchronous parallel exploration process. The results show that the performance of the GPU-B&B algorithm strongly depends on the choice of the WS strategy. For a set of standard flow shop instances our IVM-based GPU-B&B algorithm outperforms its 32-threaded multi-core counterpart on average by a factor of 5.4.

As a future work we plan to investigate the use of other hardware accelerators like Intel Xeon Phi for B&B and to extend the IVM-based B&B to a cluster-based version using many-core accelerators. Solving larger problems, like Taillard's [15] flow shop instances defined by 50 jobs and 20 machines, requires indeed the combined computing power of several multi-core and many-core processors. Furthermore, we plan to validate our approach on other permutation-based optimization problems, for instance on Travelling Salesman and Quadratic Assignment problems.

References

1. Carneiro, T., Muritiba, A., Negreiros, M., Lima de Campos, G.: A new parallel schema for branch-and-bound algorithms using GPGPU. In: 23rd International Symposium on Computer Architecture and High Performance Computing (SBAC-PAD 2011), pp. 41–47
2. Chakroun, I., Mezmaz, M., Melab, N., Bendjoudi, A.: Reducing thread divergence in a GPU-accelerated branch-and-bound algorithm. Concur. Comput. Pract. Experience **25**(8), 1121–1136 (2013)
3. Garey, M.R., Johnson, D.S., Sethi, R.: The complexity of flowshop and jobshop scheduling. Math. Oper. Res. **1**(2), 117–129 (1976)
4. Harris, M., Sengupta, S., Owens, J.D.: Parallel prefix sum (scan) with CUDA. GPU Gems **3**(39), 851–876 (2007)
5. Knuth, D.E.: The Art of Computer Programming. Seminumerical Algorithms, vol. 2, 3rd edn. Addison-Wesley Longman Publishing Co., Inc., Boston (1997)
6. Lageweg, B.J., Lenstra, J.K., Kan, A.H.G.R.: A general bounding scheme for the permutation flow-shop problem. Oper. Res. **26**(1), 53–67 (1978)
7. Laisant, C.A.: Sur la numération factorielle, application aux permutations. Bull. Soc. Math. France **16**, 176–183 (1888)
8. Lalami, M., El-Baz, D.: GPU implementation of the branch and bound methodfor knapsack problems. In: IEEE 26th International Parallel and Distributed Processing Symposium' Workshops PhD Forum (IPDPSW 2012), Shanghai, CHN, pp. 1769–1777

9. Leroy, R., Mezmaz, M., Melab, N., Tuyttens, D.: Work stealing strategies for multi-core parallel branch-and-bound algorithm using factorial number system. In: Proceedings of Programming Models and Applications on Multicores and Manycores, PMAM 2014, Orlando, FL, pp. 111–119

10. McCaffrey, J.: Using permutations in.NET for improved systems security (2003)

11. Melab, N.: Contributions à la résolution de problèmes d'optimisation combinatoire sur grilles de calcul. LIFL, USTL, thesis HDR, November 2005

12. Melab, N., Chakroun, I., Bendjoudi, A.: Graphics processing unit-accelerated bounding for branch-and-bound applied to a permutation problem using data access optimization. Concur. Comput. Pract. Exp. **26**(16), 2667–2683 (2014)

13. Mezmaz, M., Melab, N., Talbi., E.G.: A grid-enabled branch and bound algorithm for solving challenging combinatorial optimization problems. In: 21th IEEE International Parallel and Distributed Processing Symposium, (IPDPS 2007), Long Beach, CA, pp. 1–9

14. Mezmaz, M., Leroy, R., Melab, N., Tuyttens, D.: A multi-core parallel branch-and-bound algorithm using factorial number system. In: 28th IEEE International Parallel & Distributed Processing Symposium, (IPDPS 2014), Phoenix, AZ, pp. 1203–1212

15. Taillard, E.: Benchmarks for basic scheduling problems. J. Oper. Res. **64**, 278–285 (1993)

Massively Parallel Construction
of the Cell Graph

Krzysztof Kaczmarski, Paweł Rzążewski$^{(\boxtimes)}$, and Albert Wolant

Faculty of Mathematics and Information Science, Warsaw University of Technology,
Koszykowa 75, 00-662 Warszawa, Poland
{k.kaczmarski,p.rzazewski}@mini.pw.edu.pl,
wolanta@student.mini.pw.edu.pl

Abstract. Motion planning is an important and well-studied field of robotics. A typical approach to finding a route is to construct a *cell graph* representing a scene and then to find a path in such a graph. In this paper we present and analyze parallel algorithms for constructing the cell graph on a SIMD-like GPU processor.

Additionally, we present a new implementation of the dictionary data type on a GPU device. In the contrary to hash tables, which are common in GPU algorithms, it uses a radix search tree in which all values are kept in leaves. With such a structure we can effectively perform dictionary operations on a set of long vectors over a limited alphabet.

Keywords: Cell graph · Motion planning · GPGPU · CUDA

1   Introduction

Motion planning is a common task in robotics and artificial intelligence. One of the aims is to find a path, which can be traversed by a rigid body (e.g. a robot) to get to the destination point and avoid collisions with obstacles [5]. Among many possible approaches to this problem, an interesting one was presented by Dobrowolski [3]. He considered the problem of motion planning in $SO(3)$ space (i.e. rotations about the origin in the Euclidean 3-space) and presented algorithms for constructing the *cell graph*, i.e. a graph representation of the configuration space.

The ideas of Dobrowolski can be generalized to work for any space, not only for $SO(3)$, and thus can be an important building block of real-life applications. However, the running times of the algorithms were not acceptable for bigger input data (although they were still significantly faster than a naive approach). In this paper we develop and extend the ideas of Dobrowolski [3]. As the main contribution of this paper, we present and analyze parallel implementation of the algorithms constructing the cell graphs, working for GPU processors. We chose this computational model, since we GPU processors perceive as relatively cheap

---

The research was funded by National Science Center, decision DEC-2012/07/D/ST6/02483.

R. Wyrzykowski et al. (Eds.): PPAM 2015, Part I, LNCS 9573, pp. 559–569, 2016.
DOI: 10.1007/978-3-319-32149-3_52

coprocessors for server, desktop, mobile and energy-efficient machines dedicated to highly parallel tasks performing computationally intensive procedures.

One of our parallel algorithms uses some variation of a *radix search tree*. Our implementation allows us for an efficient execution of dictionary operations (insert, search) on a set of long vectors over an alphabet of a constant size. As a dictionary is a fundamental data type, widely used in many applications, we believe that our solution may be interesting and important on its own.

1.1 Definitions and Basic Properties

Let $n, \ell \in \mathbb{N}$. By $[n]$ we denote the set $\{0, 1, \ldots, n-1\}$. By $[n]^\ell$ we denote the set of all vectors of length ℓ over the alphabet $[n]$. The i-th coordinate (for binary vectors called the *i-th bit*) of a vector x is denoted by $x(i)$. The coordinates are indexed in zero-based convention, i.e. $x = x(0), x(1), \ldots, x(\ell-1)$. For i, j such that $0 \leq i < j < \ell$, by $x(i; j)$ we denote the segment $x(i), x(i+1), \ldots, x(j-1)$.

The *Hamming distance* of two binary vectors $x, y \in [2]^\ell$, denoted by $\mathrm{dist}(x, y)$, is the number of positions i, such that $x(i) \neq y(i)$. Observe that dist is a metric function, so it satisfies the triangle inequality: $\mathrm{dist}(x, y) + \mathrm{dist}(y, z) \geq \mathrm{dist}(x, z)$. From this it follows that: $(\star) \quad \mathrm{dist}(x, y) \geq |\mathrm{dist}(x, z) - \mathrm{dist}(y, z)|$.

2 Problem of Cell Graph Construction

In this section we describe the notion of the cell graph in motion planning. Although we use a very simple example, similar methods can be (and actually are) used in much more complicated settings (see for example [2,5]).

2.1 Cells and Vectors

Let us consider a system of inequalities $c_0, c_1, \ldots, c_{\ell-1}$ (constraints), describing the boundaries (see Fig. 1), that partition the space into a number of pairwise disjoint regions, called *cells*. We say that two cells are neighboring (a robot can move directly from one to another) if their boundaries share some arc (one point is not enough). Our task is to unify the cells and say which of them are neighbors. Then an obstacle-free route for a robot can be determined using a graph algorithm (see for example [1]). As the scenes (i.e. the space with the arrangement of obstacles) in real-life applications tend to be very complicated, an effective construction of the cell graph is a crucial part of this approach.

We shall represent each point $P \in \mathbb{R}^2$ by an ℓ-element binary vector v_P. The i-th bit of v_P is 1 iff P satisfies the inequality c_i (see Fig. 1). Observe that such a representation identifies all points belonging to the same cell, so it can be seen as a representation of this cell. The neighboring cells are exactly the cells whose representants differ on exactly one position, which means that their Hamming distance is 1. There are several known approaches to parallel computation of the Hamming distance in various settings [4,8,11]. However, none of them benefits for the specific properties of our task.

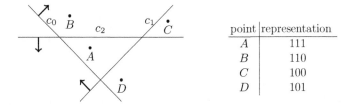

point	representation
A	111
B	110
C	100
D	101

Fig. 1. The arrangement of lines partitions the space into cells. Our object is given by three linear inequalities.

Observe that the theoretical bound for the number of different cells in the scene with ℓ constraints is 2^ℓ. However, due to the arrangement of constraints, the actual number of non-empty cells is usually much smaller. For example, there is no point represented by a vector 000 in Fig. 1. This is the reason why detecting all cells for the given scene is a hard task. However, we are usually satisfied by approximate solutions generated by a randomized procedure called a *sample generator*. This procedure generates (and possibly accumulates) some sequence of random points. The simplest one is the so-called *Shoemake's method* [12], in which the random points are generated uniformly.

2.2   Constructing the Cell Graph

Let $X = \{x_0, x_1, \ldots, x_{n-1}\}$ be a set of n binary vectors, each of length ℓ. These vectors represent the cells and are generated by a sample generator. The *cell graph* G_X for X is the graph with vertex set X, in which edges are all pairs of vectors x_i, x_j (for $0 \leq i < j < n$), such that $\mathrm{dist}(x_i, x_j) = 1$.

In this paper we are interested in solving the problem of constructing the cell graph, i.e. **finding the edge set of G_X** for the **given set $X \subseteq [2]^\ell$**. Since the degree of each vertex is at most ℓ, the graph G_X has at most $\frac{n \cdot \ell}{2}$ edges. Moreover, the total number of bits in X is $n \cdot \ell$. Thus the lower bound for the work complexity of any algorithm constructing the cell graph is $\Omega(n \cdot \ell)$.

It is also worth mentioning that many sample generators used in applications are not perfect and may output some vector more than once (so X is in fact a multiset). Observe that in this case the number of pairs i, j such that $\mathrm{dist}(x_i, x_j) = 1$ can increase to $\Theta(n^2)$. A desired property of any algorithm constructing the cell graph is to be able to deal with such a situation and output only unique pairs of neighboring vectors, without increasing the complexity.

When comparing the complexities of the algorithms we will assume that $\ell \ll n$. This is justified, since in most practical applications n is larger than ℓ by a few orders of magnitude.

3   Parallel Algorithms

Parallel algorithms presented in this section are inspired by sequential algorithms for the problem presented by Dobrowolski [3].

3.1 Heuristic Algorithm

In the naive approach we compare all pairs of vectors in total time $O(n^2 \cdot \ell)$ [3]. We improve this method by choosing a small constant $h \in \mathbb{N}$ and computing the distance between each of vectors $x_0, x_1, \ldots, x_{h-1}$ and each vector in X. Then, for each pair of vectors we compare them with each other to determine if their Hamming distance is 1. We are able to discard some pairs faster, using formula $(\star)$ and previously computed distances to vectors $x_0, x_1, \ldots, x_{h-1}$. Observe that for $h = 0$ this algorithm reduces to the naive one.

Each pair of vectors x_i, x_j is considered by one thread (this approach allows us to utilize a huge number of threads available in modern GPUs). First we compute $\mathrm{dist}(x_i, x_j)$ for $i \in [h]$ and $j \in [n]$ and store the values in a shared memory. Algorithm 1 shows the pseudo-code of this step.

Algorithm 1. ComputeDist

Input: $X = \{x_0, x_1, \ldots, x_{n-1}\} \subseteq [2]^{\ell}, h \in \mathbb{N}$
1 initialize $dist(x_i, x_j) = 0$ for all $i \in [h], j \in [n]$
2 **for** $i \in [h]$ *and* $j \in \{i+1, \ldots, n-1\}$ **do in parallel (threads)**
3   initialize $dist(x_i, x_j) = 0$
4   **for** $k \leftarrow 0$ **to** $\ell - 1$ **do**
5   **if** $x_i(k) \neq x_j(k)$ **then** $dist(x_i, x_j) \leftarrow dist(x_i, x_j) + 1$

The remaining part, shown in Algorithm 2, is analogous. The difference is that we may stop if we discover that it is greater than 1.

Algorithm 2. ParallelHeuristic

Input: $X = \{x_1, x_2, \ldots, x_n\} \subseteq [2]^{\ell}, h \in \mathbb{N}$
1 $dist \leftarrow ComputeDist(X, h)$
2 $results \leftarrow$ vector of w zeros
3 **for** $h \leq i \leq n - 1$ *and* $i < j \leq n - 1$ **do in parallel (threads)**
4   **if** $|\mathrm{dist}(x_i, x_d) - \mathrm{dist}(x_j, x_d)| \leq 1$ *for all* $d \in [h]$ **then**
5          $count \leftarrow 0$
6   **for** $k \leftarrow 0$ **to** $\ell - 1$ **do**
7   **if** $x_i(k) \neq x_j(k)$ **then** $count \leftarrow count + 1$
8   **if** $count \geq 2$ **then Break**
9   **if** $count = 1$ **then output** (x_i, x_j)

Now let us analyze the work complexity of the algorithm. Line 1 requires $O(n \cdot h \cdot \ell)$ operations. The number of iterations of the main loop (lines 3–15) is $O(n^2)$. The execution of each iteration requires $O(h + \ell)$ operations.

Thus the worst-case work complexity of this algorithm is $\Theta(n \cdot h \cdot \ell + n^2(h + \ell)) = \Theta(n^2 \cdot \ell)$, since $\ell \ll n$ and h is a constant. The space complexity is $\Theta(h \cdot n) = \Theta(n)$, as we need to store the values of $\mathrm{dist}(x_i, x_j)$.

Observe that the choice of h strongly affects the constants in the bounds for the complexity. However, the experiments show that even if h is small, the effect on execution time may be significant.

3.2   Tree-Based Algorithm

The main drawback of the previous approach is that it is not aware of the structure of the constructed cell graph. Thus Dobrowolski [3] presented an optimized algorithm, based on a different approach. This algorithm first constructs an auxiliary binary tree, storing all vectors in X. Using this tree we can determine if the particular vector x is in X in time $O(\ell)$.

In the parallel version of this algorithm, to improve memory accesses (see Sect. 4) we use a 2^r-ary tree T for $r \geq 1$. Let r be fixed and suppose for simplicity that r divides ℓ (otherwise the last segment of each vector is considered in a slightly different way). For $x \in X$, let $\widetilde{x}$ denote the vector in $[2^r]^{\ell/r}$ such that for every $i \in [\ell/r]$ the sequence $x(i \cdot \frac{\ell}{r}; (i+1) \cdot \frac{\ell}{r} - 1)$ is the binary encoding of $\widetilde{x}(i)$. By $\widetilde{X}$ we denote the set $\{\widetilde{x} : x \in X\}$. We can see T as the representation of the sequences in $\widetilde{X}$. The tree T has ℓ/r levels. Each level of T corresponds to i-th coordinate of $\widetilde{x}$. Each node contains 2^r pointers to nodes of the next level, each corresponding to a different element of $[2^r]$ (see Fig. 2 for an example). If a particular child does not exist, then there are no vectors with the particular prefix. If C is a node of T and C' is its child node, corresponding to the value $v \in [2^r]$, then we say that C' is a v-child of C.

The first step of our algorithm is sorting the vectors in X. As these vectors are binary, the sorting can clearly be done in $O(n \cdot \ell)$ time, using the *radix sort* algorithm. During this step we also remove all duplicates in X.

Then we proceed to constructing the search tree T. Each level of T is constructed in parallel, with synchronization of threads after finishing each level. For each node C and $v \in [2^r]$ we introduce the set $vectors(C, v)$. Let C be a node on level h. The set $vectors(C, v)$ consists of vectors $\widetilde{x} \in \widetilde{X}$, such that: (i) $\widetilde{x}(0; h-1)$ is represented by C in T (with just a little abuse of notation we assume that for $h = 0$ every vector satisfies this condition), and (ii) $\widetilde{x}(h) = v$. This means that

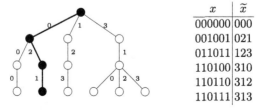

x	$\widetilde{x}$
000000	000
001001	021
011011	123
110100	310
110110	312
110111	313

Fig. 2. A search tree for $r = 2$. The search path for $\widetilde{x} = 021$ is marked.

the vectors from $vectors(C, v)$ are exactly the ones, whose search path begins with the path from the root to the v-th child of C. Observe that since the set X (and thus $\widetilde{X}$) is sorted, each set $vectors(C, v)$ can be represented by just two indices – of the first and of the last vector from this set. The Algorithm 3 shows the pseudo-code for this step. Observe that the work complexity of Algorithm 3 is $O(n \cdot \ell)$ (recall that r is a constant).

Algorithm 3. ConstructTree

Input: $\widetilde{X} = \{\widetilde{x}_0, \widetilde{x}_1, \ldots, \widetilde{x}_{n-1}\} \subseteq [2^r]^{\ell/r}$

1 create the root node (on level 0)
2 **foreach** $\widetilde{x} \in \widetilde{X}$ **do** add $\widetilde{x}$ to the set $vectors(root, \widetilde{x}(0))$
3 **for** $h \leftarrow 1$ **to** $\ell/r - 1$ **do**
4   **for** $node\ C$ $in\ level\ h - 1\ and\ v \in [2^r]$ **do in parallel (threads)**
5   **if** $vectors(C, v) \neq \emptyset$ **then**
6   create node C', being the v-child of C
7   **foreach** $\widetilde{x} \in vectors(C, v)$ **do** add $\widetilde{x}$ to the set $vectors(C', \widetilde{x}(h))$

After constructing the search tree, we can proceed to the main step– identifying neighbors. For every vector in $x \in X$ and every possible neighbor x' of x we check if $x' \in X$ (in fact we check is $\widetilde{x'} \in \widetilde{X}$). Again, we do it in parallel. For every vector $\widetilde{x}$, each bit of $\widetilde{x}$ is considered by a separate thread. Observe that each bit of $\widetilde{x}$ corresponds to a single potential neighbor of $\widetilde{x}$. Thus each thread checks if this potential neighbor exists. The Algorithm 4 shows the pseudo-code of this procedure.

Algorithm 4. ParallelTreeBased

Input: $X = \{x_0, x_1, \ldots, x_{n-1}\} \subseteq [2]^{\ell}$

1 sort X
2 $T \leftarrow ConstructTree(\widetilde{X})$
3 **for** $x \in X$ **do in parallel (blocks)**
4   **for** $k \in [\ell]$ **do in parallel (threads)**
5          $x' \leftarrow x$ with the k-th bit negated
6          $C \leftarrow$ the root of T
7   **for** $h \leftarrow 0$ **to** $\ell/r - 1$ **do**
8              $v \leftarrow \widetilde{x'}(h)$
9   **if** $there\ is\ no\ v\text{-}child\ of\ C$ **then Exit thread**
10             $C \leftarrow v$-child of C
11   **output** (x, x')

Observe that we do not have to keep the vector x' explicitly. At each step we need a segment of x' (corresponding to the current level of T), which can be found in constant time. The work complexity of the searching procedure is $O(n \cdot \ell^2)$ and so is the complexity of the whole algorithm. The space complexity of the algorithm is determined by the size of the search tree, which is $\Theta(n \cdot \ell)$.

Recall that during the sorting step we remove all duplicates. Thus this algorithm is robust in the sense that it does not assume that all input vectors are distinct and the same complexity bound holds even if X is a multiset.

The tree-based algorithm can also be adapted to implement a kind of dictionary of vectors over a limited alphabet. Using this implementation we are able to (a) construct a tree representing a given set X of n vectors of length ℓ (we use Algorithm 3) and (b) find all vectors from the given set Y, consisting of m vectors of length ℓ, which appear in our set X. For searching, we have to adapt Algorithm 4 slightly. To be specific, instead of two nested loops (starting in lines 3 and 4, respectively), we need one loop in which every thread considers a single vector from the set Y.

4   GPU Implementation Issues

In this section we discuss the implementation details of parallel algorithms described in the previous section. We shall omit an introduction to the computational model of GPGPU. The readers, who are not familiar with GPGPU programming, should refer to CUDA C literature [7,10]. There are several limitations of GPU devices which are important from the algorithmic point of view. We are interested in algorithms which are able to: (1) use coalesced memory access, (2) maximize multiprocessor occupancy, (3) hide memory latency.

4.1   Heuristic Algorithm

In order to achieve high processor occupancy we need to define the number of blocks which is at least three or four times higher than the number of streaming processors. Memory latency may be hidden if there is sufficient number of warps assigned to the same processor and memory accessing is interspersed with computations.

Algorithms 1 and 2 contain two nested loops iterating over an array of results (it is an upper-triangular square array with zeros on the main diagonal). Using blocks as the parallel computation units in the outer loop and threads in the inner one gives us a fair number of blocks and threads achieving good occupancy and hiding memory latency. Each thread reads parts of two vectors into registers and then performs comparison. Thus a significant number of computational instructions are executed between reads and writes.

Coalesced memory access is automatic if each vector is stored as a continuous array of bytes. Fragmented results of the comparison of two vectors in Algorithm 2 (one array for each block) may be stored in a shared memory and added up in parallel by threads of this block using classical parallel reduction pattern.

4.2   Tree-Based Algorithm

Tree construction in Algorithm 3 requires a synchronization after each level. Such a global synchronization can only be achieved by finishing a kernel and launching a new one. The number of threads in each kernel execution is equal to number of tree nodes in the previous level times 2^r (in our experiments we used $r = 8$, so $2^r = 256$). All threads run independently and their division into blocks may be set arbitrarily in order to achieve best processor occupancy.

Algorithm 4 again contains two nested loops. The outer one is executed for each input vector and the inner one iterates over its coordinates. Similarly as in Algorithm 2, assigning the outer loop to blocks and the inner one to threads gives good parallelism properties. Each thread performs tree searching and reads in random memory locations. Coalesced memory reads are thus not possible. However, threads may still benefit from the global memory cache since up to r threads may read the same byte from the memory performing independent searches.

5   Experimental Results and Discussion

In order to evaluate our parallel algorithms we utilized the sample generator developed by Dobrowolski [3] and some real-life input data. The number ℓ of constraints was 96. The experiment was performed on a professional computation server (Intel Xeon E5-2620 2 GHz, 15 MB cache, 6 cores, 32 GB RAM) equipped with NVIDIA Tesla K40 computational unit (2668 cores, 12 GB memory).

In all parallel algorithms there are parameters which may influence their performance, i.e. the number of blocks and threads and the value of h in the heuristic algorithm. According to NVIDIA white papers, due to the complication of the parallel processing model, the only way to find optimal values of these parameters for different devices and environments is to perform experiments. In the case of the value of h (for the heuristic algorithm) our tests indicated that the optimal value for the CPU is 3, while for K40 it is 5. The rest of the experiments for heuristic algorithm were performed with these settings. An analysis of the size of the kernel grid for the parallel tree-based algorithm (divided into three stages: sorting, building and searching) is presented in Fig. 3. The total processing time was minimal for 4096 blocks of 32 threads.

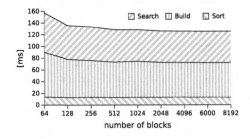

Fig. 3. The analysis of different block sizes for the tree-based algorithm.

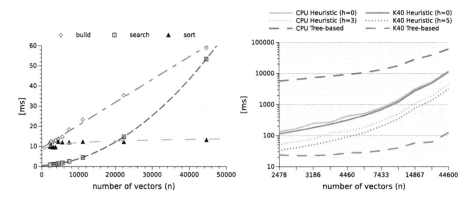

Fig. 4. The processing time of three stages of the tree-based algorithm (left). The processing time for sequential and parallel algorithms (right, note the logarithmic scale).

Figure 4 (left) presents the evaluation of the processing time in three stages of the tree-based algorithm: sorting, tree building, and neighbor searching. We can clearly see that searching time is growing faster than building, which is related to the n^2 factor in the complexity bound. However, for 44.000 vectors it is still smaller than the building part, due to high constants in the latter. Experiments show that the sorting stage does not influence the total processing time by more than 30 % in case of bigger input sets.

On the right of the Fig. 4 a comparison of several solutions is presented. Let us first analyze the heuristic methods. The sequential solution for the optimal value of h (equal to 3) is significantly faster than the solution with h set to 0, which corresponds to the naive solution. Similarly, parallel version with h set to the optimal value (5) is much faster than the naive one. Both parallel and sequential procedures show similar growth of processing time for increasing size of the input data set. This shows that the algorithm scales well. The best performance is achieved by the parallel tree-based procedure. Sequential version behaves similarly but significantly (more than two orders of magnitude) slower.

5.1 Dictionary Implemented as a Search Tree

In this section we analyze the performance of searching the dictionary implemented as a radix tree (see Sect. 3.2). Recall that we have constructed a tree representing the set X of n vectors of length ℓ and we want to find all vectors from the set Y (which has m elements), appearing in X. We consider two scenarios – the *uniform* one, in which both sets are chosen at random, with uniform distribution, and the *degenerated* one, in which most sets from X have common subsegments (i.e. the search tree has many nodes with just one child).

The results (see Fig. 5) we obtained show that our solution scales well and is very effective even for large sets Y. This makes this approach useful in real-time applications, like network control systems. Moreover, the size and the structure of

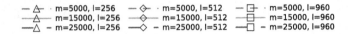

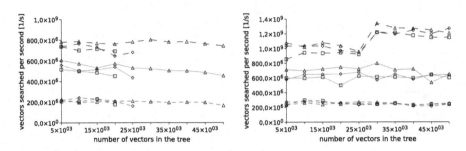

Fig. 5. The performance of searching the dictionary (uniform scenario on the left, degenerated scenario on the right). Missing entries indicate that the experiment failed due to the lack of memory. We assume that the vectors from Y are already copied to the GPU device.

the set X (uniform or degenerated scenario) and the value of ℓ do not influence the performance significantly. On the other hand, the structure of X (and of course the size of it) influences the number of nodes in the search tree and thus the amount of memory necessary to store the dictionary.

Observe that we get best performance for biggest values of m. This means that the memory is the actual limitation of our method. We expect to get significantly better efficiency, if we had a GPU device with bigger memory.

6   Conclusion and Further Research Directions

We presented two important parallel algorithms for construction of the cell graph. Our experiments show that the parallel solution based on a search tree is much faster than the heuristic approach (while the sequential tree-based algorithm is slower than its sequential heuristic counterpart). SIMD-like processors were so far mostly identified with vector-based processing. The excellent results we obtained show that creating and searching a tree structure can also be efficient. This was possible due to proper tree node construction and memory caching available in modern devices.

As a modification of the tree-based algorithm, Dobrowolski presented an algorithm, constructing the cell graph in $O(n \cdot \ell)$ time. Using an auxiliary data structure, the searching step can be performed in $O(n \cdot \ell)$ time. Unfortunately, the experiments on the real data (see Sect. 5) show that constructing the tree takes the majority of the execution time. Moreover, as this improved searching procedure requires lots of synchronization, it may actually lead to worse execution time. A very natural research direction is to design a scalable parallel algorithm, constructing the cell graph for a given set of vectors in time $O(n \cdot \ell)$.

As mentioned before, cell graphs are used in motion planning. A path in the cell graph corresponding to a given scene is equivalent to some approximate solution of the motion planning problem. Another interesting direction of research would be to extend our approach and design a whole solution to the motion planning problem, using the GP GPU setting. There are several general approaches to traversing large graphs using GP GPU (see [6,9]). It would be interesting to design an algorithm finding shortest paths in the cell graph, taking into consideration its structure.

Acknowledgement. The authors are sincerely grateful to Joanna Porter-Sobieraj and Przemysław Dobrowolski for introducing us to the problem and useful advice on possible applications in motion planning. We acknowledge the support of National Science Center, decision DEC-2012/07/D/ST6/02483.

References

1. Chen, P.C., Hwang, Y.K.: SANDROS: a dynamic graph search algorithm for motion planning. ICRA **14**, 390 (1998)
2. Canny, J.: A new algebraic method for robot motion planning and real geometry. In: Proceedings of SFCS 1987, pp. 39–48 (1987)
3. Dobrowolski, P.: An efficient method of constructing cell graph in arrangement of quadrics in 3-dimensional space of rotations. In: Proceedings of SII, pp. 671–676. IEEE (2013)
4. Grabowski, S., Fredriksson, K.: Bit-parallel string matching under hamming distance in worst case time. Inf. Process. Lett. **105**, 182–187 (2008)
5. Hwang, Y.K., Ahuja, N.: Gross motion planning - a survey. ACM Comput. Surv. **24**, 219–291 (1992)
6. Kaczmarski, K., Przymus, P., Rzążewski, P.: Improving high-performance GPU graph traversal with compression. In: Bassiliades, N., Ivanovic, M., Kon-Popovska, M., Manolopoulos, Y., Palpanas, T., Trajcevski, G., Vakali, A. (eds.) New Trends in Database and Information Systems II. AISC, vol. 312, pp. 201–214. Springer, Heidelberg (2015)
7. Kirk, D.B., Wen-mei, W.H.: Programming Massively Parallel Processors: A Hands-on Approach. Newnes, Boston (2012)
8. Liu, Y., Guo, L., Li, J., Ren, M., Li, K.: Parallel algorithms for approximate string matching with k mismatches on CUDA. In: IPDPS 2012, pp. 2414–2422 (2012)
9. Merrill, D., Garland, M., Grimshaw, A.S.: Scalable GPU graph traversal. In: PPOPP, pp. 117–128. ACM (2012)
10. NVIDIA Corporation. NVIDIA CUDA C Best Practices Guide 6.5 (2014). docs.nvidia.com/cuda/cuda-c-best-practices-guide
11. Pan, J., Lauterbach, C., Manocha, D.: Efficient nearest-neighbor computation for GPU-based motion planning. In: IEEE/RSJ International Conference on Intelligent Robots and Systems, pp. 2243–2248 (2010)
12. Shoemake, K.: Uniform Random Rotations. In: Kirk, D. (ed.) Graphics Gems III, pp. 124–132. Academic Press Professional Inc., San Diego (1992)

Benchmarking the Cost of Thread Divergence in CUDA

Piotr Bialas and Adam Strzelecki[✉]

Faculty of Physics, Astronomy and Computer Science, Jagiellonian University,
ul. Łojasiewicza 11, 30-348 Kraków, Poland
adam.strzelecki@uj.edu.pl

Abstract. All modern processors include a set of vector instructions. While this gives a tremendous boost to the performance, it requires a vectorized code that can take advantage of such instructions. As an ideal vectorization is hard to achieve in practice, one has to decide when different instructions may be applied to different elements of the vector operand. This is especially important in implicit vectorization as in NVIDIA CUDA Single Instruction Multiple Threads (SIMT) model, where the vectorization details are hidden from the programmer. In order to assess the costs incurred by incompletely vectorized code, we have developed a micro-benchmark that measures the characteristics of the CUDA *thread divergence* model on different architectures focusing on the loops performance.

Keywords: GPU · CUDA · Multithreading · SIMT · SIMD · Thread divergence

1   Introduction

Most of the current processors derive their performance from some form of SIMD instructions. This holds true for *Intel* i-Series and *Xeon* processor (8 lanes wide AVX instruction), *Intel Xeon Phi* accelerator (16 lanes wide AVX instructions), *AMD Graphics Cores Next* (64 lanes wide wavefronts) and *NVIDIA CUDA* (32 lanes wide warps). By the very nature of those instructions all the operations on a vector are performed in parallel, at least conceptually. That provides a constraint on the class of algorithms that can be efficiently implemented on such architectures. If we want to perform different operations on the different components of the vector operands we are essentially forced to issue different instructions masking the unused components in each of them which of course carries a performance penalty. The aim of this paper is to asses those cost. We will concentrate on the NVIDIA CUDA as that is the architecture that we use in our current work [2].

The CUDA and OpenCL programming models make *thread divergence* very easy to achieve as they rely on *implicit* vectorization. They model the computing environment as a great number of threads executing a single program called

R. Wyrzykowski et al. (Eds.): PPAM 2015, Part I, LNCS 9573, pp. 570–579, 2016.
DOI: 10.1007/978-3-319-32149-3_53

kernel [8], and to the programmer this may look as an multithreaded parallel execution. In fact the CUDA Programming Guide states: "For the purposes of correctness, the programmer can essentially ignore the SIMT behavior; however, substantial performance improvements can be realized by taking care that the code seldom requires threads in a warp to diverge."

On NVIDIA architectures threads are grouped warps of 32 threads that must all execute same instructions in essentially SIMD (vector) fashion. This entails that any branching instruction with condition that does not give the same result across the whole warp leads to thread divergence - while some threads take one branch other do nothing and then the other branch is taken with roles reversed. This is a picture given in the NVIDIA programming guide, which warns against thread divergence but otherwise is missing any details. Only in the Tesla architecture white-paper it is mentioned that this is achieved using a *branch synchronization stack* [8].

As the strict avoidance of thread divergence would severely limit the algorithms that could be implemented using CUDA, it would be interesting to check what are the real costs. The necessity of following two branches is one obvious performance obstacle. We will be however interested in the overhead associated with the *re-convergence* mechanism itself as manifested in loops.

Loops, which are probably the most used control statement in programming, are an interesting example. While it is easy to picture the two way branching model as in a if else statement, it is much harder to follow the execution of loops, especially nested. In this contribution we will benchmark the performance of single and double loops with bounds that are different across the threads. While such model is discouraged in CUDA programming it can nevertheless appear naturally while porting an existing multi-core algorithm to GPU. To our best knowledge such measurements were not published up to this date. We have only found some data in reference [10] but not the explicit timings of the synchronization stack operations. We will analyze our results in view of the best information on the CUDA thread divergence model that we have found.

This paper is organized as follows: In the next section we present the setup we used for benchmarking the execution of single and double loop kernel. Then we present the CUDA stack based re-convergence mechanism and in the last section we use it to analyze our results.

2 Timings

2.1 Single Loop

Our first test consisted of running single loop using the kernel from Listing 1.1. Each thread of a warp (we used only one warp) runs through the same loop but with upper limit, denoted by M, set individually for each thread. We measured the number of cycles taken by the loop using the CUDA clock64() function, repeating our loop N_UNROLL times if needed. We have also added a possibility to run same loop N_PREHEAT times before making the measurements. We have repeated each measurement 256 times and used those samples to estimate the

Table 1. Loop limits used in the single loop measurements.

n	tid																															
	0	1	2	3	4	5	6	7	8	9	10	11	12	13	14	15	16	17	18	19	20	21	22	23	24	25	26	27	28	29	30	31
	M																															
0	32	32	32	32	32	32	32	32	32	32	32	32	32	32	32	32	32	32	32	32	32	32	32	32	32	32	32	32	32	32	32	32
1	32	32	32	32	32	32	32	32	32	32	32	32	32	32	32	32	32	32	32	32	32	32	32	32	32	32	32	32	32	32	32	31
2	32	32	32	32	32	32	32	32	32	32	32	32	32	32	32	32	32	32	32	32	32	32	32	32	32	32	32	32	32	32	31	30
⋮															⋮																	
28	32	32	32	32	31	30	29	28	27	26	25	24	23	22	21	20	19	18	17	16	15	14	13	12	11	10	9	8	7	6	5	4
29	32	32	32	31	30	29	28	27	26	25	24	23	22	21	20	19	18	17	16	15	14	13	12	11	10	9	8	7	6	5	4	3
30	32	32	31	30	29	28	27	26	25	24	23	22	21	20	19	18	17	16	15	14	13	12	11	10	9	8	7	6	5	4	3	2
31	32	31	30	29	28	27	26	25	24	23	22	21	20	19	18	17	16	15	14	13	12	11	10	9	8	7	6	5	4	3	2	1

error. We have found out that when using N_PREHEAT=1 we had essentially zero errors even with N_UNROLL=1. When no "preheating" was used the results were more erratic, with difference of few cycles between the samples.

We start our measurements with all loops having same upper limit of $M = 32$ (no divergence). Next we decrease the upper limit for one of the threads and continue until all threads in warp have different upper bounds (see Table 1).

The results of the measurements as a function of n are presented in the Fig. 1. We have tested three CUDA architectures (see Table 2). We have found out that the number of cycles depended only on the architecture or Compute Capability of the card, not on the particular device. We have used CUDA 7.0 environment for all our test. We have compiled our benchmarks using the -arch=sm_20 -Xptxas -O3 flags, as sm_20 architecture did not change the results, but produced no undefined instructions in the cuobjdump disassembly listing, contrary to sm_30.

There are two interesting things on this plot to take notice of, apart from the fact that GPU are getting faster with each new architecture. Firstly although the longest loop has always same upper bound ($M = 32$) the time increases linearly with each new divergent thread. This is a clear signal of an additional cost associated with the divergence, as in fact there is altogether *less* work done by the threads. Secondly this cost of thread divergence is proportional to the number of divergent threads, up to $n = 15$, then it exhibits several "jumps" every four steps, at least for the two newest architectures (*Kepler*, *Maxwell*), visible as a deviation from the fit lines on Figs. 1 and 2. The behavior on the *Fermi* architecture is slightly different, but we will be not concerned with this architecture in this contribution, and provide the plots only for comparison.

Table 2. Devices used in test.

Architecture	Device
Fermi	GTX 480, GT 610
Kepler	GT 650M, GT 755M, GTX 770
Maxwell	GTX 850M

```
1  #define EXPR_INNER 1.3333f
2  #define EXPR_OUTER 2.3333f
3
4  __global__ void single_loop(
     int* limits, float* out,
     llong* timer) {
5
6    int tid = blockDim.x *
       blockIdx.x + threadIdx.
       x;
7    int M = limits[threadIdx.x
       ];
8    float sum = out[tid];
9
10 #pragma unroll
11   for (int k = 0; k <
       N_PREHEAT; k++)
12     for (int i = 0; i < M; i
         ++) {
13       sum += EXPR_INNER;
14     }
15
16   __syncthreads();
17   llong start = clock64();
18
19 #pragma unroll
20   for (int k = 0; k <
       N_UNROLL; ++k)
21     for (int i = 0; i < M; i
         ++) {
22       sum += EXPR_INNER;
23     }
24
25   llong stop = clock64();
26   __syncthreads();
27
28   out[tid] = sum;
29   timer[2 * tid] = start;
30   timer[2 * tid + 1] = stop;
31 }
```

Listing 1.1. Single loop kernel.

```
1  #define EXPR_INNER 1.3333f
2  #define EXPR_OUTER 2.3333f
3
4  __global__ void double_loops
     (int* limits, float* out
     , llong* timer) {
5
6    int tid = blockDim.x *
       blockIdx.x + threadIdx.
       x;
7    int M = limits[2 *
       threadIdx.x];
8    int N = limits[2 *
       threadIdx.x + 1];
9    float sum = out [0];
10
11 #pragma unroll
12   for (int k = 0; k <
       N_PREHEAT; k++)
13     for (int i = 0; i < M; i
         ++) {
14       for (int j = 0; j < N; j
           ++) {
15         sum += EXPR_INNER;
16       }
17       sum += EXPR_OUTER;
18     }
19
20   __syncthreads();
21   llong start = clock64();
22
23 #pragma unroll
24   for (int k = 0; k <
       N_UNROLL; ++k)
25     for (int i = 0; i < M; i
         ++) {
26       for (int j = 0; j < N; j
           ++) {
27         sum += EXPR_INNER;
28       }
29       sum += EXPR_OUTER;
30     }
31
32   llong stop = clock64();
33   __syncthreads();
34   out[tid] = sum;
35   timer[2 * tid] = start;
36   timer[2 * tid + 1] = stop;
37 }
```

Listing 1.2. Double loop kernel.

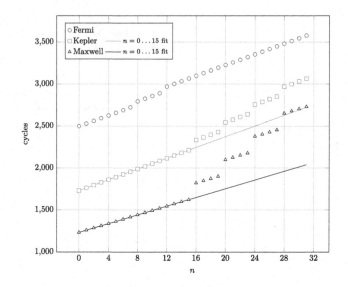

Fig. 1. Single loop timings corresponding to kernel in Listing 1.1.

2.2   Double Loop

On our second test we used a kernel with nested loops, again the loops upper bounds were set individually for each thread of the warp (see Listing 1.2). We used same patterns as for the single loop (Table 1) setting both loops bounds to the same value ($M = N$). The timings are presented in the Fig. 2. We can observe a similar pattern as in the single loop case. The number of cycles at first increases smoothly with the number of divergent threads, and then exhibits "jumps" every four steps followed by slight descents.

3   CUDA Divergence Model

To understand the observed behavior it is necessary to find out the details of the CUDA thread divergence/re-convergence model. This is not explicitly stated by NVIDIA apart from brief mention of the branch synchronization stack in [8]. A more detailed description can be found in references [1,3,4,9] and [7]. There it was established that the CUDA implementation follows the approach described in US patents [5,6]. In here we briefly describe this algorithm. Our description is based on reference [6]. We choose reference [6] over [5] because of the SSY instruction specification. While in [5] it is described as taking no arguments, cuobjdump listing shows that it expects one argument which matches the description in [6]. We concentrate only on one type of control instruction which is relevant to our example: the predicate branch instruction, ignoring all others.

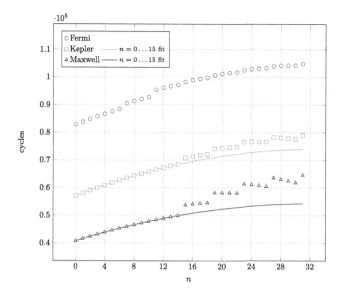

Fig. 2. Double loop timings corresponding to Listing 1.2.

Scalar Multiprocessor (SMX) processor maintains for each warp an *active mask* that indicates which threads in warp are active. When about to execute a potentially diverging instruction (branch) compiler issues one set-synchronization SSY instruction. This instruction causes new *synchronization token* to be pushed on the top of synchronization stack. Such token consist of three parts: mask, id and program counter (pc) and takes 64 bits - 32 of which are taken by the mask. In case of the SSY instruction the mask is set to active mask, id to SYNC and pc is set to the address of the synchronization point. The actual divergence is caused by the *predicated branch instruction* which has form @P0 BRA *label* which translates to the pseudocode in Algorithm 1. The P0 is a 32bit *predicate register* that indicates which threads in the warp should execute (take) the branch.

The instructions are executed according to the pseudocode in the Algorithm 2. The instruction may have a *pop-bit set*, also denoted as *synchronization command*, indicated by the suffix .s in the assembler output. The instruction which is suffixed with .s will be called a *carrier* instruction. When encountered it signals the *stack unwinding*. The token is popped from the stack and used to set the active mask and the program counter. The reference [6] does not specify exactly when carrier instructions are executed, before or after popping the stack, so we have assumed that this is done after unwinding the stack. Actually in both kernels that we have studied this carrier instruction is NOP, but this is not necessarily the case in general.

In summary: SSY and predicated branch instructions (only if some active threads diverge) push token onto the stack and the synchronization command pops it.

```
1: if None active threads take the branch then
2:      pc ← pc+1
3: else
4:      if !(All active threads take the branch) then
5:          token.mask ← active_mask && !P0
6:          token.id ← DIV
7:          token.pc ← pc+1
8:          push(token)
9:      end if
10:     active_mask ← active_mask && P0
11:     pc ← label
12: end if
```

Algorithm 1. Algorithm for executing a predicated branch instruction @P0 BRA *label*.

```
1: Fetch the instruction
2: if Instruction is a SSY label instruction then
3:      token.mask ← active_mask
4:      token.id ← SYNC
5:      token.pc ← label
6:      pc ← pc+1
7: else if Instruction is a predicated branch instruction then
8:      Execute the instruction according to Algorithm 1
9: else
10:     if Is pop-bit set in instruction then
11:         token ← pop()
12:         active_mask ← token.mask
13:         pc ← token.pc
14:         Execute the instruction
15:     else
16:         Execute the instruction
17:         pc ← pc+1
18:     end if
19: end if
20: goto 1
```

Algorithm 2. Algorithm for executing a CUDA instruction.

Using the parameters obtained from the single loop measurements we used our emulator to predict the number of spills and the timings of the double loop kernel. The results are presented in the Figs. 3 and 4. The agreement in timings while not perfect is still very good (76 cycles difference in the worst case which amounts to 0.3%). Looking at the Fig. 5 we see that the difference between the actual and predicted number of cycles follows a very regular pattern and happens only on spill occurrence, suggesting some simple mechanism contributing to the exact number of cycles needed to perform stack spill depending on previous spills history.

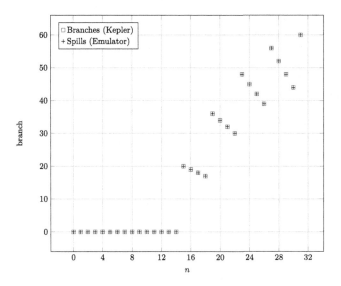

Fig. 3. Number of stack spills in the double loop kernel.

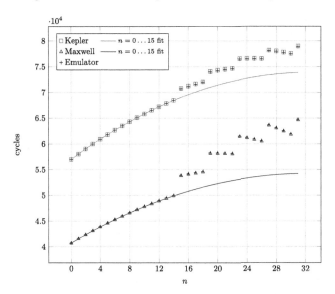

Fig. 4. Comparison of the double loop timings with emulator results.

4 Summary

The Single Thread Multiple Threads (SIMT) programming model introduced by NVIDIA with CUDA technology gives the programmer an illusion of writing a multithreaded application. In reality the threads are executed in groups of 32 threads (*warps*) performing essentially vector operations. The illusion of

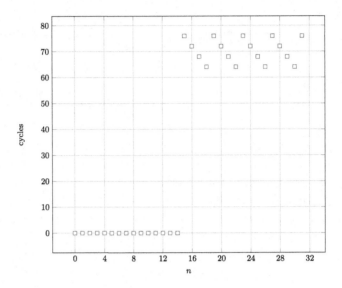

Fig. 5. Difference between the actual and predicted number of cycles for the double loop kernel on Kepler (GTX 770) architecture.

multithreading is maintained by an elaborate mechanism for managing the divergent threads within a warp. This mechanism is completely hidden from the programmer and no description is provided by NVIDIA in their programming guide. In this contribution we have presented a detailed study of two CUDA kernels leading to a large thread divergence. Following other sources, we have assumed that the management of the diverging threads is based on a synchronization stack and follows closely the idea described in reference [6] which is a NVIDIA owned patent. That assumption allowed us to fit the observed execution times with accuracy better then 1 %.

We have estimated the cost of a diverging branch instruction to be *exactly* 32 cycles on *Kepler* architecture provided that the maximum stack length did not exceed 16. After that stack entries had to be spilled into memory which carried an additional penalty of $\sim$ 84 cycles. On *Maxwell* the estimated cost of branch divergence was considerably shorter: only 24 cycles, but the cost of spilling was much higher $\sim$ 176 cycles. More detailed timings are suggesting that all of that cost, in case of no spills, can be attributed to stack unwinding.

We also performed the measurements on *Fermi* architecture, there the pattern seems to be slightly different suggesting that the physical thread synchronization stack length is shorter in this architecture.

We have analyzed only a very specific case: loops with different bounds across the threads, this is however a quite natural scenario. We would like to emphasize that in this case the additional costs reported by us are the costs of the divergence management mechanism alone, excluding additional penalty of incomplete vectorization. Nevertheless they implied an overhead of $\sim$ 20 % in case of 14 diverging threads for the double loop kernel and $\sim$ 25 % for the single loop.

In the more general case there would be additional costs associated with the serialization of the different branches.

Our work was confined to NVIDIA CUDA technology it would be however interesting how this corresponds to shader programming both in DirectX and OpenGL and we are planning to extend our work in this direction in the future.

References

1. Aamodt, T.M., et al.: GPGPU-Sim (2012). http://www.gpgpu-sim.org
2. Białas, P., Kowal, J., Strzelecki, A., et al.: GPU accelerated image reconstruction in a two-strip J-PET tomograph (2015). arXiv:1502.07478
3. Collange, S., Daumas, M., Defour, D., Parello, D.: Comparaison d'algorithmes de branchements pour le simulateur de processeur graphique Barra. In: 13'eme Symposium sur les Architectures Nouvelles de Machines, pp. 1–12 (2009)
4. Collange, S., Daumas, M., Defour, D., Parello, D.: Barra: A parallel functional simulator for GPGPU. In: 2010 IEEE International Symposium on Modeling, Analysis & Simulation of Computer and Telecommunication Systems (MASCOTS), pp. 351–360. IEEE (2010)
5. Coon, B.W., Lindholm, J.E.: System and method for managing divergent threads in a SIMD architecture, April 2008. https://www.google.com/patents/US7353369
6. Coon, B.W., Nickolls, J.R., Nyland, L., Mills, P.C., Lindholm, J.E.: Indirect function call instructions in a synchronous parallel thread processor, November 2012. https://www.google.com/patents/US8312254
7. Habermaier, A., Knapp, A.: On the correctness of the SIMT execution model of GPUs. In: Seidl, H. (ed.) Programming Languages and Systems. LNCS, vol. 7211, pp. 316–335. Springer, Heidelberg (2012)
8. Lindholm, E., Nickolls, J., Oberman, S., Montrym, J.: NVIDIA Tesla: a unified graphics and computing architecture. IEEE Micro **28**(2), 39–55 (2008)
9. Wilt, N.: The CUDA Handbook: A Comprehensive Guide to GPU Programming. Pearson Education, Upper Saddle River (2013)
10. Wong, H., Papadopoulou, M.M., Sadooghi-Alvandi, M., Moshovos, A.: DemystifyingGPU microarchitecture through microbenchmarking. In: 2010 IEEE International Symposium on Performance Analysis of Systems & Software (ISPASS), pp. 235–246. IEEE (2010)

Special Session on Efficient Algorithms for Problems with Matrix and Tensor Decompositions

Fast Algorithm for the Fourth-Order Elliptic Problem Based on Orthogonal Matrix Decomposition

Paolo Di Stolfo[1]($\boxtimes$) and Marian Vajteršic[2,3]

[1] Fachbereich Mathematik, Paris-Lodron-Universität Salzburg, Salzburg, Austria
paolo.distolfo@sbg.ac.at
[2] Fachbereich Computerwissenschaften, Paris-Lodron-Universität Salzburg,
Salzburg, Austria
marian@cosy.sbg.ac.at
[3] Institute of Mathematics, Slovak Academy of Sciences, Bratislava, Slovakia

Abstract. A fast algorithm for solving the first biharmonic boundary problem on a rectangular domain is presented. It is based on splitting the fourth-order problem into a coupled system of two-order problems, whose finite-difference approximations are solved iteratively with linear algebra routines. There, a crucial role plays the orthogonal eigenvalue decomposition of the iteration matrix, which leads to a reduction of the operational count for one iteration to the asymptotically optimal value. This approach is extensively tested from the point of view of the total computational time, number of iterations and the solution error. It is shown, that these values cope with the theoretical assumptions and scale convincingly up to a problem with 100 million grid points.

Keywords: Biharmonic equation · Orthogonal matrix decomposition · Splitting method

1 Introduction

The biharmonic boundary value problem is of importance in several areas of computational practice, e.g. in aviation industry, civil engineering, etc. [8]. An example of a concrete problem is the deflection of a loaded plate. There is a number of numerical approaches developed for solving this problem. Most of them are based on its approximation by finite differences [8], finite elements [5] and most recently, by finite volumes [9]. Our attention will be devoted to the finite-difference approximations of a 2D problem, which lead either to one block five-diagonal linear system or to a split system of two block tri-diagonal systems, both with sparse and regularly structured matrices. The latter one arises by replacing the original problem by a coupled pair of discretized Poisson equations, which need to be solved iteratively. This is the so-called splitting principle (e.g. [1,4,7,8]) and it has advantages to direct solvers in respect to the computational complexity and accuracy, particularly, when large grids are under consideration [3].

© Springer International Publishing Switzerland 2016
R. Wyrzykowski et al. (Eds.): PPAM 2015, Part I, LNCS 9573, pp. 583–593, 2016.
DOI: 10.1007/978-3-319-32149-3_54

For a rectangular domain, covered by an $N \times N$ grid, the splitting approach requires in its original form $O(N^{5/2} \log^2 N)$ operations, when the second-order approximation to the Laplacian is considered and the accuracy of the solution is set to $O(N^{-2})$ [4]. From this count, $O(N^{1/2} \log N)$ is the number of required iterations and $O(N^2 \log N)$ is the computational cost for one iteration. In paper [7], an algorithm was proposed, which reaches the optimal asymptotic complexity $O(N^2)$ for the evaluation of one iteration. This $O(\log N)$ improvement has been achieved by the exploitation of an orthogonal decomposition of the iteration matrix belonging to the splitting process. This gain was not paid off by an asymptotic increase of the number of iterations for the process with the transformed matrix. It remained unchanged, while the increase was only by a constant factor, due to a stronger convergence criterion.

In our previous work [7], this approach was tested on small grids only. A purpose of this work was to test it on modern computing platforms and to check whether the theoretical results will be confirmed also by computational experiments for large values of N. Our focus was particularly at the total computational time, the number of iterations needed and the accuracy of the obtained approximative solution.

The paper begins in Sect. 2 with a formulation of the problem, with its discretization and a description of the splitting approach. Section 3 presents the modified iteration process, where the orthogonal decomposition is exploited efficiently, together with the algorithm for its realization. The gravity point of our contribution brings Sect. 4, where the experiments are summarized and analyzed. Results for solving a problem on grids up to size 10000×10000 are presented. They manifest clear coincidence to theoretical estimations in all aspects under observation. The paper concludes with outlooks for further work.

2   Problem

The first boundary problem for the biharmonic equation (further, the biharmonic problem) on a rectangular region Ω is formulated as follows.

Find a solution $u \in C^4(\Omega) \cap C^1(\overline{\Omega})$ with piecewise continuous second derivatives on the boundary $\partial\Omega$ of Ω such that

$$\begin{aligned}
\Delta^2 u &= f \text{ in } \Omega \\
u &= g_1 \text{ on } \partial\Omega \\
u_n &= g_2 \text{ on } \partial\Omega ,
\end{aligned} \tag{1}$$

where the term u_n denotes the derivative of u in the direction of the outer normal.

The biharmonic Eq. (1) can be replaced by a coupled pair of second-order boundary value problems, introducing an unknown function v and a parameter $\gamma \neq 0$ [6]:

$$\begin{aligned}
\Delta u &= v & &\text{in } \Omega \\
u &= g_1 & &\text{on } \partial\Omega \\
\Delta v &= f & &\text{in } \Omega \\
v &= \Delta u - \gamma(u_n - g_2) & &\text{on } \partial\Omega .
\end{aligned}$$

Iterative schemes that make use of this replacement are called splitting methods. A possible iterative scheme for the above system of equations introduces new unknown functions U and V that incorporate a smoothing process for u and v:

$$
\begin{aligned}
u^{(t+1)} &= g_1 & &\text{on } \partial\Omega \\
\Delta u^{(t+1)} &= V^{(t)} & &\text{in } \Omega \\
U^{(t+1)} &= (1 - \omega_1)U^{(t)} + \omega_1 u^{(t+1)} & &\text{in } \Omega \\
v^{(t+1)} &= \Delta U^{(t+1)} - \gamma\big(\partial_n U^{(t+1)} - g_2\big) & &\text{on } \partial\Omega \\
\Delta v^{(t+1)} &= f & &\text{in } \Omega \\
V^{(t+1)} &= (1 - \omega_2)V^{(t)} + \omega_2 v^{(t+1)} & &\text{in } \Omega.
\end{aligned}
\tag{2}
$$

For this scheme, starting values $U^{(0)}$, $V^{(0)}$, and suitable smoothing parameters ω_1, ω_2 have to be defined. Given suitable smoothing parameters and $\gamma \neq 0$, the scheme converges to the solution of the biharmonic Eq. (1) (see e.g. [6]).

In our further explanations, the rectangular region Ω will be simplified to the unit square $[0, 1]^2$. Using the mesh size $h = 1/(N + 1)$ for some positive integer N, $\Omega \cup \partial\Omega$ is replaced by the set $\Omega_h \cup \partial\Omega_h$ of grid points. The discrete function $U^{(t+1)}$ on Ω_h is given as $U_{i,j}^{(t+1)} = U^{(t+1)}(ih, jh)$ for $1 \leq i, j \leq N$. The values of $U^{(t+1)}$ are arranged as a vector $U^{(t+1)} = \left(U_1^{(t+1)}, \dots, U_N^{(t+1)}\right)^\top$, where $U_j^{(t+1)} = \left(U_{j,1}^{(t+1)}, \dots, U_{j,N}^{(t+1)}\right)^\top$.

Let us consider the discretized version of the functions u, U, v, V on $\Omega_h \cup \partial\Omega_h$ and discrete analogues for Δ, ∂_n in (2) by applying the standard five-point difference approximation for the Laplacian. This yields the discretization error of $O\big(N^{-3/2}\big)$ for sufficiently large N. From these approximations, the following iterative system studied by Ehrlich [4] can be derived:

$$
\begin{aligned}
Gu^{(t+1)} &= -h^2 V^{(t)} + b \\
U^{(t+1)} &= (1 - \omega_1)U^{(t)} + \omega_1 u^{(t+1)} \\
Gv^{(t+1)} &= \tfrac{2}{h^2} M U^{(t+1)} + c \\
V^{(t+1)} &= (1 - \omega_2)V^{(t)} + \omega_2 v^{(t+1)}\ .
\end{aligned}
\tag{3}
$$

Here, G denotes the block-tridiagonal matrix $G = (-I, A, -I) \in \mathbb{R}^{N^2 \times N^2}$, where $A = (-1, 4, -1) \in \mathbb{R}^{N \times N}$ is a tridiagonal matrix. Moreover, the matrix $M = (I + D, D, \dots, D, I + D) \in \mathbb{R}^{N^2 \times N^2}$ is block-diagonal, where $D = \text{diag}(1, 0, \dots, 0, 1) \in \mathbb{R}^{N \times N}$ and $I \in \mathbb{R}^{N \times N}$ is the identity matrix. The vectors b and c are constant and depend only on the boundary conditions and the right-hand side.

3   Algorithm

Instead of iterating the scheme in (3), the equations may be combined into a single three-term recursion formula for $U^{(t+1)}$, eliminating the vectors $u^{(t)}, v^{(t)}$, and $V^{(t)}$ [7]:

$$U^{(t+1)} = (2 - \omega_1 - \omega_2)U^{(t)} \ -2\omega_1\omega_2 G^{-2}\Big(MU^{(t)}\Big) - (1 - \omega_1)(1 - \omega_2)U^{(t-1)}$$
$$+\omega_1\omega_2 d, \tag{4}$$

where $U^{(0)} = O$, $U^{(1)} = \omega_1 G^{-1}b$, and $d = G^{-1}b - h^2 G^{-2}c$.

As the computation of $G^{-2}MU^{(t)}$ in (4) is the most expensive step, it requires special attention and is handled using the matrix decomposition (MD) algorithm which takes advantage of the eigendecomposition of a symmetric matrix [2].

The matrix $A \in \mathbb{R}^{N \times N}$ is real and symmetric and, therefore, can be factored into $A = V^T \Lambda V$, where V is an orthogonal matrix. The diagonal values of $\Lambda = \mathrm{diag}(\lambda_1, \ldots, \lambda_N)$ are the eigenvalues of A, i.e.,

$$\lambda_j = 4 - 2\cos\frac{j\pi}{N+1} \ , \text{ for } j = 1, \ldots, N.$$

The corresponding eigenvectors form the columns of the symmetric matrix $V = (V_{i,j}) \in \mathbb{R}^{N \times N}$, where

$$V_{i,j} = \sqrt{\frac{2}{N+1}}\sin\frac{ij\pi}{N+1} \ , \text{ for } i, j = 1, \ldots, N. \tag{5}$$

In the following, we apply the MD algorithm to compute $y = G^{-2}z$ with $z = MU^{(t)}$ and include the computational complexity of each step:

1. Compute the vectors $\tilde{z}_j = V z_j$, for $j = 1, \ldots, N$. Due to the sparse structure of z, this requires only $O(N^2)$ operations in total.
2. Create the column vectors $\hat{z}_i$, i.e., taking the ith column of $\tilde{z}$.
3. Solve the systems $M_i \hat{y}_i = \hat{z}_i$, for $i = 1, \ldots, N$, where $M_i = (-1, \lambda_j, -1) \in \mathbb{R}^{N \times N}$. Each tridiagonal system may be solved in $O(N)$ operations. Thus, this step takes $O(N^2)$ operations in total.
4. Solve the systems $M_i \hat{\hat{y}}_i = \hat{y}_i$, $i = 1, \ldots, N$ in $O(N^2)$ operations.
5. Create the vectors $\tilde{y}_j$, $j = 1, \ldots, N$ that are the rows of $\tilde{y}$ by using $\hat{y}_i$ as their columns.
6. Compute the solution vectors $y_j = V\tilde{y}_j$, $j = 1, \ldots, N$. Note that y_j are dense vectors in general. Using the matrix $F = \mathrm{diag}(V, \ldots, V) \in \mathbb{R}^{N^2 \times N^2}$, one may also write $\tilde{y} = Fy$. While the classical matrix-vector multiplication takes $O(N^3)$ steps, the FFT algorithm reduces this number to $O(N^2 \log N)$.

Note that the vectors $\tilde{b} = FG^{-1}b$ and $\tilde{\tilde{c}} = FG^{-2}c$ may be computed analogously using the above algorithm. For $\tilde{b}$, steps 1–3 need to be performed using b instead of $MU^{(t)}$ for $\tilde{\tilde{c}}$, steps 1–5 need to be performed using c instead of $MU^{(t)}$.

We refer to the combination of (4) with the above steps for computing $G^{-2}MU^{(t)}$ as modified Ehrlich's algorithm (ME). In total, the ME algorithm costs $O(N^2 \log N)$ computational steps per iteration. To improve the computational complexity of each iteration to an asymptotically optimal value of $O(N^2)$, the final multiplication of F in step 6 needs to be eliminated. Therefore, a new iteration sequence $\overline{U}^{(t+1)}$ is defined by $\overline{U}^{(t+1)} = \frac{1}{\omega_1} FU^{(t+1)}$. Premultiplying the recurrence formula for $U^{(t+1)}$ by $\frac{1}{\omega_1} F$, one gets the equivalent formula

$$\overline{U}^{(t+1)} = (2 - \omega_1 - \omega_2)\overline{U}^{(t)} - 2\omega_2 FG^{-2}(MU^{(t)}) - (1 - \omega_1)(1 - \omega_2)\overline{U}^{(t-1)}$$
$$+\omega_2 Fd, \tag{6}$$

in which $\overline{U}^{(0)} = 0$, $\overline{U}^{(1)} = \widetilde{b}$, and $\widetilde{d} = \widetilde{b} + h^2\widetilde{c}$. Since $\widetilde{d}$ is assumed to be constant, each iteration can be performed in $O(N^2)$ operations. We refer to the combination of (6) and the steps 1–5 from above as the algorithm FV.

As a test for convergence, one may check for a given $\epsilon > 0$ whether

$$\max_{i,j} \left| U_{i,j}^{(t+1)} - U_{i,j}^{(t)} \right| < \epsilon \tag{7a}$$

for ME, and

$$\max_{i,j} \left| \overline{U}_{i,j}^{(t+1)} - \overline{U}_{i,j}^{(t)} \right| < \frac{\epsilon}{\|V\|_\infty \omega_1} \tag{7b}$$

for FV. Since the iteration matrices corresponding to the algorithms ME and FV are similar, the same optimal smoothing parameters $\omega_1 = \omega_2 = O(N^{-1/2})$ as developed in [4] can be used. From this, one gets the same rate of convergence $R_\infty = O(N^{-1/2})$ for FV, so that the number of iterations required for $\epsilon = O(h^2)$ is

$$\log \frac{\omega_1 \|V\|_\infty}{\epsilon} \frac{1}{R_\infty} = O\left(N^{1/2} \log N\right). \tag{8}$$

Due to the stronger convergence criterion (7b), FV may need a higher number of iterations than ME up to a constant multiple.

In total, the computational complexity amounts to $O(N^{5/2} \log N)$ for FV and to $O(N^{5/2} \log^2 N)$ for the ME algorithm (called *ME-FFT* in short) when $\epsilon = O(h^2)$. When using classical matrix-vector multiplications as a replacement for the FFT, the computational complexity is $O(N^{7/2} \log N)$. We refer to this variant as *ME-Matmul*.

4   Experiments

In this section, we study the performance of the algorithm FV with various mesh sizes h. Main measures of interest are the total running time and the resulting error. From this perspective, we also compare it to ME.

4.1   Experimental Setup

The algorithm FV (explained in detail in the previous section) was implemented in the C programming language. The ME algorithm was implemented in the variants ME-FFT and ME-Matmul. Experiments comparing all three algorithms are described in Sect. 4.2.

The computations were performed on the Doppler cluster at the University of Salzburg that runs mostly AMD Opteron processors of clock rates between 2.2 GHz and 2.8 GHz. (Clearly, comparative experiments were conducted on nodes having the same clock rates.)

As an experimental problem, we use the biharmonic Eq. (1) with the solution (see e.g. [7])

$$u(x, y) = x^3 - 3y^2 + 2xy \text{ on } [0, 1]^2 . \tag{9}$$

All figures and tables presented in this section are related to problem (1) with solution (9). The parameter N in figures ranges from 7 to 5000 and the logarithmic scale is used on both axes.

4.2   Performance

In this section, detailed results for running times are given. Firstly, the relationship between the running times for the FV algorithm and its computational complexity is investigated. Later on, FV is compared to the two variants ME-FFT and ME-Matmul of the modified Ehrlich's algorithm.

Total Computational Time. For sizes $N \leq 31$, the problem (1) with solution (9) was analyzed in [7]. These results are reproduced and results for larger-sized systems are given in the following.

As a tolerance, the typical choice of $\epsilon = h^2$ is used. Figure 1 depicts the running time for several problem sizes confirming the expected complexity of $O(N^{5/2} \log N)$. The constant for the dominant complexity term, which results from our experiments, can be taken equal to 10^{-6}.

Number of Iterations. In order to achieve a factor of error reduction β, the expected number of iterations is $t(N) \geq \dfrac{\log 1/\beta}{\log 1/(1-\sqrt{h})}$, where $h = \dfrac{1}{N+1}$ [3].

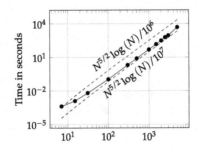

Fig. 1. Running times for $\epsilon = h^2$. The actual running time (solid line) is compared to the theoretical complexity estimation with the constants 10^{-6} and 10^{-7} (dashed lines).

In comparison to the actual number of iterations used, the curve $t(N)$ is plotted in Fig. 2. The curve $t(N)$ conforms to the actual number of iterations, slightly overestimating it for smaller N. In addition, the curve $N^{1/2} \log N$ is plotted which is an approximation to $t(N)$ in this case. For a tolerance $\epsilon = h^2 \approx N^{-2}$, the number of iterations is at least $2N^{1/2} \log N$. In the present case, this is a fairly good approximation for the actual number of iterations. Moreover, the approximation $2N^{1/2} \log N$ can be shown to be almost identical to the curve $t(N)$ for larger N. Thus, it seems that the estimate $O(N^{1/2} \log N)$ might be satisfactory in practice.

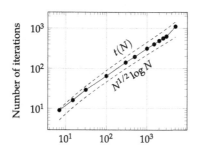

Fig. 2. The number of iterations for $\epsilon = h^2$ (solid line). The dashed curves $t(N)$ and $N^{1/2} \log N$ depict estimates for the number of iterations.

Comparison of FV with ME-FFT and ME-Matmul

- FV: This algorithm has a complexity of $O(N^2)$ per iteration, but a possibly higher number of iterations than the variants ME-FFT and ME-Matmul due to different criteria for convergence (see Sect. 2).
- ME-FFT: This variant, as described in (4), performs with a complexity of $O(N^2 \log N)$ per iteration.
- ME-Matmul: Instead of optimizing the matrix-vector multiplications by means of FFT, the classical algorithm is used. Its drawback is the higher complexity of $O(N^3)$ per iteration.

For our comparisons, the same problem (1) for (9) is solved for $\epsilon = h^2$ by all three variants. We note that for achieving this demanded tolerance, the number of iterations is $O(N^{1/2} \log N)$. Therefore, the total computational complexity amounts to $O(N^{5/2} \log N)$ for FV, to $O(N^{5/2} \log^2 N)$ for ME-FFT, and to $O(N^{7/2} \log N)$ for ME-Matmul. The actual results for these algorithms are depicted in Fig. 3.

From Fig. 3, it is evident that the running times of FV and ME-FFT follow similar patterns since their theoretical computational complexities differ only by a factor of $O(\log N)$. When comparing FV and ME-FFT, the possibly larger number of iterations performed by FV has to be taken into account. Nonetheless,

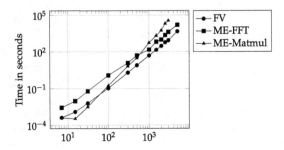

Fig. 3. The total running time of each of the three algorithms for $\epsilon = h^2$

even for large problem sizes, FV is approximately three times faster than ME-FFT.

Table 1 contains a detailed list of results per algorithm for some representative sizes of N. Two immediate conclusions can be drawn from the figures. Firstly, the number of iterations for FV is indeed higher than the number of iterations for ME-FFT due to the more strict termination criterion, which may cause an iteration count increased by a constant factor. Secondly, although competitive for small N, the variant ME-Matmul using classical matrix-vector multiplications becomes infeasible for large N.

Table 1. The number of iterations and running times for $\epsilon = h^2$

	FV		ME-FFT		ME-Matmul	
N	Iterations	Time [s]	Iterations	Time [s]	Iterations	Time [s]
7	9	0.0004	7	0.0028	7	0.0004
31	29	0.0064	21	0.0605	21	0.0032
500	195	7.9508	152	50.4387	152	30.7433
1,000	314	49.5345	239	151.2021	239	560.1155
2,000	488	306.3109	369	1,015.1991	369	5,689.0126
3,000	630	883.7871	488	4,248.0112	488	36,534.9337
5,000	1,240	4,778.6709	662	16,376.5583	–	–

For a second series of tests, the same setups are used as before except the tolerance ϵ is fixed at $\epsilon = 10^{-6}$. Table 2 lists corresponding results for representative values of N. In comparison to the results for $\epsilon = h^2$, it can be seen that for smaller N, the number of iterations is higher. It is because for small N, the tolerance h^2 is larger than 10^{-6} and thus, more iterations have to be performed until an iteration error of 10^{-6} is reached. However, starting at a size of $N = 1000$, the solution takes less time for $\epsilon = 10^{-6}$. Additionally to Table 1, results for $N = 10000$ are provided, showing that FV is still faster than ME-FFT for large systems (in this case, by a factor of approximately 5).

Table 2. The number of iterations and running times for the tolerance $\epsilon = 10^{-6}$

	FV		ME-FFT		ME-Matmul	
N	Iterations	Time [s]	Iterations	Time [s]	Iterations	Time [s]
7	18	0.0005	17	0.0061	17	0.0002
31	42	0.0079	37	0.1038	37	0.0075
500	216	8.9500	162	53.6371	162	41.7075
1,000	314	56.0913	239	152.0935	239	537.9254
2,000	466	294.5305	336	922.8362	336	6,769.0631
3,000	575	1,122.8605	423	3,678.5816	423	28,381.2390
5,000	791	3,436.3500	575	13,916.6963	–	–
10,000	1,164	19,156.2446	824	106,102.6260	–	–

4.3   Errors

In this subsection, the error values related to FV are examined. These are the termination error, given as the maximum norm of the difference of two consecutive iterations defined by (7b), and the solution error, which is the maximum norm of the difference between the computed solution and the actual solution. It is known that the approximations used yield a discretization error of $O(N^{-3/2})$. Thus, if the actual error is below the discretization error, the solution is sufficiently accurate.

In the following plots and tables, the same set-up configurations are used as in the previous subsection, which was devoted to examinations of running times. In Fig. 4, a comparison of the actual error to the function $N^{-3/2}$ is plotted.

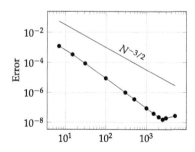

Fig. 4. The solution error compared to the theoretical estimate $N^{-3/2}$ with $\epsilon = 10^{-6}$

Figure 4 shows, that the error caused by the approximation attains the order of the discretization error, as expected. For the sake of completeness, the termination and solution errors are listed in Table 3.

Table 3. The termination errors and solution errors for $\epsilon = 10^{-6}$

N	Termination error	Solution error
7	0.00000040	0.00120771
31	0.00000042	0.00008502
500	0.00000031	0.00000035
1,000	0.00000025	0.00000009
2,000	0.00000030	0.00000002
3,000	0.00000030	0.00000002
5,000	0.00000038	0.00000003

5   Conclusions and Outlooks

An efficient algorithm for solving the model biharmonic problem has been presented. It has been shown, that it is possible to get a numerical solution on an $N \times N$ grid with the accuracy of $O(N^{-2})$ in a time proportional to $O(N^{5/2} \log^2 N)$. It was experimentally confirmed that the actual times for N ranging from 7 up to 5000 scale perfectly with this estimate and lie in a corridor marked by lines, which correspond to the dominant complexity term with constants 10^{-6} and 10^{-7} respectively. Similarly, the estimation $O(N^{1/2} \log N)$ for the number of required iterations coincides in a scalable way with the reality, whereby the absolute error of the solution is always below the discretization error.

A comparison of this algorithm with a modification of the original approach of Ehrlich [4] confirms, that the estimated theoretical $O(\log N)$ gain in total time is valid also for large values of N, despite the number of iterations needed is slightly higher. Experiments for N up to 10000 show that against Ehrlich's approach with FFT (used for matrix-multiplication) the presented algorithm needs up to 5-times less total time and when the product of matrices is computed classically, then this factor is even higher, making this approach for large parameters on serial computer not feasible.

There is a space left for further research. Another interesting comparison of FV against the modification of Ehrlich's approach offers an application of a multigrid Poisson solver, because it would also lead to the optimal complexity count for computing one iteration. While our presentation is restricted to $N \times N$ grids on a rectangle, a further extension would encompass grids with different number of points in each direction and also the case of non-rectangular domains. Here, the estimation of smoothing parameters would be an important issue. The algorithm bears potential for a parallelization. Therefore, the investigation of its behavior when examined on large HPC platforms is also an open question.

Acknowledgments. The second author was supported by the VEGA grant no. 2/0026/14.

References

1. Bialecki, B.: A fourth order finite difference method for the Dirichlet biharmonic problem. Numer. Algor. **61**(3), 351–375 (2012)
2. Buzbee, B.L., Golub, G.H., Nielson, C.W.: On direct methods for solving Poisson's equations. SIAM J. Numer. Anal. **7**(4), 627–656 (1970)
3. Di Stolfo, P.: Splitting-based fast algorithm for solving the rectangular biharmonic problem: a sequential implementation. Master's Thesis, Paris-Lodron-Universität Salzburg (2015)
4. Ehrlich, L.W.: Solving the biharmonic equation as coupled finite difference equations. SIAM J. Numer. Anal. **8**(2), 278–287 (1971)
5. Falk, R.S.: Approximation of the biharmonic equation by a mixed finite element method. SIAM J. Numer. Anal. **15**(3), 556–567 (1978)
6. McLaurin, J.W.: A general coupled equation approach for solving the biharmonic boundary value problem. SIAM J. Numer. Anal. **11**(1), 14–33 (1974)
7. Vajteršic, M.: A fast algorithm for solving the first biharmonic boundary value problem. Computing **23**(2), 171–178 (1979)
8. Vajteršic, M.: Algorithms for Elliptic Problems. Kluwer Academic Publisher, Dordrecht (1993)
9. Wang, T.: A mixed finite volume element method based on rectangular mesh for biharmonic equations. J. Comput. Appl. Math. **172**(1), 117–130 (2004)

Performance of the Parallel One-Sided Block Jacobi SVD Algorithm on a Modern Distributed-Memory Parallel Computer

Shuhei Kudo[1](✉), Yusaku Yamamoto[1], Martin Bečka[2],
and Marian Vajteršic[2,3]

[1] The University of Electro-Communications, Tokyo 182-8585, Japan
138x211x@stu.kobe-u.ac.jp, yusaku.yamamoto@uec.ac.jp
[2] Institute of Mathematics, Slovak Academy of Sciences, Bratislava, Slovakia
Martin.Becka@savba.sk
[3] Department of Computer Sciences, University of Salzburg, Salzburg, Austria
marian@cosy.sbg.ac.at

Abstract. The one-sided block Jacobi (OSBJ) method is known to be an efficient algorithm for computing the singular value decomposition. In this paper, we evaluate the performance of the most recent variant of the OSBJ method, the one with dynamic ordering and variable blocking, on the Fujitsu FX10 parallel computer. By analyzing the performance results, we identified two bottlenecks, namely, weight computation for ordering and diagonalization of 2×2 block matrices. To resolve the problem, we propose new implementations for these two tasks. Experimental results show that they are effective and can achieve speedup of up to 1.6 times in total. As a result, our OSBJ solver can compute the SVD of matrices of order 2048 to 8192 on 12 to 48 nodes of FX10 more than three times faster than ScaLAPACK PDGESVD.

Keywords: One-sided block Jacobi method · Dynamic ordering · Variable blocking · Singular value decomposition · SVD

1 Introduction

Singular value decomposition (SVD) is one of the most fundamental matrix computations and has applications in such diverse fields as signal processing, information retrieval and machine learning. This paper deals with the computation of SVD of an $m \times n$ matrix A $(m \geq n)$ on a distributed-memory parallel computer.

The standard algorithm for SVD consists of three steps, namely, transformation of the input matrix to bidiagonal form, SVD of the bidiagonal matrix, and back-transformation of the singular vectors [1]. Common matrix libraries such as LAPACK and ScaLAPACK adopt this approach. However, from the viewpoint of high performance computing, this approach has two drawbacks. First, the bidiagonalization step has fine-grained parallelism and requires as many as

© Springer International Publishing Switzerland 2016
R. Wyrzykowski et al. (Eds.): PPAM 2015, Part I, LNCS 9573, pp. 594–604, 2016.
DOI: 10.1007/978-3-319-32149-3_55

$O(n)$ interprocessor communications. This often causes performance bottleneck. Second, half of the computational work in the bidiagonalization step is done in the form of level-2 BLAS (matrix-vector multiplication), which cannot use cache memory efficiently. This lowers the performance, especially when the matrix size is large.

As an alternative to the bidiagonalization-based method, the one-sided block Jacobi (OSBJ) method [2–5] attracts attention recently. In this method, one first partitions the input matrix logically into column blocks as

$$A = [A_1, A_2, \ldots, A_\ell], \tag{1}$$

where ℓ is even, A_i has n_i columns ($1 \leq i \leq \ell$) and $n_1 + n_2 + \cdots + n_\ell = n$. The method starts with

$$A^{(0)} = A \tag{2}$$

and computes for $r = 0, 1, \ldots$ the iterates

$$A^{(r+1)} = A^{(r)} V^{(r)}. \tag{3}$$

Here, $V^{(r)}$ is an $n \times n$ orthogonal matrix designed to make the column vectors of the submatrix $[A_{i_r}^{(r)}, A_{j_r}^{(r)}]$ ($1 \leq i_r, j_r \leq \ell$) orthogonal to each other. This process is repeated until all the column vectors of $A^{(r)}$ are orthogonal to working accuracy. Then one can compute the singular values as the norms of the column vectors of $A^{(r)}$, the left singular vectors $U = [\mathbf{u}_1, \mathbf{u}_2, \ldots, \mathbf{u}_n] \in \mathbf{R}^{m \times n}$ as the normalized column vectors of $A^{(r)}$, and the right singular vectors $V = [\mathbf{v}_1, \mathbf{v}_2, \ldots, \mathbf{v}_n] \in \mathbf{R}^{n \times n}$ by solving $AV = A^{(r)}$. Although this process requires more computational work than the bidiagonalization-based method, it has larger grain parallelism and requires smaller number of interprocessor communications. Moreover, most of the computational work is performed in the form of level-3 BLAS, which is a cache-friendly operation. Thanks to these features, the OSBJ method has the potential to outperform the bidiagonalization-based method on modern distributed-memory parallel machines.

In this paper, we evaluate the performance of the OSBJ method on the Fujitsu FX10 distributed-memory parallel computer. There are many variants of the OSBJ method which differ in the order $\{(i_r, j_r)\}_{r=0,1,\ldots}$ in which the column blocks are orthogonalized and in the way the matrix is distributed across the computing nodes. As for the ordering, we adopt the so-called *parallel dynamic ordering strategy* [3] proposed by Bečka et al. In this strategy, at each step r, one chooses $\ell/2$ independent pairs $(i_{r,1}, j_{r,1}), \ldots, (i_{r,\ell/2}, j_{r,\ell/2})$ by taking into account the mutual perpendicularity of all pairs of column blocks, and then orthogonalize the columns of the $\ell/2$ submatrices $[A_{i_{r,1}}^{(r)}, A_{j_{r,1}}^{(r)}], \ldots, [A_{i_{r,\ell/2}}^{(r)}, A_{j_{r,\ell/2}}^{(r)}]$ in parallel. It has been shown that this strategy is effective in reducing the number of iterations of the OSBJ method for a wide class of random matrices. As for the matrix distribution, we divide each block $A_i^{(r)}$ ($i = 1, 2, \ldots, \ell$) into column subblocks and allocate each subblock to one computing node. This approach, in which a pair of blocks $[A_i^{(r)}, A_j^{(r)}]$ is distributed across multiple nodes, is named *variable blocking* in [4] and is shown to achieve better parallel efficiency than the conventional fixed-size blocking, where the pair $[A_i^{(r)}, A_j^{(r)}]$ is allocated to one node. We will evaluate the performance of this variant of the OSBJ method, which was originaly used in [4], on the FX10 with up to 48 nodes and analyze the results in detail. In particular, we study how the performance changes depending on the parameter ℓ, identify performance bottleneck, and propose possible improvements. We also compare the performance of the OSBJ method with that of ScaLAPACK.

The rest of this paper is organized as follows. In Sect. 2, we describe the algorithm of the parallel OSBJ method with dynamic ordering and variable blocking. Performance results on the FX10 are presented in Sect. 3. In Sect. 4, we propose possible improvements and investigate their effectiveness again by numerical experiments. Section 5 gives some concluding remarks.

2   The Parallel One-Sided Block Jacobi Algorithm with Dynamic Ordering and Variable Blocking

In this section, we describe the algorithm and implementation of the parallel OSBJ method with dynamic ordering and variable blocking. We assume that at the beginning of the algorithm, the matrix A is logically partitioned as given in (1), whereby each submatrix A_i is divided into k column subblocks and each subblock is allocated to one node. Thus the total number of computing nodes is $p = \ell k$. In the following, we assume $n_1 = n_2 = \cdots = n_\ell = n/\ell$ for simplicity.

2.1   The Algorithm

The algorithm consists of preprocessing, iteration and postprocessing. Here we explain each of these stages without going into the details of parallelization and data distribution, which will be discussed in the next subsection.

Preprocessing. The preprocessing is intended to concentrate the Frobenius norm of $A^\top A$ near the diagonal, thereby accelerating the convergence of the OSBJ method (see [2,5] for details). The preprocessing consists of the following two tasks. Both of them are computed using ScaLAPACK.

QR preprocessing. The input matrix A is decomposed as $A = Q_1 R$, where $Q_1 \in \mathbf{R}^{m \times n}$ is a matrix with orthonormal columns and $R \in \mathbf{R}^{n \times n}$ is an upper triangular matrix.

LQ preprocessing. The matrix R obtained in the previous step is decomposed as $R = L Q_2$, where $L \in \mathbf{R}^{n \times n}$ is a lower triangular matrix and $Q_2 \in \mathbf{R}^{n \times n}$ is an orthogonal matrix. After the decomposition, we set $A^{(0)} = L$.

Iteration. After preprocessing, iteration process (3) gets started with following tasks performed in one iteration.

Weight computation (WC) [4]. For every column block pair $[A_i^{(r)}, A_j^{(r)}]$, where $1 \leq i < j \leq \ell$, we compute its weight w_{ij} by

$$w_{ij} \equiv \frac{\|(A_i^{(r)})^\top A_j^{(r)} \mathbf{e}\|}{\|\mathbf{e}\|}, \tag{4}$$

where $\mathbf{e} = (1, 1, \ldots, 1)^\top \in \mathbf{R}^{n/\ell}$. The value of w_{ij} can be viewed as indicator how mutually inclined the two subspaces $\mathrm{Im}(A_i^{(r)})$ and $\mathrm{Im}(A_j^{(r)})$ are, and will be used to define the ordering.

Greedy ordering [3,4]. Let $\mathcal{I} = \{1, 2, \ldots, \ell\}$. We choose the first column block pair by $(i_{r,1}, j_{r,1}) = \mathrm{argmax}_{i,j \in \mathcal{I}, i<j} w_{ij}$, so that the most mutually inclined pair is orthogonalized. Then we set $\mathcal{I} := \mathcal{I} \backslash \{i_{r,1}, j_{r,1}\}$ and choose the second pair again by $(i_{r,2}, j_{r,2}) = \mathrm{argmax}_{i,j \in \mathcal{I}, i<j} w_{ij}$. This process is repeated until $\mathcal{I}$ becomes empty and $\ell/2$ column block pairs $\{(i_{r,s}, j_{r,s})\}_{s=1}^{\ell/2}$ are obtained. It is clear from the construction that these $\ell/2$ pairs can be orthogonalized independently. Note that the greedy ordering is an inherently sequential operation, so it will not be parallelized but will be executed by all nodes redundantly.

Computation of the Gramian (GRAM). For each pair $[A_{i_{r,s}}^{(r)}, A_{j_{r,s}}^{(r)}]$ $(1 \leq s \leq \ell/2)$, we compute the 2×2 block Gram matrix $G_s \in \mathbf{R}^{(2n/\ell) \times (2n/\ell)}$ by

$$G_s = [A_{i_{r,s}}^{(r)}, A_{j_{r,s}}^{(r)}]^\top [A_{i_{r,s}}^{(r)}, A_{j_{r,s}}^{(r)}]. \tag{5}$$

2×2 block EVD (EVD). For $1 \leq s \leq \ell/2$, diagonalize the Gram matrix as $G_s = V_s D_s V_s^\top$, where V_s is an orthogonal matrix and D_s is a diagonal matrix.

Matrix-matrix multiplication (MM). For $1 \leq s \leq \ell/2$, compute the column blocks of $A^{(r+1)}$ at the next iteration by the matrix-matrix multiplication

$$[A_{i_{r,s}}^{(r+1)}, A_{j_{r,s}}^{(r+1)}] = [A_{i_{r,s}}^{(r)}, A_{j_{r,s}}^{(r)}] V_s. \tag{6}$$

Then it is easy to see that the column vectors of $[A_{i_{r,s}}^{(r+1)}, A_{j_{r,s}}^{(r+1)}]$ are mutually orthogonalized.

The iteration process is repeated until all columns of $A^{(r)}$ are mutually orthogonal to working accuracy.

Postprocessing. After finishing the iteration process at iteration r, we compute the singular values and singular vectors of A as follows.

Computation of the singular values and left singular vectors of $A^{(0)}$. The singular values of $A^{(0)}$ are computed as the norms of the column vectors of $A^{(r)}$. Of course, they are also the singular values of A. The left singular vectors $U^{(0)} = [\mathbf{u}_1^{(0)}, \ldots, \mathbf{u}_n^{(0)}]$ of $A^{(0)}$ are computed as the normalized column vectors of $A^{(r)}$.

Computation of the right singular vectors of $A^{(0)}$. The right singular vectors of $A^{(0)}$ are computed by solving the equation $A^{(0)}V^{(0)} = A^{(r)}$ [4]. Note that this can be solved very easily because $A^{(0)}$ is a lower triangular matrix.

QR postprocessing. Compute the left singular vectors of A by $U = Q_1 U^{(0)}$.

LQ postprocessing. Compute the right singular vectors of A by $V = Q_2^\top V^{(0)}$.

2.2 Parallelization and Data Distribution

In the algorithm described above, three types of data distribution are used.

At the beginning of the algorithm, each logical block A_i $(1 \leq i \leq \ell)$ is partitioned into k column subblocks and each subblock is allocated to one computing node. We call this data distribution *DIST1*.

In the preprocessing and postprocessing stages, the QR decomposition, the LQ decomposition, the QR postprocessing and the LQ postprocessing are done using ScaLAPACK. Because these are operations on the entire matrix, all computing nodes are involved at. Thus the matrix is distributed among all the nodes using block cyclic data layout. We call this data distribution *DIST2*.

In the weight computation (WC) task, we again use DIST1. The computation of w_{ij} is performed by those $2k$ nodes, to which the block columns $A_i^{(r)}$ and $A_j^{(r)}$ are allocated. Thus $\ell/2$ weights can be computed in parallel.

In the GRAM, EVD and MM tasks, operations on the disjoint $\ell/2$ subblock pairs $[A_{i_{r,s}}^{(r)}, A_{j_{r,s}}^{(r)}]$ $(1 \leq s \leq \ell/2)$ can be done independently. Operations on each subblock pair are performed by $2k$ nodes using ScaLAPACK. Hence, the subblock pair is distributed among these nodes using block cyclic data distribution. We call this data distribution *DIST3*.

From the data distribution stated above, it is clear that redistribution of the matrix data is necessary before and after the preprocessing and postprocessing stages, before the GRAM task, and after the MM task.

2.3 Computational Work of Each Task

In the variable blocking scheme, ℓ can be chosen arbitrarily as long as ℓ is even, $\ell \geq 2$ and p is divisible by ℓ. Here we consider how the computational work and granularity of each task change depending on ℓ. Since the computations in the preprocessing and postprocessing stages do not depend on ℓ, we consider only the tasks in the iteration. We also exclude the greedy ordering, which requires far less computational work than other tasks.

When the number of column blocks is ℓ, there are $\ell(\ell - 1)/2$ block pairs and $\ell/2$ of them can be orthogonalized in one iteration in parallel. Hence, for performing $\ell(\ell-1)/2$ orthogonalizations, there are $\ell-1$ iterations needed, which define as one *sweep*. For each of the WC, GRAM, EVD and MM tasks, its order of computational work per sweep and per node, computational pattern and the size of matrices appearing in it are summarized in Table 1. Here, DGEMV, DSYRK and DGEMM denote routines for matrix-vector multiplication, multiplication of a matrix with its transpose, and multiplication of two matrices, respectively. It can be seen that the complexity of WC increases with ℓ, while that of EVD decreases with ℓ. Thus it is expected that the optimal value of ℓ is determined from the trade-off between these two effects.

Table 1. Computational work, pattern and granularity of each task.

Task	Work	Routine	Matrix size
WC	$n^2\ell^2/p$	DGEMV	$n \times (n/\ell)$
GRAM	n^3/p	DSYRK	$n \times (2n/\ell)$
EVD	$n^3/(\ell p)$	EVD	$(2n/\ell) \times (2n/\ell)$
MM	n^3/p	DGEMM	$n \times (2n/\ell)$, $(2n/\ell) \times (2n/\ell)$

3   Performance Results on the Fujitsu FX10

To evaluate the performance of the parallel OSBJ method with dynamic ordering and variable blocking, we implemented the algorithm using FORTRAN and MPI. As a substance for that, we used a code which was presented in [6]. In the GRAM, EVD and MM tasks, we used ScaLAPACK and PBLAS routines PDSYRK, PDSYEVD and PDGEMM, respectively. All experiments were done on the Fujitsu PRIMEHPC FX10 distributed memory parallel computer installed at Kobe University. It has 96 computing nodes connected via 6-dimensional torus network and each node consists of a SPARC IXfx processor with 16 cores, 32GB of main memory and 12MB of secondary cache.

For the performance evaluation, we varied the number of nodes p from 12 to 48, by running one MPI process per node. Each MPI process consisted of 16 threads, which were used by multi-threaded BLAS routines called from ScaLA-PACK and PBLAS. To construct the test matrices, we first generated random diagonal matrices whose diagonal elements follow normal distribution $N(0,1)$ and then multiplied them by random orthogonal matrices from both sides. The order of matrices was $n = 2048, 4096$ and 8192. In the case of $p = 48$, the blocking factor considered was $\ell = 4, 12, 24, 48$ and hence, the corresponding number of nodes in charge of one column block pair was $p/(\ell/2) = 96/\ell$, i.e. 24, 8, 4 and 2 respectively. For each of these cases, the process grid used in PDSYEVD to distribute the matrix data in 2-dimensional block cyclic fashion was 4×6, 2×4, 2×2 and 1×2, respectively.

In Figs. 1 and 2, we show the execution times for the test matrices of order 4096 and 8192, respectively. In both cases, the number of nodes was fixed to 48 and ℓ varied from 4 to 48. The number of iterations and sweeps (defined as the number of iterations divided by $\ell - 1$) for each case are shown in Table 2.

Table 2. Number of iterations and sweeps of the OSBJ method.

n		$\ell = 48$	$\ell = 24$	$\ell = 12$	$\ell = 4$
4096	# of iterations	246	115	50	10
	# of sweeps	5.23	5.00	4.55	3.33
8192	# of iterations	246	115	52	10
	# of sweeps	5.23	5.00	4.73	3.33

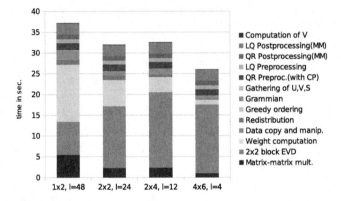

Fig. 1. Execution time for the case of $n = 4096$ and $p = 48$ (before optimization).

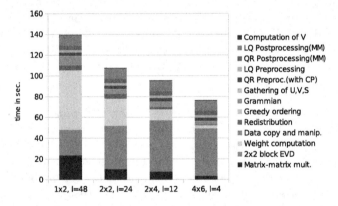

Fig. 2. Execution time for the case of $n = 8192$ and $p = 48$ (before optimization).

4   Possible Improvements and Their Effectiveness

4.1   Improvements

Figures 1 and 2 show that most of the computation time is spent for two tasks, namely for the weight computation and 2×2 block EVD. This is the same for other cases, which are omitted due to space limitations. As expected from Table 1, the time for WC increases with ℓ, while that for EVD decreases with ℓ.

On the other hand, the number of sweeps increases with ℓ, as can be seen from Table 2. The optimal value of ℓ for given n and p are determined from these factors.

To further speed up the OSBJ method, we propose improvements on the implementations of WC and EVD in the following.

Improvement of weight computation. In the current implementation [6], the weights $\{w_{ij}\}$ in (4) are computed in the form of dot products. However, it is well known that this operation cannot use the cache memory efficiently. Looking at (4), the computation of $\{w_{ij}\}$ can be divided into three parts: computation of $\mathbf{b}_j \equiv A_j^{(r)}\mathbf{e}/\|\mathbf{e}\|$ ($1 \leq j \leq \ell$), computation of $\mathbf{c}_{ij} = (A_i^{(r)})^\top \mathbf{b}_j$ ($1 \leq i < j \leq \ell$) and computation of $w_{ij} = \|\mathbf{c}_{ij}\|$ ($1 \leq i < j \leq \ell$). Among them, the second part is the heaviest, since it requires $O(n^2\ell/(2p))$ work per iteration. We therefore propose to aggregate these computations for $1 \leq j \leq \ell$ and compute them using matrix-matrix multiplication as

$$[\mathbf{c}_{i1}, \mathbf{c}_{i2}, \ldots, \mathbf{c}_{i\ell}] = (A_i^{(r)})^\top [\mathbf{b}_1, \mathbf{b}_2, \ldots, \mathbf{b}_\ell] \quad (1 \leq i \leq \ell). \tag{7}$$

Thus the cache usage of this part is greatly enhanced. In the actual implementation, the matrix multiplication (7) is distributed among several nodes, since each node has only a subset of $\{\mathbf{b}_j\}_{j=1}^{\ell}$.

Improvement of 2×2 block EVD. In the current implementation, diagonalization of the Gram matrix G_s is executed by $2k$ nodes in charge of the column block pair $[A_{i_{r,s}}^{(r)}, A_{j_{r,s}}^{(r)}]$ using ScaLAPACK PDSYEVD. However, the tridiagonalization step used in PDSYEVD requires as many interprocessor communications as the order of the input matrix. Hence, when the size $2n/\ell$ of this matrix is small, the portion belonging to communication becomes to be prevealing in the total computational time, thus the performance of PDSYEVD can become lower than that of the sequential routine DSYEVD. Meanwhile, we recently developed an optimized

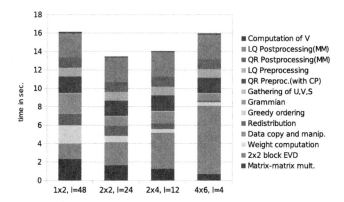

Fig. 3. Execution time for the case of $n = 4096$ and $p = 48$ (after optimization)

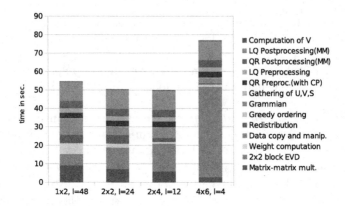

Fig. 4. Execution time for the case of $n = 8192$ and $p = 48$ (after optimization).

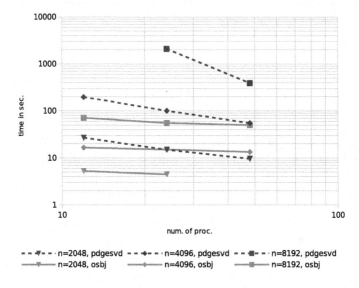

Fig. 5. Comparison of the OSBJ method with ScaLAPACK PDGESVD.

version of DSYEVD for a *single* node of FX10 which adopts a new cache-efficient implementation of the tridiagonalization algorithm, SIMD optimization and a new cache-friendly matrix storage scheme [7]. Thus it might be advantageous to replace PDSYEVD with this routine. To this end, it is required to gather the matrix G_s distributed across $2k$ nodes to one node prior to the EVD, and distribute the eigenvector matrix V_s to the $2k$ nodes after the EVD.

4.2   Performance Evaluation

We incorporated the two improvements proposed in the previous subsection into our parallel OSBJ code [6] and evaluated its performance on the FX10.

The experimental conditions are the same as those explained in Sect. 3. The results for the matrices of order 4096 and 8192 on 48 nodes are shown in Figs. 3 and 4, respectively. It can be seen that the times for both WC and EVD have been greatly reduced. Additionally, our experiments show new optimal values for ℓ, which becomes $\ell = 24$ and 12 for the cases of $n = 4096$ and 8192, respectively.

Finally, we compare the performance of our parallel OSBJ code with that of ScaLAPACK SVD routine PDGESVD. For PDGESVD, we used block size of MB=NB=50 and process grid of 3×4, 4×6, and 6×8 for the cases of 12, 24 and 48 nodes, respectively. The execution times of the OSBJ method and ScaLAPACK PDGESVD for three matrix sizes, $n = 2048$, 4096 and 8192, are shown in Fig. 5. In all cases, the optimized version of the OSBJ method was more than three times faster than PDGESVD. When $n = 8192$ and $p = 48$, the execution time of the OSBJ method and PDGESVD was 49.91 seconds and 392.38 seconds, respectively, so the former was 7.8 times faster.

5 Conclusion

In this paper, we evaluated the performance of the one-sided block Jacobi SVD method with dynamic ordering and variable blocking on the Fujitsu FX10 parallel computer. We identified two performance bottlenecks, namely, weight computation and 2×2 block EVD, proposed improvements for them and investigated their effectiveness. Thanks to these improvements, the bottlenecks were resolved and the solver achieved performance more than three times higher than that of ScaLAPACK. Experiments with the new version of OSBJ on other computers to evaluate the performance on much large number of nodes and effectivenes of the improvements on different architectures, would be a matter of a future work.

Acknowledgements. The authors thank Professor Mitsuo Yokokawa at the Education Center on Computational Science and Engineering, Kobe University, for providing the computational environments used in our experiments. The present study is supported in part by the Ministry of Education, Science, Sports and Culture, Grant-in-Aid for Scientific Research (B)(No. 26286087), Core Research for Evolutional Science and Technology (CREST) Program "Highly Productive, High Performance Application Frameworks for Post Petascale Computing" of Japan Science and Technology Agency (JST). This work is supported also by grant No. 2/0026/14 of the VEGA grant agency.

References

1. Golub, G.H., van Loan, C.F.: Matrix Computations, 4th edn. Johns Hopkins University Press, Baltimore (2012)
2. Drmač, Z., Veselić, K.: New fast and accurate Jacobi SVD algorithm: I. SIAM J. Matrix Anal. Appl. **29**, 1322–1342 (2007)
3. Bečka, M., Okša, G., Vajteršic, M.: Dynamic ordering for a parallel block-Jacobi SVD algorithm. Parallel Comput. **28**, 243–262 (2002)

4. Bečka, M., Okša, G.: Parallel One – sided Jacobi SVD algorithm with variable blocking factor. In: Wyrzykowski, R., Dongarra, J., Karczewski, K., Waśniewski, J. (eds.) PPAM 2013, Part I. LNCS, vol. 8384, pp. 57–66. Springer, Heidelberg (2014)
5. Okša, G., Vajteršic, M.: Efficient preprocessing in the parallel block-Jacobi SVD algorithm. Parallel Comput. **31**, 166–176 (2005)
6. Bečka, M., Okša, G., Vajteršic, M.: Parallel code for one-sided Jacobi-method. Technical report 2015–02, Department of Computer Sciences, University of Salzburg, April 2015
7. Kudo, S., Yamamoto, Y., Yokokawa, M.: Performance analysis and implementation method of Dongarra-Wilkinson tridiagonalization method (To appear in Proceedings of HPCS2015) (in Japanese)

New Approach to Local Computations in the Parallel One–Sided Jacobi SVD Algorithm

Martin Bečka[✉] and Gabriel Okša

Institute of Mathematics, Slovak Academy of Sciences, Bratislava, Slovakia
{Martin.Becka,Gabriel.Oksa}@savba.sk

Abstract. One sided block Jacobi algorithm for the singular value decomposition (SVD) of matrix can be a method of choice to compute SVD efficiently and accurately in parallel. A given matrix is logically partitioned into block columns and is subjected to an iteration process. In each iteration step, for given two block columns, their Gram matrix is generated, its symmetric eigenvalue decomposition (EVD) is computed and the update of block columns by matrix-matrix multiplication is performed. Another possibility is to omit the computation of Gram matrix and the update, so that there is no matrix-matrix multiplication at all. A local matrix is formed by two block columns and its QR decomposition is computed first to reduce the dimension and decrease the off-diagonal norm. Then, the one-sided serial Jacobi SVD is called (either for a local matrix or its R-factor). No update is necessary, since the result from the serial one-sided Jacobi SVD algorithm is the same as the original matrix after update. Crucial for this new approach is an efficient implementation of the QR decomposition for tall and skinny matrices, as well as a fast and accurate (serial) one-sided Jacobi SVD algorithm. Another improvement of the algorithm would be to replace the static (fixed) local stopping criterion in the inner EVD or SVD computations by the dynamic (flexible) one to reduce the work done by these routines in one parallel iteration step. Since the orthogonality of columns is crucial at the end of the algorithm, one could progressively set the local stopping criterion as the computation proceeds. We tried to implement the proposed approaches and compare the achieved results with the standard algorithm.

Keywords: Singular value decomposition · Parallel one–sided block–Jacobi algorithm · Gram matrix

1 Introduction

The one-sided Jacobi methods for computing the singular value decomposition (SVD) of a rectangular matrix are more efficient than their two-sided counterparts mainly due to the halving of the number of matrix-matrix products needed for updating the left and right singular vectors; see [1,7]. Moreover, some of them were also proved to be accurate in the relative sense and at least as accurate as the two-sided Jacobi methods; see [3–5,7]. When the preprocessing for positive

© Springer International Publishing Switzerland 2016
R. Wyrzykowski et al. (Eds.): PPAM 2015, Part I, LNCS 9573, pp. 605–617, 2016.
DOI: 10.1007/978-3-319-32149-3_56

definite matrices is used in form of the Cholesky decomposition, the one-sided Jacobi methods are very accurate eigensolvers [3]. Recently, it has been reported in [4,5] that the clever implementation of a serial one-sided Jacobi algorithm reached the efficiency of the QR method. However, the QR method uses a bidi-agonalization as a pre-processing step, which means that the relative accuracy is lost and cannot be recovered using the subsequent one-sided Jacobi method. Consequently, the QR method should not be used in such applications where the high relative accuracy of computed singular values is required (e.g., the compu-tation of energies of atoms and molecules in quantum physics and quantum chemistry), and the one-sided Jacobi method is the choice.

Next section contains a brief description of the one-sided block-Jacobi algo-rithm (OSBJA) for computing the SVD of a general rectangular matrix A. We mention two interesting approaches for the reduction of each local matrix. The first approach consists of computing the symmetric, positive definite Gram matrix from two block columns in each processor and then to use some EVD procedure for symmetric matrices for its reduction to diagonal form. Another approach uses the direct SVD of two matrix blocks inside each processor using the serial Jacobi procedure from LAPACK. We also describe the most efficient dynamic parallel ordering from [2]. For local SVD computations, we propose the dynamic local stopping criterion which enables to accelerate the computations in a parallel iteration step.

The last section is devoted to numerical experiments and comparison of five variants of the OSBJA which differ in using the static or dynamic local stopping criterion and in QR pre- and post-processing.

2   One-Sided Block-Jacobi Algorithm

2.1   Computation of Gram Matrices

The OSBJA is suited for the SVD computation of a general complex matrix A of order $m \times n$, $m \geq n$. However, we will restrict ourselves to real matrices with obvious modifications in the complex case.

We start with the block-column partitioning of A in the form

$$A = [A_1, A_2, \ldots, A_r],$$

where the width of A_i is n_i, $1 \leq i \leq r$, so that $n_1 + n_2 + \cdots + n_r = n$. The most natural choice is $n_1 = n_2 = \cdots = n_{r-1} = n_0$, so that $n = (r-1)n_0 + n_r$, $n_r \leq n_0$. Here n_0 can be chosen according to the available cache memory, which is up to 10 times faster than the main memory; this connection will be clear later on.

The OSBJA can be written as an iterative process:

$$A^{(0)} = A, \quad V^{(0)} = I_n,$$
$$A^{(k+1)} = A^{(k)}U^{(k)}, \quad V^{(k+1)} = V^{(k)}U^{(k)}, \quad k \geq 0. \tag{1}$$

Here the $n \times n$ orthogonal matrix $U^{(k)}$ is the so-called *block rotation* [6] of the form

$$U^{(k)} = \begin{pmatrix} I & & & \\ & U_{ii}^{(k)} & & U_{ij}^{(k)} \\ & & I & \\ & U_{ji}^{(k)} & & U_{jj}^{(k)} \\ & & & & I \end{pmatrix}, \tag{2}$$

where the unidentified matrix blocks are zero. The purpose of matrix multiplication $A^{(k)}U^{(k)}$ in (1) is to mutually orthogonalize the columns between column blocks i and j of $A^{(k)}$. The matrix blocks $U_{ii}^{(k)}$ and $U_{jj}^{(k)}$ are square of order n_i and n_j, respectively, while the first, middle and last identity matrix is of order $\sum_{s=1}^{i-1} n_s$, $\sum_{s=i+1}^{j-1} n_s$ and $\sum_{s=j+1}^{r} n_s$, respectively. The orthogonal matrix

$$\hat{U}^{(k)} = \begin{pmatrix} U_{ii}^{(k)} & U_{ij}^{(k)} \\ U_{ji}^{(k)} & U_{jj}^{(k)} \end{pmatrix} \tag{3}$$

of order $n_i + n_j$ is called the *pivot submatrix* of $U^{(k)}$ at step k. During the iterative process (1), two index functions are defined: $i = i(k)$, $j = j(k)$ whereby $1 \le i < j \le r$. At each step k of the OSBJA, the pivot pair (i, j) is chosen according to a given *pivot strategy* that can be identified with a function $\mathcal{F} : \{0, 1, \ldots\} \to \mathbf{P}_r = \{(l, m) : 1 \le l < m \le r\}$. If $\mathbf{O} = \{(l_1, m_1), (l_2, m_2), \ldots, (l_{N(r)}, m_{N(r)})\}$ is some ordering of $\mathbf{P}_r$ with $N(r) = r(r - 1)/2$, then the *cyclic* strategy is defined by:

  If $k \equiv r - 1 \mod N(r)$ then $(i(k), j(k)) = (l_s, m_s)$ for $1 \le s \le N(r)$.

  The most common cyclic strategies are the *row-cyclic* one and the *column-cyclic* one, where the orderings are given row-wise and column-wise, respectively, with regard to the upper triangle of A. The first $N(r)$ iterations constitute the first *sweep* of the OSBJA. When the first sweep is completed, the pivot pairs (i, j) are repeated during the second sweep, and so on, up to the convergence of the entire algorithm.

  Notice that in (1) only the matrix of right singular vectors $V^{(k)}$ is iteratively computed by orthogonal updates. If the process ends at iteration t, say, then $A^{(t)}$ has mutually highly orthogonal columns. Their norms are the singular values of A, and the normalized columns (with unit 2-norm) constitute the matrix of left singular vectors.

  One (serial) step of the OSBJA can be described in three parts:

1. For the given pivot pair (i, j), the symmetric, positive semidefinite Gram matrix is computed:

$$\hat{A}_{ij}^{(k)} = [A_i^{(k)} \ A_j^{(k)}]^T [A_i^{(k)} \ A_j^{(k)}] = \begin{pmatrix} A_i^{(k)T} A_i^{(k)} & A_i^{(k)T} A_j^{(k)} \\ A_j^{(k)T} A_i^{(k)} & A_j^{(k)T} A_j^{(k)} \end{pmatrix}. \tag{4}$$

This requires $(n_i + n_j)(n_i + n_j - 1)/2$ dot products or $m(n_i + n_j)(n_i + n_j - 1)/2$ flops. As will be soon clear, except for a part of the first sweep, the two diagonal blocks of $\hat{A}^{(k)}$ will be always diagonal. This reduces the flop count to $m(n_i n_j + n_i + n_j)$ where $m(n_i + n_j)$ comes from the computation of the diagonal elements of $\hat{A}^{(k)}$.

2. $\hat{A}_{ij}^{(k)}$ is diagonalized, i.e., the eigenvalue decomposition of $\hat{A}_{ij}^{(k)}$ is computed:

$$\hat{U}^{(k)T} \, \hat{A}_{ij}^{(k)} \, \hat{U}^{(k)} = \hat{\Lambda}_{ij}^{(k)} \tag{5}$$

and the eigenvector matrix $\hat{U}^{(k)}$ is partitioned according to (3). The matrix $\hat{U}^{(k)}$ defines the orthogonal transformation $U^{(k)}$ in (2) and (1), which is then applied to $A^{(k)}$ and $V^{(k)}$. This diagonalization requires, on average, around $8(n_i + n_j)^3$ flops.

3. Finally, an updating of two block-columns of $A^{(k)}$ and $V^{(k)}$ is required, which requires $2m(n_i + n_j)^2$ flops.

Notice that the Gram matrix $\hat{A}^{(k)}$ in the second phase is symmetric and positive definite. Hence, its SVD is equal to its EVD, and we can use, for example, the LAPACK procedure for the EVD of symmetric matrices. Of course, also a Jacobi procedure (two-sided or one-sided) can be a choice.

2.2 Direct SVD of (A_i, A_j)

Instead of computing the local Gram matrix, one can proceed by direct SVD of (A_i, A_j), i.e. of a local matrix of order $m \times (n_i + n_j)$. Notice that for $m \gg n_i + n_j$ this matrix is tall and skinny and it can be advantageous to perform its QR factorization first and subsequently to work with a 'small' square R-factor of order $n_i + n_j$. However, then the post-processing step is required which consists of the update of left singular vectors by the Q-factor.

For this local SVD computation, one can use the one-sided Jacobi procedure from LAPACK. Moreover, it is *not necessary* to accumulate the individual rotations for local right singular vectors, and the computation can proceed *without* any updates of block columns. At the end of the global iteration process, say, at iteration k, the columns of $[A_1^{(k)}, A_2^{(k)}, \ldots, A_r^{(k)}]$ are all mutually almost orthogonal and, after normalization, represent the left singular vectors. Their norms estimate singular values. If one has a copy of the original matrix A, the right singular vectors V can be computed *a posteriori* by solving the linear system

$$AV = [A_1^{(k)}, A_2^{(k)}, \ldots, A_r^{(k)}], \tag{6}$$

where the right hand side is equal to $U\Sigma$.

The problem with this approach lies in the accuracy of computed U and Σ at the end of the global iteration process. This accuracy is directly connected to the local stopping criterion of the inner Jacobi SVD procedure for each local SVD of (A_i, A_j) in each parallel iteration step.

The computational complexity per one parallel iteration step is given essentially by the complexity of the inner SVD of (A_i, A_j) which is of order $O(m(n_i + n_j)^2)$ for a version with no QR pre- and post-processing.

2.3 Dynamic Parallel Ordering

Having p processors, the above OSBJA can be parallelized with the blocking factor $r = 2p$ and, for simplicity, assume $n_1 = n_2 = \ldots = n_{2p} = n/(2p)$. Hence, each processor contains two block columns and a parallel dynamic ordering has to define which pairs of block columns will meet in a given processor in each parallel iteration step.

The computation can be organized in such a way that after the first parallel iteration step (initialization), each block column consists of orthogonal columns. Let us suppose that all $k = n/(2p)$ columns in each block column are *normalized* to the unit Euclidean norm, so that each block column is the *orthonormal basis* of the k-dimensional subspace.

The main idea is to mutually orthogonalize those block columns first which are maximally *inclined* to each other, i.e., their mutual position differs maximally from the orthogonal one. In [2], we have described four new variants of dynamic ordering that are based on estimates of principle angles between two linear subspaces of the same dimension k. Here we mention the most efficient variant 3.

Assume that the original matrix A is of full column rank, $e \equiv (1, 1, \ldots, 1)^T$ is from $\mathbb{R}^{k \times 1}$, and for each column block A_j define its *representative vector*,

$$c_j \equiv \frac{A_j\, e}{\|e\|}, \quad 1 \le j \le 2p. \tag{7}$$

Recall that all block columns of A have linearly independent (orthogonal) columns. Hence, $A_j\, e \ne 0$ for all j throughout the computation. Moreover, assume that the columns in each A_j are orthonormalized so that $\|c_j\| = 1$ for all j. The choice of e ensures the uniform participation of all k one-dimensional subspaces, which constitute $\mathrm{span}(A_j)$, in the definition of c_j.

In variant 3 (see [2]), the weight $w_{ij}^{(3)}$ describes the mutual position of the whole subspace $\mathrm{span}(A_i)$ with respect to the representative vector c_j defined in previous subsection. Hence,

$$w_{ij}^{(3)} \equiv \|A_i^T c_j\| = \frac{\|A_i^T A_j e\|}{\|e\|}. \tag{8}$$

Notice that the orientation of c_j with respect to the *whole* orthonormal basis of $\mathrm{span}(A_i)$ is taken into account.

There is a simple upper bound for $w_{ij}^{(3)}$:

$$w_{ij}^{(3)} = \frac{\|A_i^T A_j e\|}{\|e\|} \le \|A_i^T A_j\|_2.$$

Therefore, if the global Jacobi process converges with respect to the iteration number r then the positive sequence $\{\max_{i,j} w_{ij,r}^{(3)}\}_{r \ge 1}$ converges to zero.

Conversely, if $w_{ij}^{(3)} = 0$, the representative c_j is perpendicular to *all* basis vectors stored in A_i, i.e., it is perpendicular to the whole subspace $\mathrm{span}(A_i)$. Moreover, since $\sigma_k(A_i^T A_j) = \min_{\|x\|=1} \|A_i^T A_j x\| \ge 0$, this also means that at least the largest principal angle is $\pi/2$.

2.4   Stopping Criteria

Each weight $w_{ij}^{(3)}$ needs k scalar products of length m for its computation. Neglecting the length m, we propose the global stopping criterion for variant 3 as

$$\max_{i,j} w_{ij}^{(3)} < k\,\epsilon. \tag{9}$$

Locally, two block columns are not mutually orthogonalized if

$$w_{ij}^{(3)} < k\,\epsilon. \tag{10}$$

In the case of the inner (local) computation *without* Gram matrices, i.e., when the SVD of (A_i, A_j) is computed by the one-sided Jacobi procedure DGESVJ from LAPACK, one can use a *static or dynamic local* stopping criterion inside this procedure. Note that the stronger local stopping criterion, the longer time spent inside the local SVD. Therefore, one can accelerate the whole computation if, at the beginning of the global iteration process, local SVDs will be computed with *less* accuracy, but this accuracy will increase towards the end of the global iteration process. The switch to higher local accuracy can be controlled using the weights $w_{ij}^{(3)}$.

When Gram matrices are computed locally, one can use any procedure from LAPACK designed for the EVD of symmetric matrices. However, in our implementation we use again the one-sided Jacobi procedure DGESVJ which does not take into account the matrix symmetry. On the other side, using DGESVJ one has control over the local stopping criterion that can be again either static or dynamic.

Let us denote a local matrix by B_{ij}, i.e. $B_{ij} = [A_i, A_j]$ in the case of direct computation or $B_{ij} = [A_i, A_j]^T [A_i, A_j]$ in the case of the Gram matrix. At the end of local SVD, the columns of B_{ij} should be orthogonal. Write $B_{ij} = \hat{B}_{ij} D_{ij}$ where D_{ij} is a diagonal matrix containing the norms of columns in B_{ij}. Hence, $\hat{B}_{ij}$ is orthonormal.

The static local stopping criterion tests the orthogonality of computed columns of $\hat{B}_{ij}$ against some fixed value. The local computation is finished if

$$\frac{\|\hat{B}_{ij}^T \hat{B}_{ij} - I\|_{\mathrm{F}}}{\sqrt{n_i + n_j}} \le \frac{\sqrt{m}}{5}\,\epsilon_{\mathrm{M}}. \tag{11}$$

Here, $(n_i + n_j)$ is the order of $\hat{B}_{ij}^T \hat{B}_{ij}$ and I, m is the number of rows in the original matrix A and ϵ_{M} is the machine precision.

A dynamic version of the local stopping criterion takes into account the maximal weight encountered in a given parallel iteration step. Denote by RHS the right-hand side of (11) and define

$$maxw \equiv \max_{i,j} w_{ij}^{(3)}$$

in each parallel iteration step. Then we use the following rule for computing RHS:

$$\textbf{if} \quad (maxw \geq 10^{-6}) \quad \textbf{then} \quad \text{RHS} = 10^{-4} \times \sqrt{m} \times maxw \qquad (12)$$

$$\textbf{else} \quad \text{RHS} = \frac{\sqrt{m}}{5} \, \epsilon_{\text{M}}.$$

Hence, when the weights are 'large' at the beginning of the OSBJA, RHS depends only on $maxw$ and some constants but it does not depend on the machine precision at all. Conversely, towards the end of computation when the weights converge to zero, the local stopping criterion becomes 'rigid' and depends on ϵ_{M} so that the local SVDs are computed practically to full machine precision.

3   Numerical Experiments

Five variants of the OSBJA with dynamic ordering were implemented on the Geisberg Cluster at the University of Salzburg, Austria. The Doppler Cluster consists of 32 nodes where each node has 16 or 64 cores of type Opteron Series 6200, 2.2 GHz, and with 2–8 GB RAM per core. The parallel system is equipped with a variety of GPU accelerator hardware and uses the QDR Infiniband/Mellanox interconnection network. We used one node in a stand-alone mode of computation, so that measured execution times are quite reliable. All computations were performed using the MPI and the IEEE standard double precision floating point arithmetic with the machine precision $\epsilon \approx 2.22 \times 10^{-16}$.

The blocking factor $r = 2p$ was used so that all block columns were of the same width, $n_i = n/(2p)$, $1 \leq i \leq 2p$. One quality measure was consistently computed to estimate the accuracy of all algorithms that describes the relative error in the orthogonality of computed left singular vectors:

$$\tilde{Q} \equiv \frac{\|U^T U - I\|_{\text{F}}}{\sqrt{n}}.$$

The results are presented in Tables 1 and 2 for a square, random matrix of order $n = 4000$ and $n = 8000$, respectively. In both cases, the matrix singular values were randomly distributed. Five different variants of the OSBJA with dynamic ordering were tested and the abbreviations in tables have the following meaning:

– NGS: computation without Gram matrices, static local stopping criterion.
– NGD: computation without Gram matrices, dynamic local stopping criterion.
– NGQS: computation without Gram matrices, QR pre- and post-processing, static local stopping criterion,
– NGQD: computation without Gram matrices, QR pre- and post-processing, dynamic local stopping criterion,
– G: computation with Gram matrices, static local stopping criterion. This is our original implementation from previous papers.

The algorithms were run on $p = 8$, 16 and 32 cores and we provide the number of parallel iteration steps needed for the global convergence n_{it}, the total parallel execution time T_{p}, the time spent in local SVDs and pre- plus post-processing

Table 1. $n = 4000$

p		NGS	NGD	NGQS	NGQD	G
8	n_{it}	81	107	79	107	82
	T_p	1040	695	**622**	770	671
	$T_{(SVD,PP)}$	(701,-)	(458,-)	(84,428)	(61,590)	(121,301)
	$\tilde{Q}$	3e–13	7e–14	3e–13	1e–13	4e–14
16	n_{it}	181	219	177	220	177
	T_p	704	425	455	641	**247**
	$T_{(SVD,PP)}$	(440,-)	(246,-)	(28,281)	(18,412)	(32,131)
	$\tilde{Q}$	4e–13	7e–14	4e–13	9e–14	5e–14
32	n_{it}	722	770	724	778	724
	T_p	859	427	674	817	**343**
	$T_{(SVD,PP)}$	(383,-)	(191,-)	(14,362)	(8,375)	(17,144)
	$\tilde{Q}$	7e–13	8e–14	8e–13	2e–13	8e–14

Table 2. $n = 8000$

p		NGS	NGD	NGQS	NGQD	G
8	n_{it}	81	119	81	120	80
	T_p	7708	6472	6634	6912	**3848**
	$T_{(SVD,PP)}$	(5533,-)	(4484,-)	(692,3792)	(589,5417)	(856,2036)
	$\tilde{Q}$	6e–13	1e–13	7e–13	5e–13	7e–14
16	n_{it}	178	235	176	236	177
	T_p	5401	3263	4706	5604	**2711**
	$T_{(SVD,PP)}$	(3494,-)	(2070,-)	(227,2955)	(150,3578)	(249,1393)
	$\tilde{Q}$	7e–13	1e–13	7e–13	4e–13	8e–14
32	n_{it}	394	480	384	477	384
	T_p	3138	2193	2739	3747	**1532**
	$T_{(SVD,PP)}$	(1827,-)	(1087,-)	(639,1565)	(39,2010)	(78,808)
	$\tilde{Q}$	8e–13	1e–13	8e–13	4e–13	8e–14

$T_{(SVD,PP)}$ and the relative error in the orthogonality of computed left singular vectors $\tilde{Q}$. Note that in the variant G the pre- and post-processing denotes the computation of a Gram matrix and the subsequent update of block columns, respectively. All timings are in seconds and best total parallel execution times are written in bold.

Considering the NG variants, the implementation with the static local stopping criterion (LSC) requires always less n_{it} than the dynamic LSC. When no QR pre-processing is used, the complexity of a local SVD is $O(m(n/p)^2)$, whereas for NG variants with QR pre-processing it is $O((n/p)^3)$. This explains relatively small values of time spent in local SVDs for variants NGQS and NGQD, especially for larger values of p. However, the time spent in QR decompositions and subsequent updates is significantly larger than that of local SVDs. This is a drawback of all implementations which use the QR decomposition procedure from LAPACK for tall, skinny matrices.

The aim was to achieve a good performance for variant NGD. When compared to NGS, the dynamic LSC helps to decrease the time spent in local SVDs. The number of parallel iteration steps increases as compared to NGS but it is practically equal to that of NGQD (the same is true for variants NGS and NGQS). In summary, variant NGD is almost always the best one among variants NG with respect to the total parallel execution time T_p. The only exception is for $n = 4000$, $p = 8$ when NGQS is better.

Now we compare the variant NGD with the variant G. Mainly due to the smaller value of n_it, variant G is always *faster* than the variant NGD despite the extra matrix-matrix multiplications required for computing Gram matrices and updates of two block columns per iteration. Note that square, symmetric Gram matrices are 'small', just of order n/p, and the inner SVD is computed again by the one-sided Jacobi procedure DGESVJ from LAPACK which does *not* utilize the symmetry [4,5]. Our new variant NGD would be competitive with variant G if its values of n_it were only slightly decreased.

With respect to the orthogonality of computed left singular vectors, the variant G seems to be the best one. One reason could be that it works with small, symmetric matrices B_{ij} of order n/p in local SVDs. After finishing the inner SVD resulting in the non-normalized matrix of left singular vectors U_{ij}, the matrix of right singular vectors V_{ij} is computed *a posteriori* by solving the linear system $B_{ij}V_{ij} = U_{ij}$. Then, V_{ij} is used in orthogonalization of two local block columns (A_i, A_j) by matrix-matrix multiplication $(A_i, A_j)V_{ij}$ (this is an orthogonal update). It is remarkable that this approach gives practically the same values of $\tilde{Q}$ than the NGD variant which applies the procedure DGESVJ directly to (A_i, A_j). It would be interesting to explain this fact using the perturbation theory for both cases.

Notice that the static LSC in variants NGS and NGQS, which is quite stringent during the whole computation, results in equal (or very close) values of $\tilde{Q}$. On the other hand, variants NGD and NGQD with dynamic LSC (also mutually comparable) compute the left singular vectors with *better* level of orthogonality (up to one order of magnitude) although, at the beginning of the OSBJA, the local SVDs are not computed with high accuracy! We have no explanation for this interesting behavior at this moment.

Next two tables depict some interesting results of how the length of one iteration step T_it (in seconds) and the number of parallel iterations steps n_it depend on the order of maximal weight *maxw* in the dynamic ordering. Tables 3 and 4 contain results for $n = 8000$, $p = 32$ and $n = 8000$, $p = 16$, respectively.

Some trends are interesting. First of all, we should mention that although the weight $w_{ij}^{(3)}$ is only an estimate of mutual geometric position of two linear subspaces we observe almost monotonic convergence of *maxw* to zero in all numerical experiments, regardless to the variant of OSBJA.

Table 3. $n = 8000$, $p = 32$

maxweight		NGS	NGD	QRS	QRD	G
10^{-1}	T_{it}	12	7	7	10	4
	n_{it}	31	28	32	28	30
10^{-2}	T_{it}	10	6	7	8	4
	n_{it}	157	158	150	157	159
10^{-3}	T_{it}	7	3	7	7	4
	n_{it}	51	51	53	55	47
10^{-4}	T_{it}	7	2	7	7	3
	n_{it}	27	45	31	44	30
10^{-5}	T_{it}	5	2	7	7	3
	n_{it}	24	43	21	39	24
10^{-6}	T_{it}	4	2	7	7	3
	n_{it}	20	50	19	48	15
10^{-7}	T_{it}	4	4	7	7	3
	n_{it}	14	35	13	33	16
10^{-8}	T_{it}	2	2	7	7	3
	n_{it}	12	33	15	35	12
10^{-9}	T_{it}	2	-	7	-	3
	n_{it}	8	0	8	0	11
10^{-10}	T_{it}	2	-	6	-	3
	n_{it}	6	0	11	0	9
10^{-11}	T_{it}	2	2	6	-	3
	n_{it}	16	1	11	0	10
10^{-12}	T_{it}	2	2	6	7	3
	n_{it}	5	7	5	7	7
10^{-13}	T_{it}	2	2	6	7	3
	n_{it}	7	17	6	19	9
10^{-14}	T_{it}	2	2	4	5	2
	n_{it}	16	17	9	12	5

With respect to the length of one parallel iteration step T_{it}, there are clearly two types of behavior. For variants with no pre- and post-processing (QR decomposition or Gram matrices plus updates), i.e. for NGS and NGD, the decrease of T_{it} with the decrease of *maxw* is observed in the interval $10^{-1} \geq maxw \geq 10^{-8}$, whereas for smaller values of maximal weight T_{it} remains constant. Recall that for these two variants the computation in one parallel iteration step is dominated by the SVD of tall, skinny matrix constructed from two block columns. As *maxw* decreases all block columns are mutually more and more orthogonal so that the LAPACK procedure DGESVJ is faster up to some level. On the other hand, other

Table 4. $n = 8000$, $p = 16$

maxweight		NGS	NGD	QRS	QRD	G
10^{-1}	T_{it}	49	27	26	23	14
	n_{it}	26	26	33	26	26
10^{-2}	T_{it}	38	19	25	23	14
	n_{it}	59	60	51	62	61
10^{-3}	T_{it}	27	10	25	22	13
	n_{it}	21	26	22	22	19
10^{-4}	T_{it}	22	7	25	22	13
	n_{it}	16	23	15	25	13
10^{-5}	T_{it}	17	6	25	22	13
	n_{it}	10	29	11	27	13
10^{-6}	T_{it}	16	6	24	22	13
	n_{it}	7	18	7	20	6
10^{-7}	T_{it}	13	11	24	22	13
	n_{it}	9	24	10	25	8
10^{-8}	T_{it}	9	6	24	22	12
	n_{it}	4	9	4	9	8
10^{-9}	T_{it}	8	-	24	-	11
	n_{it}	7	0	5	0	3
10^{-10}	T_{it}	7	-	23	-	11
	n_{it}	4	0	6	0	7
10^{-11}	T_{it}	8	6	23	-	11
	n_{it}	3	1	3	0	2
10^{-12}	T_{it}	7	6	23	22	11
	n_{it}	3	5	3	5	2
10^{-13}	T_{it}	7	6	21	21	10
	n_{it}	6	9	3	10	6
10^{-14}	T_{it}	5	5	18	15	6
	n_{it}	3	5	3	5	3

three variants that use some sort of pre- and post-processing (NGQS, NGQD, G) have an almost constant length of T_{it} in the whole range of $maxw$ because the amount of this extra work is *constant* and does not depend on $maxw$.

With respect of the number of parallel iteration steps n_{it} spent at given $maxw$, the general trend is again the decrease of n_{it} with decrease of $maxw$— see the results for variant G. However, there are some irregularities especially for smallest orders of $maxw$, $maxw \approx 10^{-13}$, 10^{-14}. Here n_{it} can suddenly increase—see its values in Table 3 for all variants except G. Apparently, in this case there is a problem to meet the *global* convergence criterion (9). It seems that for very small values of $maxw$ it can be a problem to push the LAPACK procedure DGESVJ 'behind its limit' in requiring 'more orthogonal' left singular vectors—we are somewhere near the perturbation limit of computations with a limited (although double) precision.

Finally, notice the sudden jump of three orders of magnitude in the value of $maxw$ in variants with the dynamic LSC (NGD, NGQD). Namely, there are no parallel iterations (or there is just one iteration) when $10^{-9} \geq maxw \geq 10^{-11}$. This is connected with the switch from dynamic to static LSC in (12) which occurs for $maxw < 10^{-6}$. Hence, for $maxw \approx 10^{-7}$, the more stringent, static LSC was applied for the first time according to (12). Then, for NGD with $maxw \approx 10^{-8}$ and $p = 16$ (Table 4), T_{it} was almost halved and n_{it} decreased almost three times. For $p = 32$, the same is true for T_{it} but n_{it} remained the same (see Table 3). For NGQD, there is no reduction of T_{it} due to the QR pre- and post-processing but for $p = 16$ the value of n_{it} decreases almost three times (see Table 4). However, there is no decrease of n_{it} for $p = 32$ (see Table 4). And then, suddenly, we have a jump in $maxw$ downwards over three orders of magnitude for both dynamic algorithms and both values of p where there are *no* iterations (or just one iteration for NGD)! At the moment, this remarkable property is not well understood. Intuitively, the static LSC achieves 'more' orthogonal columns in p pairs of block columns at the end of each parallel iteration step than the dynamic LSC. However, it is 'the delay' between applying the static LSC at $maxw \approx 10^{-7}$ and the jump occurring at $maxw \approx 10^{-9}$ which is difficult to explain.

4   Conclusions

The idea of a dynamic LSC is quite interesting and results in the acceleration of the OSBJA as compared to the static LSC. However, when the local SVDs are computed with less accuracy at the beginning of a global iteration process, one needs more parallel iteration steps for a given global stopping criterion (notice that the global stopping criterion was *the same* for all algorithms). Perhaps one can gently tune the constants in (12) so that the variant NGD becomes competitive with variant G. Providing guidelines for such tuning remains an open problem.

Acknowledgment. Authors were supported by the VEGA grant no. 2/0026/14.

References

1. Anderson, A., et al.: LAPACK Users' Guide, 2nd edn. SIAM, Philadelphia (1999)
2. Bečka, M., Okša, G., Vajteršic, M.: New dynamic orderings for the parallel one-sided block-Jacobi SVD algorithm. Parallel Proc. Lett. **25**, 1–19 (2015)
3. Demmel, J., Veselić, K.: Jacobi's method is more accurate than QR. SIAM J. Matrix Anal. Appl. **13**, 1204–1245 (1992)

4. Drmač, Z., Veselić, K.: New fast and accurate Jacobi SVD algorithm: I. SIAM J. Matrix Anal. Appl. **29**, 1322–1342 (2007)
5. Drmač, Z., Veselić, K.: New fast and accurate Jacobi SVD algorithm: II. SIAM J. Matrix Anal. Appl. **29**, 1343–1362 (2007)
6. Hari, V.: Accelerating the SVD block-Jacobi method. Computing **75**, 27–53 (2005)
7. Veselić, K., Hari, V.: A note on a one-sided Jacobi algorithm. Numer. Math. **56**, 627–633 (1989)

Author Index

Printed in the United States
By Bookmasters